Exploring University Mathematics with Python

Siri Chongchitnan

Exploring University Mathematics with Python

 Springer

Siri Chongchitnan
Mathematics Institute
University of Warwick
Coventry, UK

ISBN 978-3-031-46269-6   ISBN 978-3-031-46270-2   (eBook)
https://doi.org/10.1007/978-3-031-46270-2

Mathematics Subject Classification (2020): 00-01

This Springer imprint is published by the registered company Springer Nature Switzerland AG
The registered company address is: Gewerbestrasse 11, 6330 Cham, Switzerland

Paper in this product is recyclable.

To my teachers

Preface

Motivation

Python is now arguably the world's most popular programming language, thanks to its gentle learning curve, clear syntax, wide range of open-source libraries and active online support community. Over the past decade, Python programming has become a highly desirable skill to employers not only in the STEM and ICT sectors, but also in any industry involving data. It comes to no surprise then that Python has now been integrated into school and university curriculums around the world.

A typical mathematics university curriculum would include some element of programming, usually in a standalone module. However, in my experience teaching at several UK universities, students often regard programming as just another module that is disparate from a typical 'pen-and-paper' module.

In my opinion, this is an extremely unhealthy viewpoint, because programming can often help us gain a more solid understanding of mathematics in comparison to a purely pen-and-paper approach. It is true that much of university mathematics is driven by theorems and proofs, and it is also true that Python does not prove theorems. However, Python gives us the power and freedom to glimpse into the unknown, leading us towards insightful conjectures that would have been difficult to formulate otherwise.

Hence, I was motivated to write a mathematics textbook that is richly interwoven with Python, rather than another Python textbook with some mathematical examples. The spirit of this book is one of mathematical exploration and investigation. I want to show students that Python can hugely enrich our understanding of mathematics through:

- *Calculation*: Performing complex calculations and numerical simulations instantly;
- *Visualisation*: Demonstrating key theorems with graphs, interactive plots and animations;
- *Extension*: Using numerical findings as inspiration for making deeper, more general conjectures.

Who is this book for?

I wrote this book for all learners of mathematics, with the primary audience being mathematics undergraduates who are curious to see how Python can enhance their understanding of core university material. The topics I have chosen represent a mathematical tour of what students typically study in the first and second years at university. As such, this book can also serve as a preview for high-school students who are keen to learn what mathematics is like at university.

In addition, I hope this book will also benefit mathematics lecturers and teachers who want to incorporate programming into their course. I hope to convince educators that programming can be a meaningful part of any mathematical module.

Structure

The topics covered in this book are broadly *analysis, algebra, calculus, differential equations, probability and statistics*. Each chapter begins with a brief overview of the subject and essential background knowledge, followed by a series of questions in which key concepts and important theorems are explored with the help of Python programs which have been succinctly annotated. All code is available to download online.

At the end of each section, I present a *Discussion* section which dives deeper into the topic. There are also a number of exercises (most of which involve coding) at the end of each chapter.

Assumed knowledge

In terms of programming knowledge, this book does not assume that you are a highly experienced user of Python. On the other hand, complete beginners to Python might struggle to follow the code given. I would suggest that the reader should have the most basic knowledge of programming (*e.g.* you know what a *for* loop does).

For completeness, I have included a section called *Python 101* (Appendix A) which gives instructions on installing Python and signposts to references that can help anyone pick up Python quickly.

In terms of mathematical knowledge, I do not assume any university-level mathematics. Students who are familiar with the material in the standard A-Level mathematics (or equivalent) should be able to follow the mathematical discussions in this book.

Acknowledgements

I am extremely grateful to have received extensive comments from my early reviewers, many of whom are students at my home department, Warwick Mathematics Institute. They are:

- Maleeha Ahmad
- Michael Cavaliere
- Carl Beinuo Guo
- Rachel Haddad
- Trinity Jinglan Hu
- Rachel Siyi Hong
- Kornel Ipacs
- Kit Liu
- William Mau
- Ben Middlemass
- Kush Patel
- Payal Patel
- Safeeyah Rashid
- Danny Robson
- Zac Ruane
- Reduan Soroar
- Kyle Thompson
- Ben Wadsworth
- Jiewei Xiong

I would also like to thank the team at Springer for their support. Special thanks to Richard Kruel for his belief in my idea and his unwavering encouragement.

Siri Chongchitnan
Coventry, UK

The code

a) Downloading and using the code

All code is available to download from

https://github.com/siriwarwick/book

We will be coding in **Python 3** (ideally 3.9 or higher). To find out which version you have, see the code box below. If you don't have Python installed, see Appendix A.

There are many ways to run Python programs, but by default, I will assume that you are working in **JupyterLab** (or **Jupyter Notebook**). You will be working with files with the .ipynb extension. For more information on the Jupyter IDE (integrated development environment), see https://jupyter.org.

There are alternative IDEs to Jupyter, for example:

- *IDLE* (which comes as standard with your Python distribution);
- *Spyder* (https://www.spyder-ide.org);
- *PyCharm* (https://www.jetbrains.com/pycharm);
- *Replit* (https://replit.com).

If you prefer these IDEs, you will be working with files with the .py extension.

Code and annotations will be given in grey boxes as shown below. The code is given on the right, whilst the explanation is given on the left.

filename.ipynb (for checking Python version)	
Let's check your version of Python	`from platform import python_version` `python_version()`

b) About %matplotlib

We will often use the line

%matplotlib

to make any plot interactive (rather the usual static 'inline' plot). The zoom and pan buttons in the GUI window will be particularly useful. To return the static plot, use the command:

%matplotlib inline

%matplotlib is one of the so-called 'magic' commands[1] that only work in the Jupyter environment and not in standard Python.

If you have difficulty running with just %matplotlib, try running this line of code in an empty cell

%matplotlib -l

(that's a small letter L after the dash). This should list all the graphics backends available on your machine. Choose one that works for you. For example, if qt is on the list, replace %matplotlib by

%matplotlib qt

---

[1] For more on magic commands, see https://ipython.readthedocs.io/en/stable/interactive/magics.html

c) Coding style

Everyone has their own style of coding. The code that I present in this book is just one of many ways of approaching the problems. Keeping the purpose and the audience in mind, I tried to minimise the use of special packages and clever one-liners, but instead I placed a greater emphasis on readability. In particular, I do not use the `if __name__ == "__main__":` idiom. My goal in this book is to show how basic knowledge Python goes a long way in helping us delve deeply into mathematics.

In short, *every piece of code shown can be improved in some ways.*

A caveat: some code may produce warnings or even errors as new versions of Python roll out since the publication of this book. If you feel something isn't quite right with the code, check the book's GitHub page for possible updates.

d) Getting in touch

Comments, corrections and suggestions for improvement can be posted on the discussion section on the book GitHub

$$\texttt{https://github.com/siriwarwick/book/discussions}$$

or emailed to `siri.chongchitnan@warwick.ac.uk`. I will be happy to hear from you either way.

CONTENTS

All code and brief descriptions

All code (in `.ipynb` and `.py`) is available to download from `https://github.com/siriwarwick/book`.

	Section	Filename	What does the code do?
1. Analysis	1.3	sequence-convergence.ipynb	Plots terms in a sequence
	1.4	series-convergence.ipynb	Plots partial sums of a series
	1.5	harmonic.ipynb	Demonstrates that the harmonic number H_n is approximately $\ln n + \gamma$
	1.6	fibonacci.ipynb	Demonstrates that the Fibonacci sequence F_n scales like ϕ^n (where ϕ is the Golden ratio)
	1.7		Explores the ε-δ definition of continuity using
		continuity.ipynb	- `ipywidgets` slider (Jupyter only)
		continuityslider.ipynb	- *Matplotlib* slider
	1.8	thomae.ipynb	Plots Thomae's function
	1.9	bisection.ipynb	Performs bisection root finding
	1.10	differentiation.ipynb	Performs numerical differentiation (forward Euler)
	1.12	counterexample.ipynb	Shows a counter-intuitive property of the function $f(x) = x + 2x^2 \sin(1/x)$, showing its behaviour near the origin
2. Calculus	2.2	Eh.ipynb	Plots the error $E(h)$ in numerical differentiation (forward difference) as a function of step size h
	2.3	taylor.ipynb	Plots a function and its Taylor series of various degrees
	2.4	taylorhm.ipynb	Plots the error R_N in the Taylor-series approximation as a function of the degree N
	2.5	weierstrass.ipynb	Plots the Weierstrass function
	2.6	trapezium.ipynb	Plots the error $E(h)$ in numerical integration (Trapezium Rule) as a function of step size h
	2.7	simpson.ipynb	Plots the error $E(h)$ in numerical integration (Simpson's Rule) as a function of step size h
	2.8	improper.ipynb	Plots the error in numerical integration of an improper integral
	2.9	fourier.ipynb	Plots the Fourier series (up to n terms) of a given function
3. Vector Calculus	3.2	cycloid.ipynb	Generates a cycloid from a rolling wheel
	3.3	ellipse.ipynb	Calculates the perimeter of an ellipse and compares with Ramanujan's approximation
	3.4	curvature.ipynb	Plots a parametric curve in $\mathbb{R}^2$ and its curvature κ
	3.5	torsion.ipynb	Plots a parametric curve in $\mathbb{R}^3$ and its torsion τ
	3.6	quadrics.ipynb	Visualises a family of quadric surfaces
	3.7	surfacearea.ipynb	Calculates the area of (part of) an ellipsoidal surface
	3.8	grad.ipynb	Plots a 3D surface and its contour lines
	3.9	div.ipynb	Demonstrates the divergence theorem
	3.10	curl.ipynb	Demonstrates Stokes' theorem for a family of surfaces

	Section	Filename	What does the code do?
4. Differential Equations	4.3	odesolver.ipynb	Solves a first-order ODE numerically and plots the result
	4.4	pendulum.ipynb	Animates a pendulum by solving a 2nd order ODE
	4.5	doublependulum.ipynb	Animates the (chaotic) double pendulum by solving a system of ODEs
	4.6	lorenz.ipynb	Animates the Lorenz system, showing a strange attractor
	4.7	logistic.ipynb	Plots the bifurcation diagram for the logistic map
	4.8	mandelbrot.ipynb	Plots the Mandelbrot set
		mandelbrot3D.ipynb	Visualises the connection between the Mandelbrot set and the logistic map
	4.9	heat.ipynb	Animates the solution of the 1D heat equation
	4.10	wave.ipynb	Animates the solution of the 2D wave equation
5. Linear Algebra	5.3	planes.ipynb	Visualises a system of linear equations as planes in $\mathbb{R}^3$
	5.4	solvetimes.ipynb	Plots the time taken to solve a linear system as a function of matrix dimension
	5.5	transformation.ipynb	Visualises a 2×2 matrix as a geometric transformation
	5.6	eigshow.ipynb	Visualises the eigenvectors and eigenvalues of a 2×2 matrix via a game
	5.7	diagonalise.ipynb	Visualises matrix diagonalisation as a sequence of 3 transformations
	5.8	svd.ipynb	Performs singular-value decomposition of an image
	5.9	ranknullity.ipynb	Visualises the rank-nullity theorem
	5.10	gramschmidt.ipynb	Plots the Legendre polynomials obtained via the Gram-Schmidt process
6. Algebra & Number Theory	6.3	cayley.ipynb	Produces the Cayley table for a group
	6.4	dihedral.ipynb	Visualises the dihedral group as rotation and reflection of a polygon
	6.5	permutation.ipynb	Produces the Cayley table for a permutation group
	6.6	quaternion.ipynb	Rotates a point about an axis in $\mathbb{R}^3$ using a quaternion
	6.7	inversewheel.ipynb	Produces an intriguing pattern in a circle using pairs of multiplicative inverses modulo n
	6.8	crt.ipynb	Visualises the solutions of a pair of congruences
	6.9	legendre.ipynb	Visualises the Quadratic Reciprocity Law
	6.10	pnt.ipynb	Visualises the Prime-Number Theorem
	6.12	zetaanim.ipynb	Animates the image of the critical line $\mathrm{Re}(s) = \frac{1}{2}$ under the Riemann zeta function

	Section	Filename	What does the code do?
7. Probability	7.4	pascal.ipynb	Produces Pascal's triangle modulo 2
	7.5	coin1.ipynb	Simulates multiple coin throws and plots the distribution of the number of heads observed
		coin2.ipynb	Plots the distribution of the first occurrence of a particular sequence in multiple coin throws
	7.6	birthday.ipynb	Simulates the birthday problem
	7.7	montyhall.ipynb	Simulates the Monty Hall problem
	7.8	normal.ipynb	Simulates the Galton board (visualising the normal distribution)
	7.9	poisson.ipynb	Simulates scattering dots randomly in a grid and plots the distribution of dot counts per square
	7.10	montecarlo.ipynb	Plots the fractional error for Monte Carlo integration as a function of the number of random points used
	7.11	buffon.ipynb	Simulates the Buffon's needle problem
8. Statistics	8.3	CLT.ipynb	Demonstrates the Central Limit Theorem by sampling from a given distribution
	8.4	ttest.ipynb	Plots the distribution of the population mean from a small sample and performs a one-sample t-test
	8.5	chi2test.ipynb	Plots the observed and expected frequencies for categorical data and performs a χ^2 test
	8.6	regression.ipynb	Plots a regression line through data points and demonstrates Simpson's paradox
	8.7	bivariate.ipynb	Plots the bivariate normal distribution and its contour ellipses
	8.8	randomwalk.ipynb	Generates 1D random walks and plots the mean distance travelled as a function of time step
	8.9	bayesian.ipynb	Plots the prior and posterior for Bayesian inference
	8.10	clustering.ipynb	Performs k-means clustering
		classification.ipynb	Performs k-nearest neighbour (kNN) classification

Visualisation recipes

Recipe	Section	Code
Plots		
Plot with different types of lines	1.3	sequence-convergence.ipynb
Plot a large number of curves, with a gradual colour change	2.3	taylor.ipynb
Two-panel plot in $\mathbb{R}^2$	1.5	harmonic.ipynb
Plot parametric curves and surfaces in $\mathbb{R}^2$ and $\mathbb{R}^3$	3.1	see text
Plot polar curves	3.1	see text
Plotting in $\mathbb{R}^3$ with *Plotly*	5.3	planes.ipynb
Three-panel plot in $\mathbb{R}^3$	5.9	ranknullity.ipynb
Plot one panel in $\mathbb{R}^2$ next to one in $\mathbb{R}^3$	3.8	grad.ipynb
Plot a 3D surface and its contour lines	3.8	grad.ipynb
Shade the area under a graph	8.4	ttest.ipynb
Draw a filled polygon	6.4	dihedral.ipynb
Sliders		
Slider controlling a plot in $\mathbb{R}^2$	1.7	continuityslider.ipynb
Slider controlling a plot in $\mathbb{R}^3$	3.6	quadrics.ipynb
One slider controlling two plots in $\mathbb{R}^2$	3.4	curvature.ipynb
One slider controlling two plots, one in $\mathbb{R}^2$ and one in $\mathbb{R}^3$	3.5	torsion.ipynb
	8.7	bivariate.ipynb
Two sliders controlling one plot in $\mathbb{R}^2$	2.5	weierstrass.ipynb
Heatmaps		
Heatmap + vector field + slider (in $\mathbb{R}^2$)	3.9	div.ipynb
Heatmap + slider (polar coordinates)	3.10	curl.ipynb
Animations		
Animation in $\mathbb{R}^2$	4.4	pendulum.ipynb
	4.9	heat.ipynb
Animation with two panels in $\mathbb{R}^2$	6.12	zetaanim.ipynb
Animation in $\mathbb{R}^3$	4.6	lorenz.ipynb
	4.10	wave.ipynb
Visualising matrices		
Display a matrix with colour-coded elements	6.3	cayley.ipynb
	4.8	mandelbrot.ipynb
Read in an image file	5.8	svd.ipynb
Visualising data		
Plot a histogram with *Matplotlib*	7.5	coin1.ipynb
Plot a histogram with *Seaborn*	8.3	CLT.ipynb
Fit a line through a scatter plot	7.10	montecarlo.ipynb
Read data from a file and store as *Pandas* dataframes	8.6	regression.ipynb
Write data to a file	8.7	see text
Shade 2D regions according to classifications with *Scikit-learn*	8.10	classification.ipynb

Analysis

Real analysis is the study of real numbers and maps between them (*i.e.* functions). Analysis provides a solid foundation for calculus, which in turn gives rise to the development of other branches of mathematics. The main feature of analysis is its *rigour*, meaning that every concept in analysis is defined precisely with logical statements without the need for pictures.

Fig. 1.1: *Augustin-Louis Cauchy* (1789–1857), one of the founders of modern analysis. Cauchy is commemorated on a French stamp shown on the right. (Image source: [137].)

Analysis is a subject which many new university students find to be most different from how mathematics is taught in school. The proof-driven nature of the subject can be overwhelming to some students. In this chapter, we will see how Python can help us visualise and understand key concepts in analysis.

In addition to real numbers, the main objects in analysis are sequences, series and functions. We will first see how these objects can be represented and manipulated in Python. We then give a survey of some key theorems in analysis and see how Python can help us understand these theorems more deeply. We will focus on understanding and visualising the theorems themselves, rather than the proofs. The proofs of the theorems discussed in this chapter can be found in good analysis textbooks, amongst which we recommend [10, 20, 91, 143, 189].

1.1 *Basics of NumPy and Matplotlib*

NumPy is an indispensable Python library for mathematical computing. Most mathematical objects and operations that you are familiar with are part of NumPy. Equally indispensable is the *Matplotlib* library which is the key to visualising mathematics using graphs and animations. It is standard practice to import NumPy and Matplotlib together using the following two lines at the beginning of your code.

```
import numpy as np
import matplotlib.pyplot as plt
```

1.2 *Basic concepts in analysis*

Sequences

A sequence (a_n) is simply a succession of numbers with the subscript $n = 1, 2, 3 \ldots$ labelling the order of appearance of each term. For example, the sequence $(a_n) = \left(\frac{1}{n^2}\right)$ contains the terms $a_1 = 1$, $a_2 = \frac{1}{4}$, $a_3 = \frac{1}{9}$ and so on. The bracketed notation (a_n) denotes the entire sequence, whereas a_n denotes a single term in the sequence.

Sequences can be finite, for example, $(a_n)_{n=1}^5 = 1, \frac{1}{4}, \frac{1}{9}, \frac{1}{16}, \frac{1}{25}$. However, in analysis, we are mainly interested in the behaviour of infinite sequences $(a_n)_{n=1}^\infty$.

Sequences are easily represented as NumPy *arrays*, which are one of the most useful objects in Python (another method is to use *lists* which we will discuss later). Naturally, we cannot store infinitely long arrays in Python. But take the array to be long enough and it will usually be sufficient to reveal something about the behaviour of the infinite sequence.

It is often necessary to create a sequence of equally spaced numbers between two given numbers (in other words, an arithmetic progression). Such a sequence can be easily created using NumPy's `linspace` command:

$$\texttt{np.linspace}(\textit{first element, last element, how many elements})$$

The third argument is optional. If not provided, the default is 50 elements, equally spaced between the first and the last elements. It is worth noting the terminology here: each term in a sequence can be identified with an *element* in a NumPy array. `linspace` is especially useful for plotting graphs, since we usually want equally spaced points along the *x*-axis.

The same sequence can also be created using the `arange` command of the form:

$$\texttt{np.arange}(\textit{first element, >last element, step})$$

where >*last element* means any number greater than the last element, but not exceeding *last element* + *step*. This is because, by convention, the 2nd argument in `arange` is not included in an `arange` array. In practice, just choose a number that is slightly larger than the final element desired. The third argument is optional, with the default = 1.

If you want to create a short sequence by manually specifying every element, use the following syntax:

$$\texttt{np.array}([\textit{element1, element2, \ldots, last element}])$$

Here are some examples of how to create sequences in Python and obtain new ones using array operations.

Creating sequences as arrays	
Standard header (for all code on this page)	`import numpy as np`
100 evenly spaced values from 0 to 5	`np.linspace(0, 5, 100)`
$(a_n) = 3, 4, 5, \ldots 99$	`a_n = np.arange(3, 100)`
$(b_n) = 3, 5, 7, \ldots 99$	`b_n = np.arange(3, 99.1, 2)`
$(c_n) = \frac{1}{3}, \sqrt{2}, \pi$	`c_n = np.array([1/3, np.sqrt(2), np.pi])`
$(d_n) = 100$ terms, all 0.	`d_n = np.zeros(100)`
$(e_n) = 100$ terms, all 0.7.	`e_n = np.full(100, 0.7)`

Array operations	
Addition and scalar multiplication $(x_n) = (2a_n - 5)$ $(y_n) = (a_n + x_n)$	`x_n = 2*a_n - 5` `y_n = a_n + x_n`

Multiplication and division
$(x_n y_n)$
$(2/x_n)$
(x_n/y_n)
	`x_n*y_n`
	`2/x_n`
	`x_n/y_n`

Exponentiation $\left(x_n^3\right)$ (3^{x_n})	`x_n**3` `3**x_n`

Other useful array commands	
The number of elements in an array x_n Shape of the array	`len(x_n)` `x_n.shape`

Calling elements of (x_n) by its index
The first element
The fifth element
The last element
The first 9 elements (*slicing*)
	`x_n[0]`
	`x_n[4]`
	`x_n[-1]`
	`x_n[0:9]`

$(z_n) = $ zero array, same shape as x_n $(w_n) = $ array of 0.7, same shape as x_n	`z_n = np.zeros_like(x_n)` `w_n = np.full_like(x_n, 0.7)`

Here are some Python tips to help avoid errors.

- **Initial index** In Python, the index of each array starts from 0, not 1.
- **** means power** In some programming languages (*e.g.* MATLAB), x^2 is coded as `x^2`. However, in Python, the operation `a^b` is a *bitwise addition modulo 2* (*i.e.* a and b are converted to binary and added bitwise modulo 2). This is something that we will seldom need in mathematics problems, so be careful!
- **List vs array** Instead of a NumPy array, one can also work with a Python *list*. There are subtle differences between the two data types. For most numerical purposes, we will be using arrays, although there will be occasions where we will also use lists, or mix the two data types. We compare arrays and lists in detail in the Appendix A.3.

Series

A *series* is the sum of terms in a sequence. If the sequence is infinitely long, the sum of the terms is an *infinite series* denoted $\sum_{n=1}^{\infty} x_n$. For example, the sequence $\left(\frac{1}{n^2}\right)_{n=1}^{\infty}$ gives rise to the series

$$\sum_{n=1}^{\infty} \frac{1}{n^2} = 1 + \frac{1}{4} + \frac{1}{9} + \frac{1}{16} + \dots$$

Of course, Python cannot deal with infinitely many terms, but we can still compute the *partial sums*

$$S_N = \sum_{n=1}^{N} x_n = x_1 + x_2 + \dots + x_N.$$

for large values of N. These partial sums should give us a good idea about whether the series *converges* to a finite number, or *diverges* (*e.g.* the sum becomes arbitrarily large, or fluctuates without settling down to a value). We will study the convergence of a number of interesting series in §1.4.

A useful command for evaluating series is sum(*an array*). Another useful operator in Python 3 is the @ operator, which is equivalent to the *dot product* of two arrays, *i.e.*

$$\texttt{x@y} = \texttt{np.dot(x, y)} = \texttt{sum(x*y)}$$

Of course there is also the option of using the for loop. This is the most appropriate method if you need to plot the partial sums. Here are two examples of series evaluation using various methods.

Series evaluation	
$\sum_{n=1}^{100} n$	```import numpy as np``` `x= np.arange(1,101)` `sum(x)` `# or` `S = 0` `for n in np.arange(1,101):` `    S += n`
$\sum_{n=1}^{100} n(n+3)$	`x@(x+3)          # or` `sum(x*(x+3))    # or` `np.dot(x, x+3)  # or` `S = 0` `for n in np.arange(1,101):` `    S += n*(n+3)`

Functions

A function f is a mapping which takes as its input an object $x \in A$ (where A is called the *domain*), and produces an output $f(x) \in B$ (where B is called the *codomain*). We denote such a mapping by the notation $f : A \rightarrow B$.

Functions in Python are defined using the `def` command, which has a rich structure. A function expects brackets in which arguments (if any) are placed. In most cases, you will also want your function to return an output.

Instead of `def`, another way to define a function in a single line is to use the *lambda function* (also called *anonymous function*).

For example, let $F(x) = 2x + 1$ and $G(x, y) = x + y$. In the box below, the left side shows the `def` method. The right side shows the lambda-function method.

Defining functions	
Method I: `def`	**Method II**: Lambda functions
`def F(x):` `    return 2*x+1`	`F = lambda x : 2*x+1`
`def G(x,y):` `    return x+y`	`G = lambda x,y : x+y`

We often use lambda functions when creating very simple functions, or when feeding a function into another function. In the latter case, the lambda function does not even need to have a name (hence 'anonymous'). This will be useful when we discuss integration in Python – for a preview, take a look at the end of §3.1.

Now that we can define sequences, series and functions in Python, we are ready to investigate some key definitions and theorems in analysis.

1.3 The ε, N definition of convergence for sequences

A sequence of real numbers (x_n) is said to *converge* to $x \in \mathbb{R}$ (and we write $\lim\limits_{n \to \infty} x_n = x$) if,

For any $\varepsilon > 0$, there exists an integer $N \in \mathbb{N}$ such that, for all $n > N$, $|x_n - x| < \varepsilon$.

Consider sequences defined by the following expressions for each term.

a) $x_n = 1 - \dfrac{1}{n}$ b) $y_n = \dfrac{\sin n}{n}$ c) $z_n = \dfrac{\ln n}{n}$ d) $E_n = \left(1 + \dfrac{1}{n}\right)^n$.

For each sequence, find the limit as $n \to \infty$. Investigate the relationship between ε and N.

This is one of the most important definitions in analysis. It gives a rigorous way to express an intuitive concept of convergence as $n \to \infty$ very precisely. Let's first try to unpack the definition.

Let's call $|x_n - x|$ the *error* (*i.e.* the absolute difference between the nth term of the sequence and the limit x). The definition above simply says that *down the tail of the sequence, the error becomes arbitrarily small.*

If someone gives us ε (typically a tiny positive number), we need to find how far down the tail we have to go for the error to be smaller than ε. The further we have to go down the tail, the larger N becomes. In other words, N depends on ε.

Now let's consider the first sequence $x_n = 1 - \frac{1}{n}$. The first step is to guess the limit as $n \to \infty$. Clearly, the $\frac{1}{n}$ term becomes negligible and we guess that the limit is 1. Now it remains to *prove* this conjecture formally. We go back to the definition: for any given $\varepsilon > 0$, the error is given by

$$|x_n - 1| = \frac{1}{n} < \frac{1}{N} < \varepsilon,$$

where the final inequality holds if we let N be any integer greater than $1/\varepsilon$. For instance, if someone gives us $\varepsilon = 3/10$, then any integer $N \geq 10/3$ would work (*e.g.* $N = 4$). We have shown that this choice of N implies that the definition holds, hence we have proved that $\lim\limits_{n \to \infty} x_n = 1$.

The definition holds if there exists one such integer N, so that in the proof above, $N + 1$ or $3N^2 + 5$ are also equally good candidates (one just goes further down the tail of the sequence). However, it is still interesting to think about the smallest N (let's call it $N_{\min}$) that would make the definition of convergence work. We could tabulate some values of $N_{\min}$ for the sequence x_n.

ε	0.3	0.2	0.1	0.05	0.03
$N_{\min}$	4	5	10	20	34

We can easily confirm these calculations with the plot of the error $|x_n - 1|$ against n (solid line in fig. 1.2).

Next, let's consider the sequences y_n. What does it converge to? Since $\sin n$ is bounded by -1 and 1 for all $n \in \mathbb{N}$, whilst n can be arbitrarily large, we guess that y_n converges to 0. The proof can be constructed in the same way as before. For any $\varepsilon > 0$, we find the error:

$$|y_n - 0| = \frac{|\sin n|}{n} \leq \frac{1}{n} < \frac{1}{N} < \varepsilon, \tag{1.1}$$

where the final inequality holds if we let N be any integer greater than $1/\varepsilon$. However, the latter may not be $N_{\min}$, because we have introduced an additional estimate $|\sin n| \leq 1$.

It seems that $N_{\min}$ cannot be obtained easily by hand, and the error plot (dashed line in fig. 1.2) also suggests this. For example, we find that when $\varepsilon = 0.1$, $N_{\min} = 8$ rather than 10 as we might have expected from Eq. 1.1.

A similar analysis of the relationship between ε and N can be done for remaining sequences z_n and E_n. The sequence z_n converges to 0 because, intuitively, $\ln n$ is negligible compared with n when n is large (we discuss how this can be proved more rigorously in the Discussion section).

Finally, the famous sequence E_n can be regarded as the definition of Euler's number e

$$e = \lim_{n \to \infty} \left(1 + \frac{1}{n}\right)^n.$$

The errors of all these sequences approach zero, as shown in fig. 1.2 below. Matplotlib's `plot` function renders the errors as continuous lines, but keep in mind that the errors form sequences that are defined only at integer n, and the lines are for visualisation purposes only.

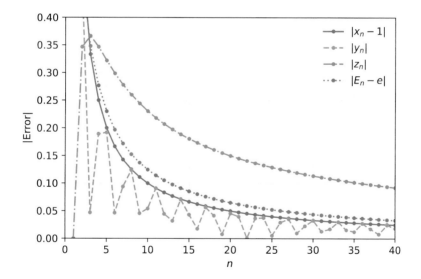

Fig. 1.2: The error, *i.e.* the absolute difference between the sequences x_n, y_n, z_n, E_n and their respective limits.

The code `sequence-convergence.ipynb` plots fig. 1.2. Remember that all code in this book can be downloaded from the GitHub page given in the Preface.

sequence-convergence.ipynb (for plotting fig. 1.2)	
	```import numpy as np``` ```import matplotlib.pyplot as plt```
Consider up to $n = 40$	```n = np.arange(1,41)```

Define x_n
Error = $\|x_n - x\|$ where $x = \lim\limits_{n\to\infty} x_n = 0$
	```recp = 1/n```
	```xn = 1-recp```
	```err_xn = recp```

$\lim\limits_{n\to\infty} y_n = 0$	```yn = np.sin(n)/n``` ```err_yn = abs(yn)```
$\lim\limits_{n\to\infty} z_n = 0$	```zn = np.log(n)/n``` ```err_zn = abs(zn)```
$\lim\limits_{n\to\infty} E_n = e$	```En = (1+recp)**n``` ```err_En = abs(En-np.e)```

Plotting with dots (markersize = 3) joined by various line styles	```plt.plot(n, err_xn, 'o-' , ms=3)```
	```plt.plot(n, err_yn, 'o--', ms=3)```
	```plt.plot(n, err_zn, 'o-.', ms=3)```
	```plt.plot(n, err_En, 'o:' , ms=3)```

r allows typesetting in LaTeX	```plt.xlim(0, 40)```
	```plt.ylim(0, 0.4)```
	```plt.xlabel(r'n')```
	```plt.ylabel('|Error|')```
	```plt.legend([r'$|x_n-1|$', r'$|y_n|$',```
	``` r'$|z_n|$', r'$|E_n-e|$'])```
	```plt.grid('on')```
	```plt.show()```

Suppose that at some point, the error shrinks monotonically (*i.e.* it does not fluctuate like the error for y_n), we can use the code below to search for $N_{\min}$ for a given value of epsilon. Take E_n for example.

Finding N given ε	

For any given ε, say, 10^{-5}	```epsilon = 1e-5```
	```n = 1```
	```err = np.e - (1 + 1/n)**n```

Continue searching until the error $< \varepsilon$
Increasing n by 1 per iteration
	```while err >= epsilon:```
	``` n += 1```
	``` err = np.e - (1 + 1/n)**n```

Report $N_{\min}$	```print("N_min = ", n-1)```

In this case, we find that for the error to satisfy $\|E_n - e\| < 10^{-5}$, we need $n > N_{\min} = 135912$.

DISCUSSION

- **Floor and ceiling.** Instead of saying "N is the smallest integer greater than or equal to x", we can write

$$N = \lceil x \rceil,$$

which reads "the *ceiling* of x". For example, in our proof that the sequence $x_n = 1 - 1/n$ converges, given any $\varepsilon > 0$, we can take $N = \lceil \varepsilon^{-1} \rceil$.
Similarly, one can define $\lfloor x \rfloor$ (the *floor* of x) as the smallest integer less than or equal to x.

- **Squeeze Theorem.** Consider $y_n = \sin n / n$. The numerator is bounded between -1 and 1, so we see that

$$\frac{-1}{n} \le y_n \le \frac{1}{n}.$$

As n becomes large, y_n is *squeezed* between 2 tiny numbers of opposite signs. Thus, we have good reasons to believe that y_n also converges to 0. This idea is formalised by the following important theorem.

Theorem 1.1 *(Squeeze Theorem) Let a_n, b_n, c_n be sequences such that $a_n \le b_n \le c_n$ for all $n \in \mathbb{N}$. If $\lim_{n \to \infty} a_n = \lim_{n \to \infty} c_n$, then $\lim_{n \to \infty} a_n = \lim_{n \to \infty} b_n = \lim_{n \to \infty} c_n$.*

The Squeeze Theorem can also be used to prove that $z_n = \ln n / n \to 0$. Using the inequality $\ln x < x$ for $x > 0$, observe that

$$0 \le \frac{\ln n}{n} = \frac{2 \ln \sqrt{n}}{n} < \frac{2\sqrt{n}}{n} = \frac{2}{\sqrt{n}}.$$

Now take the limit as $n \to \infty$ to find that $\lim_{n \to \infty} \ln n / n = 0$.

- **Monotone convergence.** Why is the sequence $E_n = (1 + \frac{1}{n})^n$ convergent? First, we will show that $E_n < 3$. The binomial expansion gives:

$$E_n = \left(1 + \frac{1}{n}\right)^n = 1 + 1 + \sum_{k=2}^{n} \binom{n}{k} \frac{1}{n^k}$$

Observe that:

$$\binom{n}{k} \frac{1}{n^k} = \frac{1}{k!} \frac{n(n-1)(n-2)\ldots(n-(k-1))}{n \cdot n \cdot n \ldots \cdot n}$$

$$= \frac{1}{k!} \cdot 1 \cdot \left(1 - \frac{1}{n}\right)\left(1 - \frac{2}{n}\right)\cdots\left(1 - \frac{k-1}{n}\right)$$

$$< \frac{1}{k!}$$

$$= \frac{1}{1 \cdot 2 \cdot 3 \ldots \cdot k}$$

$$\le \frac{1}{1 \cdot 2 \cdot 2 \ldots \cdot 2}$$

$$= \frac{1}{2^{k-1}}.$$

Therefore, the sequence is bounded above by the sum of a geometric series.

$$E_n < 2 + \sum_{k=2}^{n} \frac{1}{2^{k-1}} < 2 + \sum_{k=2}^{\infty} \frac{1}{2^{k-1}} = 3$$

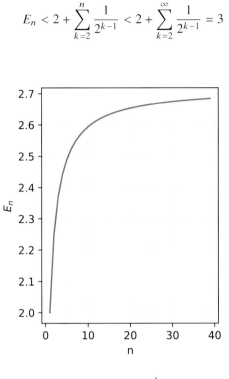

Fig. 1.3: $E_n = (1 + \frac{1}{n})^n$.

In addition, the graph of the sequence E_n (fig. 1.3) shows that the sequence is strictly increasing (*i.e.* $E_n < E_{n+1}$). The proof is an exercise in inequalities (see [20]). These facts imply that E_n converges due to the following theorem for *monotone* (*i.e.* increasing or decreasing) sequences.

Theorem 1.2 *(Monotone Convergence Theorem) A monotone sequence is convergent if and only if it is bounded.*

1.4 Convergence of series

> By plotting the partial sums, conjecture whether each of following series converges or diverges.
>
> a) $\displaystyle\sum_{k=1}^{\infty} \frac{1}{k^2}$,   b) $\displaystyle\sum_{k=1}^{\infty} \frac{(-1)^{k+1}}{k}$,   c) $\displaystyle\sum_{k=1}^{\infty} \frac{1}{\sqrt{k}}$,   d) $\displaystyle\sum_{k=1}^{\infty} \frac{1}{1+\ln k}$.

Let's have a look at one way to plot the partial sum for (a) where $S_n = \sum_{k=1}^{n} \frac{1}{k^2}$.

series-convergence.ipynb (for plotting the blue solid curve in fig. 1.4)

	```import numpy as np``` ```import matplotlib.pyplot as plt```
Calculate partial sums up to S_{10}	```x = np.arange(1,11)```

Initialise S to be an array with 10 zeros. We will use S to collect the partial sums
Define each term as a function
	```S = np.zeros(len(x))```
	
	```def afunc(k):```
	``` return 1/k**2```

The first partial sum	```S[0] = afunc(1)```
$S_i = S_{i-1} + (i\text{th term})$	```for i in range(1, len(x)):``` ``` S[i] = S[i-1] + afunc(x[i])```

Plot the partial sum with dots joined by solid line	```plt.plot(x, S, 'o-')```
	```plt.xlim(1, 10)```
	```plt.xlabel('Number of terms')```
	```plt.ylabel('Partial sum')```
	```plt.legend(['a'])```
	```plt.grid('on')```
	```plt.show()```

The graphs for (b), (c) and (d) can be obtained similarly by augmenting the above code. Fig. 1.4 shows the result. We are led to conjecture that series (a) and (b) converge, whilst (c) and (d) diverge.

DISCUSSION

- **p-series.** Whilst the conjectures are correct, the graphs do not constitute a *proof*. In analysis, we usually rely on a number of *convergence tests* to determine whether a series converges or diverges. These tests yield the following very useful result in analysis:

  Theorem 1.3 *The p-series* $\sum \frac{1}{n^p}$ *converges if* $p > 1$ *and diverges if* $p \leq 1$.

  where the shorthand $\sum$ here means $\sum_{n=n_0}^{\infty}$ for some integer n_0. This theorem explains why series (a) converges and (c) diverges.

- **Euler's series.** The exact expression for the sum in series (a) was found by Euler in 1734. The famous result

$$\sum_{n=1}^{\infty} \frac{1}{n^2} = \frac{\pi^2}{6}, \tag{1.2}$$

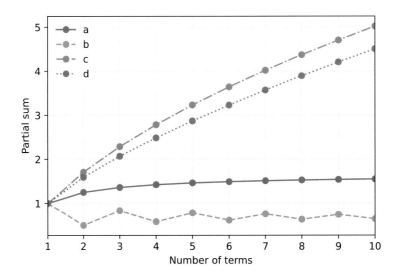

Fig. 1.4: The partial sums for the series (a)-(d) up to 10 terms

(sometimes called the *Basel problem*) has a special place in mathematics for its wide-ranging connections, especially to number theory.

Euler's series (1.2) is often interpreted as a particular value of the *Riemann zeta function* ($\zeta(2) = \pi^2/6 \approx 1.6449$). See reference [4] for a number of accessible proofs of this result. A proof using Fourier series will be discussed later in §2.9. We will study the zeta function more carefully later in §6.11.

- **Taylor series.** The value of series (b) can be derived from the well-known Taylor (Maclaurin) series

$$\ln(1+x) = x - \frac{x^2}{2} + \frac{x^3}{3} - \frac{x^4}{4} \cdots$$

valid for $x \in (-1, 1]$. Substituting $x = 1$ shows that series (b) converges to $\ln 2 \approx 0.6931$. We will discuss why this holds on the interval $(-1, 1]$ in §2.4.

- **Comparison test.** Finally, we can understand why series (d) diverges by observing that, because $n > \ln n$ for all $n \in \mathbb{N}$, we have

$$\frac{1}{1 + \ln n} > \frac{1}{1 + n}.$$

Thus, series (d) is greater than $\sum \frac{1}{n}$, which is divergent (as it is a p-series with $p = 1$). This technique of deducing if a series converges or diverges by considering its magnitude relative to another series (usually a p-series) can be formally expressed as follows.

Theorem 1.4 (Comparison Test) *Let x_n and y_n be real sequences such that (eventually) $0 \leq x_n \leq y_n$. Then a) $\sum x_n$ converges if $\sum y_n$ converges,   b) $\sum y_n$ diverges if $\sum x_n$ diverges.*

1.5 The Harmonic Series

The Harmonic Series is given by

$$\sum_{n=1}^{\infty} \frac{1}{n} = 1 + \frac{1}{2} + \frac{1}{3} + \dots$$

Show that the partial sum of the series grows logarithmically (*i.e.* increases at the same rate as the log function).

We wish to calculate the partial sums of the Harmonic Series, where each partial sum of N terms is given by

$$\sum_{n=1}^{N} \frac{1}{n}.$$

As in the previous section, to calculate the partial sum of $N + 1$ terms in Python, we simply add one extra term to the partial sum of N terms.

Another thing to note is that, because the sum grows very slowly, we will get a more informative graph if we plot the x-axis using log scale, whilst keeping the y-axis linear. This is achieved using the command `semilogx`.

The question suggests that we might want to compare the partial sums with the (natural) log curve. We will plot the two curves together on the same set of axes. In fact, if the partial sums up to N terms grow like $\ln N$, then it might even be interesting to also plot the difference between the two.

The code `harmonic.ipynb` produces two graphs, one stacked on top of the other. The top panel shows the growth of the harmonic series in comparison with the log. The difference is shown in the bottom panel. The calculation itself is rather short, but, as with many programs in this book, making the plots informative and visually pleasing takes a little more work.

The resulting plots, shown in fig. 1.5 shows a very interesting phenomenon: the upper plot shows that the partial sums grows very slowly just like the log, but offset by a constant. When we plot the difference between the two curves, we see that the difference is between 0.57 and 0.58.

These graphs lead us to conjecture that there is a constant γ such that

$$\gamma = \lim_{N \to \infty} \left(\sum_{n=1}^{N} \frac{1}{n} - \ln N \right). \tag{1.3}$$

It is useful to express this as an approximation:

$$\text{For large } N, \quad \sum_{n=1}^{N} \frac{1}{n} \approx \ln N + \gamma. \tag{1.4}$$

harmonic.ipynb (for plotting fig. 1.5)	
	```python\nimport numpy as np\nimport matplotlib.pyplot as plt\n```
How many terms in the sum? $n = 1, 2, \ldots$ Nmax Initialise hplot, using it to collect the partial sums. harmo is the running total	```python\nNmax = 1e5\nn = np.arange(1, Nmax+1)\nhplot = np.zeros_like(n)\nharmo = 0\n```
Collecting the partial sums $\texttt{hplot[N]} = \sum_{n=1}^{N+1} 1/n$	```python\nfor N in np.arange(0, int(Nmax)):\n harmo += 1/n[N]\n hplot[N] = harmo\n```
Create 2 stacked plots, specifying the dimension in inches The upper plot (log scale on the x-axis) showing the Harmonic Series and $\ln n$ Adjust legend location	```python\nfig, (ax1, ax2)=plt.subplots(2, figsize=(5,6))\nax1.semilogx(n, hplot, n, np.log(n) , '--')\nax1.set_xlim([10, Nmax])\nax1.set_ylim([2, 12])\nax1.legend(['Harmonic series', r'$\ln n$'],\n loc = 'lower right')\nax1.grid('on')\n```
The lower plot in red (log x-axis) showing Harmonic Series $- \ln n$	```python\nax2.semilogx(n, hplot-np.log(n), 'r')\nax2.set_xlim([10, Nmax])\nax2.set_ylim([0.57, 0.63])\nax2.set_xlabel(r'n')\nax2.legend([r'Harmonic - $\ln n$'])\nax2.grid('on')\nplt.show()\n```

DISCUSSION

- **The Euler-Mascheroni constant.** The constant γ is known as the *Euler-Mascheroni* constant (not to be confused with Euler's number e), where

$$\gamma = \texttt{np.euler\_gamma} = 0.5772\ldots$$

consistent with our findings. The convergence can be proved by showing that the difference is monotone decreasing (as seen in the lower panel of fig. 1.5) and bounded below. Hence the limit exists by the monotone convergence theorem. A comprehensive account of the history and mathematical significance of γ can be found in [121].

- **The Harmonic Series is divergent.** Another observation from the graphs is that the Harmonic Series diverges to ∞, just like the log. In fact, we can deduce the divergence by the following proof by contradiction. Suppose that the Harmonic series converges to S, then, grouping the terms pairwise, we find

$$S = \left(1 + \frac{1}{2}\right) + \left(\frac{1}{3} + \frac{1}{4}\right) + \left(\frac{1}{5} + \frac{1}{6}\right) + \ldots$$

$$> \left(\frac{1}{2} + \frac{1}{2}\right) + \left(\frac{1}{4} + \frac{1}{4}\right) + \left(\frac{1}{6} + \frac{1}{6}\right) + \ldots$$

$$= S.$$

Several other accessible proofs can be found in [113].

See exercise 1 for an intriguing physical situation in which the Harmonic Series appears.

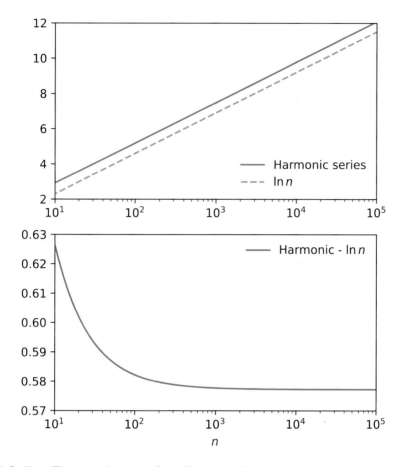

Fig. 1.5: *Top:* The partial sums of the Harmonic Series grows like $\ln n$. *Bottom:* The difference between the two curves approach a constant $\gamma \approx 0.577$.

1.6 The Fibonacci sequence

The Fibonacci sequence is given by the recursive relation:

$$F_1 = 1, \quad F_2 = 1, \quad F_n = F_{n-1} + F_{n-2}.$$

Investigate the growth of the following quantities: a) F_n b) $R_n := F_n/F_{n-1}$.

The Italian mathematician *Fibonacci* (1170–1250) mentioned the sequence in his *Liber Abaci* ('book of calculations') published in 1202, although the sequence was already known to ancient Indian mathematicians as early as around 200BC. The sequence is ubiquitous in nature and has surprising connections to art and architecture. See [169] for a readable account of the Fibonacci sequence.

A quick calculation of a handful of terms shows that

$$F_n = (1, 1, 2, 3, 5, 8, 13, 21, 34, 55, 89, 144, \ldots),$$

and R_n is the ratio of consecutive terms. The growth of F_n and R_n is shown in fig. 1.6. The figure is producing by the code `fibonacci.ipynb`.

Let's consider the growth of F_n. The semilog plot in fig. 1.6 is, to a good approximation, a straight line. This suggests that, for large n at least, we have an equation of the form

$$\ln F_n \approx \alpha n + \beta \implies F_n \approx A\phi^n,$$

where α and β are the gradient and y-intercept of the linear graph, and the constants $A = e^\beta$ and $\phi = e^\alpha$. We are only interested in the growth for large n so let's focus on the constant ϕ for now.

The gradient α of the line can be calculated using two consecutive points at n and $n - 1$. Joining these points gives a line with gradient

$$\alpha = \frac{\ln F_n - \ln F_{n-1}}{1} = \ln \frac{F_n}{F_{n-1}} = \ln R_n.$$
$$\implies \phi = e^\alpha = R_n.$$

Thus we conclude that F_n grows like ϕ^n for large n, where $\phi = \lim_{n \to \infty} R_n$, which according to the other plot in fig. 1.6, appears to be just below 1.62. In fact, Python tells us that $R_{25} \approx 1.618034$.

Finally, we estimate the y-intercept from the expression

$$\beta = \ln F_n - n \ln R_n.$$

Using $n = 25$, Python tells us that $A = e^\beta \approx 0.4472136$.

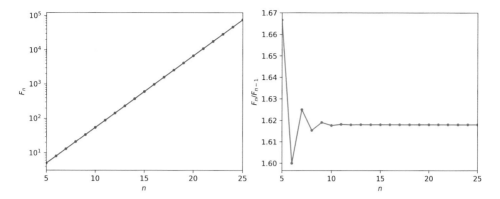

Fig. 1.6: *Left:* The Fibonacci sequence F_n plotted against n. The vertical scale is logarithmic. *Right:* The ratio R_n between consecutive Fibonacci numbers appears to approach a constant $\phi \approx 1.618$.

fibonacci.ipynb (for plotting fig. 1.6)	
	`import numpy as np` `import matplotlib.pyplot as plt`
Plot up to F_{25}	`Nend = 25`
Initialise the sequences F_n and $R_n = F_n/F_{n-1}$	`F = np.zeros(Nend+1)` `R = np.zeros(Nend+1)`
Define F_1 and F_2	`F[1] = 1` `F[2] = 1`

Iterate the recurrence	`for i in np.arange(3,Nend+1):`
	`    F[i] = F[i-1] + F[i-2]`
	`    R[i] = F[i]/F[i-1]`

Plot two figures side by side	`fig,(ax1,ax2)=plt.subplots(1,2,figsize=(11,4))`
Set smallest N value on the x-axis Plotting on domain [Nmin, Nend]	`Nmin = 5` `Nplt = range(Nmin, Nend+1)`

Use vertical log scale to see the growth of F_n	`ax1.semilogy(Nplt, F[Nmin:Nend+1], 'bo-', ms=3)`
	`ax1.set_xlabel(r'$n$')`
	`ax1.set_ylabel(r'$F_n$')`
	`ax1.set_xlim(Nmin, Nend)`

Manual adjustment of tick frequency	`ax1.set_xticks(range(Nmin, Nend+1, 5))` `ax1.grid('on')`

Use linear scales for plotting R_n	`ax2.plot(Nplt, R[Nmin:Nend+1], 'ro-', ms=3)`
	`ax2.set_xlabel(r'$n$')`
	`ax2.set_ylabel(r'$F_{n}/F_{n-1}$')`
	`ax2.set_xlim(Nmin, Nend)`
	`ax2.set_xticks(range(Nmin, Nend+1, 5))`
	`ax2.grid('on')`

	`plt.show()`

DISCUSSION

- **The Golden Ratio**. ϕ in fact corresponds to the *Golden Ratio*

$$\phi = \frac{1 + \sqrt{5}}{2}.$$

In fact, our estimate R_{25} is remarkably accurate to 9 decimal places. The connection between R_n and ϕ can be seen by observing that, from the recurrence relation, we find

$$\frac{F_n}{F_{n-1}} = 1 + \frac{F_{n-2}}{F_{n-1}} \implies R_n = 1 + \frac{1}{R_{n-1}}. \tag{1.5}$$

Thus if the $\lim_{n \to \infty} R_n$ exists and equals ϕ, then it satisfies the equation

$$\phi = 1 + \frac{1}{\phi},$$

which defines the Golden Ratio.

- **Binet's formula**. Python allowed us to discover part of a closed-form expression for F_n called Binet's formula

$$F_n = \frac{\phi^n - (1 - \phi)^n}{\sqrt{5}}.$$

We will derive this formula when we study matrix diagonalisation in §5.7 . For large n, we see that $F_n \approx \frac{1}{\sqrt{5}}\phi^n$. More precisely, for any positive integer n, the Fibonacci number F_n is the integer closest to the number $\frac{1}{\sqrt{5}}\phi^n$. This follows from the fact that the second term in Binet's formula is small. To see this, let $r = 1 - \phi$ and note that $|r|^n = (0.618\ldots)^n < 1$ for all n. Therefore,

$$\frac{1}{\sqrt{5}}(1 - \phi)^n < \frac{1}{\sqrt{5}} < \frac{1}{\sqrt{4}} = \frac{1}{2}.$$

This shows that F_n is the integer nearest to $\frac{1}{\sqrt{5}}\phi^n$.

- **Contractive sequences**. It remains to explain why the limit of R_n exists. This is a consequence of the following theorem (see [20]).

Theorem 1.5 *If a sequence x_n eventually satisfies the relation*

$$|x_{n+1} - x_n| \leq C|x_n - x_{n-1}|$$

where the constant $C \in [0, 1)$, then x_n is convergent.

Such a sequence is said to be *contractive*. We show that the sequence R_n is contractive as follows. From Eq. 1.5, it is clear that for $n \in \mathbb{N}$, $R_n \in [1, 2]$, which implies $R_{n+1} = 1 + \frac{1}{R_n} \geq \frac{3}{2}$. Using (1.5) again we find that for $n \geq 4$,

$$\begin{aligned} |R_{n+1} - R_n| &= \left| \frac{1}{R_n} - \frac{1}{R_{n-1}} \right| \\ &= \frac{|R_n - R_{n-1}|}{|R_n R_{n-1}|} \\ &\leq \frac{2}{3} \cdot \frac{2}{3}|R_n - R_{n-1}| = \frac{4}{9}|R_n - R_{n-1}|. \end{aligned}$$

1.7 The ε, δ definition of continuity

Students are taught in school that a continuous function is one for which the graph can be drawn without lifting the pen. However, at university, the focus of mathematics is shifted away from drawings, descriptions and intuition to mathematics that is based purely on logic. In this spirit, we want to be able to define continuity logically (*i.e.* using symbols, equations, inequalities...) without relying on drawings.

A function $f: A \to \mathbb{R}$ is said to be *continuous at a point* $x_0 \in A$ if, for all $\varepsilon > 0$, there exists $\delta > 0$ such that, for all $x \in A$, we have

$$|x - x_0| < \delta \implies |f(x) - f(x_0)| < \varepsilon.$$

For the following functions, investigate the relationship between ε and δ at $x = x_0$.

a) $f(x) = \dfrac{2}{x}$ $(x_0 = 1)$,

b) $g(x) = \ln(x + 1)$ $(x_0 = 1)$, c) $h(x) = \begin{cases} \dfrac{\sin x}{x} & x \neq 0, \\ 1 & x = 0. \end{cases}$ $(x_0 = 0)$.

The ε-δ definition given above was first published in 1817 by the Bohemian mathematician and philosopher *Bernard Bolzano* (1781–1848). It expresses precisely what it means for a function to be continuous at a point purely in terms of inequalities. This is one of the most important definitions in real analysis, and is also one which many beginning undergraduates struggle to understand. Let's first try to unpack what the definition says.

If f is continuous at x_0, it makes sense to demand that we can always find some y values *arbitrarily* close to (or equal to) $y_0 = f(x_0)$. Symbolically, the set of "*all y values arbitrarily close to y_0*" can be expressed as follows.

Definition Let $\varepsilon > 0$ and $y_0 \in \mathbb{R}$. The *ε-neighbourhood* of y_0 is the set of all $y \in \mathbb{R}$ such that $|y - y_0| < \varepsilon$.

Now, we want the y values inside the ε-neighbourhood of y_0 to "come from" some values of x. In other words, there should be some x values such that $y = f(x)$. It is natural to demand that those values of x should also be in some neighbourhood of x_0. We are satisfied as long as "*there exists such a neighbourhood*" in the domain A. This statement can be written symbolically as:

$$\exists \delta > 0 \text{ such that } \forall x \in A, |x - x_0| < \delta \ldots$$

The symbol $\exists$ reads "*there exists*", and $\forall$ reads "*for all*".

Combining what we have discussed so far, we say that a function f is continuous at x_0 if there exists a neighbourhood of x_0 which gets mapped by f into an arbitrarily small neighbourhood of $y_0 = f(x_0)$. Symbolically:

$$\forall \varepsilon > 0, \exists \delta > 0 \text{ such that } \forall x \in A, |x - x_0| < \delta \implies |f(x) - f(x_0)| < \varepsilon.$$

This explains the intuition behind the definition. One could think about this as a game in which we are given a value of ε, and our job is to find a suitable value of δ.

In fact, if we can find a certain value of δ that works, then so will any other values δ' such that $0 < \delta' < \delta$. Often we may find that there is a largest value, $\delta_{\max}$, that works for each choice of ε.

Now let's see how Python can help us visualise this ε-δ game for the function a) $f(x) = 2/x$, at $x_0 = 1$. The code `continuity.ipynb` produces a GUI (graphical user interface) as shown in fig. 1.7. Warning: This code only works in a Jupyter environment[1].

continuity.ipynb (for plotting the top panel of fig. 1.7). Jupyter only	
`ipywidgets` make the plot interactive	```import matplotlib.pyplot as plt
import numpy as np
from ipywidgets import interactive```
The function f	```def f(x):
return 2/x```
The inverse function f^{-1}	```def finverse(x):
return 2/x```
Domain We will study the continuity of f at x_0	```x = np.linspace(0.5, 2)
x0 = 1
y = f(x)
y0 = f(x0)```
Now define a function of ε to feed into the slider $f(x_0) + \varepsilon$ $f(x_0) - \varepsilon$ The x values of the above two points (We use them to calculate δ)	```def plot(eps):
y0p = y0+eps
y0m = y0-eps
x0p = finverse(y0p)
x0m = finverse(y0m)```
Where to draw vertical and horizontal dotted lines	``` vertical = [x0, x0p, x0m]
horizontal = [y0, y0p, y0m]```
Plot $y = f(x)$ in red An easy way to plot horizontal lines (black dotted) ...and vertical lines (cyan dotted)	``` plt.plot(x, y, 'r')
for Y in horizontal:
plt.axhline(y = Y, color = 'k',
linestyle = ':')
for X in vertical:
plt.axvline(x = X, color = 'c',
linestyle = ':')
plt.show()```
The largest viable value of δ	``` delta= min(abs(x0-x0p), abs(x0-x0m))```
Report values (using unicode for Greek letters)	``` print(f'Given \u03B5 = {eps:.2}')
print(f'Found \u03B4 = {delta:.4}')```
Finally, set slider for $\varepsilon \in [0.01, 0.4]$ in steps of 0.01	```interactive(plot, eps=(0.01, 0.4, 0.01))```

In the code, we use the `ipywidgets`[2] library to create a slider for the value of ε. Drag the blue slider to change the separations of the horizontal dotted lines which correspond to the ε-neighbourhood of $f(x_0)$.

---

[1] If you are running a non-Jupyter IDE, Matplotlib's slider (to be discussed shortly) can be used to to produce fig. 1.7. The code is given on GitHub as `continuity.py`.

[2] https://ipywidgets.readthedocs.io

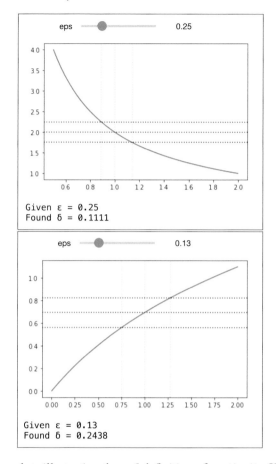

Fig. 1.7: Interactive plots illustrating the ε-δ definition of continuity. Here we investigate the continuity at $x_0 = 1$ of functions defined by $f(x) = 2/x$ (top), and $g(x) = \ln(x + 1)$ (bottom). In each graph, you can use the slider to adjust the value of ε, and a viable value of δ is displayed.

Given this ε, the largest δ-neighbourhood of x_0 can be found by taking the smallest separation between the vertical dotted lines. In this example, we can write

$$\delta_{\max} = \min \left\{ \left| f^{-1}(y_0 + \epsilon) - x_0 \right|, \left| f^{-1}(y_0 - \epsilon) - x_0 \right| \right\}.$$

Conveniently, f is decreasing and so $f^{-1}(x)$ can be evaluated easily (in fact $f = f^{-1}$). Thus, it is also possible to calculate precisely the value of $\delta_{\max}$. Given $\varepsilon = 1/4$ as shown in the top panel of fig. 1.7, our formula gives

$$\delta_{\max} = \min \left\{ \left| f^{-1}(9/4) - 1 \right|, \left| f^{-1}(7/4) - 1 \right| \right\}$$
$$= \min \{1/9, 1/7\}$$
$$= 1/9.$$

This agrees with the value of δ displayed by the code at the bottom of the GUI.

We can easily modify the code to illustrate the continuity of $g(x) = \ln(x + 1)$ at $x_0 = 1$, as shown in the lower panel of fig. 1.7. We leave it as an exercise for you to show that the exact expression for δ shown is given by

$$\delta_{max} = 2(1 - e^{-0.13}) \approx 0.2438.$$

The previous code relies on the fact that the function is either strictly decreasing or increasing and that the expression for the inverse function can be found explicitly. If these conditions do not hold, the positions of the vertical lines may not be so easy to obtain. This is the case for the function h shown in fig. 1.8, where

$$h(x) = \begin{cases} \dfrac{\sin x}{x} & x \neq 0, \\ 1 & x = 0. \end{cases}$$

One way to get around this is to ask the GUI to read out the coordinates of where the horizontal lines intersect the curve $y = h(x)$. In fig. 1.8, with $\varepsilon = 0.1$, we find that $|h(x) - h(0)| = \varepsilon$ at $x \approx \pm 0.79$, so taking any δ to be less than this value would make the ε-δ definition work. Indeed, $h(x)$ is continuous at $x = 0$.

The code `continuityslider.ipynb` produces the interactive GUI shown in fig. 1.8. We use Matplotlib's own `Slider` widget to create an interactive plot (which you can pan and zoom in to get a more accurate value of δ).

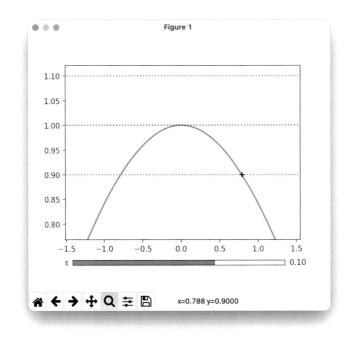

Fig. 1.8: An interactive GUI showing the graph of $h(x) = \sin x / x$ around $x = 0$ (with $h(0) := 1$). The dotted lines show $y = 1$ and $1 \pm \varepsilon$, where the value of ε can be adjusted using the slider. The coordinates of the cursor are given at the bottom. The readout shows that for $\varepsilon = 0.1$, we need $\delta \approx 0.79$.

continuityslider.ipynb (for plotting fig. 1.8)

See Preface "b) About %matplotlib"

```python
import numpy as np
import matplotlib.pyplot as plt
from matplotlib.widgets import Slider
%matplotlib
```

The function takes an array

enumerate gives pairs of numbers:
the indices (ind) and elements (x) in the array
The function maps 0 to 1...

and maps nonzero x to $\frac{\sin x}{x}$

```python
def f(xarray):
    y = np.zeros_like(xarray)
    for ind, x in enumerate(xarray):
        if x==0:
            y[ind] = 1
        else:
            y[ind] = np.sin(x)/x
    return y
```

Choose domain to plot

Initial value of ε

```python
xarray = np.linspace(-2, 2,200)
y = f(xarray)
x0 = 0
y0 = f([x0])
eps = 0.1
```

y arrays for 3 horizontal lines at y_0
and $y_0 \pm \varepsilon$

```python
harray0 = np.full_like(xarray, y0)
harrayP = np.full_like(xarray, y0+eps)
harrayM = np.full_like(xarray, y0-eps)
```

Leave a space at the bottom for a slider

```python
fig,ax = plt.subplots()
plt.subplots_adjust(bottom = 0.2)
```

Plot the function with a thick line

```python
plt.ylim(0.5,1.15)
plt.plot(xarray, f(xarray) , lw=2)
```

Plot the 3 horizontal dotted lines:
$y = y_0$ in blue, $y = y_0 \pm \varepsilon$ are in red
Use commas to unpack the lists - see Discussion

```python
h0, = plt.plot(xarray, harray0, 'b:')
hP, = plt.plot(xarray, harrayP, 'r:')
hM, = plt.plot(xarray, harrayM, 'r:')
```

The slider's dimensions and location
Create a slider
specify range of values, step size and...
the initial value of ε

```python
axeps = plt.axes([0.15, 0.1, 0.7, 0.02])
eps_slide = Slider(axeps, '\u03B5',
            0, 0.15, valstep = 0.001,
            valinit = eps)
```

Update the plot if slider is changed
Take new value of ε
The location of the horizontal lines $y_0 \pm \varepsilon$ are
updated

```python
def update(val):
    eps = eps_slide.val
    hP.set_ydata(y0 + eps)
    hM.set_ydata(y0 - eps)
    fig.canvas.draw_idle()
```

Redraw the graph

```python
eps_slide.on_changed(update)
```

```python
plt.show()
```

- **Proof of continuity**. The graphs shown in this section do not *prove* continuity. They only allow us to visualise the ε-δ game. Writing a rigorous ε-δ proof is an important topic that will keep you very much occupied in your analysis course at university. To give a flavour of what is involved, here is a rigorous proof that $f(x) = 2/x$ is continuous at $x = 1$.

  Proof: For all $\varepsilon > 0$, take $\delta = \min\{1/2, \varepsilon/4\}$, so that $\forall x \in \mathbb{R} \setminus \{0\}$,

  $$|x - 1| < \delta \implies |f(x) - f(1)| = \left|\frac{2}{x} - 2\right|$$
  $$= \frac{2|x - 1|}{|x|} \qquad (*)$$

  Since $|x - 1| < 1/2$, the reverse triangle inequality gives

  $$-1/2 < |x| - 1 < 1/2 \implies 2/3 < \frac{1}{|x|} < 2.$$

  Substituting this into $(*)$, we find

  $$|f(x) - f(1)| < 4\delta \leq \varepsilon. \qquad \square$$

- **Limits**. A closely related related concept to continuity is that of continuous *limits*. Given a function $f : (a, b) \to \mathbb{R}$, and a point $c \in (a, b)$. We write

  $$\lim_{x \to c} f(x) = L,$$

  if $\forall \varepsilon > 0, \exists \delta > 0$ such that $\forall x \in (a, b)$, we have

  $$0 < |x - c| < \delta \implies |f(x)| - L| < \varepsilon.$$

  The only difference between this definition and that of continuity is that for limits, there is no mention of what happens at $x = c$, but only what happens close to c.
  Using this definition, it can be shown that the following important theorem holds.

  Theorem 1.6 *Let $A \subseteq \mathbb{R}$ and define $f : A \to \mathbb{R}$. f is continuous at $c \in A$ if and only if* $\lim_{x \to c} f(x) = f(c)$.

  This rigorous definition of the continuous limit lays a strong foundation for differentiation and integration, both of which can be expressed as continuous limits, as we will see later.

- **The sinc function**. In the language of limits, our plot of $h(x)$ shows that

  $$\lim_{x \to 0} \frac{\sin x}{x} = 1. \qquad (1.6)$$

  The proof of this limit based on the Squeeze Theorem (for continuous limits) can be found in [20]. The function $h(x)$ is sometimes written $\text{sinc}(x)$. It has many real-world applications, particularly in signal processing.

• **Why comma?** In the code, you may be wondering why we used a comma on the LHS of the assignment

```
hP, = plt.plot(xarray, harrayP, 'r:')
```

rather than

```
hP  = plt.plot(xarray, harrayP, 'r:')
```

Indeed, had we removed the comma, Python would report errors when the slider is moved. So what is going on here? Well, let's ask Python what type the object hP is. With the comma, the command `type(hP)` tells us that hP is a `Line2d` object. This object has many properties including the x and y coordinates of the lines (called `xdata` and `ydata`) and optional colours and line thickness attributes (type `dir(hP)` to see the full list of attributes). Indeed, when the slider is moved, the `update` function updates the y coordinates of the dashed line.

Without the comma, however, we find that hP is a *list*. Furthermore, `len(hP)=1`, meaning that the object `plt.plot(...)` is in fact a list with one element (namely, the `Line2d` object). When we move the slider, the y coordinates (`ydata`) is not a property of this list, but rather the object *within* the list. This explains why Python reports an error.

To put this in another way, we want to change the filling of the sandwich within the box, rather than put new filling onto the box itself.

In summary, the comma tells Python to *unpack* the list (*i.e.* take out the sandwich), so we can update its content.

1.8 Thomae's function

Thomae's function $f : (0, 1) \to \mathbb{R}$ is defined by

$$f(x) = \begin{cases} \frac{1}{q} & \text{if } x \in \mathbb{Q} \text{ and } x = \frac{p}{q} \text{ in lowest form, where } p, q \in \mathbb{N}, \\ 0 & \text{otherwise.} \end{cases}$$

Plot this function. For how many values $x \in (0, 1)$ is $f(x) > \frac{1}{10}$? or $\frac{1}{100}$?
Deduce that Thomae's function is continuous at irrational x and discontinuous at
rational x.

This function, named after the German mathematician *Carl Johannes Thomae* (1840–1921), serves as a classic illustration of how university mathematics differs from school mathematics. Few school students would have seen functions defined in such an exotic way.

Let's try to make sense of the function, for example, by first looking for values of x that would be mapped to the value $1/8$. A little experiment reveals that there are 4 such values:

$$\frac{1}{8}, \frac{3}{8}, \frac{5}{8}, \frac{7}{8}.$$

Note that $f(2/8) = f(6/8) = 1/4$ and $f(4/8) = 1/2$.

It is clear that if $f(x) = 1/8$ then x must be a rational number of the form $p/8$ where p and 8 have no common factors apart from 1. Another way to say this is that p has to be *coprime* to 8. Yet another way to say this is that the *greatest common divisor* (gcd) of p and 8 is 1, *i.e.*

$$\gcd(p, 8) = 1.$$

(More about the gcd when we discuss number theory in chapter 6.)

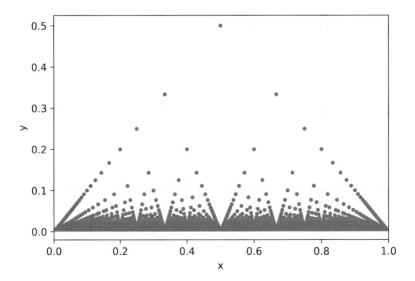

Fig. 1.9: Thomae's function

The graph of Thomae's function is shown in fig. 1.9. To plot this graph, we can conveniently use NumPy's gcd function as shown in the code thomae.ipynb. The code also counts how many values of x satisfy $f(x) > 1/10$.

thomae.ipynb (for plotting fig. 1.9)	
	``import numpy as np
import matplotlib.pyplot as plt```
Initialising x and y as empty lists	```xlist = []
ylist = []```
Search for fractions with denominators from 2 to 199 If p and q are coprime... append the lists of x and y coordinates	```for q in range(2,200):
for p in range(1,q):
if np.gcd(p,q) == 1:
xlist.append(p/q)
ylist.append(1/q)```
Plot the points with big red dots	```plt.plot(xlist, ylist, 'or', ms=3)
plt.xlabel('x')
plt.ylabel('y')
plt.xlim(0,1)
plt.grid('on')
plt.show()```
Count how many points are above 0.1 Use list comprehension to do this and report	```lim = 0.1
num = sum(y > lim for y in ylist)
print(f'Found {num} points above y={lim}')```

In the code, we use the *for* loops to append values to empty lists. This creates 2 growing lists of the x and y coordinates. Instead of lists, one could also use NumPy arrays (using np.empty and np.append).

Running the code tells us that there are 27 values. As an exercise, try to write them all down. As for the case $f(x) > 1/100$, the code gives us 3003 values.

What do these results mean? Well, they imply that given any number $\varepsilon > 0$, there are a *finite* number of x such that $f(x) > \varepsilon$. This means that at any $x_0 \in (0, 1)$, we can find a neighbourhood of x_0 sufficiently small that it does not contain any values of x such that $f(x) > \varepsilon$. In other words, $|f(x)| < \varepsilon$ for all x sufficiently close to x_0.

Now let x_0 be any irrational number in $(0, 1)$. Since $f(x_0) = 0$, the previous paragraph gives the following result:

$$\forall \varepsilon > 0, \exists \delta > 0 \text{ such that, } \forall x \in (0, 1), |x - x_0| < \delta \implies |f(x) - f(x_0)| < \varepsilon.$$

This is precisely the ε-δ definition for the continuity of Thomae's function at any irrational $x_0 \in (0, 1)$.

DISCUSSION

- **The Archimedean property**. In deducing the fact that there are a finite number of x such that $f(x) > \varepsilon$, we implicitly used the following property of real numbers.

Theorem 1.7 (*The Archimedean property*) *For any $\varepsilon > 0$, there exists an integer $n \in \mathbb{N}$ such that $\varepsilon > 1/n$.*

(Can you see precisely where this has been used?) This property sounds like an obvious statement, but, like many results in introductory analysis, it has to be proven. It turns out that the Archimedean property follows from other *axioms*, or properties of real numbers which are satisfied by decree. These axioms consist of the usual rules of addition an multiplication, plus some axioms on inequalities which allow us to compare sizes of real numbers. In addition, we also need:

The Completeness Axiom. Every bounded nonempty set of real numbers has a least upper bound.

For example, the least upper bound of real numbers $x \in (0, 1)$ is 1. The axiomatic approach to constructing $\mathbb{R}$ is a hugely important topic in introductory real analysis which you will encounter at university.

- **Sequential criterion**. The ε-δ definition for continuity is equivalent to the following.

Theorem 1.8 *(**Sequential criterion for continuity**) The function $f : A \to \mathbb{R}$ is continuous at $x_0 \in A$ if and only if, for every sequence $x_n \in A$ converging to x_0, we have $f(x_n) \to f(x_0)$.*

This theorem can be used to prove that Thomae's function is discontinuous at any rational number $x_0 \in (0, 1)$. Consider the sequence

$$x_n = x_0 - \frac{\sqrt{2}}{n},$$

which must necessarily be a sequence of irrational numbers (here we use the well-known result that $\sqrt{2}$ is irrational). The sequence clearly converges to $x_0 \in \mathbb{Q}$. Thus, we have found a sequence $x_n \to x_0$ such that $f(x_n) = 0 \nrightarrow f(x_0)$. This proves that f cannot be continuous at $x_0 \in \mathbb{Q} \cap (0, 1)$.

1.9 The Intermediate Value Theorem and root finding

Solve the equation $e^{-x} - x = 0$.

The Intermediate Value Theorem (IVT) is an important result concerning functions which are continuous on a closed bounded interval $[a, b]$. Continuity on an interval simply means that the ε-δ definition applies to all points in the interval. Such functions have special properties such as the following 'intermediate value' property.

Theorem 1.9 (Intermediate Value Theorem) *Let f be a function which is continuous on $[a, b]$. If v is a value strictly between $f(a)$ and $f(b)$, then there exists $c \in (a, b)$ such that $f(c) = v$.*

In other words, a continuous function f on $[a, b]$ takes *all possible values* between $f(a)$ and $f(b)$.

Now consider the function $f(x) = e^{-x} - x$, which comprises the exponential function and a linear function. Both functions (and their difference) can all be shown to be continuous at every $x \in \mathbb{R}$ using the ε-δ definition. In particular, we note that f is continuous on the interval $[0, 1]$ and that f takes opposite signs at the end points of this interval, *i.e.*

$$f(0) = 1, \quad f(1) = 1/e - 1 < 0.$$

Therefore, the IVT tells us that $\exists c \in (0, 1)$ such that $f(c) = 0$. The graph below illustrates that the root of $f(x) = 0$ indeed lies in this interval.

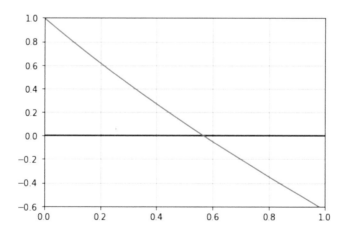

Fig. 1.10: The graph $y = e^{-x} - x$ intersects the x-axis somewhere in the interval $(0,1)$. This is consistent with the Intermediate Value Theorem.

One way to proceed is to consider the sign of $f(x)$ at the midpoint $x = 0.5$. We find that

$$f(0.5) = 1/\sqrt{e} - 0.5 > 0.$$

Since $f(x)$ has opposite signs at the endpoints of the 'bisected' interval $(0.5, 1)$, the IVT again implies that f has a root in $(0.5, 1)$.

We can carry on bisecting this interval and study the sign of $f(x)$ at the midpoint and repeat until we achieve a root c in an interval that is as small as the desired accuracy.

The code below illustrates how this process (called the *bisection method* of root finding) can be iterated as many times as required to achieve a root with accuracy `acc`. The core of the code is a `while` loop which repeats the iteration until the size of the bisected interval shrinks below `acc`.

bisection.ipynb (for performing root-finding using the bisection method)	
	`import numpy as np` `import matplotlib.pyplot as plt`
Define $f(x)$	`def f(x):` `    return np.exp(-x)-x`
Specify accuracy required Specify interval $[a, b]$	`acc = 0.5e-5` `a, b = 0, 1`
Function values at interval endpoints	`fa, fb = f(a), f(b)`
If $f(a)$ and $f(b)$ don't have opposite signs, report error and abort mission	`if fa*fb>=0:` `    raise ValueError('Root is not bracketed.')`

Iteration counter
Repeat as long as the error is too large
This step is the bisection
	`n=0`
	`while (b-a)/2 > acc:`
	`    x = (a+b)/2`
	`    fx = f(x)`

If lucky (found exact root),
. . .jump out of while loop
If root lies in $[a, x]$,
. . .make x the new b
Otherwise, root lies in $[x, b]$
	`    if fx == 0:`
	`        break`
	`    if fx*fa < 0:`
	`        b = x`
	`    else:`
	`        a = x`

Report result of every iteration Increase iteration count and repeat	`    print(f'Iteration number {n}, x={x}')` `    n += 1`
Final bisection Report answer	`x = (a+b)/2` `print(f'Final iteration number {n}, x= {x}')`

Running the code above with $\text{acc} = 0.5 \times 10^{-5}$ (to ensure 5 dec. pl. accuracy) shows that

```
Final iteration number 17, x= 0.5671424865722656
```

Thus, we can report that the root of the equation $e^{-x} - x = 0$ is approximately 0.56714 (5 dec. pl.).

DISCUSSION

- **Newton-Raphson method**. The bisection method is a relatively slow but reliable method of root-finding for most practical applications. A faster root-finding method called the *Newton-Raphson method* is discussed in the exercise 11. Although the Newton-Raphson method is faster, it requires additional information, namely, the expression for the derivative $f'(x)$, which is not always available in real applications.

- **Bracketing**. For the bisection algorithm to start, we first need to *bracket* the root, *i.e.* find an interval $[a, b]$ in which $f(a)$ and $f(b)$ have opposite signs . However, this may not be possible, for instance, with $f(x) = x^2$. In this case another root-finding method must be used. See [131, 132] for other root finding algorithms in Python.

- **Throwing**. The Python command `raise` is useful for flagging a code if an error has occurred. This practice is also known as *throwing* an error. The simplest usage is:

  `if` (some conditions are satisfied):
  ```
  raise ValueError('Your error message')
  ```

  It is good practice to be specific in your error message about what exactly has gone wrong.

- **Numerical analysis** is the study of the accuracy, convergence and efficiency of numerical algorithms. This field of study is essential in understanding the limitation of computers for solving mathematical problems. We will explore some aspects of numerical analysis in this book, particularly in the next chapter. For further reading on numerical analysis, see, for example, [40, 170, 182].

1.10 Differentiation

> For each of the following functions, plot its graph and its derivative on the interval
> $[-1, 1]$.
>
> a) $f(x) = \sin \pi x$,   b) $g(x) = \sqrt{|x|}$,   c) $H(x) = \begin{cases} x^2 \sin(1/x^2) & x \neq 0, \\ 0 & x = 0. \end{cases}$
>
> Which functions are differentiable at $x = 0$?

Let $f: (a, b) \to \mathbb{R}$ and let $c \in (a, b)$. The **derivative** of f at $x = c$, denoted $f'(c)$ is
defined as:

$$f'(c) = \lim_{h \to 0} \frac{f(c + h) - f(c)}{h}. \tag{1.7}$$

A function is said to be **differentiable** at $x = c$ if the limit above is finite.

In school mathematics, the derivative is often defined as the *rate of change* of a function,
or the gradient of the tangent to $y = f(x)$ at $x = c$. However, in university *analysis*,
pictorial definitions are not only unnecessary, but must also be avoided in favour of rigorous
logic-based definitions. The limit (1.7) has a precise definition in terms of ε-δ (see §1.7).

First let's consider the derivative of $f(x) = \sin \pi x$. You will probably have learnt how
to differentiate this type of function at school. But let's see how we can also work this out
from *first principles* using the definition above. Recall the trigonometric identity:

$$\sin \alpha - \sin \beta = 2 \cos \left(\frac{\alpha + \beta}{2} \right) \sin \left(\frac{\alpha - \beta}{2} \right).$$

Using this identity in the limit, we have:

$$\begin{aligned} f'(c) &= \lim_{h \to 0} \frac{\sin \pi(c + h) - \sin \pi c}{h} \\ &= \lim_{h \to 0} \frac{2 \cos \pi(c + \frac{h}{2}) \sin \frac{\pi h}{2}}{h} \\ &= \lim_{h \to 0} \cos \pi(c + \frac{h}{2}) \cdot \lim_{h \to 0} \pi \frac{\sin \frac{\pi h}{2}}{\frac{\pi h}{2}}. \end{aligned} \tag{1.8}$$

where we have broken up the limit into a product of two limits. The first limit in (1.8) is
simply $\cos \pi c$ (technical note: this step requires the continuity of cos). The second limit can
be evaluated using the result from Eq. 1.6 in §1.7, giving us π. Therefore, we have

$$f'(c) = \pi \cos \pi c,$$

as you might have expected. Note that f' is defined at all values of $c \in \mathbb{R}$. In other words, f
is differentiable on $\mathbb{R}$. In particular, it is certainly differentiable at $x = 0$ with $f'(0) = \pi$.

The next function g can be written in piecewise form (using the definition of the modulus)
as:

$$g(x) = \begin{cases} \sqrt{x} & x \geq 0, \\ \sqrt{-x} & x < 0. \end{cases}$$

This can be differentiated in the usual way for $x \neq 0$, giving

$$g'(x) = \begin{cases} \frac{1}{2\sqrt{x}} & x > 0, \\ -\frac{1}{2\sqrt{-x}} & x < 0. \end{cases}$$

But that at $x = 0$, the limit definition gives

$$g'(0) = \lim_{h \to 0} \frac{\sqrt{|h|}}{h},$$

which becomes arbitrarily large near $h = 0$, so the limit does not exist. Thus, g is not differentiable at $x = 0$.

Before we consider $H(x)$, let's pause to consider how derivatives can be calculated on the computer. Of course, one could simply differentiate the function by hand, then simply code the result. However, in real applications, we may have limited information on the function to be differentiated. Sometimes the expression for the function itself cannot be easily written down. This means that it is often impractical to rely on the explicit expression for the derivative.

Instead, we can work with a numerical approximation of the limit definition (1.7). For example, we could say:

$$f'(x) \approx \frac{f(x+h) - f(x)}{h}, \tag{1.9}$$

for a small value of h. This is called the *forward-difference* or *forward-Euler* estimate of the derivative. The word '*forward*' comes from the fact that the gradient at x is approximated as the slope of the line joining the points on the curve at x and $x + h$, a little "forward" from x.

Figure 1.11 shows the graphs of f, g (dashed blue lines) and their approximate derivatives (solid orange lines) calculated using the forward-difference approximation (1.9) with $h = 10^{-6}$. Note in particular that the lower panel shows that the derivative of g blows up near $x = 0$, where $y = g(x)$ has a sharp point (similar to that the graph of $y = |x|$). The graph indeed confirms that g is not differentiable at $x = 0$.

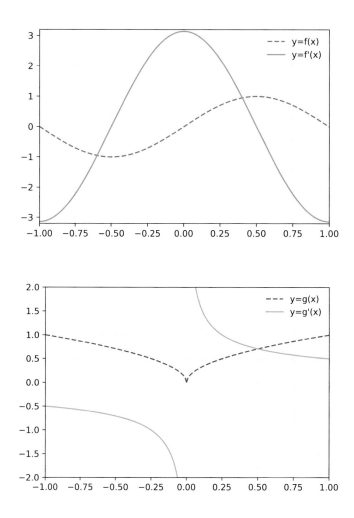

Fig. 1.11: The graphs of functions $f(x) = \sin \pi x$ (top) and $g(x) = \sqrt{|x|}$ (bottom) – in dashed blue lines – and their derivatives (solid orange lines) calculated using the forward-difference approximation (1.9) with $h = 10^{-6}$.

The code `differentiation.ipynb` plots the graphs of g and g'. Note that we work on the positive and negative x values separately because, otherwise, Matplotlib would join up points around $x = 0$, creating a false visual impression of the values of $g'(x)$ near 0.

differentiation.ipynb (for plotting the bottom panel of fig. 1.11)	
| | ```
import numpy as np
import matplotlib.pyplot as plt
``` |
| Function to be differentiated | ```
def g(x):
    return np.sqrt(np.abs(x))
``` |
| Pick a small $h$<br>Work with $x > 0$ and $x < 0$ separately | ```
h=1e-6
xp= np.linspace(0,1)
xn= np.linspace(-1,-h)
gxp = g(xp)
gxn = g(xn)
``` |
| $g'(x)$ approximation (forward Euler) | ```
dgp = (g(xp+h) - gxp)/h
dgn = (g(xn+h) - gxn)/h
``` |
| | ```
plt.plot(xp, gxp, 'b--',
         xp, dgp, 'orange',
         xn, gxn, 'b--',
         xn, dgn, 'orange')
plt.legend(["y=g(x)" ,"y=g'(x)"])
plt.grid('on')
plt.xlim([-1,1])
plt.ylim([-2,2])
plt.show()
``` |

Now let's return to $H(x)$. With simple modifications of the code, we can plot H and H' as shown in fig. 1.12. It appears that the derivative fluctuates wildly around $x = 0$. One might even be tempted to conclude from the graph that H is not differentiable at $x = 0$.

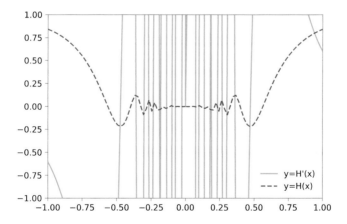

Fig. 1.12: The graphs of function $H(x)$ (blue dashed lines) and its derivative (solid orange lines) calculated using the approximation (1.9) with $h = 10^{-6}$.

But graphs can be misleading! In fact, H *is* differentiable at $x = 0$, as we now prove using the limit definition:
```

$$H'(0) = \lim_{h \to 0} \frac{H(h) - H(0)}{h}$$
$$= \lim_{h \to 0} h \sin(1/h^2).$$

Note that we replaced $H(h)$ by the expression for $h \neq 0$. This follows from the fact that the limit as $h \to 0$ is defined without requiring us to know what happens *at* $h = 0$. Observe then that for any $h \in \mathbb{R}$, we have

$$-|h| \leq h \sin(1/h^2) \leq |h|,$$

since $-1 \leq \sin(1/h^2) \leq 1$. Taking the limit as $h \to 0$, we conclude that $H'(0) = 0$ by the Squeeze Theorem.

DISCUSSION

- **How small should h be?** In the forward-difference approximation 1.9, it appears as though the smaller h is, the more accurate the estimate for the derivative becomes. Surprisingly, this is not the case! You could try this yourself by changing h in the code from 10^{-6} to, say 10^{-20}. What do you observe?

  In fact, there is an optimal value of h which gives the most accurate answer for the derivative. Larger or small values of h would give answers that are less accurate. We will explore this in chapter 2, in which we will also see that there are many other derivative approximations that are more accurate than the forward-difference formula (but they take more resources to compute).

- **Differentiable means continuous**. The following useful theorem establishes the link between continuity and differentiability.

  Theorem 1.10 *If $f : (a, b) \to \mathbb{R}$ is differentiable at a point $c \in (a, b)$, then f is continuous at c.*

- **Can computers really do maths?** It is worth reminding ourselves that whilst computers can help us understand mathematics more deeply, they cannot *think* mathematically. It is our job to check and interpret results that Python tells us. Often the answers we get are not what we expect (*e.g.* when the step-size h is too small). Sometimes the answers are just plain wrong (*e.g.* $H'(0)$). So one should never treat a computer like an all-knowing black box which always gives us the correct answers all the time.

1.11 The Mean Value Theorem

> Show that the function $f(x) = (x - 1)\sin x$ has a turning point in the interval $(0,1)$. Find the x coordinate of the turning point.

Method 1
It is natural for students to associate the phrase 'turning point' with where $f'(x) = 0$. This means that we have to solve the equation

$$\sin x + (x - 1)\cos x = 0. \tag{1.10}$$

Suppose $\cos x = 0$, then Eq. 1.10 gives $\sin x = 0$, but this is impossible because $\sin x$ and $\cos x$ cannot be zero at the same time, so we conclude that $\cos x \neq 0$. We can then safely divide Eq. 1.10 by $\cos x$, giving

$$\tan x = 1 - x. \tag{1.11}$$

A quick sketch reveals that there are infinitely many solutions, but only one in $(0, 1)$, as shown in fig. 1.13.

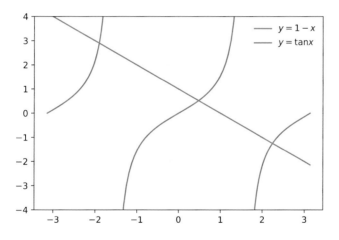

Fig. 1.13: The curves $y = \tan x$ and $y = 1 - x$ intersect infinitely many times on $\mathbb{R}$, but only once on $(0, 1)$.

One way to locate this root is to do a bisection search for the solution of (1.11) in $(0, 1)$ using the code in §1.9. Whilst this method will yield the x-coordinate of the turning point, we needed the explicit expression for the derivative $f'(x)$ that we just found. However, as discussed earlier, such an expression may not always be available.

Method 2
Let's take another look at the problem. This time, suppose we *don't* know how to differentiate the function $f(x) = (x - 1)\sin x$, nor do we know about its graph. Can we still deduce that there is a turning point in the interval $(0, 1)$? Yes, according to the following theorem:

Theorem 1.11 *(Mean Value Theorem) If $f : [a, b] \to \mathbb{R}$ is continuous on $[a, b]$ and differentiable on (a, b), then there exists $c \in (a, b)$ such that*

$$f'(c) = \frac{f(b) - f(a)}{b - a}.$$

With $(a, b) = (0, 1)$, we find that $f(0) = f(1) = 0$, so by the Mean Value Theorem, $\exists c \in (0, 1)$ such that $f'(c) = 0$, *i.e.* there exists a turning point in $(0, 1)$

The Mean Value Theorem (MVT) is a very powerful theorem in analysis. The word '*mean*' refers to the fact that at c, the gradient is simply the average trend on the interval (a, b), *i.e.* the slope of the straight line joining the two endpoints of the curve.

To find the location of the turning point without manual differentiation, we can employ the forward-difference estimate (1.9)

$$f'_{\mathrm{est}}(x) = \frac{f(x + h) - f(x)}{h}, \tag{1.12}$$

for some small h (say 10^{-6}). It is then just a matter of finding where $f'_{\mathrm{est}}(x) = 0$ numerically using, say, the bisection code in §1.9.

Both methods give us, after 17 bisections, the following answer for the x coordinates of the turning point

$$x = 0.47973 \quad \text{(5 dec. pl)}.$$

In summary, the existence of the turning point is guaranteed by the MVT, and its location can be estimated without knowing the derivative explicitly. The plots below (of f and its exact derivative) are consistent with our numerical answer for the turning point.

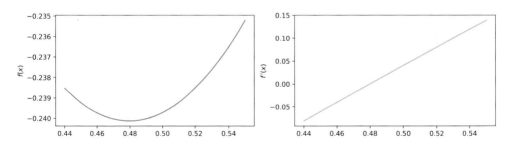

Fig. 1.14: The curve $y = (x - 1) \sin x$ (left) and its exact derivative (right) around the turning point at $x \approx 0.48$.

DISCUSSION

- **Rolle's Theorem**. It is a useful exercise to show that the MVT is a consequence of the following theorem.

  Theorem 1.12 *(Rolle's Theorem) If $f : [a, b] \to \mathbb{R}$ is continuous on $[a, b]$ and differentiable on (a, b), with $f(a) = f(b)$, then there exists $c \in (a, b)$ such that $f'(c) = 0$.*

  As a consistency check, putting $f(a) = f(b)$ in the MVT gives Rolle's Theorem, so we see that the MVT is a more general result.

  Michel Rolle (1652–1719) was a French self-taught mathematician. Apart from Rolle's Theorem, he is also credited with introducing the notation $\sqrt[n]{x}$ for the nth root of x.

- **Cauchy's Mean Value Theorem**. A generalised version of the MVT is the following.

  Theorem 1.13 *(Cauchy's Mean Value Theorem) Let f and g be functions that are continuous on $[a, b]$ and differentiable on (a, b). Suppose that $g'(x) \neq 0$ for all $x \in (a, b)$. Then there exists $c \in (a, b)$ such that*

  $$\frac{f'(c)}{g'(c)} = \frac{f(b) - f(a)}{g(b) - g(a)}.$$

  As a consistency check, putting $g(x) = x$ gives the MVT, so we see that Cauchy's MVT is a more general result.

1.12 A counterexample in analysis

In mathematics, a counterexample is a specific example which proves that a statement is *false*. For instance, to prove that the statement "*every prime number is odd*", one only needs to supply the counterexample: "*2 is an even prime.*"

In analysis, counterexamples are typically used to demonstrate the falsehood of statements that seem *intuitively true*. Such counterexamples are very instructive. They warn us that we cannot always rely on our intuition, and that functions are a more nuanced entity than we might have thought when studying mathematics in school.

We will study one such counterexample in this section, with the help of Python for visualisation.

> Is the following statement true or false?
> *Let f be differentiable at $c \in \mathbb{R}$ with $f'(c) > 0$. Then there exists a neighbourhood of c in which f is strictly increasing.*

f is said to be *strictly increasing* on an interval I if, $\forall x_1, x_2 \in I$ such that $x_1 < x_2$, we have $f(x_1) < f(x_2)$. f is said to be *increasing* on an interval I if, $\forall x_1, x_2 \in I$ such that $x_1 \leq x_2$, we have $f(x_1) \leq f(x_2)$.

The property $f'(c) > 0$ tells us that the gradient of the curve $y = f(x)$ at c is positive. Thus, it makes intuitive sense to think that at points very close to $x = c$, the curve should also have positive gradients that are not drastically different from $f'(c)$, suggesting that f is increasing around $x = c$. Besides, since f is differentiable at c, it is also continuous there (this is theorem 1.10). So it makes sense that we should not have any wild jumps around $x = c$ that might invalidate continuity at c.

Yet the statement is *false*. Consider the following counterexample.

$$f(x) = \begin{cases} x + 2x^2 \sin\left(\frac{1}{x}\right) & x \neq 0, \\ 0 & x = 0. \end{cases} \tag{1.13}$$

We will show that $f'(0) > 0$, yet there exists no neighbourhood of 0 in which f is strictly increasing.

Firstly, to prove that $f'(0) > 0$, we use the limit definition (1.7).

$$f'(0) = \lim_{h \to 0} \frac{f(0+h) - f(0)}{h}$$

$$= 1 + \lim_{h \to 0} 2h \sin\left(\frac{1}{h}\right)$$

$$= 1,$$

where the last step follows from the Squeeze Theorem as before (try justifying this by following the calculation at the end of §1.10).

We now show that f is *not* increasing around $x = 0$. It suffices to show that in any neighbourhood of 0, we can find a point with negative gradient. Symbolically, we want to show that $\forall \delta > 0$, $\exists c \in (-\delta, \delta)$ such that $f'(c) < 0$ (we will say more about this in the Discussion section).

By applying the usual differentiation techniques, we find that for $x \neq 0$,

$$f'(x) = 4x \sin\left(\frac{1}{x}\right) - 2\cos\left(\frac{1}{x}\right) + 1. \tag{1.14}$$

For all $\delta > 0$, there exists an integer $n \in \mathbb{N}$ such that $0 < \frac{1}{n} < 2\pi\delta$. This follows from the Archimedean property discussed in §1.8. Note that the point $x = \frac{1}{2\pi n}$ is within the neighbourhood $(-\delta, \delta)$. However,

$$f'\left(\frac{1}{2\pi n}\right) = -1.$$

Hence, we have successfully used the counterexample to disprove the given statement.

Let's use Python to visualise the curve $y = f(x)$. The code below produces fig. 1.15, which plots the curve in two neighbourhoods of 0 (the neighbourhood on the right panel is 10 times smaller than the left). We see a sinusoidal behaviour in both plots. Try increasing the zoom level by a small modification of the code, or by using %matplotlib and using the zoom button on the GUI (see §1.7). In any case, you should see a sinusoidal behaviour no matter how much you zoom in (of course you should increase the plotting resolution in the code accordingly). The figure suggests that within any δ neighbourhood of 0, we can find points at which $f'(x)$ is positive, negative or zero!

One way to understand this result is to see that the graph for the function f gets increasingly wiggly towards the origin, whilst being constrained to bounce between the parabolas $y = x + 2x^2$ and $y = x - 2x^2$ (where $\sin\left(\frac{1}{x}\right) = \pm 1$). These parabolic boundaries intersect at 0, hence forcing $f'(0) = 1$. Try adding these parabolas to the plots in fig. 1.15.

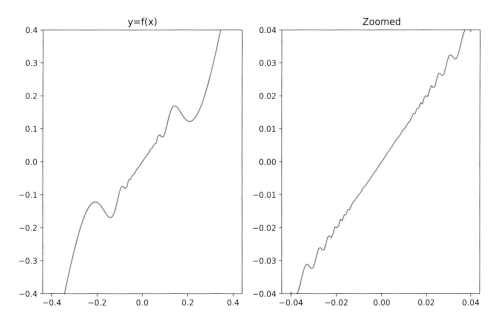

Fig. 1.15: The curve $y = f(x)$ where $f(x) = x + 2x^2 \sin\left(\frac{1}{x}\right)$ $(x \neq 0)$ and $f(0) = 0$. The right panel is the same plot but $10\times$ zoomed in. The sinusoidal behaviour is seen no matter how much we zoom in towards the origin.

---

counterexample.ipynb (for plotting fig.1.15)

| | ```
import numpy as np
import matplotlib.pyplot as plt
``` |
| Define the function | ```
def f(xarray):
``` |
| Given $x$ array...<br>... map 0 to 0 | ```
    y=np.zeros_like(xarray)
    for i, x in enumerate(xarray):
        if x==0:
            y[i] = 0
        else:
``` |
| and map the rest to $x + 2x^2 \sin \frac{1}{x}$<br>then return an array | ```
            y[i] = x*(1+2*x*np.sin(1/x))
    return y
``` |
| | ```
fig, (ax1, ax2) = plt.subplots(1,2,
                        figsize=(10,6))
x1 = np.linspace(-0.4, 0.4, 100)
x2 = np.linspace(-0.04, 0.04, 100)
``` |
| Left panel<br>$\delta = 0.4$ | ```
ax1.plot(x1, f(x1))
ax1.set_ylim(-0.4, 0.4)
ax1.title.set_text('y=f(x)')
ax1.grid('on')
``` |
| Right panel<br>$\delta = 0.04$ | ```
ax2.plot(x2, f(x2))
ax2.set_ylim(-0.04,0.04)
ax2.title.set_text('Zoomed')
ax2.grid('on')
``` |
| | ```
plt.show()
``` |

DISCUSSION

- **Derivative of a monotone function**. There is a subtle connection between the sign of the derivative and the monotonicity of f. The MVT can be used to show that, on an interval I,

$$f'(x) \geq 0 \text{ on } I \iff f \text{ is increasing on } I,$$
$$f'(x) > 0 \text{ on } I \implies f \text{ is strictly increasing on } I.$$

The converse to the second statement does not hold. Can you think of a simple counterexample? A dramatic one is given in exercise 14.

- **A counterexample of historical importance**. Perhaps the most notorious counterexample in analysis is a function which is continuous everywhere but is *nowhere differentiable*. The discovery of such a function by Karl Weierstrass in 1872 sent a shockwave through the mathematical world, leading to the reform and development of analysis into the rigorous subject that we know today. We will meet this function in the next chapter.

- **More counterexamples** in analysis have been compiled by Gelbaum and Olmsted [72], a highly recommended book full of surprising and enlightening counterexamples.

1.13 Exercises

1 (*Book-stacking problem*) Here is an interesting physical situation in which the Harmonic Series appears. The problem was published by Paul Johnson in 1955 [101]. The code `harmonic.ipynb` may be helpful in this question.

I wish to create a leaning tower of books using multiple identical copies of a book. Using n books arranged in a tower perpendicular to the edge of the table, I push the top book as far out as I can, and do the same for the next book below, working my way towards the bottom of the tower. See the figure below when $n = 4$. We can assume that the books and the table are rigid enough that there are no vertical movements.

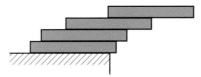

a. Show that using n books, the overhang (in units of books) can be written as the Harmonic Series

$$\frac{1}{2}\left(1 + \frac{1}{2} + \frac{1}{3} \ldots + \frac{1}{n}\right).$$

Deduce that using 4 books, the overhang exceeds the length of a book.

b. Plot the overhang (in unit of books) against the number of books used to build the tower. Consider up to 1000 books. Suggestion: use log scale on the horizontal axis.

c. On the same plot, plot the result when the eq. 1.4 (logarithmic approximation of the Harmonic Series) is used to calculate the overhang.

d. Using the log approximation:
   i. estimate the overhang when 10^6 books are used to create the tower. (Ans: around 7-book long overhang.)
   ii. estimate the number of books needed to build a leaning tower with a 10-book long overhang. (Your answer is probably greater than the estimated number of physical books that exist in the world.)

2 (*Famous approximations for π*) Below are three historically important approximations for π.

- *Madhava series* (14th century), sometimes called the *Gregory-Leibniz approximation* (1671–1673)

$$\pi = 4\left(1 - \frac{1}{3} + \frac{1}{5} - \frac{1}{7} + \cdots\right)$$

- *Wallis product* (1656)

$$\pi = 2\left(\frac{2}{1} \cdot \frac{2}{3}\right) \cdot \left(\frac{4}{3} \cdot \frac{4}{5}\right) \cdot \left(\frac{6}{5} \cdot \frac{6}{7}\right) \cdots$$

- *Viète's formula* (1593)

$$\pi = 2 \cdot \frac{2}{\sqrt{2}} \cdot \frac{2}{\sqrt{2 + \sqrt{2}}} \cdot \frac{2}{\sqrt{2 + \sqrt{2 + \sqrt{2}}}} \cdots$$

Let n denote the number of iterations in each approximation scheme. For example, the zeroth iteration ($n = 0$) gives 4, 2 and 2 for the three approximations respectively.

a. On the same set of axes, plot the results of the three approximations against the number of iterations up to $n = 10$.

b. On a set of logarithmic axes, plot the absolute fractional error

$$\left| \frac{\text{Estimate after } n \text{ iterations } - \pi}{\pi} \right|$$

for the three approximations up to $n = 100$. This gives us an idea of how fast the approximations converge to π.

You should find that the error for Viète's formula does not appear to go smaller than a minimum limit of around 10^{-16}. The reason for this is the *machine epsilon* which will be discussed in §2.2.

c. Recall that an estimate x of π is accurate to p decimal places if $|x - \pi| < 0.5 \times 10^{-p}$. For each of the three approximations of π, calculate how many iterations are needed to obtain π accurate to 5 decimal places.

(Answers: 200000, 157080 and 9.)

3 (*Ramanujan's formula for π*) In 1910, Ramanujan gave the following formula for π.

$$\pi = \left[\frac{2\sqrt{2}}{9801} \sum_{n=0}^{\infty} \frac{(4n)!(1103 + 26390n)}{(n!)^4 396^{4n}} \right]^{-1}.$$

(Famously, he simply 'wrote down' many such formulae.) Calculate the first 3 iterations of this approximation. How many decimal places is each approximation accurate to? Try writing a code that calculates the series up to n terms. Can your code accurately evaluate the result, say, when $n = 10$? If not, explain why.

Suggestion: In Python, we can calculate the factorial, say 15!, using the following syntax:

```
import math
math.factorial(15)
```

The factorials in Ramanujan's formula give rise to huge numbers. Think about what can go wrong.

4 (*Reciprocal Fibonacci number*) Use `fibonacci.ipynb` as a starting point.
The reciprocal Fibonacci constant is given by

$$\psi = \sum_{n=1}^{\infty} \frac{1}{F_n} = 3.35988566624317755\ldots$$

a. Calculate the sum to 20 terms.
Suggestion: Try to do this in a vectorised way using arrays rather than a loop.

b. Calculate how many terms are needed to obtain ψ to 10 decimal places. (Answer: 49.)

5 (*Generalised Fibonacci sequences*) Use `fibonacci.ipynb` as a starting point.

    a. Suppose we use different initial values F_0 and F_1. Use Python to investigate the behaviour of the ratio $R_n := F_n/F_{n-1}$. Show that R_n always converges to the Golden Ratio.

      (If you like a challenge, you could demonstrate this behaviour using sliders.)

    b. (*Lucas sequences*) Let P and Q be fixed integers (not both zero). Define the *Lucas sequence*, $U_n(P,Q)$, by the following rules:

$$U_0 = 0, \quad U_1 = 1, \quad U_n = PU_{n-1} - QU_{n-2}.$$

      Note that the Fibonacci sequence corresponds to $U_n(1, -1)$.

      Write a code that plots the large-n behaviour of the ratio of consecutive terms $R_n := U_{n+1}/U_n$ for the following values of P and Q.

        i. $(P,Q) = (3, 1)$
        ii. $(P,Q) = (2, 1)$
        iii. $(P,Q) = (1, 1)$

      Make a conjecture on the range values of P such that R_n is a convergent sequence.

      (Answer: see `https://mathworld.wolfram.com/LucasSequence.html`)

6 (*Order of convergence*) Given a convergent sequence, it is natural to ask: *how fast does the sequence converge?* In this question, we explore how to quantify the speed of convergence.

Suppose a sequence (x_n) converges to x. Define the *absolute error*, E_n, as the sequence

$$E_n = |x_n - x|.$$

The speed of convergence can be quantified by two positive constants: the *rate* (C) and the *order* (q) of convergence, defined via the equation

$$\lim_{n \to \infty} \frac{E_{n+1}}{(E_n)^q} = C > 0.$$

    a. Verify by hand that the sequence $\left(\frac{1}{n}\right)$ converges with $q = C = 1$.
    b. For most applications, only the order of convergence is of interest as it is a better indicator of how fast the sequence converges. It can be shown that we can estimate q using the formula

$$q \approx \frac{\ln(E_{n+1}/E_n)}{\ln(E_n/E_{n-1})},$$

      where n should be large enough so that q stabilises, but not so large that the denominator in the formula is zero.

        i. Using this approximation, verify (using Python) that the ratio of consecutive Fibonacci numbers $R_n = F_{n+1}/F_n$ converges with order 1.

          Suggestion: it might help to plot q as a function of n. You should find that the graph settles on $q = 1$ before becoming undefined when n is too large.

        ii. Consider the Madhava series for π in Question 2. Show that $q < 1$.

          Technically, we say that the convergence of the series is *sublinear* (*i.e.* terribly slow).

        iii. Conjecture a sequence that converges *superlinearly* ($q > 1$) and demonstrate this property graphically..

7 In `continuityslider.ipynb` (§1.7), add a slider in the GUI so that the value of x_0 can be changed.

8 (*Euclid's orchard*) Thomae's function (§1.8) has an interesting connection to the following problem, sometimes called *Euclid's orchard* or the *visible points* problem.

Consider grid points with positive integer coordinates (m, n). A grid point P is said to be *visible* from $(0,0)$ if a straight line joining $(0,0)$ to P does not pass through any other grid points. For example, as shown in the figure below, the point $(1, 2)$ is visible from $(0,0)$, but $(2, 4)$ is not.

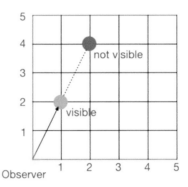

You should first try to experiment on a piece of grid paper to come up with a conjecture on which points are visible from $(0, 0)$.

   a. Write a code that produces a grid and marks each point that is visible from $(0,0)$ with a red dot (you will need to impose a sensible cutoff).

   b. For each visible point (m, n), plot its image under the mapping

$$F(m, n) = \left(\frac{m}{m+n}, \frac{1}{m+n} \right).$$

What do you see? Can you explain how this mapping works?
(Answer: You see Thomae's function!)

9 (*Root-finding*) Use the bisection code to solve the following problems. Give your answers to 4 decimal places.

   a. Solve the equation $x^3 - x^2 - x - 1 = 0$.
      Suggestion: Start by plotting.
   b. Solve $\frac{\sin x}{x} = 0.9$ (hence verifying the intersection point seen in fig. 1.8)
   c. Find the numerical value of $\sqrt{3}$ using only the four basic operations $+ - \times \div$.
      Suggestion: Start by defining $f(x) = x^2 - 3$.

10 (*Generalised Golden Ratio*) Suppose we generalise the Golden Ratio to ϕ_n defined as the positive root of the following order-n polynomial

$$x^n - x^{n-1} - x^{n-2} \cdots - 1 = 0.$$

For example, $\phi_1 = 1$ and $\phi_2 \approx 1.618$.
Write a code that plots the sequence ϕ_n (obtained by bisection). Make a conjecture for the value of $\lim_{n \to \infty} \phi_n$.

11 (*Newton-Raphson method*) Let f be differentiable function on $\mathbb{R}$. The *Newton-Raphson method* for finding the root of the equation $f(x) = 0$ comprises the following steps.

- Start with an initial guess x_0.
- Calculate

$$x_{n+1} = x_n - \frac{f(x_n)}{f'(x_n)},$$

where the expression for $f'(x)$ must be explicitly known.
- Repeat the above iteration as long as $|x_{n+1} - x_n| < \varepsilon$ for a fixed tolerance ε specified by the user (e.g. 0.5×10^p for p-decimal place accuracy). This step ensures that we stop only when the first p-decimal places are stabilised.

a. Write a Python code for this method and use it to solve the problems in question 9.
b. Let x_n be the estimate of the solution of $f(x) = x^2 - 3$ after n iterations using the Newton-Raphson methods.
Calculate the absolute error $E_n = |x_n - \sqrt{3}|$ and verify that the Newton-Raphson with order 2. Do this by calculating q defined in question 6).

12 (*Creating a pretty plot*) Plot the graph of the sinc function $f : [0, 2\pi] \to \mathbb{R}$ defined by

$$f(x) = \begin{cases} \frac{\sin x}{x} & \text{if } x \neq 0, \\ 1 & \text{if } x = 0. \end{cases}$$

On the same set of axes, plot its first and second derivatives, estimated using the forward Euler method (*i.e.* do not differentiate anything by hand in this question). Try to use different types of lines in your plot (think about colour-blind readers). Here is an example output.

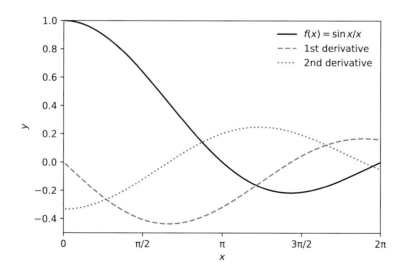

Fig. 1.16: A pretty plot of the sinc function and its first and second derivatives.

13 (*Mean Value Theorem*) Consider the sinc function f defined in the previous question.
Let's apply the Mean Value Theorem (theorem 1.11) to f on the interval $[0, \pi]$.
In particular, let's determine the value(s) of c in the statement of the theorem. In other
words, we wish to find $c \in [0, \pi]$ such that

$$f'(c) = \frac{f(\pi) - f(0)}{\pi}.$$

   a. Show (by hand) that one solution is $c = \pi$.
   b. Use a root-finding algorithm to find another solution to 4 decimal places.
      Your answer should be consistent with the dashed line in fig. 1.16 above.

14 (*Another counterexample*) Here is another interesting counterintuitive example in
analysis. Use counterexample.ipynb as a template.

Consider the function $f : [0, 1] \to \mathbb{R}$ defined by

$$f(x) = \begin{cases} x\,(2 - \cos \ln x - \sin \ln x) & \text{if } x \in (0, 1], \\ 0 & \text{if } x = 0. \end{cases}$$

It can be shown that f is strictly increasing, yet there are *infinitely* many points where
$f'(x) = 0$ in $(0, 1]$.

   a. Use Python to plot the graph of the function. Your graph should demonstrate the
      self-similar structure when we zoom in towards the origin.
   b. Show (by hand) that there are points of inflection at $x = e^{-2n\pi}$ where $n \in \mathbb{N}$.
      Indicate them on your plot.

# CHAPTER TWO

Calculus

Calculus broadly encompasses the study of *differentiation* and *integration*. At school, these subjects are often taught in a rather algorithmic way with a focus on employing various techniques of differentiation and integration using stock formulae and tricks. This approach only scratches the surface of what has to be one of the most profound mathematical inventions that have helped us understand the physical world and the laws of nature.

At university, the focus of calculus shifts dramatically from *how* to *why* you can differentiate or integrate a function, and calculus becomes a much more rigorous subject with close links to analysis. A good review of these links is given in [161, 189], for example.

Fig. 2.1: (L-R) *Sir Isaac Newton* (1642–1726) and *Gottfried Leibniz* (1646–1716) formulated calculus independently, although the question of who came to it first was a highly contentious intellectual feud of the late 17th century. (Image source: [137].)

How can we use Python to help us understand calculus? Since Python is primarily a numerical tool, we will be interested in the numerical values of derivatives and integrals rather than their algebraic expressions. For example, we are <u>not</u> interested in the problems

$$\text{If } f(x) = e^{-x^2}, \text{ find } f'(x) \qquad \text{OR} \qquad \text{Find } \int \sin x \, dx,$$

which are symbolic in nature, and so must be solved by specialised packages that have been taught the *rules* of calculus (*e.g.* the *SymPy* library). Instead, we are interested in problems that require numerical answers, such as

$$\text{If } f(x) = e^{-x^2}, \text{ find } f'(2) \qquad \text{OR} \qquad \text{Find } \int_0^\pi \sin x \, dx,$$

which can be solved with basic binary arithmetic on any computer.

Calculus on the computer is a surprisingly subtle topic. Because Python does not know even the most basic rules of differentiation and integration, we need to approximate those quantities numerically using the gradient of a tangent for derivatives, and area under a graph for integrals. The surprise here, as we will see, is that there is a limit to how accurate these numerical answers can be, stemming from the inevitable fact that computers operate on a system of floating-point binary numbers (this will be explored in §2.2) .

We will also explore some applications of calculus, including applying Taylor's Theorem to quantify the error in approximating a function as a polynomial (§2.4) and approximating a function as a Fourier series, comprising sine and cosine waves of various frequencies (§2.9).

2.1 Basic calculus with SciPy

Another useful Python library for mathematicians is *SciPy* (https://www.scipy.org), which contains ready-made functions and constants for advanced mathematics. We will use often use SciPy throughout this book.

You will need SciPy version at least 1.10. To check this, run the following lines:

```
import scipy
scipy.__version__
```

If you have an older version of SciPy, update it using pip. See Appendix A.

In this chapter, we will need the SciPy module `scipy.integrate`, which contains several integration routines such as the Trapezium and Simpson's Rules. The go-to workhorse for computing integrals in Python is the function **quad** (for *quadrature*, a traditional term for numerical integration). The **quad** function itself is not a single method but a set of routines which work together to achieve an accurate answer efficiently. The algorithm behind **quad** was initially conceived in Fortran in a package called *QUADPACK*, and is described in detail by [165].

Here's the basic syntax for the **quad** function.

Integration with SciPy	

For integration with SciPy	`import numpy as np`
	`import matplotlib.pyplot as plt`
	`import scipy.integrate as integrate`

Define $f(x) = e^{-x^2}$	`f = lambda x: np.exp(-x**2)`
$\int_0^\infty e^{-x^2} \, dx$	`integral, error = integrate.quad(f, 0, np.inf)`

The output for (`integral, error`) is

(0.8862269254527579, 7.101318390472462e-09)

Note that **quad** returns a pair of numbers (called a *tuple* of length 2). The first number is the value of the definite integral, and the second is an estimate of the absolute error (which should be tiny for a reliable answer). In this case, the exact answer is $\sqrt{\pi}/2$, which agrees with SciPy's answer to 16 decimal places.

The mathematical details of how functions like SciPy's `integrate` work are usually hidden from users behind a 'Wizard-of-Oz' curtain that users rarely look behind. This is contrary to the spirit of this book whose emphasis is on mathematics and not commands. Therefore, in this chapter, we will only use SciPy's `integrate` function occasionally, and only to confirm what we can do by other, more transparent methods.

As for differentiation, SciPy had a `derivative` routine, but it is now obsolete. In the next section, we will discuss how to differentiate a function numerically.

2.2 Comparison of differentiation formulae

Consider the following 3 formulae for approximating the derivative of a function f at point x.

Let $h > 0$ be a small number (we call this the *step size*).

The *forward-difference* formula: $f'(x) \approx \dfrac{f(x + h) - f(x)}{h}$.

The *backward-difference* formula: $f'(x) \approx \dfrac{f(x) - f(x - h)}{h}$.

The *symmetric-difference* formula: $f'(x) \approx \dfrac{f(x + h) - f(x - h)}{2h}$.

Let $f(x) = x^3$. Compare the actual value of the derivative $f'(1)$ to the above approximations for a range of step sizes h. Do this by plotting the absolute error $E(h)$ for each formula, where

$E(h) = |$actual value of the derivative$-$numerical approximation using step size $h|$.

The forward, backward and symmetric-difference formulae all approach the gradient of the tangent to the graph $y = f(x)$ at point x as the step size $h \to 0$, so it is reasonable to expect that the smaller the h, the more accurate the expression will be. However, we show in this section that on the computer, this is not the case. This may come as a surprise to many beginners. It is very important to be aware that when coding derivatives, hugely inaccurate answers could result if h is too large or too small.

Let's study the accuracy of these approximations at a fixed value $x = 1$ and calculate the absolute difference between $f'(1)$ and the approximations as we vary h. The code below produces the graph of $E(h)$, which in this case is

$$E(h) = |3 - \text{approximation}|$$

for h in the range $[10^{-20}, 1]$.

Eh.ipynb (for plotting fig. 2.2)	
	`import numpy as np` `import matplotlib.pyplot as plt`

Range of h from 10^{-20} to 10^0 (log scale)
Point of interest
Define the function
Actual expression for $f'(x)$
	`h = np.logspace(-20, 0, 300)`
	`x = 1`
	`f = lambda x: x**3`
	`actual = 3*x**2`

	`fx = f(x)`
	`fxp = f(x+h)`
	`fxm = f(x-h)`

Forward difference Symmetric difference	`est1 = (fxp-fx)/h` `est2 = (fxp-fxm)/(2*h)`
$E(h)$	`err1 = abs(actual-est1)` `err2 = abs(actual-est2)`

Plot the errors on log axes	`plt.loglog(h,err2, 'k', lw=2)`
	`plt.loglog(h,err1, 'r', lw=1)`
	`plt.legend(['Symmetric difference',`
	`              'Forward difference'])`
	`plt.xlabel(r'$h$')`
	`plt.ylabel(r'Absolute error $E$')`
	`plt.xlim([1e-20, 1])`
	`plt.grid('on')`
	`plt.show()`

Fig. 2.2 shows the logarithmic plot of the two approximations (forward and symmetric differences). The graphs show a number of distinctive and surprising features:

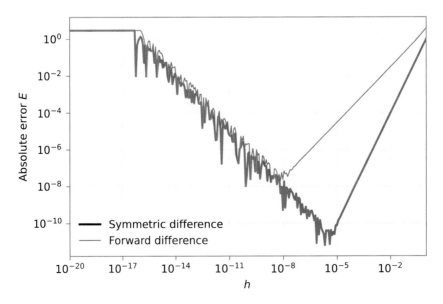

Fig. 2.2: $E(h)$ defined as $|f'(1) - \text{approximation}|$ is plotted for the forward-difference approximation (thin red line) and the symmetric-difference approximation (thick black line).

- For each approximation, there appears to be an optimal value of h which minimises the error, namely:

$$h_{\text{opt}} \sim \begin{cases} 10^{-8} & \text{(Forward difference)} \\ 10^{-6} & \text{(Symmetric difference)} \end{cases}$$

The symbol $\sim$ is used in this context to mean a rough order-of-magnitude estimate.
- The minimum error (ignoring small-scale fluctuations) are roughly:

$$E(h_{\text{opt}}) \sim \begin{cases} 10^{-8} & \text{(Forward difference)} \\ 10^{-11} & \text{(Symmetric difference)} \end{cases}$$

- For $h \gtrsim h_{\text{opt}}$, we see a linear behaviour in both cases. Since this is a log-log plot, a line with gradient m actually corresponds to the power-law relation $E(h) \sim h^m$. From the graph we see that

$$E(h \gtrsim h_{\text{opt}}) \sim \begin{cases} h & \text{(Forward difference)} \\ h^2 & \text{(Symmetric difference)} \end{cases}$$

- For $10^{-16} \lesssim h \lesssim h_{\text{opt}}$, we see rapid fluctuations, but the overall trend for both approximations is

$$E(10^{-16} \lesssim h \lesssim h_{\text{opt}}) \sim h^{-1}.$$

- For $h \lesssim 10^{-16}$, both approximations give the same constant.

You should try changing the function and the point x at which the derivative is calculated. You should see that there are no effects on any of the observations summarised above. You should also verify that the backward-difference approximation gives essentially the same graph as the forward-difference graph (the fluctuations will be slightly difference).

The key takeaway here is that numerical derivatives do not behave in the way we might expect: smaller h does not produce a more accurate estimate. Always use $h \approx h_{\text{opt}}$ whenever we code a derivative.

DISCUSSION

- **The machine epsilon**. The reason behind the behaviour of $E(h)$ that we saw is the fact that computers use binary numbers. A computer uses 64 binary digits to represent any real number as a *floating-point number* (or simply *float*). It can be shown that this accuracy is dictated by a number called the *machine epsilon*, $\varepsilon_{\text{mach}}$, defined as the distance between 1 and the next floating-point number greater than 1. For double-precision floats (which is what most computers today use by default), we have

$$\varepsilon_{\text{mach}} = 2^{-52} \approx 2.2 \times 10^{-16}.$$

See [182] for an explanation of how $\varepsilon_{\text{mach}}$ is calculated. The machine epsilon is one of the main reasons why numerical results are sometimes very different from theoretical expectations.

For instance, $\varepsilon_{\text{mach}}$ is the reason why we see the flat plateau for very small h in fig. 2.2. If $h \lesssim \varepsilon_{\text{mach}}$, the floating-point representation of $1 + h$ is 1, and the approximation formulae give zero because $f(x + h) = f(x - h) = f(x)$. Therefore $E(h) = 3$ if h is too small, as seen in the graph.

- **Truncation and rounding errors** The V-shape of the graphs is due to two numerical effects at play. One effect is the *rounding error*, E_R, which occurs when a number is represented in floating-point form and rounded up or down according to a set of conventions. It can be shown that

$$E_R \sim h^{-1},$$

and therefore E_R dominates at small h. The tiny fluctuations are due to different rounding rules being applied at different real numbers.

In addition, there is also the *truncation error*, E_T, which is associated with the accuracy of the approximation formula. In fact, *Taylor's Theorem* (see §2.4) tells us that the error in the forward and symmetric difference formulae can be expressed as follows.

$$\text{Forward difference: } f'(x) = \frac{f(x+h) - f(x)}{h} - \frac{h}{2} f''(\xi_1), \tag{2.1}$$

$$\text{Symmetric difference: } f'(x) = \frac{f(x+h) - f(x-h)}{2h} - \frac{h^2}{6} f'''(\xi_2), \tag{2.2}$$

for some numbers $\xi_1, \xi_2 \in (x, x+h)$. The truncation error is simply each of the error terms above, arising from the truncation of the infinite Taylor series for f. Note that the powers of h in these remainder terms are exactly what we observed in the $E(h)$ graph. In exercise 2, you will explore the $E(h)$ graph of a more exotic formula for the derivative.

- **Big O notation** Another way to express the accuracy of the approximations (2.1)-(2.2) is to use the $O(h^n)$ notation, where n is determined from the scaling of the error term. We say that the forward difference formula is an $O(h)$ approximation, and the symmetric difference formula is $O(h^2)$. The higher the exponent n, the faster the error shrinks, so one does not have to use a tiny h to achieve a good accuracy.

2.3 Taylor series

> Evaluate and plot the partial sums of the Taylor series for:
> a) $\sin x$,   b) $\ln(1 + x)$ c) $1/(1 + x)$.
> In each case, at what values of x does the Taylor series converge?

Recall that the Taylor series for a smooth function f expanded about $x = 0$ (also known as Maclaurin series) is given by:

$$f(x) = f(0) + f'(0)x + \frac{f''(0)}{2!}x^2 + \frac{f'''(0)}{3!}x^3 + \ldots = \sum_{n=0}^{\infty} \frac{f^{(n)}(0)}{n!}x^n. \qquad (2.3)$$

For the given functions, we find

$$\sin x = \sum_{n=1}^{\infty} \frac{(-1)^{n+1}}{(2n-1)!}x^{2n-1}, \qquad (2.4)$$

$$\ln(1 + x) = \sum_{n=1}^{\infty} \frac{(-1)^{n+1}}{n}x^n, \qquad (2.5)$$

$$\frac{1}{1 + x} = \sum_{n=0}^{\infty}(-x)^n. \qquad (2.6)$$

The code `taylor.ipynb` plots the graph $y = \ln(1 + x)$ along with partial sums of the Taylor series up to the x^{40} term (fig. 2.3). As there are so many lines, it might help to start plotting only, say, the 5th partial sum onwards. The first few partial sums do not tell you much more than the fact that they are terrible approximations.

It might also help to systematically colour the various curves. In our plot, we adjust the `(r,g,b)` values gradually to make the curves 'bluer' as the number of terms increases (*i.e.* by increasing the `b` value from 0 to 1 and keeping `r=g=0`).

From the graphs (and further experimenting with the number of terms in the partial sums), we can make the following observations.

- $\sin x$ – The Taylor series appears to converge to $\sin x$ at *all* $x \in \mathbb{R}$.
- $\ln(1 + x)$ – The Taylor series appears to converge to $\ln(1 + x)$ for $x \in (-1, 1)$, possibly also at $x = 1$. For $x > 1$, the graphs for large n show a divergence (*i.e.* the y values become arbitrarily large in the right neighbourhood of $x = 1$).
- $\frac{1}{1+x}$ – The Taylor series also appears to converge to $\frac{1}{1+x}$ for $x \in (-1, 1)$, similar to $\ln(1 + x)$. At $x = 1$, the Taylor series gives alternating values of ± 1, so clearly does not converge to the function value of $\frac{1}{2}$. For $|x| > 1$, the Taylor series blows up.

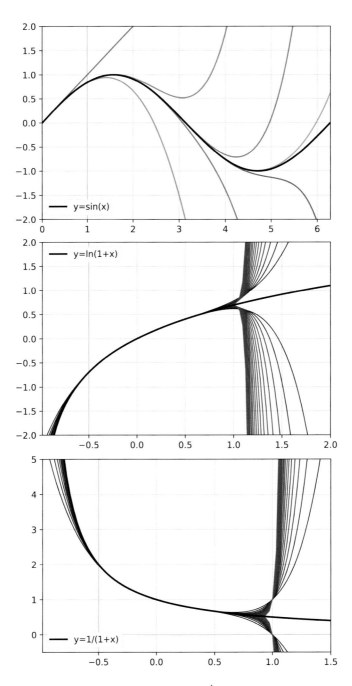

Fig. 2.3: The graphs of $y = \sin x$, $\ln(1 + x)$ and $\frac{1}{1+x}$ (thick black lines) and their Taylor series. The top panel shows up to 7 terms in the series (up to the term x^{13}). The lower panels show the series up to the x^{40} term in the series, with bluer lines indicating higher number of terms.

taylor.ipynb (for plotting the middle panel in fig. 2.3)	
| | ```python
import numpy as np
import matplotlib.pyplot as plt
``` |
| $n$th term of the Taylor series | ```python
def nth_term(x,n):
        return -(-1)**n*x**n/n
``` |
| Specify maximum number of terms<br>Domain of the function | ```python
n_max = 40
x = np.linspace(-0.99, 2, 100)
S = np.zeros_like(x)
``` |
| With every pass of the *for* loop<br>Add a term<br>Adjust the colour (b value) of the curve<br>Start plotting from $n = 5$ | ```python
for n in np.arange(1,n_max+1):
    S = S + nth_term(x,n)
    b=n/n_max
    if (n>=5):
        plt.plot(x,S,label='_nolegend_',
        color = (0,0,b), lw=1)
``` |
| Plot the function in thick black line | ```python
plt.plot(x, np.log(1+x),lw=2,color='k')
plt.legend([r'$y=ln(1+x)$'], loc='upper left')
plt.xlim([-0.99, 2])
plt.ylim([-2,2])
plt.grid('on')
plt.show()
``` |

DISCUSSION

- **Radius of convergence** The radius of convergence, R, of a power series is a real number such that the series $\sum a_n x^n$ converges for $|x| < R$, and diverges for $|x| > R$. The interval $(-R, R)$ is called the *interval of convergence* of the series.

  For example, the series (2.6) could be regarded as a geometric series with common ratio x. We know that the geometric series converges to the sum to infinity $\frac{1}{1+x}$ if $|x| < 1$ and diverges when $|x| > 1$, as can be seen graphically in fig. 2.3. We say that the radius of convergence of the series is 1.

- **Ratio Test**. We discussed the comparison test in the previous chapter. Another useful convergence test is the following result known as *d'Alembert's Ratio Test*:

  Theorem 2.1 (Ratio Test) *Let T_n be a sequence such that $T_n \neq 0$ eventually. Let*

  $$L = \lim_{n \to \infty} \left| \frac{T_{n+1}}{T_n} \right|.$$

  If $L < 1$ then the series $\sum T_n$ converges. If $L > 1$ then the series $\sum T_n$ diverges.

  Applying the Ratio Test to the series (2.4) gives

  $$L = \lim_{n \to \infty} \frac{x^2}{2n(2n + 1)} = 0, \text{ for all } x \in \mathbb{R}.$$

  This proves our conjecture that the Taylor series converges for all x.

  However, we have not proved that the series converges to $\sin x$ (we only proved that it converges to some function). We will come back to this in the next section.
```

- **Differentiating and integrating power series**. There is a relation between the Taylor series for $\frac{1}{1+x}$ and $\ln(1 + x)$. Here is a very useful theorem from analysis:

Theorem 2.2 *The series $\sum a_n x^n$ can be differentiated and integrated term by term within the interval of convergence. The resulting power series has the same radius of convergence.*

Using this result, we can integrate both sides of Eq. 2.6 with respect to x as long as $|x| < 1$, yielding exactly Eq. 2.5. This proves our conjecture that the two series share the same interval of convergence.

However, we have not established the convergence at the end points $x = \pm 1$. We will discuss this in the next section.

2.4 Taylor's Theorem and the Remainder term

Suppose we approximate $\sin x$ and $\ln(1 + x)$ by the following polynomials (obtained by truncating their Taylor series $P(x)$):

$$\sin x \approx P_{2N-1}(x) := x - \frac{x^3}{3!} + \frac{x^5}{5!} \cdots + (-1)^{N+1} \frac{x^{2N-1}}{(2N-1)!}$$

$$= \sum_{n=1}^{N} \frac{(-1)^{n+1}}{(2n-1)!} x^{2n-1}, \tag{2.7}$$

$$\ln(1+x) \approx P_N(x) := x - \frac{x^2}{2} + \frac{x^3}{3} \cdots (-1)^{N+1} \frac{x^N}{N}$$

$$= \sum_{n=1}^{N} \frac{(-1)^{n+1}}{n} x^n. \tag{2.8}$$

Quantify the accuracy of these approximations by investigating the difference between the function and its order-k polynomial approximation:

$$R_k(x) = f(x) - P_k(x).$$

You will probably be familiar with the approximation of a function $f(x)$ as an order-k polynomial $P_k(x)$ by truncating its Taylor series after a finite number of terms. At university, we are not only concerned with the series itself, but also the error term (the 'remainder') in the expansion, defined by $R_k(x)$. The following theorem gives a useful expression for the remainder term.

Theorem 2.3 (Taylor's Theorem) *Let* $I = [a, b]$ *and* $N = 0, 1, 2 \ldots$. *Suppose that* f *and its derivatives* $f', f'', \ldots f^{(N)}$ *are continuous on* I, *and that* $f^{(N+1)}$ *exists on on* (a, b). *If* $x_0 \in I$, *then,* $\forall x \in I \setminus \{x_0\}$, $\exists \xi$ *between* x *and* x_0 *such that*

$$f(x) = f(x_0) + f'(x_0)(x - x_0) + \cdots + \frac{f^{(N)}(x_0)}{N!}(x - x_0)^N + R_N(x)$$

where $\quad R_N(x) = \frac{f^{(N+1)}(\xi)}{(N+1)!}(x - x_0)^{N+1}$

Note that when $x = x_0$, the equality is trivial.

Although the theorem bears the name of the English mathematician *Brook Taylor* (1685–1731) who studied the polynomial expansion of functions, it was *Joseph-Louis Lagrange* (1736–1813) who provided the expression for the remainder term. R_n is often called the *Lagrange form of the remainder.*

In this form, the remainder is not completely determined due to the appearance of an unknown ξ. However, in practice, this expression is sufficient for us to place a bound on R, giving us some idea of the magnitude of the error.

Applying Taylor's Theorem to $f(x) = \ln(1 + x)$, the remainder R_N is found to be

$$R_N(x) = \frac{(-1)^N}{N+1} \left(\frac{x}{1 + \xi} \right)^{N+1}, \qquad \text{for some } \xi \in (0, x). \tag{2.9}$$

On the domain $0 < x \leq 1$ (where the Taylor series converges), we find that $|R_N(x)|$ is bounded by:

$$\frac{1}{N+1}\left(\frac{x}{1+x}\right)^{N+1} < |R_N(x)| < \frac{1}{N+1}x^{N+1}. \tag{2.10}$$

As the polynomial order N increases, we expect the quality of the approximation to improve, and so $R_N(x)$ should shrink to 0. Indeed, as $N \to \infty$, the RHS of (2.10) goes to zero. This means that for all $x \in (0, 1]$, the Taylor series for f converges to $f(x)$ (and not any other functions)[1].

Interestingly, this also explains why we can evaluate the Taylor series at $x = 1$, giving us the familiar series for $\ln 2$ which we saw in §1.4.

The code `taylorthm.ipynb` calculates $R_N(x)$ at a fixed $x \in (0, 1]$ and plots it as a function of N. Fig. 2.4 shows the graph of $R_N(x = 0.4)$ (solid red line) when $\ln(1 + x)$ is approximated by the Taylor polynomial of order N. The upper and lower bounds for $|R_N|$ (eq. 2.10) are shown in dotted lines. The graph shows that Taylor's Theorem holds, and that the constant ξ must be close to 0.

Now let's turn to the function $f(x) = \sin x$. Since the coefficients of the even powers of x vanish, the $2N$-th order approximation is the same as the $(2N - 1)$th order approximation. In other words, $R_{2N}(x) = R_{2N-1}(x)$. Calculating the remainder using Taylor's Theorem, we find

$$R_{2N-1}(x) = R_{2N}(x) = \frac{(-1)^N \cos \xi}{(2N+1)!}x^{2N+1}, \qquad \text{for some } \xi \in (0, x).$$

Note that the cosine is positive and decreasing on the interval $(0, 0.4)$. Thus, we find the upper and lower bounds for $|R_N|$:

$$\frac{\cos x}{(2N+1)!}x^{2N+1} < |R_{2N-1}(x)| = |R_{2N}(x)| < \frac{1}{(2N+1)!}x^{2N+1}. \tag{2.11}$$

In fig. 2.4, we can see that the remainder and its bounds are almost indistinguishable, up until around $N = 12$ where the $|R_N|$ violates the bounds when it shrinks below a small number of order $\varepsilon_{\text{mach}} \approx 10^{-16}$. This is due to an additional error arising from the subtraction of two nearly equal numbers (namely, $f(x)$ and $P_N(x)$).

In both cases, our conclusion is that Taylor's Theorem is verified. In addition, we also saw that the Taylor series for $\sin x$ converges much more rapidly than that of $\ln(1 + x)$. By the time we reach the order-12 polynomial, the approximation for $\sin x$ is so good that the error is smaller than $\varepsilon_{\text{mach}}$.

---

[1] Keen-eyed readers might remember from fig. 2.3 that the Taylor series of $\ln(1 + x)$ also converges to $\ln(1 + x)$ when $-1 < x < 0$. However, our proof does not work in this case, since it no longer follows from eq. 2.9 that $R_N(x) \to 0$. What is needed in that case is a different technique (for example, integrating another Taylor series).

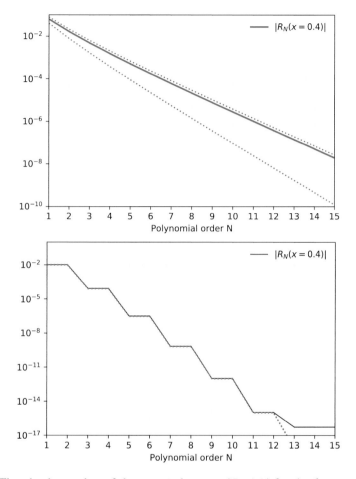

Fig. 2.4: The absolute value of the remainder term $|R_N(x)|$ for the function $\ln(1 + x)$ (top) and $\sin x$ (bottom), evaluated at $x = 0.4$, plotted as a function of N (the order of the approximating polynomial). As N increases, the remainder approaches zero. The bounds for $|R_N|$ are plotted in dotted lines.

DISCUSSION

- **Ratio Lemma**. We showed in the previous Section that the Taylor series for $\sin x$ converges, but it remains to show that it converges to $\sin x$. Here is a very useful result which will help us prove this.

  Theorem 2.4 (Ratio Lemma) *Let a_n be a sequence such that $a_n > 0$. Suppose L is a constant such that $0 < L < 1$ and $a_{n+1}/a_n \leq L$ eventually. Then, a_n converges to 0.*

  Letting a_N be the sequence $|x|^{2N+1}/(2N + 1)!$ which appears on the RHS of Eq. 2.11 (where $x \neq 0$). Then the ratio $a_{N+1}/a_N = \frac{x^2}{(2N+3)(2N+2)} \to 0$ regardless of the value of x. The Ratio Lemma says that $a_N \to 0$, and therefore the remainder $R_N \to 0$. This proves that the Taylor series of $\sin x$ converges to $\sin x$ for all $x \in \mathbb{R}$.

- **A pathological function**. You may be wondering if it is possible for the Taylor series of a function to converge to a different function. The answer is yes! Here is a classic counterexample. Define $f : \mathbb{R} \to \mathbb{R}$ by

$$f(x) = \begin{cases} e^{-1/x^2} & \text{if } x \neq 0, \\ 0 & \text{if } x = 0. \end{cases}$$

It can be shown that $f^{(n)}(0) = 0$ for all $n \in \mathbb{N}$ (see, for example, [91] for calculation details). Therefore, the Taylor series of f converges to 0 everywhere, but only coincides with f at a single point where $x = 0$.

---

taylorthm.ipynb (for plotting the top panel of fig. 2.4)

	```import numpy as np``` ```import matplotlib.pyplot as plt```
Term of degree N in the Taylor series	```def Nth_term(x,N):``` ``` return -(-1)**N*x**N/N```

Choose any $x \in (-1, 1]$
Specify the maximum polynomial order
	```x = 0.4```
	```N_max = 15```
	```Nlist = np.arange(1,N_max+1)```
	```P = 0```

Value of the polynomial
Lower bound for the remainder
Upper bound for the remainder
	```PNlist = []```
	```lowlist = []```
	```hilist = []```

Append the lists with every iteration
$P_N(x)$
 	
Lower and upper bounds in Eq. 2.10
	```for N in Nlist:```
	``` P = P + Nth_term(x,N)```
	``` PNlist.append(P)```
	``` Np = N+1```
	``` low = (x/(1+x))**Np/Np```
	``` lowlist.append(low)```
	``` hi = x**Np/Np```
	``` hilist.append(hi)```

The remainder $|R_N(x = 0.4)|$	```RN = abs(PNlist-np.log(1+x))```

Plot $|R_N|$ (thick red line), log y-axis
 	
Upper and lower bounds in dotted lines
	```plt.semilogy(Nlist, RN, lw=2,```
	``` color = 'r')```
	```plt.semilogy(Nlist,lowlist,'r:',```
	``` Nlist,hilist, 'r:')```
	```plt.legend([r'$|R_N(x=0.4)|$'])```
	```plt.xticks(Nlist)```
	```plt.xlim([1,N_max])```
	```plt.xlabel('Polynomial order N')```
	```plt.ylim([1e-10,0.1])```
	```plt.grid('on')```
	```plt.show()```

2.5 A continuous, nowhere differentiable function

In 1872, Karl Weierstrass announced the discovery of a function which is continuous on $\mathbb{R}$, but is nowhere differentiable. The *Weierstrass function* $f : \mathbb{R} \to \mathbb{R}$ is given by

$$f(x) = \sum_{n=0}^{\infty} a^n \cos(b^n \pi x),$$

where $a \in (0, 1)$, b is an odd integer, and $ab > 1 + \frac{3\pi}{2}$.
Plot the Weierstrass function for some choices of a and b.

A function is continuous wherever it is differentiable. However, it came as a surprise to the late 19th-century mathematical world that the converse is not true. The Weierstrass function, arguably the most famous counterexample in mathematics, led to the development of a rigorous foundation for analysis. Although Weierstrass's counterexample was the first to be published, and certainly the most impactful, such a pathological function was discovered much earlier in 1830 by Bolzano. There are now many known examples of such a function.

An accurate plot of the function had to wait until the advent of computers (surely frustrating mathematicians over the decades). Luckily for us, we can use Python to visualise Weierstrass's original function. The code below produces an interactive GUI with two sliders for adjusting the values of a and b (fig 2.5). The code looks long, but it is simply a straightforward extension of the single-slider code that we used in §1.7.

Using this GUI, we see that the parameter a adjusts the amplitude of the small-scale fluctuations (as $a \to 0$, f reduces to a regular cosine curve). The parameter b adjusts the density of the small-scale fluctuations (higher b gives more substructures with high-frequency oscillations).

The GUI also allow us to zoom into the curve, revealing a self-similar (or fractal-like) oscillating structure. This self-similarity is the key intuitive reason why the curve is not differentiable: zooming in always reveals the same oscillations, meaning that there are no smooth segments on the curve, despite the fact that the function is continuous on $\mathbb{R}$.

Whilst these properties may be easy to understand intuitively, Weierstrass's original proof is rather technical. We refer interested readers to [161] for a readable walkthrough of the proof. It should be mentioned that the conditions $ab > 1 + 3\pi/2$ and b an odd integer are both specific to Weierstrass's own working. It was shown by Hardy that the only conditions required are that $ab > 1$ and $b > 1$ (not necessarily an integer).

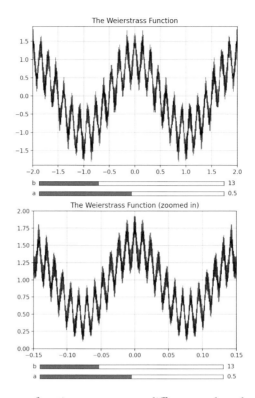

Fig. 2.5: The Weierstrass function on two very different scales, showing its self-similar structure.

DISCUSSION

- **Other monsters**. Poincaré famously described Weierstrass's function as a "monster". Here are two other famous monsters. Both these functions will be explored in the exercises.

  – The *Blancmange function*. For $m = 0, 1, 2 \ldots$, let $f_m : \mathbb{R} \to \mathbb{R}$ be defined by

  $$f_0(x) = \min\{|x - k|, k \in \mathbb{Z}\}, \qquad f_m(x) = \frac{1}{2^m} f_0(2^m x).$$

  The Blancmange function, $g : \mathbb{R} \to \mathbb{R}$, is defined by $g(x) = \sum_0^\infty f_m(x)$. It is continuous on $\mathbb{R}$ but is nowhere differentiable.

  – *Riemann's function*, $R : \mathbb{R} \to \mathbb{R}$ is defined by

  $$R(x) = \sum_{k=1}^{\infty} \frac{\sin(k^2 x)}{k^2}. \tag{2.12}$$

  Riemann conjectured that R was nowhere differentiable, and Weierstrass drew inspiration from this function in the construction of his counterexample. However, it is now known that R is differentiable only at points of the form $\pi a/b$ where a, b are odd coprime integers.

- **Further reading**. For historical accounts and accessible material on continuous, nowhere differentiable functions, see [118, 202]. A more comprehensive treatment can be found in [98].

weierstrass.ipynb (for plotting fig. 2.5)	
	```import numpy as np```
	```import matplotlib.pyplot as plt```
	```from matplotlib.widgets import Slider```
Create an interactive GUI	```%matplotlib```
Initial values of *a* and *b*	```a, b = 0.5, 13```
Maximum number of term in the series	```m_max = 25```
x values (need high resolution)	```x = np.linspace(-2, 2, 2500)```
Each term in the series. Pass *a* and *b* as arguments so they can be updated with the slider	```def fn(x, n, a, b):``` ``` return a**n*np.cos(np.pi*x*b**n)```
The Weierstrass function	```def g(x, a, b):```

Sum the terms from *n* = 0 to m_max	``` S = np.zeros_like(x)```
	 ``` for i in np.arange(0,m_max+1):```
	 ``` S = S + fn(x,i,a,b)```
	 ``` return S```

	```fig,ax = plt.subplots()```
Leave a space at the bottom for sliders	```plt.subplots_adjust(bottom=0.2)```

Plot Weierstrass function (thin black line)	```Wfunc, = plt.plot(x, g(x,a,b),'k', lw=0.5)```
	 ```plt.xlim([-2,2])```
	 ```plt.ylim([-2,2])```
	 ```plt.grid('on')```
	 ```plt.title('The Weierstrass Function')```

Create a slider for $a \in (0, 1)$	```axa = plt.axes([0.15, 0.05, 0.7, 0.02])```
	 ```a_slide = Slider(axa, 'a', 0, 1,```
	 ``` valstep=0.01, valinit=a)```

Create a slider for *b*	```axb = plt.axes([0.15, 0.1, 0.7, 0.02])```
	 ```b_slide = Slider(axb, 'b', 1, 25,```
	 ``` valstep=0.01, valinit=b)```

Update the plot if slider is changed
 Take *a* and *b* from sliders
	```def update(val):```
	 ``` a = a_slide.val```
	 ``` b = b_slide.val```
	 ``` Wfunc.set_ydata(g(x,a,b))```
	 ``` fig.canvas.draw_idle()```

Redraw	```a_slide.on_changed(update)``` ```b_slide.on_changed(update)```
	```plt.show()```

2.6 Integration with Trapezium Rule

Partition the interval $[a, b]$ into n subintervals $[x_0, x_1], [x_1, x_2] \ldots [x_{n-1}, x_n]$ of equal length (where $x_0 = a$ and $x_n = b$). Let $h = (b-a)/n$. The *Trapezium Rule* (or Trapezoidal Rule) states that the integral $\int_a^b f(x)\,dx$ can be approximated as:

$$\int_a^b f(x)\,dx \approx \frac{h}{2}\left[y_0 + y_1 + 2\sum_{i=1}^{n-1} y_i\right], \qquad (2.13)$$

where $y_i = f(x_i)$. We say that the RHS is the approximation of the integral using the Trapezium Rule with n strips.

Use the Trapezium Rule to approximate the integral $\int_1^2 \ln x\,dx$.
How does the accuracy of answer vary with the width of each strip h?

The Trapezium Rule approximates the area under the graph $y = f(x)$ as the sum of the area of n equally spaced trapezia. The left panel of fig. 2.6 demonstrates this idea for $f(x) = \ln x$, using 10 trapezia, or *strips*. The concavity of the curve suggests that the Trapezium-Rule estimate will be less than the actual answer.

Similar to the error analysis of derivatives in §2.2 , we can study the error in numerical integration by defining $E(h)$ as the absolute difference between the Trapezium-Rule estimate and the actual answer. The latter can be obtained by integration by parts, yielding $\int_1^2 \ln x\,dx = [x \ln x - x]_1^2 = 2 \ln 2 - 1$. Hence, we find

$$E(h) = (2 \ln 2 - 1 - \text{Trapezium-Rule estimate}) > 0.$$

The graph of $E(h)$ (produced by the code `trapezium.ipynb`) is shown on the right of fig. 2.6. Since $E(h)$ appears to be a straight line on the logarithmic plot, we can approximate this behaviour as $E \propto h^k$ where k is the gradient of the straight line. Python tells us that

`Gradient of line = 1.99999.`

Thus, we conjecture that the Trapezium Rule is an $O(h^2)$ approximation.

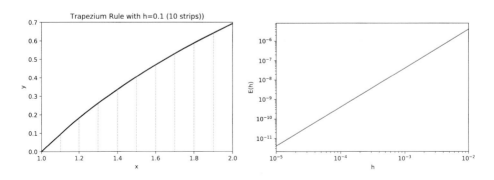

Fig. 2.6: Trapezium Rule. *Left*: The integration scheme for $\int_1^2 \ln x\,dx$ using 10 trapezoidal strips. *Right*: The error $E(h)$ plotted on log scales, showing that $E(h) \propto h^2$.

DISCUSSION

- **The error term**. It can be shown (*e.g.* [182]) that the error term in the Trapezium Rule can be written as

$$\int_a^b f(x)\,dx = \frac{h}{2}\left[y_0 + y_1 + 2\sum_{i=1}^{n-1} y_i\right] - \frac{(b-a)h^2}{12}f''(\xi), \qquad (2.14)$$

for some $\xi \in (a, b)$. The exponent of h in the error term confirms our finding that the Trapezium Rule is an $O(h^2)$ approximation. Note from the formula that if f is a linear function, the error vanishes. This is consistent with the geometric picture – the area under a straight line is *exactly* a trapezium.

- **numpy.trapz** NumPy actually has a built-in Trapezium Rule. Look up the `trapz` command in NumPy's documentation and verify that it gives the same result as our own code.

trapezium.ipynb (for plotting fig. 2.6)	
| | ```python
import numpy as np
import matplotlib.pyplot as plt
``` |
Integration limits $\int_a^b$	`a, b = 1, 2`
Create N evenly spaced values (number of strips) on log scale. **round** gives nearest integer.	`N = np.round(np.logspace(2,5))`
Exact answer for the integral	`actual= 2*np.log(2)-1`
h (width of each strip)	`hlist = (b-a)/N`
$E(h)$ (to be filled in using the *for* loop)	`error = []`
| Eq. 2.13 | ```python
def trapz(y,h):
    return h*(sum(y)-(y[0]+y[-1])/2)
``` |
| Given a fixed number of strips,<br>Create partition $x_i$<br>$y_i$<br><br>Apply the Trapezium Rule<br>and collect the value of $E(h)$ | ```python
for n in N:
    x = np.linspace(a,b,int(n+1))
    y = np.log(x)
    h = (b-a)/n
    estim = trapz(y,h)
    error.append(actual-estim)
``` |
| | ```python
plt.loglog(hlist, np.abs(error))
plt.xlim([1e-5, 1e-2])
plt.xlabel('h')
plt.ylabel('E(h)')
plt.grid('on')
plt.show()
``` |
| Calculate the gradient from the first and last points on the line...<br>and report | ```python
k=(np.log(error[0])-np.log(error[-1]))/\
    (np.log(hlist[0])-np.log(hlist[-1]))
print(f'Gradient of line = {k:.5f}.')
``` |
```

2.7 Integration with Simpson's Rule

Partition the interval $[a, b]$ into $2n$ equal subintervals $[x_0, x_1], [x_1, x_2] \ldots [x_{2n-1}, x_{2n}]$ (where $x_0 = a$ and $x_{2n} = b$). Let $h = (b - a)/2n$. *Simpson's Rule* states that the integral $\int_a^b f(x)\,dx$ can be approximated as:

$$\int_a^b f(x)\,dx \approx \frac{h}{3}\left[y_0 + y_{2n} + 4\sum_{i=1}^n y_{2i-1} + 2\sum_{i=1}^{n-1} y_{2i}\right], \qquad (2.15)$$

where $y_i = f(x_i)$.

Use Simpson's Rule to approximate the integral $\int_1^2 \ln x\,dx$.

How does the accuracy of answer vary with the width of each strip?

Thomas Simpson (1710–1761) was an English self-taught mathematician known today for his integration rule (although, as he himself acknowledged, the result was already known to Newton and a number of previous mathematicians).

In Simpson's Rule, the curve $y = f(x)$ is approximated using a parabola drawn over each subinterval $[x_i, x_{i+2}]$. Each parabola goes through three points, namely, (x_i, y_i), (x_{i+1}, y_{i+1}) and (x_{i+2}, y_{i+2}). For easier comparison with the Trapezium Rule, we say that the formula (2.15) uses n strips, each with width $H \equiv 2h$, where h is the width of each substrip. This means that one parabola is drawn per strip. Fig. 2.7 (left panel) demonstrates this scheme with 10 strips: the thick red vertical dashed lines show the boundaries of strips at $x_0, x_2, \ldots x_{20}$. The thin blue dashed lines are the centres of the strips at $x_1, x_3, \ldots x_{19}$.

As before, we define the absolute error in the approximation as

$$E(H) = |2\ln 2 - 1 - \text{Simpson's-Rule estimate}|.$$

The graph of $E(H)$ is shown in the right panel of fig. 2.7. For easy comparison with the result for Trapezium Rule, we plot this graph over the same domain as that in fig. 2.6. We can make a few interesting observations from this graph.

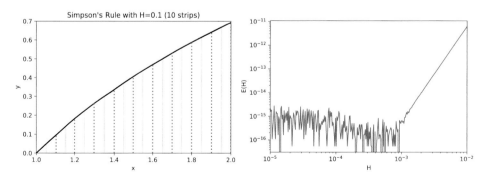

Fig. 2.7: Simpson's Rule. *Left*: The integration scheme for $\int_1^2 \ln x\,dx$ using 10 strips (20 substrips). *Right*: The error $E(H)$ plotted on log scales. The domain is the same as that in fig. 2.6 for easier comparison.

- The values of the error, given the same number of strips, is many orders of magnitude smaller than the error using Trapezium Rule. For example, using the strip width 10^{-2} gives a surprisingly accurate answer with $E \sim 10^{-11}$ for Simpson's Rule, but for the Trapezium Rule, we find a much larger $E \sim 10^{-5}$.
- There appears to be two distinct regimes of the curve. The straight part where $H \gtrsim 10^{-3}$, and the oscillating part for smaller H. Python tells us that the value of the gradient of the straight part is

$$\text{Gradient of line} = 3.99997.$$

Thus, we conjecture that Simpson's Rule is an $O(h^4)$ approximation. Together with the previous point, we conclude that Simpson's Rule is a far superior method in terms of accuracy (at the expense of an increased number of calculations).

- The oscillating part of the curve occurs around $E(H) \sim 10^{-15}$. This is a magnitude comparable to $\varepsilon_{\text{mach}}$. Indeed, we are just seeing the numerical artefact when subtracting two nearly equal numbers in the calculation of $E(H)$.

Here is the code for plotting $E(H)$ and calculating the gradient of the straight part.

simpson.ipynb (for plotting fig. 2.7)	
| | ```python
import numpy as np
import matplotlib.pyplot as plt
``` |
| Integration limits $\int_a^b$ Create $N$ evenly spaced values (number of strips) on log scale. round gives nearest integer. Exact answer for the integral $H$ (width of each strip) $E(H)$ (to be filled in using the *for* loop) | ```python
a, b = 1, 2
N = np.round(np.logspace(2,5,300))

actual= 2*np.log(2)-1
Hlist = (b-a)/N
error = []
``` |
| Eq. 2.15 | ```python
def simp(y,h):
    return (h/3)*(y[0]+y[-1]+\
    4*sum(y[1:-1:2])+2*sum(y[2:-1:2]))
``` |
| Given a fixed number of strips Number of substrips The partition $x_i$ $y_i$ Width of each substrip Apply Simpson's Rule and collect the value of $E(H)$ | ```python
for n in N:
    n2 = 2*n
    x = np.linspace(a,b,int(n2+1))
    y = np.log(x)
    h = (b-a)/n2
    estim = simp(y,h)
    error.append(actual-estim)
``` |
| | ```python
plt.loglog(Hlist , np.abs(error))
plt.xlim([1e-5,1e-2])
plt.xlabel('H')
plt.ylabel('E(H)')
plt.grid('on')
plt.show()
``` |
| Calculate the gradient of the straight part Note that the first 100 elements of these lists are the right third of the graph shown in fig. 2.7 | ```python
k=(np.log(error[0])-np.log(error[50]))/\
    (np.log(Hlist[0])-np.log(Hlist[50]))
print(f'Gradient of line = {k:.5f}.')
``` |

DISCUSSION

- **Smaller h is not always better**. Just as we saw in numerical differentiation, we have a similar phenomenon for numerical integration: smaller h does not always guarantee a more accurate answer. Fig. 2.7 shows that the minimum strip width (below which the roundoff error dominates) occurs at around 10^{-3}.

- **The error term**. It can be shown (*e.g.* [182]) that the error term in Simpson's Rule can be written as

$$\int_a^b f(x)\,dx = \frac{h}{3}\left[y_0 + y_{2n} + 4\sum_{i=1}^{n} y_{2i-1} + 2\sum_{i=1}^{n-1} y_{2i} \right] - \frac{(b-a)h^4}{180} f^{(4)}(\xi),$$

for some $\xi \in (a, b)$. The derivation of the error term is surprisingly much more difficult than that of the Trapezium Rule.

The exponent of h in the error term confirms our conjecture that Simpson's Rule is an $O(h^4)$ approximation.

Note from the formula that if f is a cubic function, the error vanishes. This too is surprising given that Simpson's Rule is based on parabolas. If we consider the simplest case of a single strip on $[x_0, x_2]$, the formula suggests that given a cubic curve on this interval, we can draw a parabola which intersects the cubic at 3 points (where $x = x_0, x_1, x_2$) such that both the cubic and the parabola have exactly the same area under the curves! See exercise 9.

Another interesting observation is that the error term allows us to estimate the minimum strip width by setting magnitude of the error term to be roughly $\varepsilon_{\text{mach}}$. We find

$$H_{\min} = 2h_{\min} \sim 2\left(\frac{180}{(b-a)f^{(4)}(\xi)}\varepsilon_{\text{mach}} \right)^{1/4}.$$

With $f(x) = \ln x$ on $[1, 2]$ and $|f^{(4)}(\xi)| \geq \frac{3}{8}$, we find a conservative estimate of $H_{\min} \lesssim 1.1 \times 10^{-3}$. This is in good agreement with what we see in fig. 2.7. A similar calculation of $h_{\min}$ for the Trapezium Rule is left as an exercise.

- **Romberg integration**. In fact, there is a beautiful connection between the Trapezium and Simpson's Rules. Each method approximates the function $y = f(x)$ on a strip with a degree-n polynomial ($n = 1$ and 2 respectively). One could clearly extend this approximation by using a higher-order polynomial on each strip. This iterative construction is formalised in the technique called *Romberg integration*, frequently used in numerical applications as it has the advantage of being easy to implement using an iterative scheme.

The Trapezium Rule is the lowest-order ($O(h^2)$) Romberg integration, and Simpson's Rule is the next order ($O(h^4)$) Romberg integration. The next order ($O(h^6)$) Romberg integration is based on dividing each strip into 4 substrips, and approximating $f(x)$ over the strip using an order-4 polynomial. This approximation, known as *Boole's Rule*, will be explored in exercise 11.

Werner Romberg (1909–2003), was a German mathematician who escaped Nazi Germany and settled in Norway. His integration method was based on previous work by Maclaurin and Huygens. *George Boole* (1815–1864) was an English mathematician and inventor of symbolic logic which we now call *Boolean algebra*.

2.8 Improper integrals

Evaluate the following integrals numerically.

$$I_1 = \int_0^\infty e^{-x^2}\, dx, \qquad I_2 = \int_0^\infty x^2 e^{-x^2}\, dx, \qquad I_3 = \int_0^\infty \frac{\sin x}{x}\, dx.$$

An *improper integral* is an integral that either has an integration limit involving ∞ or one in which the integrand is not well-defined at one of the integration limits (*e.g.* the integrand in I_3 is not well defined at $x = 0$).

When calculating an integral numerically, it is always useful to start by plotting the integrand to survey the behaviour of the function over the integration domain. This gives us a clue as to the kind of answer we might expect. Any discontinuities or divergence would be revealed in this step and we can then conclude that either the integral diverges, or plan to work strategically around the discontinuities. The graphs of the integrands of I_1 to I_3 are shown in fig. 2.8.

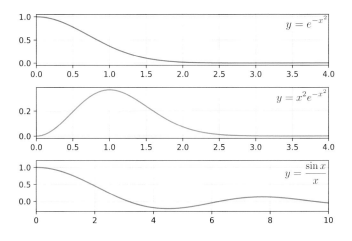

Fig. 2.8: Graphs of the integrands of I_1 to I_3 (top to bottom).

We see that the integrands all appear to converge to $y = 0$ as $x \to \infty$. However, this asymptotic behaviour alone does not guarantee the convergence of the integrals (after all, $\int_1^\infty \frac{1}{x}\, dx$ is divergent). But our graphical investigation has not revealed any obviously divergent integrals.

You might also notice that for I_3, the integrand does not diverge as $x \to 0$, but approaches 1 (due to the limit $\lim_{x\to 0}(\sin x/x) = 1$, see eq. 1.6). To avoid zero division, it might be useful for numerical purposes to define the integrand of I_3 as

$$f(x) = \begin{cases} \dfrac{\sin x}{x} & x \neq 0, \\ 1 & x = 0. \end{cases}$$

This makes f is continuous on $\mathbb{R}$.

Now let's see how these improper integrals can be evaluated in Python.

First solution: quad

The quickest method for tackling improper integrals is use `scipy.integrate.quad` which accepts ∞ (`np.inf`) in the argument. The code for evaluating I_1 was shown in the beginning of this chapter (§2.1). Using quad, we obtain the following values of the integrals, as well as the error estimates.

```
I1, err1 = (0.8862269254527579, 7.101318390472462e-09)
I2, err2 = (0.4431134627263801, 8.053142703522972e-09)
I3, err3 = (2.247867963468921, 3.2903230524472544)
```

The error estimates for I_1 and I_2 look reassuringly tiny, but the error estimate for I_3 is alarmingly large (even larger than the answer itself!). Python also gives a warning message that *"The integral is probably divergent, or slowly convergent"*. Further investigation is needed.

The quad method is really easy, but it gives us very little understanding of what is happening mathematically. When coding becomes an opaque black box, it makes the interaction between mathematics and computing less meaningful, and the problem becomes less mathematically enlightening.

Second solution: substitution

Let's take a look at a more transparent method, this time without using quad.

All the integrals involve infinity in the limit. But *infinity* is not a binary number, so to compute it numerically we must replace infinity with a finite limit. But simply replacing infinity with a large number is not always going to work. After all, it's not clear how large the large number should be.

Here is a more sophisticated strategy. To work out $I = \int_0^\infty f(x)\,dx$, let's first break it up into two integrals:

$$I = \int_0^\alpha f(x)\,dx + \int_\alpha^\infty f(x)\,dx,$$

where the break point α is a positive number to be determined. For the second integral, use a substitution to turn the limit ∞ to a finite number. Let's try $u = 1/x$ (a different substitution might work even better, depending on the behaviour of the integrand). This yields

$$I = \int_0^\alpha f(x)\,dx + \int_0^{1/\alpha} \frac{f(1/u)}{u^2}\,du. \tag{2.16}$$

When $\alpha = 1$, the two terms can also be combined into a single integral (upon renaming the dummy variable u as x)

$$I = \int_0^1 \left[f(x) + \frac{f(1/x)}{x^2} \right] dx. \tag{2.17}$$

Let's try using formula (2.16) to tackle I_1 using Simpson's Rule to evaluate each integral. You can use our own Simpson's Rule code, or, equivalently, `scipy.integrate.simpson`.

Supposing for now that the exact answer is

$$I_1 = \frac{\sqrt{\pi}}{2}.$$

(We will explain this in the Discussion section.) Let's vary the value of the break point α and see how the error behaves. Using $h = 10^{-3}$ in Simpson's Rule, the code `improper.ipynb` produces fig. 2.9.

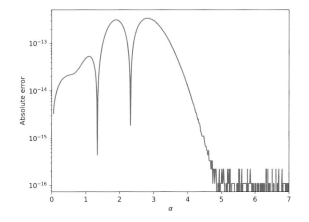

Fig. 2.9: The accuracy when $I_1 = \int_0^\infty e^{-x^2}\, dx$ is evaluated by splitting it into two integrals (eq. 2.16) and using Simpson's Rule with $h = 10^{-3}$. The graph shows the absolute error as the break point α varies. The absolute error is minimised when $\alpha \gtrsim 5$.

Fig 2.9 shows that the best accuracy is obtained when α is around 5 (where the absolute error is machine-epsilon limited). Nevertheless, for any α the worst accuracy is still a respectable 13 decimal places. With $\alpha = 5$, Python gives the following excellent estimate which is accurate to 15 decimal places:

```
integ = 0.886226925452758
exact = 0.8862269254527579
```

A couple of highlights from the code:

- In evaluating the second integral in eq. 2.16 , we choose a tiny cutoff for the lower limit (10^{-8} in the code) in place of zero, which would have produced a division-by-zero error. It is always a good idea to check that the answer for the integral is insensitive to the choice of the tiny cutoff. Indeed, varying the cutoff by a few orders of magnitude does not change the shape of the graph.
- Simpson's Rule requires an even number of strips, *i.e.* an odd number of points N in `np.linspace(a, b, N)`. We ensure that N is odd using the function

$$\texttt{makeodd(N) = N + 1 - (N\%2)}.$$

The syntax N%2 gives the remainder when N is divided by 2. Therefore, the function gives N when N is odd, and N+1 when N is even.
- Nevertheless, SciPy's `integrate.simpson` would still work without our `makeodd` function, but one should be aware of how the routine deals with an odd number of half-strips. The result can be different depending on your version of SciPy. For more on this point, consult SciPy's documentation[2].

---

[2] https://docs.scipy.org/doc/scipy/reference/generated/scipy.integrate.simpson.html

<table>
<tr><td colspan="2">improper.ipynb (for producing fig. 2.9)</td></tr>
</table>

```python
                                          import numpy as np
SciPy's Simpson's Rule                    import matplotlib.pyplot as plt
                                          from scipy.integrate import simpson
Range of break point α in eq. 2.16
Each run of the for loop will fill this list   alpha = np.linspace(0.05, 7, 500)
f (the first integrand of (2.16)          error = []
g (the second integrand)                  f = lambda x: np.exp(-x**2)
                                          g = lambda x: np.exp(-1/x**2)/x**2
Exact value of the integral
Width of each half-strip                  exact = 0.5*np.sqrt(np.pi)
                                          h = 1e-3
Function   which   always   gives   an
odd integer                               def makeodd(N):
(see text for explanation)                    return N + 1 - (N%2)

Loop over α values                        for a in alpha:
No. of half-strips in each integrand (int     N1 = int(a/h)
converts a float to an integer)               N2 = int(1/(a*h))
Domain of the first integration               x1 = np.linspace(0, a, makeodd(N1))
Lower limit in the second integral cannot     x2 = np.linspace(1e-8, 1/a, makeodd(N2))
be 0, so we introduce a tiny cutoff 10⁻⁸.
Use Simpson's Rule to evaluate (2.16)         integ = simpson(f(x1),x1)+simpson(g(x2),x2)
(note the syntax: y values then x values)     err = abs(exact-integ)
Collect the absolute errors                   error.append(err)

Plot the result with a green line   plt.semilogy(alpha,error,'g')
(log vertical axis)                 plt.xlim([0,max(alpha)])
                                    plt.xlabel(r'$\alpha$')
                                    plt.ylabel('Absolute error')
                                    plt.grid('on')
                                    plt.show()
```

Applying the same splitting trick to integral I_2, you should find a similar behaviour in the absolute error. To work out the exact answer, one can integrate by parts and use the previous exact result to deduce that

$$I_2 = \frac{\sqrt{\pi}}{4}.$$

With $\alpha = 5$, Python gives an excellent estimate accurate to 15 decimal places.

```
integ = 0.44311346272637947
exact = 0.44311346272637897
```

Finally, for the integral I_3, our trick produces

$$I_3 = \int_0^\alpha \frac{\sin x}{x}\,dx + \int_0^{1/\alpha} \frac{\sin(1/x)}{x}\,dx.$$

For small x, the $\sin(1/x)$ term fluctuates very rapidly. The resulting error from integrating this term numerically is enormous (try it!).

We could try neglecting the second troublesome integral and evaluating the first integral up to a large finite number $\alpha = A$, *i.e.* we make the approximation

$$I_3 \approx \int_0^A \frac{\sin x}{x}\,dx.$$

The integral converges very slowly as shown in fig. 2.10. Here we have used the Trapezium Rule with strip width $h = 10^{-3}$, and we have assumed that the exact answer is

$$I_3 = \frac{\pi}{2}.$$

(This will be explained in the Discussion). From fig. 2.10, it is clear that we have to integrate to quite a large number to achieve high accuracy. More precisely, the error fluctuates more and more as A increases, but there is an envelope $A \propto (\text{error})^{-1}$, which guarantees a minimum achievable accuracy. We can deduce, for instance, that to achieve an answer that is accurate to p decimal places (*i.e.* the error must be at most 0.5×10^{-p}), this can be achieved when $A \approx 2 \times 10^p$ (limited by rounding error).

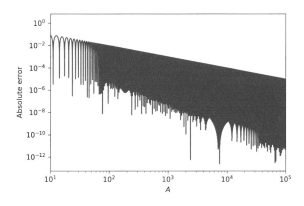

Fig. 2.10: The absolute error defined by $\left| \int_0^A \frac{\sin x}{x} \, dx - \frac{\pi}{2} \right|$ plotted as the upper limit A varies. The integral is evaluated using Trapezium Rule with $h = 10^{-3}$.

Although we have attacked these integrals using a variety of different numerical methods, it is also possible to calculate these integrals using Python's library for symbolic mathematics called *SymPy*. We will explore SymPy in chapter 5, but if you are impatient, go straight to the code box in the very last exercise in that chapter.

Finally, a rather different approach to numerical integration based on probability is *Monte Carlo integration*, which will be discussed in §7.10.

DISCUSSION

- **Key to success with numerical integration**. The main takeaway here is that there are no magical Python functions which can deal with every integral accurately. The key to success is to work with Python by using our mathematical understanding (*e.g.* introduce a suitable transformation). Always investigate the behaviour of the integrands, and where possible, avoid the black-box approach (where you leave everything to Python without knowing what it does). Numerical integration is an art which takes experience to master, so be patient!

- **A Gaussian integral and a trick**. Here is a neat integration trick to show that $I_1 = \sqrt{\pi}/2$. Let $I = \int_{-\infty}^{\infty} e^{-x^2}\, dx$. Note that by symmetry, $I = 2I_1$. Since we can also write $I = \int_{-\infty}^{\infty} e^{-y^2}\, dy$, we can multiply the two expressions for I and, assuming that we can move terms around, we have

$$I^2 = \int_{-\infty}^{\infty} e^{-x^2}\, dx \int_{-\infty}^{\infty} e^{-y^2}\, dy = \iint_{\mathbb{R}^2} e^{-(x^2+y^2)}\, dx\, dy.$$

Here we need some knowledge of multivariable calculus. If double integration is unfamiliar to you, come back to revisit this point after studying the next chapter. Note that the domain in the final integral is the whole of $\mathbb{R}^2$. We can evaluate this integral in polar coordinates (r, θ). We have $x^2 + y^2 = r^2$, and the area element $dx\, dy = r\, dr\, d\theta$. Thus,

$$I^2 = \int_{\theta=0}^{2\pi} \int_{r=0}^{\infty} r e^{-r^2}\, dr\, d\theta = \pi.$$

Therefore $I = \sqrt{\pi}$ and $I_1 = \sqrt{\pi}/2$.

This kind of integral, known as *Gaussian integral*, occurs frequently in university mathematics, especially in probability and statistics (you may recognise that the integrand is the normal distribution). But if we change the integration limits to $\int_a^b e^{-x^2}\, dx$ for arbitrary real numbers a, b, then there are no elementary expressions for the answer. This is why it is important for us to perform this kind of numerical integration accurately.

- **The sine integral and another trick**. The integral I_3, which we found tricky to handle numerically due to its slow convergence, is in fact a special value of the following function called the *sine integral*:

$$\text{Si}(x) = \int_0^x \frac{\sin t}{t}\, dt.$$

This function occurs frequently in physical and engineering applications. The quickest way to evaluate this is to use the following SciPy command:

```
scipy.special.sici(x)[0]
```

Note that only first element of `sici(x)` is the sine integral. The second element is the *cosine integral* which will be explored in exercise 13.

In general, there are no elementary expressions for this integral, except at $x = 0$ (clearly $\text{Si}(0) = 0$) and $x \to \infty$ where the integral becomes $I_3 = \int_0^{\infty} \frac{\sin x}{x}\, dx = \pi/2$ as we saw earlier. At university, you will learn a number of different methods that can help you evaluate I_3 exactly. One technique is to use *contour integration* (in a topic called *complex analysis*). Another technique the *Laplace transform*, which is a mathematical tool with a huge range of engineering applications.

Even without these advanced techniques, there is a clever trick which can help us evaluate the integral. Again this relies on some elementary knowledge of multivariable calculus. This trick is based on the following simple observation:

$$\int_0^\infty e^{-xy} \sin x \, dy = \frac{\sin x}{x},$$

(where x is held constant in the integral). Therefore, we can write the original integral as

$$I_3 = \int_0^\infty \left(\int_0^\infty e^{-xy} \sin x \, dx \right) dy,$$

where we have assumed that the order of integration can be switched (thanks to a result in analysis called *Fubini's theorem*). You would most likely have come across the inner integral presented as an exercise in integration by parts with the use of a *reduction formula* (whereby the original integral appears upon integrating by part twice). You should verify that:

$$\int_0^\infty e^{-xy} \sin x \, dx = \frac{1}{1 + y^2}.$$

Returning to the original integral, we then find:

$$I_3 = \int_0^\infty \frac{dy}{1 + y^2} = \left[\tan^{-1} y \right]_0^\infty = \frac{\pi}{2}.$$

2.9 Fourier series

Let $f : \mathbb{R} \to \mathbb{R}$ be a 2π-periodic function defined by:

$$f(x) = \begin{cases} 1 & x \in [0, \pi) \\ 0 & x \in [\pi, 2\pi) \end{cases}$$

and continued outside $[0, 2\pi)$ in the same way, so that the graph of $y = f(x)$ looks like a *square wave*.
Show that f can be written as a series in $\sin nx$ as

$$f(x) = \frac{1}{2} + \frac{2}{\pi} \left(\sin x + \frac{1}{3} \sin 3x + \frac{1}{5} \sin 5x \dots \right).$$

Plot the partial sums of this series.

The topic of *Fourier series* is a vital part of university mathematics. We give a brief summary here. The goal is to write a 2π-periodic function f (defined on $\mathbb{R}$) as a sum of sines and cosines of different frequencies and amplitudes. In other words, we want to express $f(x)$ as

$$f(x) = \frac{a_0}{2} + \sum_{n=1}^{\infty} (a_n \cos nx + b_n \sin nx), \tag{2.18}$$

(the constant term a_0 is traditionally written out separately to make the formula for a_n and b_n easier to remember). The French mathematician *Joseph Fourier* (1768–1830) proposed such a series to study the problem of heat conduction, and showed that the coefficients of the series are given by:

$$a_n = \frac{1}{\pi} \int_{-\pi}^{\pi} f(x) \cos nx \, dx, \qquad b_n = \frac{1}{\pi} \int_{-\pi}^{\pi} f(x) \sin nx \, dx. \tag{2.19}$$

We can think of equation 2.18 as a decomposition of a function into different resolutions. Large n are high-frequency sinusoidals, so they capture the small-scale fluctuations of f at fine detail. Similarly, small n sinusoidals would capture the low-frequency, broad-brush behaviour of f.

In fact, this composition can be generalised to a continuous range of frequency n, in which case the decomposition is called a *Fourier transform*. Fourier series and Fourier transform constitute a topic called *Fourier analysis*, which is an indispensable tool in signal and image processing, geoscience, economics and much more. See [65, 94, 190] for some excellent books on Fourier analysis.

Back to the square-wave function. Performing the integrations in Eqs. 2.19, we find:

$$a_0 = 1, \qquad a_n = 0,$$

$$b_n = \frac{1}{n\pi} [1 - (-1)^n] = \begin{cases} \frac{2}{n\pi} & \text{for } n \text{ odd} \\ 0 & \text{for } n \text{ even} \end{cases}$$

where $n = 1, 2, 3 \dots$ (you should verify these results). Putting these into the Fourier series (2.18) gives:

$$f(x) = \frac{1}{2} + \frac{2}{\pi} \sum_{n \text{ odd} \geq 1}^{\infty} \frac{\sin nx}{n}. \tag{2.20}$$

The code below produces an interactive plot with a slider which changes the number of terms in the truncated Fourier series ($n = 0, 1, 2, \ldots n_{\max}$) . Some snapshots for different values of $n_{\max}$ are shown in fig. 2.11.

Here are some interesting observations from fig. 2.11.

- The more terms we use, the closer we are to the square wave. However, the series, even with infinitely many terms, will never be exactly the same as the square wave. For example, at the points of discontinuities of the square wave function, the Fourier series equals $\frac{1}{2}$ (just put $x = k\pi$ in (2.20)), but the square wave never takes this value. This means that one has to take the equal sign in Eq. 2.20 with a pinch of salt. Some people write $\sim$ instead of $=$ to emphasise the difference.
- Near each discontinuity, there appears to be an overshoot above 1 (and undershoot below 0). Try zooming into an overshoot and read off the maximum value (choosing a large value of $n_{\max}$). You should find that Fourier series overshoots the square wave by just under 9%. The undershoot below 0 is also by the same amount.

DISCUSSION

- **Jump discontinuities**. It can indeed be shown that if there is a jump discontinuity in the function f at $x = a$, then its Fourier series converges to the average of the left and right limits, *i.e.*

$$\frac{1}{2} \left(\lim_{x \to a^-} f(x) + \lim_{x \to a^+} f(x) \right). \tag{2.21}$$

- **Gibbs phenomenon**. An overshoot (and an undershoot) always occurs when using the truncated Fourier series to approximate a discontinuous function. For large $n_{\max}$, the overshoot moves closer to the point of discontinuity but does not disappear. In fact, the overshoot can be shown to be around 9% of the magnitude of the jump. More precisely, the fractional overshoot can be expressed as

$$\frac{1}{\pi} \int_0^\pi \frac{\sin x}{x} \, dx - \frac{1}{2} \approx 0.0894898722 \quad (10 \text{ dec. pl.}) \tag{2.22}$$

Note the appearance of the sine integral!

The overshoot and undershoot of Fourier series near a discontinuity are known as *Gibbs phenomenon* after the American mathematician *Josiah Gibbs* (1839–1903) who gave a careful analysis of the overshoot magnitude. However, it was the English mathematician *Henry Wilbraham* (1825–1883) who first discovered the phenomenon.

- **Parseval's theorem and $\zeta(2)$**. There is an elegant relation between the function f and its Fourier coefficients a_n and b_n (without the sine and cosines).

Parseval's Theorem, named after the French mathematician *Marc-Antoine Parseval* (1755–1836), states that

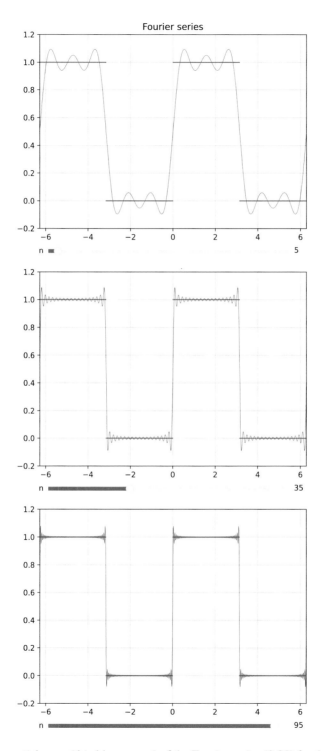

Fig. 2.11: The partial sums (thin blue curves) of the Fourier series (2.20) for the square-wave function (shown in red) with $n_{\max} = 5, 35$ and 95.

$$\frac{1}{2\pi} \int_{-\pi}^{\pi} |f(x)|^2 \, dx = \left(\frac{a_0}{2}\right)^2 + \frac{1}{2} \sum_{1}^{\infty} a_n^2 + \frac{1}{2} \sum_{1}^{\infty} b_n^2. \qquad (2.23)$$

Here is a brief interpretation of why this holds. The LHS is the average of $|f(x)|^2$ over the period. The terms on the RHS are the averages of terms when the Fourier expansion is squared. We are left with the averages of $(a_0/2)^2$, $(a_n \cos nx)^2$ and $(b_n \sin nx)^2$ (the cross terms average to zero).

Applying Parseval's identity to the square wave function, Eq. 2.23 becomes:

$$\frac{1}{2} = \frac{1}{4} + \frac{2}{\pi^2} \left(1 + \frac{1}{3^2} + \frac{1}{5^2} + \cdots\right).$$

Another way to express this interesting result is:

$$\sum_{k=1}^{\infty} \frac{1}{(2k-1)^2} = \frac{\pi^2}{8}. \qquad (2.24)$$

We end this chapter with a calculation of $\zeta(2) = \sum_{n=1}^{\infty} \frac{1}{n^2}$ that follows from Eq. 2.24. Let $S = \zeta(2)$. Observe that:

$$S = \sum_{n=1}^{\infty} \frac{1}{n^2}$$

$$= \sum_{n \text{ odd}} \frac{1}{n^2} + \sum_{n \text{ even}} \frac{1}{n^2}$$

$$= \sum_{k=1}^{\infty} \frac{1}{(2k-1)^2} + \sum_{k=1}^{\infty} \frac{1}{(2k)^2}$$

$$= \frac{\pi^2}{8} + \frac{S}{4}$$

$$\implies S = \frac{4}{3} \cdot \frac{\pi^2}{8} = \frac{\pi^2}{6}.$$

This beautiful result[3] also verifies our numerical calculation of $\zeta(2)$ in §1.4.

---

[3] A subtle point: in the second line of the working, we assumed that the sum S remains unchanged by the rearrangement of the terms in the series. Although it is safe to do here, a rearrangement can change the sum of series like the alternating harmonic series. Look up *Riemann's rearrangement theorem*.

fourier.ipynb (for producing fig. 2.11)

| | ```python
import numpy as np
import matplotlib.pyplot as plt
from matplotlib.widgets import Slider
``` |
| Create an interactive GUI | ```python
%matplotlib
``` |
| Initial values of $n_{max}$ (an odd number) | ```python
nmax = 5
``` |
| A shorter way to call $\pi$ | ```python
pi = np.pi
``` |
| Domain $[-2\pi, 2\pi]$ at high resolution | ```python
x = np.linspace(-2*pi, 2*pi, 1001)
``` |
| The square-wave function | ```python
def f(xarray):
``` |
| Pre-populate output with 0's | ```python
    y = np.zeros_like(xarray)
    for ind, x in enumerate(xarray):
``` |
| If $x$ modulo $2\pi$ is... | ```python
        xmod = x%(2*pi)
``` |
| in the right domain then... | ```python
        if xmod<pi:
``` |
| change the output from 0 to 1 | ```python
            y[ind] = 1
``` |
| Insert discontinuities at multiples of $\pi$ | ```python
        if x%pi==0:
            y[ind]= np.nan
    return y
``` |
| The Fourier series | ```python
def Fourier(x, nmax):
``` |
| $\sum_1^{n_{max}} \frac{\sin nx}{n}$ with $n$ odd | ```python
    S = np.zeros_like(x)
    for n in np.arange(1,nmax+1,2):
        S += np.sin(n*x)/n
    return 0.5+ 2*S/pi
``` |
| | ```python
fig,ax = plt.subplots()
``` |
| Leave a space at the bottom for a slider | ```python
plt.subplots_adjust(bottom=0.15)
``` |
| Plot the square wave in red | ```python
plt.plot(x, f(x),'r',lw=1.5)
``` |
| Plot the Fourier series in thin blue line | ```python
Ffunc,= plt.plot(x, Fourier(x,nmax),'b',
        lw=0.5)
plt.xlim([-2*pi, 2*pi])
plt.ylim([-0.2, 1.2])
plt.grid('on')
plt.title(r'Fourier series')
``` |
| Create a slider for $n_{max}$ from 1 to 101 | ```python
axn = plt.axes([0.15, 0.05, 0.7, 0.03])
n_slide = Slider(axn, 'n', 1, 101,
``` |
| Keep $n_{max}$ odd by setting the step size to 2 | ```python
        valstep = 2, valinit = nmax)
``` |
| | ```python
def update(val):
``` |
| Take $n_{max}$ from slider | ```python
    nmax = n_slide.val
``` |
| Recalculate the Fourier series | ```python
    Ffunc.set_ydata(Fourier(x,nmax))
    fig.canvas.draw_idle()
``` |
| Update the plot if slider is changed | ```python
n_slide.on_changed(update)
``` |
| | ```python
plt.show()
``` |

2.10 Exercises

1 Perform the following modifications on the code Eh.ipynb which we used to calculate
the derivative of $f(x) = x^3$ at $x = 1$ (§2.2).

- Change the point x at which the derivative is calculated.
- Change the function $f(x)$ to any differentiable function.
- Change the approximation to the backward-difference approximation.

Verify that the qualitative behaviour of $E(h)$ does not change by these modifications.

2 (*Five-point stencil*) Use Eh.ipynb to help you with this question.
Apart from the forward, backward and symmetric-difference formulae for the derivative
discussed in §2.2, there are infinitely many other similar differentiation formulae. Those
involving more points typically produce a more accurate answer (given the same step
size h) at the expense of increased calculation time.
Here is an exotic one called the *five-point stencil* formula:

$$f'(x) \approx \frac{1}{12h}(-f(x + 2h) + 8f(x + h) - 8f(x - h) + f(x - 2h)).$$

a. Plot the graph of the absolute error $E(h)$ for the derivative of $f(x) = \cos x$ at $x = 1$.
You should see a V-shape similar to fig. 2.2.
b. On the same set of axes, plot the symmetric difference formula. How do the two
graphs compare?
c. If the five-point stencil formula is an $O(h^k)$ approximation, use your graph to show
that $k = 4$.

Note: Such formulae are studied in a topic called *finite-difference methods*, which we
will explore in §4.9.

3 (*Taylor series and convergence*) Consider the following Taylor series.

$$\sinh x = \sum_{n=0}^{\infty} \frac{x^{2n+1}}{(2n + 1)!},$$

$$\tan^{-1} x = \sum_{n=0}^{\infty} (-1)^n \frac{x^{2n+1}}{(2n + 1)},$$

$$\sqrt{1 + x} = \sum_{n=0}^{\infty} \frac{(-1)^{n+1}(2n)!}{4^n (n!)^2 (2n - 1)} x^n.$$

a. Modify the code taylor.ipynb (§2.3) to produce the graph of each of the above
functions and its Taylor series, similar to fig. 2.3.
Experiment with the domain of the plot, and conjecture the radius of convergence
for each Taylor series.
(For example, you should find that for $\sqrt{1 + x}$, the radius of convergence is 1.)
b. Choose a value of x for which all three Taylor series converge to the respective
function (for example, $x = 0.3$).
Plot the absolute error

$$|R_N| = |f(x) - \text{Taylor series of order } N|$$

as a function of the polynomial order N, similar to fig. 2.4 (but forget about the upper and lower bound in dotted lines). Plot the absolute error for all 3 Taylor series on the same set of axes.

From the plot, deduce which Taylor series converges fastest and slowest. (Answer: $\sinh x$ is fastest, $\sqrt{1+x}$ is slowest.)

4 (*Taylor's Theorem*) Consider $f(x) = \ln(1 + x)$. We showed using Taylor's Theorem (theorem 2.3) that the remainder R_N is given by eq. 2.9.

   a. Fix $x = 0.4$. Using the Taylor polynomial of order 1, show that the constant ξ is approximately 0.1222 (to 4 dec. pl).
   b. Plot a graph of ξ as a function of N. You should obtain a decreasing function. Verify that ξ always lies in the interval $(0, 0.4)$, in accordance with Taylor's Theorem.

5 (*The Blancmange function*) For $m = 0, 1, 2 \ldots$, let $f_m : \mathbb{R} \to \mathbb{R}$ be defined by

$$f_0(x) = \min\{|x - k|, k \in \mathbb{Z}\}, \qquad f_m(x) = \frac{1}{2^m} f_0(2^m x).$$

   a. Plot the graph of the function $f_m(x)$ for a few values of m. Show that the graphs are straight-line segment with sharp (non-differentiable) points similar to $y = -|x|$.
   b. Plot the *Blancmange function* defined by

$$g(x) = \sum_{m=0}^{\infty} f_m(x).$$

   On the domain $[0, 1]$, you should obtain the graph in fig. 2.12 (which looks like the eponymous dessert). Clearly you will need to impose some cutoff to approximate the infinite sum.
   c. Generalise the Blancmange function by redefining $f_m(x)$ as

$$f_{m,K}(x) = \frac{1}{K^m} f_0(K^m x).$$

   Plot the generalised Blancmange function $g_K(x) = \sum_{m=0}^{\infty} f_{m,K}(x)$ for $K = 2, 3, 4 \ldots 20$ on the same set of axes.
   (Better yet, vary the plot using a slider for K. Use weierstrass.ipynb as a template).
   Conjecture the shape of the function when $K \to \infty$. Can you prove it?
   d. For $K = 2$, create a figure which shows the self-similarity structure of the graph (in the style of fig. 1.15). Note also that the graph is periodic on $\mathbb{R}$.
   Conjecture the values of $K \in \mathbb{R}$ for which the graph shows i) periodicity, or ii) self-similarity.

For an accessible account of the Blancmange function and its properties, see [200].

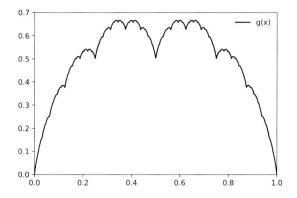

Fig. 2.12: The Blancmange function ($K = 2$). It is continuous on $\mathbb{R}$ but is nowhere differentiable.

6 (*Riemann's function*) Consider Riemann's non-differentiable function, $R(x)$, defined by

$$R(x) = \sum_{k=1}^{\infty} \frac{\sin(k^2 x)}{k^2}.$$

a. Plot the graph $y = R(x)$. What is the period of this function? (Prove it.)
b. The function R is continuous on $\mathbb{R}$ but is only differentiable at certain points, including those of the form $x = a\pi/b$ where a and b are odd, coprime integers. It is known that the derivative at such points takes the same value. Let's use Python to investigate what this value is.
   i. Use the symmetric-difference formula (see §2.2), with $h = 10^{-6}$, to find the numerical derivative of R at $x_0 = 3\pi/5$. To do this, you will have to approximate the infinite sum by the partial sum with $N_{\max}$ terms:

$$R(x_0) \approx \sum_{k=1}^{N_{\max}} \frac{\sin(k^2 x_0)}{k^2}. \tag{$*$}$$

   ii. Plot the approximation of the derivative ($*$) as a function of the cutoff $N_{\max}$. You should choose the values of $N_{\max}$ such that a convergent behaviour emerges. Hence, conjecture the value of the derivative at $x_0 = 3\pi/5$.
   iii. Change x_0 to see if your conjecture still holds.
   (Answer: See [73].)

7 (*The error function*) The *error function* erf: $\mathbb{R} \to \mathbb{R}$ is defined by the integral:

$$\text{erf } x = \frac{2}{\sqrt{\pi}} \int_0^x e^{-t^2} \, dt.$$

We often encounter this function in the context of probability, statistics and in simple models of diffusion.

a. Plot the graph of the error function. Suggestion: Use `scipy.special.erf`.
b. The error function can be expressed as a power series

$$\text{erf } x = \frac{2}{\sqrt{\pi}} \sum_{n=0}^{\infty} \frac{(-1)^n x^{2n+1}}{(2n+1)n!}.$$

Suppose we want to use a power series in x to approximate erf(x) for $|x| < 2$, accurate to 4 decimal places.

Show that we need 15 terms in the partial sum.

Suggestion: Plot the absolute error (semi-log scale) for $x \in [0, 2]$. Increase the number of terms in the partial sum until the error falls below 0.5×10^{-4}.

8. (*Trapezium Rule*) Use `trapezium.ipynb` as a starting point for this question.
Consider the evaluation of $\int_1^2 \ln x \, dx$.

a. Evaluate the integral using Trapezium Rule, extending the graph of $E(h)$ (fig. 2.6) to smaller values of h. Note that when h becomes sufficiently tiny, rapid fluctuations start to appear in the plot of $E(h)$. This is where roundoff error starts to dominate. Take note of the value of h around which the fluctuations occur. (We can regard this value of h as the optimal value of the strip-size for the numerical integration.) How does your graph compare with fig. 2.7 (right panel)?
b. Using the expression for the error term in eq. 2.14, estimate the value of h below which the roundoff error dominates. Do this by hand by setting the error term to $\varepsilon_{\text{mach}}$. You should find that this value roughly agrees with the graph in part (a).

9. (*Simpson's Rule*) Use `simpson.ipynb` as a starting point in this question.
Recall that Simpson's Rule uses parabolas to approximate the area under a graph. In this question, we will demonstrate the (somewhat surprising) observation that Simpson's Rule gives an exact answer when integrating a *cubic* polynomial. (The reason for this was discussed at the end of §2.7.)
Consider the cubic polynomial defined by $f(x) = (2x - 1)^3$.

a. Verify that $\int_0^4 f(x) \, dx = 300$.
b. Evaluate the integral $\int_0^4 f(x) \, dx$ using Simpson's Rule with 1 strip (*i.e.* 2 substrips). You should also obtain 300.
c. Consider the parabola $P(x) = 36x^2 - 58x - 1$. Verify that the parabola agrees with the cubic at $x = 0, 2$ and 4.
d. Verify that $\int_0^4 P(x) \, dx = 300$.
e. Plot the cubic and the parabola on the same set of axes over the domain $[0, 4]$. This graph shows the parabola that is drawn when Simpson's Rule is applied.
(Clearly both the cubic and the parabola should look like they have the same signed area under the graph.)

10. (*Midpoint Rule*) Use `trapezium.ipynb` to help you with this question.
The *Midpoint Rule* is another useful integration scheme, in addition to the Trapezium and Simpson's Rules.
We start as before by partitioning the domain of integration into n equal subintervals $[x_0, x_1], [x_1, x_2] \ldots [x_{n-1}, x_n]$. Let $h = (b - a)/n$. The Midpoint Rule states that the integral $\int_a^b f(x) \, dx$ can be approximated as:

$$\int_a^b f(x)\, dx \approx h \sum_{i=0}^{n-1} f(m_i), \quad \text{where } m_i = \frac{x_i + x_{i+1}}{2} \qquad (2.25)$$

In other words, at the midpoint m_i of each strip, we draw a rectangle of height $f(m_i)$, and sum the area of these rectangles to approximate the integral. The figure below demonstrates the Midpoint Rule with $n = 5$.

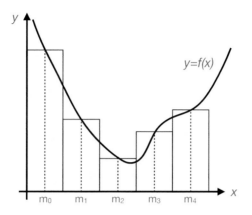

Fig. 2.13: The Midpoint Rule with 5 strips.

a. Write a code which evaluates the integral $\int_1^2 \ln x\, dx$ using the Midpoint Rule with $h = 10^{-5}$.

b. Plot the graph of the absolute error $E(h)$ on log scales. On the same set of axes, plot $E(h)$ for the Trapezium Rule. You should see two parallel lines.
Verify that the Midpoint Rule is an $O(h^2)$ approximation.

c. Plot the graph of the ratio of the errors:

$$\frac{E_{\text{Midpoint}}(h)}{E_{\text{Trapezium}}(h)}.$$

You should find that this ratio is approximately constant over a large range of h. Hence, make a conjecture on the exact form of the error term for the Midpoint Rule.

d. Use the Midpoint Rule to verify the result

$$\int_0^1 \frac{\ln x}{x - 1} = \frac{\pi^2}{6}.$$

Note: The Midpoint Rule allows us to avoid the singularities at the endpoints. This makes it a useful method for evaluating integrals in which the integrand is not well-defined at one or both of the endpoints. The combination of a well-judged substitution and the Midpoint Rule is a powerful arsenal for tackling improper integrals (see [170] for details of such techniques).

11 (*Boole's Rule*) Use `simpson.ipynb` to help you with this question.
   Another interesting higher-order integration scheme called *Boole's Rule*.
   Partition the domain of integration into $4n$ equal subintervals $[x_0, x_1], [x_1, x_2] \ldots [x_{4n-1}, x_{4n}]$.
   Let $h = (b - a)/4n$ (*i.e.* divide each strip into 4 substrips). Boole's Rule states that the
   integral $\int_a^b f(x)\, dx$ can be approximated as:

$$\int_a^b f(x)\, dx \approx \frac{2h}{45} \left[7(y_0 + y_{4n}) + 32 \sum_{i=1,3,5\ldots}^{4n-1} y_i + 12 \sum_{i=2,6,10\ldots}^{4n-2} y_i + 14 \sum_{i=4,8,12\ldots}^{4n-4} y_i \right].$$

   a. Write a code that evaluates the integral $\int_1^2 \ln x\, dx$ using Boole's rule with $h = 10^{-2}$.
   b. Plot the graph of the absolute error $E(h)$ on log scales. Hence verify that Boole's
      Rule is an $O(h^6)$ approximation.

12 (*Numerical integration challenge*) Use numerical integration techniques studied in this
   chapter to verify the following results. Remember to always plot the integrand first and
   plan how to deal with any singularities.

   a) $\displaystyle\int_0^1 \frac{1}{\sqrt{-\ln x}}\, dx = \sqrt{\pi}$ b) $\displaystyle\int_0^\infty \left(\frac{\sin x}{x}\right)^2 dx = \frac{\pi}{2}$

   c) $\displaystyle\int_0^\infty \frac{dx}{(1 + x^\phi)^\phi} = 1$ (where ϕ is the Golden Ratio).

   For more interesting integrals against which you can test your numerical integration
   skills, see the classic encyclopaedic reference by [78], and also [153, 208] for deep
   mathematical insights into many fascinating (and mind-blowing) integrals.

13 (*Sine and cosine integrals*) The sine and cosine integrals are defined by

$$Si(x) = \int_0^x \frac{\sin t}{t}\, dt, \quad Ci(x) = \gamma + \ln x + \int_0^x \frac{\cos t - 1}{t}\, dt,$$

   where γ is the Euler-Mascheroni constant (see §1.5). Plot $Si(x)$ and $Ci(x)$ on the domain
   $[0, 4\pi]$, performing the integration numerically. (Suggestion: Use `quad`.) Check your
   results using `scipy.special.sici`.
   Conjecture the asymptotic values of $Si(x)$ and $Ci(x)$ as $x \to \infty$.

14 (*Fourier series recap*) Use `fourier.ipynb` to help you with this question.
   Consider the 2π-periodic function, $f : \mathbb{R} \to \mathbb{R}$, defined by

$$f(x) = (x + \pi)^2, \qquad x \in [-\pi, \pi).$$

   Its Fourier series (which you may like to verify by hand) is given by

$$f(x) = \frac{4\pi^2}{3} + \sum_{n=1}^\infty 4(-1)^n \left(\frac{1}{n^2} \cos nx - \frac{\pi}{n} \sin nx \right).$$

   a. Substitute $x = 0$ and show that $\zeta(2) = \pi^2/6$.
   b. Substitute $x = \pi$ and verify the convergence property at jump discontinuities (eq.
      2.21).
   c. Apply Parseval's Theorem (eq. 2.23) to show that $\zeta(4) = \pi^4/90$.
   d. Modify `fourier.ipynb` to plot the function and the partial sums of its Fourier
      series. Your output should resemble fig. 2.14 shown below.

e. Verify numerically the magnitude of Gibbs phenomenon (eq. 2.22). Suggestion: You will need a large number of terms in the partial sum.

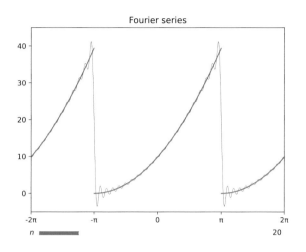

Fig. 2.14: The Fourier series (thin blue line) for $y = (x + \pi)^2$ defined on $[-\pi, \pi]$ (extended periodically outside the interval).

15 (*Fourier series with period L*) Fourier series can be used to approximate a function $f : \mathbb{R} \to \mathbb{R}$ with arbitrary period $L > 0$ (not necessarily 2π). In this case, we have the Fourier series:

$$f(x) = \frac{a_0}{2} + \sum_{n=1}^{\infty} \left(a_n \cos\left(\frac{2n\pi x}{L}\right) + b_n \sin\left(\frac{2n\pi x}{L}\right) \right), \quad \text{where}$$

$$a_n = \frac{2}{L} \int_{-L}^{L} f(x) \cos\left(\frac{2n\pi x}{L}\right) dx, \quad b_n = \frac{2}{L} \int_{-L}^{L} f(x) \sin\left(\frac{2n\pi x}{L}\right) dx.$$

Consider the function, $f : \mathbb{R} \to \mathbb{R}$, with period $L = 2$, defined by

$$f(x) = \cosh x, \quad x \in [-1, 1).$$

You may like to verify that the Fourier coefficients are given by:

$$a_n = \frac{2(-1)^n \sinh 1}{n^2 \pi^2 + 1}, \quad b_n = 0.$$

a. Modify the code `fourier.ipynb` to plot the function and the partial sums of its Fourier series.
b. By substituting a suitable value of x in the Fourier series, find a closed-form expression for

$$\sum_{n=0}^{\infty} \frac{1}{n^2 \pi^2 + 1}.$$

Verify your answer numerically.

THREE

Vector Calculus and Geometry

Vector calculus is a generalisation of calculus concepts (*i.e.* differentiation and integration) from $\mathbb{R}$ to $\mathbb{R}^n$. Sometimes called *multivariable calculus* or *differential geometry*, this is one of the core topics in any mathematics course.

Fig. 3.1: *Carl Friedrich Gauss* (1777–1855) discovered profound results in a myriad of topics in mathematics and physics. His legacy is that of an intellectual titan whose work dramatically revolutionised mathematics. Many results in this chapter are due to him. (Image source: [137].)

One of the most useful applications of vector calculus is the study of curves and surfaces in $\mathbb{R}^3$. Python's *Matplotlib* is especially useful in helping us visualise 3-dimensional curves and surfaces. Double and triple integrals (of well behaved functions) can also be performed quickly as we will demonstrate below. We will use these visualisation tools to create interactive graphics to help us understand the length, curvature and torsion of curves in $\mathbb{R}^3$, as well as the surface area and volume bounded by surfaces. The chapter concludes with two important results in vector calculus: the Divergence Theorem and Stokes' Theorem.

Excellent reviews of vector calculus can be found in [28, 93, 174, 189, 192]. For good references on differential geometry of curves and surfaces, see [37, 171, 201]. We will introduce many more references along the way.

We now present a lightning review of some essential concepts in vector calculus, with accompanying code snippets where appropriate. For full mathematical explanations and plenty more examples, see the textbooks on vector calculus recommended above.

S. Chongchitnan, *Exploring University Mathematics with Python*,

3.1 *Basic concepts in vector calculus*

All code snippets in §3.1 will require the following header.

```
import numpy as np
import matplotlib.pyplot as plt
[code snippet]
```

Parametric curves

A parametric curve $\mathbf{r} : (a, b) \rightarrow \mathbb{R}^n$ is defined by

$$\mathbf{r}(t) = (f_1(t), f_2(t), \ldots f_n(t)),$$

where we assume that each component f_i is a smooth function of t (*i.e.* infinitely differentiable on $(a, b) \subseteq \mathbb{R}$). In this chapter, we will only encounter curves in $\mathbb{R}^2$ or $\mathbb{R}^3$.

Here are Python commands for plotting curves in $\mathbb{R}^2$ and $\mathbb{R}^3$. We have given two of several command variations that can be used to plot a parametric curve in $\mathbb{R}^3$. They require the keyword argument `projection='3d'`.

Plotting parametric curves in $\mathbb{R}^2$ and $\mathbb{R}^3$	
$t \in [0, 2\pi]$	`t = np.linspace(0,2*np.pi)`

Plot $\mathbf{r}(t) = (\cos t, \sin t)$ (a circle)
 This prevents the circle from appearing stretched
	`plt.plot(np.cos(t), np.sin(t))`
	 `plt.axis('equal')`
	 `plt.show()`

Plot $\mathbf{r}(t) = (\cos t, \sin t, t)$
 (a helix)
	`ax = plt.axes(projection='3d')`
	 `ax.plot(np.cos(t), np.sin(t), t)`
	 `plt.show()`
	
	 `# OR`
	
	 `fig = plt.figure()`
	 `ax = fig.add_subplot(projection='3d')`
	 `ax.plot(np.cos(t), np.sin(t), t)`
	 `plt.show()`

Polar coordinates

Polar coordinates (r, θ) are related to Cartesian coordinates (x, y) by

$$x = r \cos \theta, \qquad y = r \sin \theta.$$

To plot a polar curve of the form $r = f(\theta)$, use either of the following methods.

Plotting a curve in polar coordinates	
$\theta \in [0, 2\pi]$ Example: plot $r = 2\cos 3\theta$	```t = np.linspace(0, 2*np.pi, 100)``` ```r = 2*np.cos(3*t)```

Method I: Convert to Cartesian
and use θ as the parameter
Prevent shape distortion
	```# Method I```
	```plt.plot(r*np.cos(t), r*np.sin(t))```
	```plt.axis('equal')```
	```plt.show()```

Method II: Plot on polar axes	```# Method II```
	```ax = plt.axes(projection='polar')```
	```ax.plot(t, r)```
	```plt.show()```

Both methods allow the possibility that $r < 0$, however Method II displays the curve in an unconventional way when $r < 0$. Try this, and notice how the centre of the plot does not correspond to $r = 0$.

If you'd like to exclude the part $r < 0$ when using Method I, insert the following command before plotting.

$$r = r.clip(min=0)$$

Parametric surfaces

A parametric surface $\mathbf{S} : A \to \mathbb{R}^3$ is defined by

$$\mathbf{S}(u, v) = (f_1(u, v), f_2(u, v), f_3(u, v)),$$

where $(u, v) \in A \subseteq \mathbb{R}^2$. We assume that each component f_i is infinitely differentiable in u and v.

A parametric curve has one parameter t, whilst a parametric surface has two (u and v). In the code below, we use the command `meshgrid` to create all possible ordered pairs of u and v. The surface is created using the `plot_surface` command.

Plotting a parametric surface	

$u \in [-1, 1]$
$v \in [0, 2]$
Create 50×50 coordinate grid
	```u = np.linspace(-1,1)```
	```v = np.linspace(0,2)```
	```U, V = np.meshgrid(u,v)```

Plot $\mathbf{S}(u, v) = (u, 2v, 1 - u - v)$
(a plane)
	```ax = plt.axes(projection='3d')```
	```ax.plot_surface(U, 2*V, 1-U-V)```
	```plt.show()```

Alternatively, the first three lines may be equivalently written using the `mgrid` command as follows.

$$U, V = np.mgrid[-1:1:50j, 0:2:50j]$$

where `50j` specifies that we want 50 equally spaced values.

Cylindrical and spherical coordinates

A surface can be expressed in many possible coordinate choices, but picking one that exploits the symmetry of the surface can greatly simplify your calculations.

Cylindrical coordinates are useful for describing surfaces with some rotational symmetry about the z-axis. In essense, cylindical coordinates are simply polar coordinates with the z coordinate added on as the 3rd dimension. The relationship between cylindrical coordinates (r, θ, z) and Cartesian coordinates (x, y, z) is:

$$x = r\cos\theta, \quad y = r\sin\theta, \quad z = z,$$

where $r \geq 0$, $\theta \in [0, 2\pi]$ and $z \in \mathbb{R}$.

Spherical coordinates are useful for describing surfaces with some rotational symmetry about the origin. The relationship between spherical coordinates (r, θ, ϕ) and Cartesian coordinates (x, y, z) is:

$$x = r\cos\theta\sin\phi, \quad y = r\sin\theta\sin\phi, \quad z = r\cos\phi,$$

where $r \geq 0$, $\theta \in [0, 2\pi]$ and $\phi \in [0, \pi]$.

In terms of *Matplotlib*, there are no special commands that allow you to plot surfaces directly in these coordinates (unlike polar coordinates). Simply use Cartesian coordinates and treat them as parametric surfaces as above. Here is an example of how to plot the unit sphere.

Plotting a sphere	

$u \in [0, 2\pi]$ (this is θ)
$v \in [0, \pi]$ (this is ϕ)
Create 50×50 grid
	```u = np.linspace(0,2*np.pi)```
	```v = np.linspace(0,np.pi)```
	```U, V = np.meshgrid(u,v)```

Plot the unit sphere
(convert spherical to Cartesian coordinates)
	```x = np.cos(U)*np.sin(V)```
	```y = np.sin(U)*np.sin(V)```
	```z = np.cos(V)```

Prevent shape distortion	```ax = plt.axes(projection='3d')```
	```ax.plot_surface(x,y,z)```
	```ax.set_box_aspect((1,1,1))```
	```plt.show()```

Partial derivatives

Suppose we differentiate a multivariable function $f(x, y, z)$ with respect to one of its variables, say x, whilst treating y and z as constants. The result is called the *partial derivative* of f with respect to x, and is denoted

$$\frac{\partial f}{\partial x} \quad \text{or} \quad f_x$$

This is in contrast to the usual (total) derivative $\frac{df}{dx}$, which does not make sense in this case.

For most vector calculus applications, the function $f(x, y, z)$ will be infinitely differentiable in all 3 variables on the given domain in $\mathbb{R}^3$. There are 3 possible first-order partial derivatives: $\frac{\partial f}{\partial x}, \frac{\partial f}{\partial y}, \frac{\partial f}{\partial z}$. There are 6 possible second-order partial derivatives:

$$\frac{\partial^2 f}{\partial x^2}, \frac{\partial^2 f}{\partial y^2}, \frac{\partial^2 f}{\partial z^2}, \frac{\partial^2 f}{\partial x \partial y}, \frac{\partial^2 f}{\partial y \partial z}, \frac{\partial^2 f}{\partial x \partial z}.$$

The last thee are partial derivatives of mixed orders, which can be done in any order (*e.g.* $\frac{\partial^2 f}{\partial x \partial y} = \frac{\partial^2 f}{\partial y \partial x}$). But be aware that there are pathological counterexamples where this does not hold (see [72]).

In this chapter, we will not perform partial derivatives numerically. As far as possible, one should try to differentiate by hand in order to avoid the truncation and rounding errors discussed in §2.2. However, if numerical differentiation is needed, the procedure is exactly the same as that described in §2.2. For example, using the forward-difference formula, we have

$$\frac{\partial f}{\partial x} \approx \frac{f(x + h, y, z) - f(x, y, z)}{h}.$$

Multiple integrals

A multivariable function may be integrated with respect to more than one variable. For example, we can integrate a 2-variable function $f(x, y)$ over a rectangle $R = [0, 1] \times [0, 2]$. This can be expressed as any the following double integrals.

$$\iint_R f(x, y)\, dx\, dy = \int_0^1 \int_0^2 f(x, y)\, dy\, dx = \int_0^2 \int_0^1 f(x, y)\, dx\, dy.$$

Note that we perform the innermost integral first and proceed outwards. This means that

$$\int_0^2 \int_0^1 f(x, y)\, dx\, dy = \int_0^2 \left(\int_0^1 f(x, y)\, dx \right) dy$$

although brackets are normally omitted.

Similarly, we can integrate a 3-variable function $f(x, y, z)$ over a cuboid $C = [0, 1] \times [0, 2] \times [-1, 1]$. This can be expressed as any the following triple integrals.

$$\iiint_C f(x, y, z)\, dx\, dy\, dz = \int_0^1 \int_0^2 \int_{-1}^1 f(x, y, z)\, dz\, dy\, dx$$

$$= \int_{-1}^1 \int_0^2 \int_0^1 f(x, y, z)\, dx\, dy\, dz$$

$$= \int_{-1}^1 \int_0^1 \int_0^2 f(x, y, z)\, dy\, dx\, dz.$$

There are 3 other possible orders of integrations. Try to write them down.

To perform double and triple integrals in Python, one can integrate one variable at a time using, say, the Trapezium or Simpson's Rules described in §2.6-2.7. Alternatively, for 'well-behaved' functions (Python will most likely warn you if your function is not well-behaved), you can use SciPy's `dblquad` and `tplquad` which are multivariable generalsations of the `quad` function described in §2.1. Two examples are given below.

Double and triple integrals

$$I = \int_0^2 \int_{x^2}^{2x} dy\,dx$$

Output: I[0] = 1.3333333333333333
(Exact answer: $I = 4/3$.)

$$J = \int_0^1 \int_0^{\sqrt{x}} \int_0^{1+x+y} xy\,dz\,dy\,dx$$

Output: J[0] = 0.38690476190095546
(Exact answer: $J = 65/168$.)

```
from scipy.integrate import dblquad, tplquad

I = dblquad(lambda y, x: 1,
            0 , 2,
            lambda x: x**2, lambda x: 2*x)
print("I = ", I[0])

J = tplquad(lambda z,y,x: x*y,
            0 , 1,
            0 , lambda x: x**(1/2),
            0 , lambda x,y: 1+x+y)
print("J = ", J[0])
```

When using `dblquad` and `tplquad`, there are a few quirky rules to keep in mind:

- The integrand must be defined as a function (even if it's 1). When defining this function, **the variables must be listed in the order of integration** (*i.e.* from in to out). Note that in the examples given, we have used lambda (anonymous) functions to define the integration limits.
- The limits of the integrals must all be defined as *functions* unless they are constants. Here's a quirky thing about `tplquad`. For the innermost limits, the variables in the functions must be listed in **reverse order of integration**.
- As with `quad`, the output of `dblquad` and `tplquad` is a tuple (pair of numbers): namely, the value of the integral and an estimate of the absolute error.

Finally, it is worth remembering the formulae for the double integral in polar coordinates:

$$\iint dx\,dy = \iint r\,dr\,d\theta \qquad\qquad \text{(polar)}$$

as well as triple integrals in cylindrical and spherical coordinates.

$$\iiint dx\,dy\,dz = \iiint r\,dr\,d\theta\,dz \qquad\qquad \text{(cylindrical)}$$

$$= \iiint r^2 \sin\phi\,dr\,d\theta\,d\phi \qquad\qquad \text{(spherical)}$$

3.2 The cycloid

> Point P lies on the circumference of a unit circle in the x-y plane. As the circle rolls horizontally without slipping, P traces out a curve. Investigate the equation and the shape of this curve.

This famous parametric curve was named the *cycloid* by Galileo (see [138] for a fascinating historical account). Let us try to derive its parametric equations.

In fig. 3.2(a), O is the origin and $\mathbf{p}$ is the position vector of a fixed reference point P on the wheel. Vector $\mathbf{a}$ is the position vector of the centre of circle, and $\mathbf{b}$ is the vector joining the centre of the circle to P.

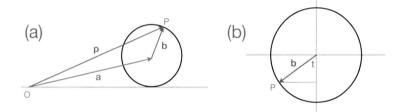

Fig. 3.2: (a) The rolling wheel with unit radius. (b) The vector $\mathbf{b}$ in the frame where the centre of the wheel is stationary.

Let's parametrise the curve traced out using t which equals the x coordinates of the centre of the circle (*i.e.* the distance travelled by the wheel). From the diagram, we see that at any value of t, the three vectors above always satisfy the relation

$$\mathbf{p}(t) = \mathbf{a}(t) + \mathbf{b}(t).$$

We seek a more explicit expression for $\mathbf{p}(t)$.

Since the circle has unit radius, we have $\mathbf{a}(t) = (t, 1)$.

Recall the definition of the radian: on the unit circle, an arc of length t subtends an angle of t radians. The *rolling* of the wheel implies that the horizontal distance travelled by the wheel equals the arc length that has rolled on the x-axis. Thus, we deduce that the angle swept out by the wheel also equals t.

We can then obtain the components of vector $\mathbf{b}(t)$ from fig. 3.2(b), which shows the wheel in the frame where its centre is fixed at the origin. Resolving $\mathbf{b}$ into horizontal and vertical components, we find $\mathbf{b}(t) = (-\sin t, -\cos t)$.

In conclusion, we have shown that the parametric equation of the cycloid is

$$\mathbf{p}(t) = (t - \sin t, 1 - \cos t) . \tag{3.1}$$

The code on the next page produce an interactive figure, a snapshot of which is shown in fig. 3.3. The user controls the x position of the centre of the wheel with a slider, whilst point P traces out the cycloid shown in red.

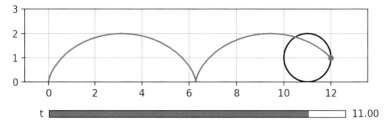

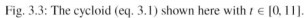

Fig. 3.3: The cycloid (eq. 3.1) shown here with $t \in [0, 11]$.

cycloid.ipynb (for producing fig. 3.3)	
| | ```python
import numpy as np
import matplotlib.pyplot as plt
from matplotlib.widgets import Slider
``` |
| Create an interactive GUI | ```python
%matplotlib
``` |
| Parametrize the unit circle with $\theta \in [0, 2\pi]$ | ```python
theta = np.linspace(0,2*np.pi)
``` |
| Parametric equation of the rolling circle at time $t$ | ```python
circ_x = lambda t: t + np.cos(theta)
circ_y = 1 + np.sin(theta)
``` |
| Coordinates of the reference point $P$ (which lies on the cycloid) at time $t$ | ```python
cycl_x = lambda t: t - np.sin(t)
cycl_y = lambda t: 1 - np.cos(t)
``` |
| Set initial time | ```python
t = 0
``` |
| Leave space for a slider | ```python
fig,ax = plt.subplots()
plt.subplots_adjust(bottom=0.2)
plt.ylim(0, 3)
plt.xlim(-1, 1+4*np.pi)
``` |
| Set equal aspect ratio to prevent shape distortion | ```python
plt.gca().set_aspect('equal')
plt.grid('on')
``` |
| Plot the circle in black<br>Plot the cycloid in red | ```python
Circ,=plt.plot(circ_x(t), circ_y, 'k')
Cycl,=plt.plot(cycl_x(t), cycl_y(t),
                'r', markersize=3)
``` |
| Mark point $P$ with a big red dot | ```python
Pnt,=plt.plot(cycl_x(t), cycl_y(t),
                'ro', markersize=5)
``` |
| Adjust the position of the slider<br>Set range and resolution of the $t$ slider | ```python
axt = plt.axes([0.18, 0.33, 0.67, 0.02])
t_slide = Slider(axt, 't', 0, 4*np.pi,
                valstep=0.001, valinit=t)
``` |
| Get the $t$ value from slider<br>Slide the circle along | ```python
def update(val):
    t = t_slide.val
    Circ.set_xdata(circ_x(t))
    T = np.linspace(0, t, int(50*t))
``` |
| Plot the cycloid from time 0 to $t$<br>Mark point $P$ only at present $t$ value | ```python
    Cycl.set_data(cycl_x(T),cycl_y(T))
    Pnt.set_data([cycl_x(t)],[cycl_y(t)])
    fig.canvas.draw_idle()
``` |
| Redraw figure when the slider is changed | ```python
t_slide.on_changed(update)
``` |
| | ```python
plt.show()
``` |
```

DISCUSSION

- **Cusps**. The cycloid has a characteristic cusp (sharp point) when t is an even multiple of π. At each cusp, the tangent vector is zero. We can check this by verifying that $\mathbf{p}'(0) = (0,0)$.

- **Regular curves**. A *regular parametrisation* of a curve $\mathbf{r}(t)$ is one in which $|\mathbf{r}'(t)| \neq 0$ on the whole curve. Thus, another way to state the previous point is that the parametrisation (3.1) is not regular. A *regular curve* is one for which there exists a regular parametrisation. Regular curves are important because it allows a change of parameter (*reparametrisation*) to be done without altering properties of the curve. Regular curves also have well-defined curvature, as we will see in §3.4.

- **Roulettes**. As a curve C_1 rolls on another curve C_2 without slipping, the locus of a reference point P on C_1 is called a *roulette*. Some readers (or their parents) may be familiar with the *spirograph*, a retro toy which generates roulettes. The cycloid is a roulette where C_1 is a circle and C_2 is a line. You will explore the case of a circle rolling on another circle in exercise 2.

  One could also formulate the roulette problem differently: suppose P traces out a straight line, what possible pairs of curves C_1, C_2 can we have? This problem is usually phrased in terms of a wheel of shape C_1 travelling on a road with shape C_2. The centre of a circular wheel of a bicycle travelling on a flat road traces out a straight line, of course. A famously outrageous wheel-road pair is that of a bicycle with *square* wheels which can travel smoothly on a road made up of precisely placed *catenary* segments! Incredibly, Macalester college in Minnesota hosts this precise setup for visitors to try. For other intriguing road-wheel pairs, see [87, 119].

3.3 Arc length of an ellipse

The parametric equation of an ellipse E with semi-major axis 1 and semi-minor axis b (where $0 < b \leq 1$) is given by

$$\mathbf{r}(t) = (\cos t, b \sin t), \qquad t \in [0, 2\pi).$$

Calculate the perimeter of the ellipse as a function of b.
In 1914, Ramanujan gave the following approximation for the perimeter of the ellipse with semi-major axis a and semi-minor axis b:

$$\pi \left(3(a + b) - \sqrt{(a + 3b)(3a + b)} \right). \tag{3.2}$$

Investigate the accuracy of this approximation for the ellipse E.

The arc length, s, of a parametric curve $\mathbf{r}(t)$ where $t \in [t_0, t_1]$ is given by

$$s = \int_{t_0}^{t_1} |\mathbf{r}'(t)| \, dt.$$

For the total arc length of the ellipse E, the integral becomes

$$s = 4 \int_0^{\pi/2} \sqrt{\sin^2 t + b^2 \cos^2 t} \, dt$$

$$= 4 \int_0^{\pi/2} \sqrt{1 - (1 - b^2) \cos^2 t} \, dt$$

where we have used the symmetry of the ellipse. Now use the substitution $x = \frac{\pi}{2} - t$. We find:

$$s = 4 \int_0^{\pi/2} \sqrt{1 - (1 - b^2) \sin^2 x} \, dx. \tag{3.3}$$

If you haven't thought about this before, it will probably come as a surprise to you that there is no analytic expression for this integral unless $b = 1$ (where the ellipse becomes a unit circle with circumference $s = 2\pi$). The fact that there are only approximations to the perimeter of the ellipse has been known since at least 1609 when the German astronomer *Johannes Kepler* (1571–1630) proposed the first approximation in connection with planetary orbits which he discovered to be elliptical. Over the centuries, many approximations have been put forward by great mathematical minds (see [199] for a comprehensive review).

The approximation (3.2) is due to the legendary mathematician *Srinivasa Ramanujan* (1887–1920). Ramanujan came from an impoverished background in south India and rose to mathematical prominence in England through a combination of innate genius, hard work and resourcefulness. There are by now volumes of books and even a few movies about Ramanujan's inspirational but tragically short life. See [25] for a thorough and enlightening account of Ramanujan's life and work.

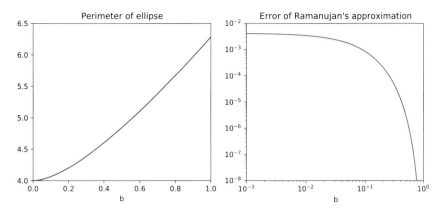

Fig. 3.4: *Left:* The perimeter of the ellipse $\mathbf{r}(t) = (\cos t, b \sin t)$ as a function of b, calculated using numerical integration (quad). *Right:* The fractional error of Ramanujan's approximation (3.2).

Back to the perimeter of the ellipse. The numerical approximation of the integral and the fractional error of Ramanujan's approximation are plotted in fig. 3.4, produced using the code ellipse.ipynb. The left panel of the figure uses the quad function for numerical integration. It shows the correct limits as $b \to 0$ (a vanishingly flat ellipse) and $b \to 1$ (a unit circle). Plotting Ramanujan's approximation of the perimeter in this figure gives an essentially identical curve.

The plot on the right shows the *fractional error* of Ramanujan's approximation plotted on log scale, so that we can examine more carefully the small b regime where the ellipse is flatter (more *eccentric*). This is the regime where most approximations do not perform well. Nevertheless, we see that Ramanujan's formula performs remarkably well here, with the worst accuracy being roughly 0.4%.

DISCUSSION

- **Working with tuples**. When using quad in the code, be very careful not to do mathematical operations on the resulting tuple. This is a very common pitfall for beginners.

  You cannot do mathematical operations on values in a tuple because values in a tuple cannot be changed (in geek speak, we say that tuples are *immutable*). In the last line of the code above, try adding 1 or multiplying by 4. The results aren't what you might expect. For example, multiplying a tuple by integer n simply produces n copies of the tuple, concatenated together.

  Can you predict what calamity would occur had we coded the parameter as s=4*integrate.quad(f, 0, np.pi/2)? (try it!) More about this in the Appendix (A.3).

  Also, recall that s[1] (the second element of the tuple s) is an estimate of the error associated with the numerical integration. You should always check that this number is tiny.

- **Eccentricity**. For the ellipse with semi-major axis a and semi-minor axis b, the shape of the ellipse can be quantified by its *eccentricity* defined as

$$e = \sqrt{1 - \frac{b^2}{a^2}}.$$

The letter e is not to be confused with Euler's constant. We see that the eccentricity is a number between 0 (a circle) and 1 (a vanishingly flat ellipse). The integral expression for the perimeter can be expressed in terms of the eccentricity as

$$s = 4a \int_0^{\pi/2} \sqrt{1 - e^2 \cos^2 t}\, dt. \tag{3.4}$$

- **Arc-length parametrisation**. A regular curve $\mathbf{r}(t)$ can be parametrised using the arc length parameter

$$s(t) = \int_{t_0}^{t} |\mathbf{r}'(u)|\, du. \tag{3.5}$$

The resulting parametrisation, $\mathbf{r}(s)$, is called the *arc-length parametrisation* (or *unit-speed* parametrisation, since $|d\mathbf{r}/ds| = 1$). Both parametrisations describe the same curve, but the arc-length parametrisation simplifies many otherwise nasty formulae and proofs in vector calculus (as we will see later). Most of the time, we will not need to know the explicit form of the arc-length parametrisation. Just the fact that it exists is enough.

A case in point is the ellipse $\mathbf{r}(t) = (a \cos t, b \sin t)$. We do not know the expression for $s(t)$ explicitly. However, $\mathbf{r}'(t) \neq \mathbf{0}$ ($\sin t$ and $\cos t$ cannot be zero at the same time), so the ellipse has a regular parametrisation, and therefore has an arc-length parametrisation $\mathbf{r}(s)$, with $|\mathbf{r}'(s)| = 1$.

- **Elliptic integrals**. The integral

$$E(\phi, k) = \int_0^{\phi} \sqrt{1 - k^2 \sin^2 x}\, dx, \tag{3.6}$$

is called the *elliptic integral of the second kind*. For the *elliptic integral of the first kind*, take the reciprocal of the above integrand:

$$F(\phi, k) = \int_0^{\phi} \frac{1}{\sqrt{1 - k^2 \sin^2 x}}\, dx. \tag{3.7}$$

These functions have analogous behaviour to trigonometric functions, and arise in physical problems involving elliptical geometry. For example, the perimeter of the ellipse with eccentricity e (eq. 3.4) can be expressed as $4a\, E(\frac{\pi}{2}, e)$.

ellipse.ipynb (for producing fig. 3.4)

```
import numpy as np
import matplotlib.pyplot as plt
import scipy.integrate as integrate
%matplotlib
```

Sample b linearly for the perimeter plot
But logarithmically for the error plot
Array for perimeter values
Array for fractional error in Ramanujan's approximation

```
b_lin = np.linspace(0,1,100)
b_log = np.logspace(-3,0,100)
perim = np.zeros(100)
error = np.zeros(100)
```

Integrand in eq. 3.3

```
def integrand(x,b):
    return np.sqrt(1-(1-b**2)*np.sin(x)**2)
```

Ramanujan's approximation (3.2) with $a = 1$

```
def Ramanujan(b):
    return np.pi*(3*(1+b) - \
            np.sqrt((1+3*b)*(3+b)))
```

Fill in array for first plot

quad gives a tuple (see Discussion)
Store perimeter values

```
for i, b in enumerate(b_lin):
    f = lambda x: integrand(x,b)
    s = integrate.quad(f, 0, np.pi/2)
    perim[i] = 4*s[0]
```

Fill in array for second plot

Store fractional error values

```
for i, b in enumerate(b_log):
    f = lambda x: integrand(x,b)
    s = integrate.quad(f, 0, np.pi/2)
    error[i]=1- Ramanujan(b)/(4*s[0])
```

Plot two figures side by side

```
fig, (ax1, ax2) = plt.subplots(1,2,
                    figsize=(10,4))

ax1.plot(b_lin,perim, 'blue')
ax1.set_xlim([0, 1])
ax1.set_ylim([4, 6.5])
ax1.set(xlabel = 'b')
ax1.set(title ='Perimeter of ellipse')
ax1.grid('on')

ax2.loglog(b_log, error, 'red')
ax2.set_xlim([1e-3, 1])
ax2.set_ylim([1e-8, 1e-2])
ax2.set(xlabel = 'b')
ax2.set(title = \
   "Error of Ramanujan's approximation")
ax2.grid('on')

plt.show()
```

3.4 Curvature

The *Lemniscate of Bernoulli* is given by the parametric equation

$$\mathbf{r}(t) = \left(\frac{\cos t}{1 + \sin^2 t}, \frac{\cos t \sin t}{1 + \sin^2 t} \right), \quad t \in [0, 2\pi].$$

Find an expression for the curvature of the Lemniscate. Identify the points on the Lemniscate where the curvature takes maximum and minimum values.

The Lemniscate (meaning ribbon) was named after *Jacob Bernoulli* (1655–1705), who showed that it is one possible locus of points such that the *product* of the distances to two given points is constant. This is analogous to the ellipse which is the locus of points such that the *sum* of the distances to two given points (called *foci*) is constant.

The Bernoullis were an influential Swiss family of eminent (and occasionally competitive) intellectuals. In particular, Jacob, his brother Johann and the latter's son Daniel made substantial contributions to many areas in mathematics and physics. At university, you will come across many results named after them.

For a curve in $\mathbb{R}^2$, it is interesting to ask: *how much does a curve curve?* A huge circle appears to curve less than a tiny circle (the outer lane of an Olympic running track is mostly straight). The quantity which measures this concept is the curvature, κ. For a unit-speed curve, $\mathbf{r}(s)$, it is defined by

$$\kappa(s) = |\ddot{\mathbf{r}}(s)|. \tag{3.8}$$

We can see why this definition makes sense by applying to the circle radius R parametrised as $\mathbf{r}_{\text{circle}}(s) = \left(R \cos \frac{s}{R}, R \sin \frac{s}{R} \right)$, $s \in [0, 2\pi R)$. You can check that this is a unit-speed parametrisation, and that the curvature is

$$\kappa_{\text{circle}}(s) = \frac{1}{R}.$$

Indeed, the larger the circle, the smaller the curvature.

From definition (3.8), it can be shown that for any regular curve $\mathbf{r}(t)$ (not necessarily unit speed), the same formula becomes dramatically more complicated, namely:

$$\kappa(t) = \frac{|\mathbf{r}'(t) \times \mathbf{r}''(t)|}{|\mathbf{r}'(t)|^3}. \tag{3.9}$$

(See recommended texts on differential geometry.) You may be wondering how we can perform the cross product with vectors in $\mathbb{R}^2$. We simply append a zero z-component, so that (x, y) becomes $(x, y, 0)$ in the cross product. The formulae (3.8)-(3.9) are also valid for regular parametric curves in $\mathbb{R}^3$. Note from the formula that we need the parametrisation to be regular - can you see why?

For our Lemniscate, substituting in the derivatives and simplifying (this gets a little messy) gives the following expression for the curvature:

$$\kappa(t) = \frac{3|\cos t|}{\sqrt{1 + \sin^2 t}}. \tag{3.10}$$

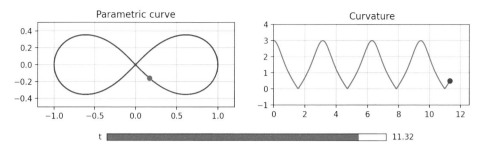

Fig. 3.5: *Left:* The Lemniscate of Bernoulli. The slider controls the value of t on both graphs. *Right:* The curvature $\kappa(t)$ at the corresponding point on the Lemniscate.

Let's use Python to visualise the Lemniscate and the curvature at each point on the curve. The code below produces two graphs, one of the Lemniscate and one for its curvature, as shown in fig. 3.5.

We make the following observations.

- The Lemniscate does indeed look like a ribbon. As t goes from 0 to 2π, the figure-of-eight loop is traced out once.
- Visually, the least curved part of the Lemniscate look like the centre of the ribbon (where the curve looks the 'straightest'). Indeed, the curvature plot tells us that the minimum $\kappa(t) = 0$ is achieved at $t = $ odd multiples of $\pi/2$.
- The maximum curvature $\kappa(t) = 3$ is achieved at the leftmost and rightmost points of the ribbon (where $t = $ multiples of π). This might be slightly more tricky to judge visually.

DISCUSSION

- **Signed curvature.** You may have noticed that the curvature plot has cusps (sharp points) where $\kappa = 0$. To understand this, note that, by definition (3.8), $\kappa \geq 0$. However, at the centre of the Lemniscate, the second derivative $\ddot{\mathbf{r}}(s)$ changes sign. As an analogy, try plotting the second derivative of $f(x) = x^3$ around the point of inflection at $x = 0$. You will see that $y = |f''(x)|$ has a similar cusp.

  In fact, for curves in $\mathbb{R}^2$, it is possible to formulate the concept of curvature more precisely by assigning a sign to the curvature at each point. This quantity is called the *signed curvature*, κ_s. Its magnitude, $|\kappa_s|$, is exactly κ. The sign of κ_s at a particular point on the curve is determined by the direction of rotation of the tangent vector at that point: anticlockwise rotation means $\kappa_s > 0$, clockwise rotation means $\kappa_s < 0$. For example, on the Lemniscate, $\kappa_s > 0$ on the right half of the ribbon, and $\kappa_s < 0$ on the left half.

  Be aware that, confusingly, some authors refer to the signed curvature as "curvature".

- **The Fundamental Theorem of Plane Curves.** The signed curvature is particularly important in the study of geometry because it turns out that for any smooth function f, there exists a unit-speed curve whose signed curvature $\kappa_s(s)$ equals $f(s)$. Furthermore, this curve is unique up to translation and rotation in the plane. In this sense, the signed curvature completely characterises a curve in $\mathbb{R}^2$.

  This result is called the *Fundamental Theorem of Plane Curves*. See [171] for proof.

- **Arc length of the Lemniscate**. Is it possible to work with the arc-length parametrisation of the Lemniscate? Using eq. 3.5, we find

$$s = \int_0^t \frac{1}{\sqrt{1 + \sin^2 x}} \, dx. \tag{3.11}$$

This can be expressed as an elliptic integral $F(k, i)$ which we saw in §3.3. In this case, the arc-length parameter is clearly too complicated for practical use, but it is still instructive to see that both the Lemniscate and its curvature can be plotted as functions of s. See exercise 5.

curvature.ipynb (for producing fig. 3.5)	
| | ```python
import numpy as np
import matplotlib.pyplot as plt
from matplotlib.widgets import Slider
``` |
| Create an interactive GUI | ```python
%matplotlib
``` |
| $x$ component of $\mathbf{r}(t)$ | ```python
def rx(t):
    return np.cos(t)/(1+np.sin(t)**2)
``` |
| $y$ component of $\mathbf{r}(t)$ | ```python
def ry(t):
    return np.sin(t)*np.cos(t)\
            /(1+np.sin(t)**2)
``` |
| The curvature $\kappa(t)$ | ```python
def kappa(t):
    return 3*np.abs(np.cos(t))\
            /(1+np.sin(t)**2)
``` |
| Set initial $t$ | ```python
t = 0
``` |
| Plot two figures side by side | ```python
fig,(ax1, ax2) = plt.subplots(1,2,
                    figsize=(10,6))
``` |
| Leave space for a slider | ```python
plt.subplots_adjust(bottom=0.2)
``` |
| Set equal respect ratio | ```python
ax1.set_aspect('equal')
ax1.set_ylim(-0.5,0.5)
ax1.set_xlim(-1.2,1.2)
ax1.title.set_text('Parametric curve')
ax1.grid('on')
``` |
| Plot the parametric curve in blue | ```python
r, = ax1.plot(rx(t), ry(t),
                    'b', markersize=3)
``` |
| with a moving red dot which the slider controls | ```python
Pnt1, = ax1.plot(rx(t), ry(t),
                    'ro', markersize=6)
``` |
| Similar settings for the second plot | ```python
ax2.set_aspect('equal')
ax2.set_ylim(-1,4)
ax2.set_xlim(0,4*np.pi)
ax2.title.set_text('Curvature')
ax2.grid('on')
``` |
| Plot the curvature values in red | ```python
kap, = ax2.plot(t, kappa(t),
                    'r', markersize=3)
``` |
| with a moving blue dot which the slider controls | ```python
Pnt2, = ax2.plot(t, kappa(t),
                    'bo', markersize=6)
``` |
| Adjust the position of the slider<br>Set range and resolution of the $t$ slider<br>from $t = 0$ to $4\pi$ | ```python
axt = plt.axes([0.25, 0.32, 0.5, 0.02])
t_slide = Slider(axt, 't',
                0, 4*np.pi, valstep=0.001,
                valinit=t)
``` |
| Get the $t$ value from slider | ```python
def update(val):
    t = t_slide.val
    T = np.linspace(0,t,200)
``` |
| Plot the parametric curve from 0 to $t$<br>Mark the dot only at time $t$<br>Do the same for the curvature plot | ```python
    r.set_data(rx(T),ry(T))
    Pnt1.set_data(rx(t),ry(t))
    kap.set_data(T, kappa(T))
    Pnt2.set_data(t,kappa(t))
    fig.canvas.draw_idle()
``` |
| Redraw figure when the slider is changed | ```python
t_slide.on_changed(update)
plt.show()
``` |

3.5 Torsion

Plot the curve

$$\mathbf{r}(t) = \left(\cos t, \sin t, \frac{t^2}{20}\right), \quad t \in [0, 6\pi]. \tag{3.12}$$

Find an expression for the *torsion* τ of the curve. The formula for the torsion is given by

$$\tau = \frac{\mathbf{r}' \times \mathbf{r}'' \cdot \mathbf{r}'''}{|\mathbf{r}' \times \mathbf{r}''|^2}. \tag{3.13}$$

Torsion quantifies how much a curve 'escapes' from a 2D plane, in the sense that a curve is contained in a plane if and only if $\tau = 0$ on the entire curve. Let us review the definition of τ and understand how it appears naturally in the study of parametric curves in $\mathbb{R}^3$.

Let $\mathbf{r}(s)$ be a unit-speed parametrisation of a curve in $\mathbb{R}^3$. We saw in the previous section that the magnitude of the second derivative $\ddot{\mathbf{r}}$ tells us about the curvature κ. Let us write the vector $\ddot{\mathbf{r}}$ as

$$\ddot{\mathbf{r}} = \kappa\, \mathbf{n}, \tag{3.14}$$

where $\mathbf{n}$ is a unit vector called the *principal normal*.

Let $\mathbf{t}$ be the unit tangent vector defined in the usual way by $\mathbf{t} = \dot{\mathbf{r}}$. Define the *binormal*, $\mathbf{b}$, as

$$\mathbf{b} = \mathbf{t} \times \mathbf{n}. \tag{3.15}$$

It is then straightforward to show the following facts about vectors $\{\mathbf{t}, \mathbf{n}, \mathbf{b}\}$ (remember that they are all functions of the parameter s).

1. The vectors $\mathbf{t}(s), \mathbf{n}(s), \mathbf{b}(s)$ are perpendicular to one another at all values of s on the curve.
2. All three are unit vectors.
3. Given a value of s on the curve, any vector $\mathbf{v} \in \mathbb{R}^3$ can be written in the form

$$\mathbf{v} = c_1\mathbf{t} + c_2\mathbf{n} + c_3\mathbf{b},$$

   where c_1, c_2, c_3 are some real constants.

Another way to express the above points is to say that $\{\mathbf{t}, \mathbf{n}, \mathbf{b}\}$ is an *orthonormal basis* of $\mathbb{R}^3$ (analogous to the standard basis $\{\mathbf{i}, \mathbf{j}, \mathbf{k}\}$). If you have not come across the concept of an orthonormal basis before, don't worry, it will all be clear when you start studying a subject called *Linear Algebra*, which we will also cover from a Python perspective in chapter 5.

The set $\{\mathbf{t}, \mathbf{n}, \mathbf{b}\}$ is called the *Frenet frame* or *Frenet-Serret frame*. Jean Frédéric Frenet (1816–1900) and *Joseph Alfred Serret* (1819–1885) were French mathematicians who independently studied space curves in this formulation. In fact, they showed that on a unit-speed curve $\mathbf{r}(s)$ with $\kappa \neq 0$, the vectors $\{\mathbf{t}, \mathbf{n}, \mathbf{b}\}$ satisfy three differential equations that can be written in an elegant matrix form called the *Frenet-Serret equations*:

$$\begin{bmatrix} \dot{\mathbf{t}} \\ \dot{\mathbf{n}} \\ \dot{\mathbf{b}} \end{bmatrix} = \begin{pmatrix} 0 & \kappa & 0 \\ -\kappa & 0 & \tau \\ 0 & -\tau & 0 \end{pmatrix} \begin{bmatrix} \mathbf{t} \\ \mathbf{n} \\ \mathbf{b} \end{bmatrix}. \tag{3.16}$$

Note that each bold component is itself a vector. The variable $\tau(s)$ appearing in this equation is called the *torsion* of the curve.

A useful example to illustrates what τ means is the *helix* shown in fig. 3.6. The helix has the unit-speed parametrisation:

$$\mathbf{r}(s) = \left(\cos \frac{s}{\sqrt{2}}, \sin \frac{s}{\sqrt{2}}, \frac{s}{\sqrt{2}} \right), \qquad s > 0. \tag{3.17}$$

The helix winds around the unit cylinder (since $x^2 + y^2 = 1$), travelling up the z-axis in an anti-clockwise direction as s increases. The Frenet-frame vectors at a point on the helix are also shown in the figure.

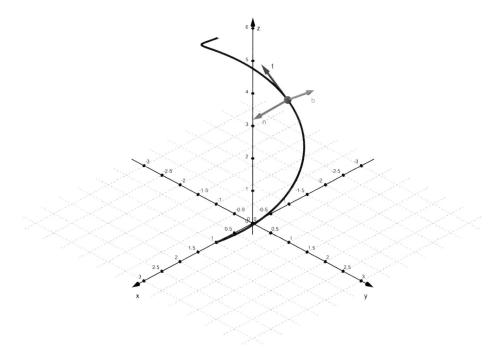

Fig. 3.6: The helix $\mathbf{r}(s) = \left(\cos \frac{s}{\sqrt{2}}, \sin \frac{s}{\sqrt{2}}, \frac{s}{\sqrt{2}} \right)$ ($s > 0$). The Frenet-frame vectors $\mathbf{t}, \mathbf{n}, \mathbf{b}$ at the grey dot on the helix are also shown. This figure was created using GeoGebra.

You can quickly verify that this parametrisation is indeed unit-speed. From the Frenet-Serret equation, we find that $\kappa = \frac{1}{2}$ and that the torsion is

$$\tau = -\dot{\mathbf{b}} \cdot \mathbf{n} \tag{3.18}$$
$$= \frac{1}{2}.$$

Intuitively, we can think of τ = constant as the result of the helix escaping from a 2D plane at a 'constant rate'.

In the previous section, we saw that for a regular curve $\mathbf{r}(t)$ which is not unit speed, the expression 3.9 for $\kappa(t)$ was rather more complicated. Similarly, the expression for $\tau(t)$ also becomes dramatically more complicated than (3.18). Instead, we now have formula 3.13 given above.

Now that we understand the meaning of τ more clearly, let us try to calculate it for the given curve (3.12). The x and y components satisfy $x^2 + y^2 = 1$, suggesting that the curve winds around the unit cylinder like the helix. However, unlike the helix, the z component suggests that it is not escaping from a 2D plane at a constant rate. Hence, we expect τ to be a nontrivial function of t.

Substituting equation 3.12 into formula 3.13 and simplifying, we find the expression for the torsion:

$$\tau(t) = \frac{10t}{t^2 + 101}. \tag{3.19}$$

By differentiating this function, we find that $\tau(t)$ attains the maximum value at $t = \sqrt{101} \approx 10.05$.

Let's visualise the curve and its torsion using Python. In fig. 3.7, the curve $\mathbf{r}(t)$ (3.12) is shown on the left panel, and the torsion $\tau(t)$ is on the right panel. A slider controls the position of a point on both plots simultaneously. The code for producing these plots is `torsion.ipynb`.

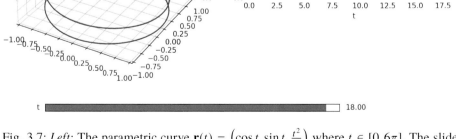

Fig. 3.7: *Left:* The parametric curve $\mathbf{r}(t) = \left(\cos t, \sin t, \frac{t^2}{20}\right)$ where $t \in [0, 6\pi]$. The slider controls the value of t. Note that 3D figures produced with *Matplotlib* can be rotated with a mouse. *Right:* The torsion $\tau(t)$ at the corresponding point on the curve.

Mixing a 3D and 2D plot with a slider is a slightly complicated affair, so the syntax is somewhat different from how we did this for the curvature plot in the previous section. In particular, note that to plot a curve in $\mathbb{R}^3$ next to one in $\mathbb{R}^2$, we use the command `add_subplot` to specify the `projection` of the 3D graph. Contrast this with the `subplots` command which we have used so far to plot multiple graphs of the same kind.

DISCUSSION

- **Why antisymmetric?** You might be wondering why the Frenet-Serret equation 3.16 features an anti-symmetric matrix.

$$\begin{pmatrix} 0 & \kappa & 0 \\ -\kappa & 0 & \tau \\ 0 & -\tau & 0 \end{pmatrix}$$

  Surely this has got nothing to do with the geometry of the curve itself since the equation describes the geometry of *any* regular curves in $\mathbb{R}^3$. One explanation is the following which relies on some knowledge of linear algebra. Feel free to skip and revisit this point after studying chapter 5.

  Consider the 3×3 matrix P whose rows are the Frenet-frame vectors $\mathbf{t}, \mathbf{n}$ and $\mathbf{b}$. We can write it in the "vector of vectors" form as before.

$$P = \begin{pmatrix} \mathbf{t} \\ \mathbf{n} \\ \mathbf{b} \end{pmatrix}.$$

  The orthonormality of the Frenet frame can be expressed as the matrix equation

$$P P^T = I.$$

  Differentiating the above equation with respect to s, we find

$$P' P^T + P(P^T)' = 0. \tag{3.20}$$

  Multiply the above on the right by P and using the fact that $P^T P = I$, we obtain a matrix equation which resembles the Frenet-Serret equation:

$$P' = AP, \quad \text{where } A := -P(P^T)'$$

  Finally, it is straightforward to show that $A^T = -A$ (using eq. 3.20 and the fact that the transpose and differentiation can be done in any order). This proves that the anti-symmetry of A is no coincidence: it follows from the orthonormality of the vectors, and the differentiation process which reveals an additional structure.

- **The Fundamental Theorem of Space Curves**. Analogous to the Fundamental Theorem of Plane Curves, we have the following theorem: let f and g be smooth functions such that $f(s) > 0$ for all $s \geq 0$. Then, there exists a unit-speed curve in $\mathbb{R}^3$ with curvature $\kappa(s) = f(s)$ and torsion $\tau(s) = g(s)$. The curve is unique up to rotation and translation in $\mathbb{R}^3$.

torsion.ipynb (for producing fig. 3.7)

| | ```python
import numpy as np
import matplotlib.pyplot as plt
from matplotlib.widgets import Slider
%matplotlib
``` |
| Create an interactive GUI<br>$x$, $y$, $z$ components of $\mathbf{r}(t)$ | ```python
rx = lambda t: np.sin(t)
ry = lambda t: np.cos(t)
rz = lambda t: t**2/20
``` |
| The torsion $\tau(t)$ | ```python
tau = lambda t: 10*t/(t**2+101)
``` |
| Set initial $t$ | ```python
t = 0
``` |
| Create a new figure size 10×6 inches | ```python
fig = plt.figure(figsize=(10,6))
``` |
| The left plot (in $\mathbb{R}^3$)<br>Plot the curve $\mathbf{r}(t)$ in blue | ```python
ax = fig.add_subplot(121, projection='3d')
r, = ax.plot(rx(t), ry(t), rz(t),
                    'b', markersize=3)
``` |
| Plot a moving red dot which the slider controls | ```python
Pnt1, = ax.plot(rx(t), ry(t), rz(t),
                    'ro', markersize=6)
ax.set_title('Parametric curve')
ax.set_xlim(-1, 1)
ax.set_ylim(-1, 1)
ax.set_zlim(0, 20)
``` |
| The right plot (in $\mathbb{R}^2$)<br>Make it a horizontally elongated figure | ```python
ax = fig.add_subplot(122)
ax.set_aspect(8)
ax.set_title('Torsion')
ax.set_xlim(0,6*np.pi)
ax.set_xlabel('t')
ax.set_ylim(0,0.6)
ax.grid('on')
``` |
| Plot the torsion values in red | ```python
torsion, = ax.plot(t, tau(t),
                    'r', markersize=3)
``` |
| Plot a moving blue dot which the slider controls | ```python
Pnt2, = ax.plot(t, tau(t),
                    'bo', markersize=6)
``` |
| Add a $t$ slider at the bottom<br>Set range and resolution of the $t$ slider<br>from $t = 0$ to $6\pi$ | ```python
axt = plt.axes([0.2, 0.1, 0.5, 0.02])
t_slide = Slider(axt, 't',
                    0, 6*np.pi, valstep=0.001,
                    valinit=t)
``` |
| Get the $t$ value from slider<br>We will plot the trace from 0 to $t$<br>The end point to be marked with a dot<br>Update the 3 components of $\mathbf{r}$<br>Similarly, update the position of the dot<br>Do the same update for the torsion plot<br>Update 2 components | ```python
def update(val):
    t = t_slide.val
    T = np.linspace(0,t,200)
    P = T[-1:]
    r.set_data_3d(rx(T),ry(T),rz(T))
    Pnt1.set_data_3d(rx(P),ry(P),rz(P))
    torsion.set_data(T, tau(T))
    Pnt2.set_data(P, tau(P))
    fig.canvas.draw_idle()
``` |
| Redraw figure when the slider is changed | ```python
t_slide.on_changed(update)
``` |
| | ```python
plt.show()
``` |

3.6 Quadric surfaces

> Plot the family of surfaces in $\mathbb{R}^3$ given by the equation
>
> $$x^2 + y^2 + az^2 = 1, \qquad (3.21)$$
>
> where $-2 \leq a \leq 2$.

Quadric surfaces are a generalisation of the quadratic equation $ax^2 + bx + c = 0$ to $\mathbb{R}^3$. A quadric is an equation of the form

$$a_1 x^2 + a_2 y^2 + a_3 z^2 + a_4 xy + a_5 yz + a_6 xz + a_7 x + a_8 y + a_9 z + a_{10} = 0, \qquad (3.22)$$

where $a_i \in \mathbb{R}$. The question asks us to plot the graph of a family of quadrics with a simple expression.

Before we start coding, let us consider what we can deduce about these surfaces. Firstly, the terms $x^2 + y^2$ suggests that this family of surfaces will have rotational symmetry about the z-axis. One might therefore consider working in cylindrical coordinates (r, θ, z). The precise range of values of r will depend on the constant a. There are 3 cases to consider.

- $a = 0$. The equation of the surface 3.21 reduces to $x^2 + y^2 = 1$. This is not just a unit circle. The equation says that we see the unit circle regardless of the value of z. This suggests that the surface is a *cylinder* with unit radius. It extends infinitely in the positive and negative z directions. The cylinder is shown in the central panel in fig. 3.8.
- $a > 0$. In terms of cylindrical coordinates, eq. 3.21 reads

$$z = \pm \sqrt{\frac{1 - r^2}{a}}. \qquad (3.23)$$

  For the term in the square root to be non-negative, we need $0 \leq r \leq 1$. To figure out the shape of surfaces in $\mathbb{R}^3$, it usually helps to consider its projections onto the x-y, y-z and x-z planes (by respectively setting z, x or y to 0 in the equation of the surface).

  - In the x-y plane, we see $x^2 + y^2 = 1$, the unit circle.
  - In the y-z plane, we see $y^2 + az^2 = 1$, an ellipse.
  - In the x-z plane, we see $x^2 + az^2 = 1$, an ellipse.

  These observations suggest that the surface is an *ellipsoid*. For example, the ellipsoidal surface with $a = 1.2$ is shown on the third panel in fig. 3.8.
- $a < 0$. Eq. 3.23 now suggests that $r \geq 1$. For the projections, instead of the ellipses, we see pairs of hyperbolae in the y-z and x-z planes. We still see the unit circle in the x-y plane. Furthermore, the projection on any $z = $ constant slice is a circle of radius $\sqrt{1 + |a|z^2}$. The circle gets bigger as $|z|$ increases.
  These observations suggest that the surface is a *hyperboloid of one sheet* as shown on the first panel in fig. 3.8, which shows the surface when $a = -1$.

In summary, as a increases from negative to zero and to positive values, the quadric surface changes from a hyperboloid to a cylinder and to an ellipsoid.

Fig. 3.8 is produced by code `quadrics.ipynb`, which uses a slider to control 3D graphics. Unlike 3D curves, surfaces do not get updated using the method that we have used so far (namely, using the command `set_data` in the `update` function to update the

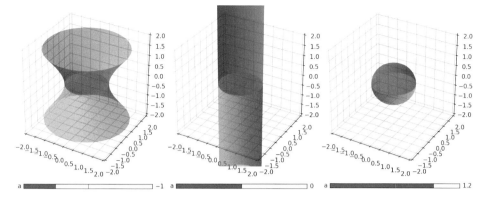

Fig. 3.8: The surfaces given by the equation $x^2 + y^2 + az^2 = 1$ with (from left to right) $a = -1, 0$ and 1.2. The slider adjusts the value of a. The portions of the surfaces with $z \geq 0$ and $z < 0$ are shown in blue and orange respectively.

plot). The workaround we have used is to simply clear the surface plot and replot everything whenever the slider is moved.

DISCUSSION

- **Other quadrics in $\mathbb{R}^3$.** A sequence of translation and rotation will bring eq. 3.22 to one of a number of standard quadrics [171]. This is sometimes called the *principle axis theorem*. We have already met the ellipsoid, hyperboloid of one sheet and the cylinder. Other quadrics will be explored in Ex. 10.

- **Quadrics in $\mathbb{R}^2$.** Quadrics in 2 dimensions are called *conic sections*. They comprise the ellipse, the parabola and the hyperbola. Properties of conic sections have been studied as far back as ancient Greece, particularly by *Menaechmus* (380–320BC) who is credited with the discovery of conic sections.

- **Quadrics in $\mathbb{R}^n$.** We can express the general equation of quadrics (eq. 3.22) in a more compact form as
$$\mathbf{x}^T A \mathbf{x} + \mathbf{b} \cdot \mathbf{x} + c = 0,$$

where $\mathbf{x} = (x \; y \; z)^T$, A is a symmetric 3×3 matrix, $\mathbf{b} \in \mathbb{R}^3$ and $c \in \mathbb{R}$. In this way, we can see how quadrics can be generalised to $\mathbb{R}^n$ simply by changing the dimension of the vectors and matrix A accordingly. Generalised quadrics beyond 3 dimensions are not easily visualised, so diagrams and figures cannot give us a full understanding of the geometry of these objects.

In general, given an algebraic expression like the generalised quadric, one can understand the geometry of the corresponding object using the tools of abstract algebra (*e.g.* groups and rings). This subject is called *algebraic geometry*, a vast topic that students normally meet in their senior undergraduate years. For the keen readers, take a peek at [24,173,185] which are classic introductory texts in algebraic geometry.

quadrics.ipynb (for plotting fig. 3.8)

| Create an interactive GUI | ```python
import numpy as np
import matplotlib.pyplot as plt
from matplotlib.widgets import Slider
%matplotlib
``` |
| $z$ coord. (positive root of eq. 3.23) | ```python
def zfunc(r,a):
    return np.sqrt((1-r**2)/a)
``` |
| Produce $(x, y, z)$ coordinates of sampled points on the surface for a given value $a$<br>**Case I**: $a = 0$ (cylinder)<br><br>All $(z, \theta)$ combinations<br>$r = 1$ (unit circle) for all $z$ values<br>**Case II**: $a > 0$ (ellipsoid)<br>$r \in [0, 1]$<br>All $(r, \theta)$ combinations<br>The $z$ coordinates for those points<br>**Case III**: $a < 0$ (hyperboloid of 1 sheet)<br>(similar)<br><br><br><br>$x$ and $y$ coordinates of all sampled points<br>(Convert cylindrical to Cartesian coordinates) | ```python
def data(a):
    theta = np.linspace(0,2*np.pi)
    if (a==0):
        z = np.linspace(0,5)
        Z, tc = np.meshgrid(z,theta)
        rc = 1
    elif (a>0):
        r = np.linspace(0,1)
        rc, tc = np.meshgrid(r, theta)
        Z = zfunc(rc,a)
    else:
        r = np.linspace(1,2)
        rc, tc = np.meshgrid(r, theta)
        Z = zfunc(rc,a)
    X = rc*np.cos(tc)
    Y = rc*np.sin(tc)
    return X, Y, Z
``` |
| Leave a space at the bottom for the slider | ```python
fig = plt.figure()
plt.subplots_adjust(bottom=0.15)
``` |
| Equal aspect ratio (so a sphere appears undistorted)<br>Plot the quadric surface given $a$<br>Get coordinates of points<br>Clear canvas<br>Plot the upper half (positive square root)..<br>and the lower half (negative square root)<br>(`alpha` adjusts the transparency) | ```python
ax = fig.add_subplot(projection='3d')
ax.set_box_aspect((1,1,1))

def plotfig(a):
    X, Y, Z = data(a)
    ax.clear()
    P1=ax.plot_surface(X,Y, Z, alpha=0.5)
    P2=ax.plot_surface(X,Y,-Z, alpha=0.5)
    ax.set_xlim(-2,2)
    ax.set_ylim(-2,2)
    ax.set_zlim(-2,2)
    return P1, P2
``` |
| Initial value of $a$ to plot | ```python
a = 0
``` |
| Set position and size of the $a$ slider<br>Set the slider's range and resolution | ```python
axa = plt.axes([0.3, 0.05, 0.45, 0.02])
a_slide = Slider(axa, 'a', -2, 2,
            valstep = 0.05, valinit = a)
``` |
| Plot the quadric for the initial $a$ | ```python
plotfig(a)
``` |
| Get the $a$ value from slider...<br>and replot | ```python
def update(val):
    a = a_slide.val
    plotfig(a)
``` |
| Redraw figure when the slider is changed | ```python
a_slide.on_changed(update)
plt.show()
``` |

3.7 Surface area

Consider the following surfaces.
i) The unit sphere $x^2 + y^2 + z^2 = 1$,
ii) The ellipsoid $x^2 + y^2/4 + z^2 = 1$.
For each surface, calculate the area of the portion of the surface bounded by the planes $z = 0$ and $z = h$ where $|h| \leq 1$.

We briefly mentioned algebraic geometry in the last section. Another way in which geometry is studied at university is *differential geometry*, where properties of curves and surfaces are analysed using the tool of calculus (*i.e.* differentiation and integration). In this example, we will need the following important result from differential geometry on how the area of a surface can be calculated (see recommended texts on differential geometry for proof).

Theorem 3.1 *The area of the portion of a surface parametrised by* $\mathbf{S}(u, v)$ *corresponding to the domain* $R \subseteq \mathbb{R}^2$ *is given by*

$$Area = \iint_R \left| \frac{\partial \mathbf{S}}{\partial u} \times \frac{\partial \mathbf{S}}{\partial v} \right| du\, dv. \tag{3.24}$$

Equivalently, the area can also be expressed as

$$Area = \iint_R \sqrt{EG - F^2}\, du\, dv, \tag{3.25}$$

$$where \ \ E = \frac{\partial \mathbf{S}}{\partial u} \cdot \frac{\partial \mathbf{S}}{\partial u}, \qquad F = \frac{\partial \mathbf{S}}{\partial u} \cdot \frac{\partial \mathbf{S}}{\partial v}, \qquad G = \frac{\partial \mathbf{S}}{\partial v} \cdot \frac{\partial \mathbf{S}}{\partial v}.$$

Let us apply Theorem 3.1 to the two surfaces given. It is useful to note that both surfaces are symmetric about the x-y plane, so it will be sufficient to consider the upper half of each surface (where $z \geq 0$). Fig. 3.9 shows the surfaces and the areas bounded by $z = 0$ and $z = 0.5$.

Starting with the unit sphere, let's use spherical coordinates to parametrise the surface as

$$\mathbf{S}(u, v) = (\cos u \sin v, \sin u \sin v, \cos v), \tag{3.26}$$

where $u \in [0, 2\pi]$ and $v \in [v_0, \pi/2]$. The angle v starts from $v_0 = \cos^{-1} h$.

Calculating the area using Theorem 3.1, we find

$$\frac{\partial \mathbf{S}}{\partial u} = (-\sin u \sin v, \cos u \sin v, 0), \qquad \frac{\partial \mathbf{S}}{\partial v} = (\cos u \cos v, \sin u \cos v, -\sin v),$$

$$E = \sin^2 v, \qquad F = 0, \qquad G = 1 \implies Area = \int_{u=0}^{2\pi} \int_{v=v_0}^{\pi/2} \sin v \, dv \, du = 2\pi h.$$

Thus, we have arrived at a remarkably simple conclusion: the area of the strip increases linearly with h.

Next, for the ellipsoidal surface, one could lightly modify the above spherical-coordinates parametrisation. Alternatively, let's try a modified cylindrical-coordinates parametrisation:

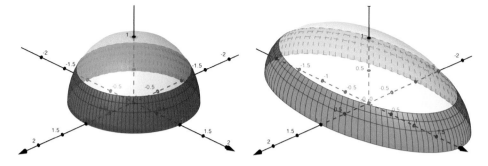

Fig. 3.9: The upper halves of the unit sphere $x^2 + y^2 + z^2 = 1$ (left) and the ellipsoid $x^2 + y^2/4 + z^2 = 1$ (right). The portions shown in solid shading are bounded by $z = 0$ and $z = 0.5$. This figure was created using GeoGebra.

$$\mathbf{S}(u, v) = \left(\sqrt{1 - u^2} \cos v, \; 2\sqrt{1 - u^2} \sin v, \; u \right), \tag{3.27}$$

where $u \in [0, h]$ and $v \in [0, 2\pi)$. Calculating the area using Theorem 3.1, you should find that we arrive at a much tougher double integral:

$$\text{Area} = \int_{v=0}^{2\pi} \int_{u=0}^{h} \sqrt{4 - 3(1 - u^2) \sin^2 v} \, du \, dv. \tag{3.28}$$

This would be a good place to bring in Python to help us evaluate this double integral. The code `surfacearea.ipynb` evaluates the integral as h goes from 0 to 1 and plots the area of the strip as a function of h.

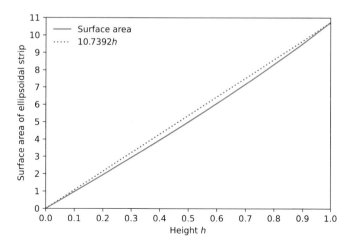

Fig. 3.10: The red solid line shows the surface area of the ellipsoid $x^2 + y^2/4 + z^2 = 1$ bounded between $z = 0$ and $z = h$, plotted as a function of h. The variation is almost linear. The dotted blue line is a straight line joining the endpoints for comparison.

We see that the variation of the surface area with h is almost linear. Python tells us that the area of the semi-ellipsoid is:

$$\text{Surface area of semi-ellipsoid} \approx 10.7392 \text{ (4 dec. pl).}$$

The error reported by `dblquad` is a tiny number of order 10^{-7}, which gives us confidence in the accuracy of the above answer.

DISCUSSION

- **Archimedes' tombstone and cartography**. Have you ever noticed that the surface area of a sphere equals the area of the curved surface of the cylinder which exactly contains it?

This result was discovered by the legendary Greek mathematician *Archimedes* (c.287–212 BC) in around 250BC using rudimentary geometric arguments (remember this was long before calculus was invented). Archimedes' tombstone (now lost) was said to have been inscribed with a figure similar to the one shown on the left.

You might also have noticed from our calculation that the surface area of a strip with height h on the sphere (found to be $2\pi h$) is exactly equal to the surface of the unit cylinder height h.

  In fact, a similar calculation shows that we can choose any region, R, on the sphere and project it horizontally to a region, $\tilde{R}$, on the wrapping cylinder. R and $\tilde{R}$ will have the same area.

  This kind of projection can be used in cartography (map-making). The resulting map of the Earth is called the *Lambert cylindrical projection* of the Earth. It does not distort the area of landmasses (unlike the more popular *Mercator projection*). However, the Lambert projection distorts lengths and angles.

- **First fundamental form**. The coefficients E, F and G in Theorem 3.1 are called the coeffcients of the *first fundamental form* of a surface. It turns out that these coefficients tell us everything about the *intrinsic* properties of the surface. By intrinsic, we mean properties that can be measured by an inhabitant on the surface. Such properties include the length of a curve, the angle between two lines, the area of a portion of the surface, and even the *curvature* at a point on the surface. If you are intrigued by this result, look up Gauss's *Theorema Egregium*.

- **Surface area of an ellipsoid**. In §3.3, we saw that the arc length of an ellipse cannot be expressed in terms of elementary functions, but only in terms of an elliptic integral. A similar misfortune befalls the surface area of the ellipsoid. There is no explicit formula for the surface area, but one can express it in terms of elliptic integrals.

  There is, however, an interesting approximation which was discovered surprisingly recently in 2004 by the Danish scientist Knud Thomsen and communicated via email to the *Numericana* website [146]. Thomsen's approximation for the surface area of an ellipsoid $(x/a)^2 + (y/b)^2 + (z/c)^2 = 1$ is

$$S \approx 4\pi \left(\frac{(ab)^p + (ac)^p) + (bc)^p}{3} \right)^{1/p}, \qquad p = 1.6075. \qquad (3.29)$$

Using this formula for our semi-ellipsoid with $a = c = 1, b = 2$, we find

$$\text{Surface area of semi-ellipsoid} \approx 10.7328 \text{ (4 dec. pl).}$$

This is within 0.06% of the value found by Python – an excellent approximation indeed.

---

surfacearea.ipynb (for producing fig. 3.10)

|  | ```python
import numpy as np
import matplotlib.pyplot as plt
from scipy.integrate import dblquad
``` |
| The integrand (3.28). Note the order of variables | ```python
def integrand(u,v):
    return np.sqrt(4-3*(1-u**2)*np.sin(v)**2)
``` |
| List of strip areas (to be filled)<br>List of strip heights, $h$<br>Initial area and height<br>Small increment in $h$ | ```python
Alist = [0]
hlist = [0]
A, h = 0, 0
dh = 1e-2
``` |
| Evaluate the double integral<br>Outer limits ($v$ from 0 to $\pi/2$)<br>Inner limits ($u$ from $h$ to $h + dh$)<br>Accumulate the area, one small strip at a<br>time (factor 4 from rotational symmetry) | ```python
while (h<1):
    dA = dblquad(integrand,
        0, np.pi/2,
        h, h+dh)
    A += 4*dA[0]
    h += dh
    Alist.append(A)
    hlist.append(h)
``` |
| Area of the semi-ellipsoid | ```python
Area = Alist[-1]
``` |
| Plot the area for varying $h$ (red)<br>Linear comparison (blue dotted line)<br><br><br>Set $x$ ticks in steps of of 0.1<br><br><br>Set $y$ ticks in steps of 1<br>Report equation of linear approximation<br>(to 4 dec. pl.) | ```python
plt.plot(hlist, Alist, 'r')
plt.plot([0,1],[0, Area], ':b')
plt.xlim(0,1)
plt.xlabel('Height $h$')
plt.xticks(np.arange(0,1.1,0.1))
plt.ylim(0,11)
plt.ylabel('Surface area of ellipsoidal strip')
plt.yticks(np.arange(0,11.1,1))
plt.legend(['Surface area', f'{Area:.4f}$h$'])
plt.grid('on')
plt.show()
``` |

3.8 Normal to surfaces and the grad operator

Consider the surface given by the equation

$$z = F(x, y) = x + 2\sin(x + y).$$

a) Plot the surface and its contour lines projected onto the x-y plane.
b) Find the unit normal to the surface at the origin.

Suppose F and G are functions such that $F : \mathbb{R}^2 \to \mathbb{R}$ and $G : \mathbb{R}^3 \to \mathbb{R}$. These functions are called *scalar fields*. Each of them associates a number to every point in space, *e.g.* the temperature or pressure at different points in a room.

The *gradient* of a function $F : \mathbb{R}^2 \to \mathbb{R}$ is defined in Cartesian coordinates as

$$\nabla F(x, y) = \left(\frac{\partial F}{\partial x}, \frac{\partial F}{\partial y} \right).$$

Similarly, if $G : \mathbb{R}^3 \to \mathbb{R}$, then

$$\nabla G(x, y, z) = \left(\frac{\partial G}{\partial x}, \frac{\partial G}{\partial y}, \frac{\partial G}{\partial z} \right).$$

The notation ∇F is read "grad F"; indeed ∇F is sometimes written as grad F.

We can view ∇ as an *operator* and write

$$\nabla := \left(\frac{\partial}{\partial x}, \frac{\partial}{\partial y}, \frac{\partial}{\partial z} \right). \tag{3.30}$$

The ∇ operator can be thought of as the generalisation of the derivative for multivariable functions. *The grad operator takes a scalar and gives us a vector.* (Note that in the context of vector calculus, a scalar means any one-dimensional object, like a function $f(x, y, z)$.)

Although the name *gradient* might suggest a link to the *tangent* of a curve or a surface, ∇F actually gives a vector that is *normal* to the surface F = constant. Let us state this more precisely.

Theorem 3.2 *Consider the functions $F : \mathbb{R}^2 \to \mathbb{R}$ and $G : \mathbb{R}^3 \to \mathbb{R}$. Let P be the point with coordinates (x_0, y_0, z_0). Then, at P,*
a) the vector $\nabla F(x_0, y_0)$ is normal to the curve F = constant.
b) the vector $\nabla G(x_0, y_0, z_0)$ is normal to the surface G = constant.

This theorem is so important in vector calculus that it is worth discussing the proof below.

First let's consider statement a). The point $P(x_0, y_0)$ clearly lies on the curve $F(x, y) = F(x_0, y_0)$. Suppose we use the variable t to parametrise this curve so that

$$F(x(t), y(t)) = F(x_0, y_0).$$

Let's also assume that at $x(0) = x_0$ and $y(0) = y_0$. Differentiating the above equation with respect to t and using the Chain Rule, we find

$$\frac{\partial F}{\partial x} x'(t) + \frac{\partial F}{\partial y} y'(t) = 0 \tag{3.31}$$

Now let's evaluate expression (3.31) at P. Recall that the vector $\mathbf{t} = (x'(0), y'(0))$ is tangent to the curve at P. Eq. 3.31 can then be expressed as the dot product:

$$\nabla F(x_0, y_0) \cdot \mathbf{t} = 0. \tag{3.32}$$

In other words, ∇F is perpendicular to the tangent of the curve $F = $ constant at P. This proves (a).

Let us demonstrate this with the given example $F(x, y) = x + 2\sin(x + y)$, and P the origin. The curve passing through P is $F(x, y) = 0$, which we can solve for y, giving

$$y = -x - \sin^{-1}\frac{x}{2}.$$

It is not terribly difficult to try to plot this by hand, but it is useful to note that around the origin (where $\sin^{-1} x \approx x$), we expect the curve passing through the origin to be roughly that of a straight line $y = -3x/2$.

Fig. 3.11 confirms this behaviour on the righthand panel, where the colour-coded contour lines are the curves corresponding to $F(x, y) = $ constant (these contour lines are also called *level curves*). When the constant is zero, the contour line is roughly straight and passes through the origin shown as the red dot.

Furthermore, the normal to the contour line at the origin can be calculated by finding ∇F:

$$\nabla F(x, y) = (1 + 2\cos(x + y), 2\cos(x + y)) \implies \nabla F(0, 0) = (3, 2).$$

This is parallel to the red arrow in the figure, which shows the vector

$$\mathbf{n} = (-3/\sqrt{14}, -2/\sqrt{14}).$$

The reason we chose to plot this vector (rather than $(3, 2)$) is that $\mathbf{n}$ is the projection of the red vector in the left-hand panel of the figure. Thus, Python has visually demonstrated part (a) of Theorem 3.2.

Now let us consider part (b) of Theorem 3.2. Consider the function $G : \mathbb{R}^3 \to \mathbb{R}$ defined by

$$G(x, y, z) = z - F(x, y) = z - x - 2\sin(x + y).$$

The equation $G(x, y, z) = $ constant represents a surface of constant temperature (sometimes called *isosurface*), which is just a 3-dimensional generalisation of the contour lines we saw above.

The origin lies on the surface $G(x, y, z) = 0$. This surface is shown in the left-hand panel of fig. 3.11 in particular. Going through the same proof as before shows us that

$$\nabla G(0, 0, 0) \cdot \mathbf{t} = 0,$$

where $\mathbf{t}$ is the tangent vector to *any* curve passing through the origin. In other words, $\nabla G(0, 0, 0)$ is perpendicular to the tangent vector of every curve on the surface passing through the origin. We conclude that at the origin, ∇G is normal to the surface $G = 0$, which is precisely the statement of the theorem.

Let's calculate this normal.

$$\nabla G(x, y, z) = (-1 - 2\cos(x + y), -2\cos(x + y), 1) \implies \nabla G(0, 0, 0) = (-3, -2, 1).$$

It is common practice to express the answer as a unit vector $\hat{\mathbf{n}}$ (as only the normal direction is of interest).

$$\hat{\mathbf{n}} = \left(-\frac{3}{\sqrt{14}}, -\frac{2}{\sqrt{14}}, \frac{1}{\sqrt{14}} \right). \tag{3.33}$$

The unit normal $\hat{\mathbf{n}}$ is shown as the red arrow in the left-hand panel of fig. 3.11. When projected onto the x-y plane, we simply neglect the z component, and hence we see the 2D vector $\mathbf{n}$ on the right-hand panel.

The code for producing (3.11) is given in `grad.ipynb`. The left-hand panel is interactive. Try to use the mouse to spin the figure around so that you see the same projection as that on the right panel. You can also see from the figure that $\hat{\mathbf{n}}$ is indeed the normal to the surface at the origin (remember that orthogonality can only been seen if we set the aspect ratio to be equal in all directions in the plotting code).

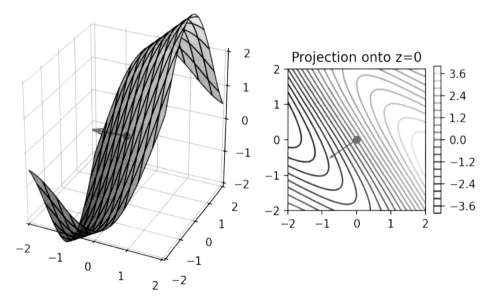

Fig. 3.11: *Left:* The surface $z = x + 2\sin(x + y)$. The dot in the middle of the surface is the origin. The vector is $\hat{\mathbf{n}}$ (eq. 3.33), the unit normal to the surface at the origin. *Right:* The contour lines on the x-y plane. The projections of the origin and the unit normal are also shown in red. The projected normal $\mathbf{n}$ is perpendicular to the contour line through the origin, as expected from theorem 3.2.

DISCUSSION

- **Directional derivative**. Suppose $G(x, y, z)$ represents the temperature at a point with coordinates (x, y, z) in a room. One might ask: how fast is the temperature changing at a certain point (x_0, y_0, z_0) if one travels in a given direction?

  For example, starting at (x_0, y_0, z_0), if we travel in the direction along the tangent plane to the surface $G = $ constant, then the rate of change of temperature is zero.

  If we specify the travel direction by a unit vector $\hat{\mathbf{u}} = (a, b, c)$, then it makes sense to define the rate of change associated with travelling in the direction $\hat{\mathbf{u}}$ as the limit:

$$D_{\hat{\mathbf{u}}}G(x_0, y_0, z_0) := \lim_{h \to 0} \frac{G(x_0 + ha, y_0 + hb, z_0 + hc) - G(x_0, y_0, z_0)}{h}.$$

This quantity is called the *directional derivative* of G in the direction $\hat{\mathbf{u}}$.

It turns out that the grad operator can also be used to calculate directional derivatives. The above limit can be shown to be equal to the scalar product:

$$D_{\hat{\mathbf{u}}}G(x_0, y_0, z_0) = \nabla G(x_0, y_0, z_0) \cdot \hat{\mathbf{u}}.$$

- **Steepest descent**. Once we know how to calculate the rate of change of temperature in a given direction, we can ask, which direction is the temperature changing most rapidly? From the previous equation, using the definition of the dot product, we have

$$D_{\hat{\mathbf{u}}}G = |\nabla G| \cos \theta,$$

where $\theta \in [0, \pi]$ is the angle between ∇G and $\hat{\mathbf{u}}$. From this simple observation, we can deduce that

  - The directional derivative is maximised when $\theta = 0$, so the temperature is increasing the fastest when we walk in the direction $\hat{\mathbf{u}} = \nabla G/|\nabla G|$ (normal to the surface $G = $ constant). This is called the direction of *steepest ascent*.
  - Similarly the direction $\hat{\mathbf{u}} = -\nabla G/|\nabla G|$ sees the temperature dropping the fastest. This is called the direction of *steepest descent*.
  - The rate of change is zero when $\theta = \pi/2$, *i.e.* the direction tangential to the surface $G = $ constant, in agreement with the previous bullet point.

The direction of steepest descent is particularly important in helping us find the minimum of a multivariable function, a common optimisation problem arising in physics, engineering and data science. For details on how to implement the steepest descent and other optimisation algorithms in Python, see [74, 100].

- **Normal vectors and computer graphics**. Light reflects off objects at an angle which can be determined from the normal vectors on the surface. Thus, normal vectors play a key role in rendering realistic 3D graphics in computer games and cinematic animations. In such applications, paths of light rays are traced between the camera and light sources in a scene, reflecting off intervening surfaces where the normal ∇F is determined numerically. This is a somewhat crude description of the highly sophisticated technique of *raytracing*. Once thought computationally prohibitive and time consuming, raytracing can now be done in real time on home computers and game consoles. For a glimpse into the fascinating mathematics behind computer graphics and games, see [127, 207].

grad.ipynb (for producing fig. 3.11)	
| | ```python
import numpy as np
import matplotlib.pyplot as plt
``` |
| gridspec allows us to control the relative size of subplots | ```python
import matplotlib.gridspec as gridspec
%matplotlib
``` |
| Define $F(x, y)$ for the surface $z = F(x, y)$ | ```python
def F(x, y):
    return x + 2*np.sin(x+y)
``` |
| Create a square grid of 50×50 points
$z$ coordinate for each point on the grid | ```python
x = np.linspace(-2,2)
y = np.linspace(-2,2)
X, Y = np.meshgrid(x, y)
Z = F(X, Y)
``` |
| Unit normal to surface $\hat{\mathbf{n}}$ | ```python
N = np.array([-3,-2,1])/np.sqrt(14)
``` |
| Create 2 figure side by side
with the left figure wider than the default size | ```python
fig = plt.figure()
gs = gridspec.GridSpec(1, 2,
      width_ratios=[1.5,1])
``` |
| Left panel (3D plot)

Equal aspect ratio to see orthogonality
Plot the surface
Adjust density or grid lines on surface
Show grid lines on surface in black
Choose your favourite colourmap
Origin = a red dot
Plot the unit normal in red
(or type *N to unpack N into 3 components) | ```python
ax1 = fig.add_subplot(gs[0],
            projection='3d')
ax1.set_box_aspect((1,1,1))
ax1.plot_surface(X,Y, Z, alpha=0.7,
            rstride=5, cstride=5,
            edgecolor='k',
            cmap='viridis')
ax1.plot(0,0,0, 'or')
ax1.quiver(0,0,0,N[0],N[1],N[2],
            length=1, color='r')
ax1.set_xlim(-2,2)
ax1.set_ylim(-2,2)
ax1.set_zlim(-2,2)
``` |
| Right panel
Plot up to 20 contour lines
Legend shown as a bar (reduced size)
Origin = a red dot
Plot the projected normal in red
in exactly the dimensions specified
Equal aspect ratio to see orthogonality | ```python
ax2 = fig.add_subplot(gs[1])
p = ax2.contour(X, Y, Z, 20)
fig.colorbar(p, shrink = 0.4)
ax2.plot(0,0, 'or')
ax2.quiver(0,0, N[0],N[1], color='r',
         scale_units='xy', scale=1)
ax2.set_aspect('equal')
ax2.set_title('Projection onto z=0')

fig.tight_layout()
plt.show()
``` |

3.9 The Divergence Theorem and the div operator

Consider the cube:

$$V = \{(x, y, z) \in \mathbb{R}^3 : -1 \leq x \leq 1, \ -1 \leq y \leq 1, \ -1 \leq z \leq 1\}.$$

Fluid flows out of the cube with flux given by the vector field

$$\mathbf{F}(x, y, z) = \left(2xz, \ z + 2\cos y, \ 2z^3\right),$$

where $(x, y, z) \in V$. Calculate the net outward flux.

A *vector field* assigns a vector to a point in space. For example, the wind direction or electric field at a point can be described by vector fields. Every point is assigned not only a magnitude (as in a scalar field) but also a direction.

In this example, we will study the *Divergence Theorem* - one of the most important results in vector calculus. It has wide ranging applications in physics, particularly in fluid mechanics and electromagnetism where vector fields are ubiquitous. We will only give an outline of the theorem below. For an accessible proof, see [192]. See also [28] for a good intuitive explanation of the theorem.

Start with a closed container with volume V containing fluid that flows outward through its surface S. At each point $P(x, y, z)$ on the surface, suppose that the fluid velocity is $\mathbf{v}(x, y, z)$ with unit $\mathrm{m\,s}^{-1}$, and the density at P is $\rho(x, y, z)$ with unit $\mathrm{kg\,m}^{-3}$. Note that the vector $\mathbf{F} = \rho\mathbf{v}$ has unit $\mathrm{kg\ m}^{-2}\,\mathrm{s}^{-1}$, meaning that it quantifies the rate at which the fluid flows through a small area element containing point P.

The vector $\mathbf{F}$ is called the *flux*. Furthermore, we can work out the total mass of fluid emptying from V per unit time by integrating the flux over the entire surface S, *i.e.*

$$\text{Flux across } S = \iint_S \mathbf{F} \cdot \hat{\mathbf{n}} \ \mathrm{d}S, \tag{3.34}$$

where $\hat{\mathbf{n}}$ is the outward-pointing unit normal at each point on S.

On the other hand, consider a volume element $\mathrm{d}V = \mathrm{d}x\,\mathrm{d}y\,\mathrm{d}z$ within V. Let us write the flux vector as $\mathbf{F} = (F_1, F_2, F_3)$. It can be shown that the rate at which the fluid flows out of this volume element is given by

$$\left(\frac{\partial F_1}{\partial x} + \frac{\partial F_2}{\partial y} + \frac{\partial F_3}{\partial z}\right) \mathrm{d}x\,\mathrm{d}y\,\mathrm{d}z = \begin{pmatrix} \frac{\partial}{\partial x} \\ \frac{\partial}{\partial y} \\ \frac{\partial}{\partial z} \end{pmatrix} \cdot \begin{pmatrix} F_1 \\ F_2 \\ F_3 \end{pmatrix} \mathrm{d}x\,\mathrm{d}y\,\mathrm{d}z = \nabla \cdot \mathbf{F}\,\mathrm{d}V,$$

where ∇ is the differential operator defined in eq. 3.30. Therefore, the total mass of fluid emptying from the entire volume per unit time can be obtained by integrating over the whole volume:

$$\iiint_V \nabla \cdot \mathbf{F}\,\mathrm{d}V. \tag{3.35}$$

Eqs. 3.34 and 3.35 are the same quantity calculated in two ways. Therefore, we have the following.

Theorem 3.3 *(The Divergence Theorem) Let* $\mathbf{F} : \mathbb{R}^3 \to \mathbb{R}^3$ *be a differentiable vector field. Let V be a volume in* $\mathbb{R}^3$ *and S its boundary surface. Then*

$$\iiint_V \nabla \cdot \mathbf{F} \, dV = \iint_S \mathbf{F} \cdot \hat{\mathbf{n}} \, dS,$$

where $\hat{\mathbf{n}}$ *is the outward-pointing unit normal to the surface S.*

The name of the theorem refers to the *divergence operator* (the 'div') defined as:

$$\text{div } \mathbf{F} := \nabla \cdot \mathbf{F}.$$

Note that the div operator takes a vector and gives us a number that quantifies the local rate of outflow per unit volume.

Discovered by Lagrange in 1764, the Divergence Theorem was proved decades later by Gauss and also by the Russian mathematician *Mikhail Vasilyevich Ostrogradsky* (1801– 1862). In some texts, the Divergence Theorem is called the Gauss or Gauss-Ostrogradsky theorem.

Now we are ready to address the given question. Fig. 3.12 shows a sketch of the given vector field defined in the cube V, plotted using the `quiver` command. The longer the arrow, the larger the magnitude of the flux (*i.e.* faster flow). The tails of the arrows are at 250 sampled points within the cube (sampled using the `meshgrid` command). The code for generating this interactive figure is given below.

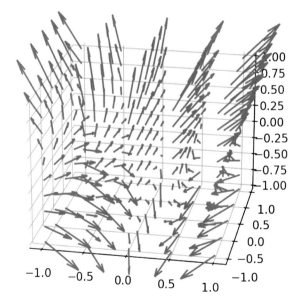

Fig. 3.12: Sketch of the vector field $\mathbf{F}(x, y, z) = \left(2xz, \; z + 2 \cos y, \; 2z^3\right)$.

Plotting a vector field (fig. 3.12)	
| Interactive plot | ```
import numpy as np
import matplotlib.pyplot as plt
%matplotlib

fig = plt.figure()
ax = plt.axes(projection='3d')
``` |
| Define 3D grid points Avoid a crowded figure - don't use too many points Components of $\mathbf{F}(x, y, z)$ | ```
x, y, z = np.meshgrid(np.linspace(-1, 1,5),
                      np.linspace(-1, 1,5),
                      np.linspace(-1, 1,10))
u = 2*x*z
v = z + 2*np.cos(y)
w = 2*z**3
``` |
| Plot vector field in 3D (length adjusts arrow length) | ```
ax.quiver(x, y, z, u, v, w , length=0.2)

plt.show()
``` |

Let's calculate the net flux (*i.e.* each side of the Divergence Theorem) in two ways. First, let's calculate the triple integral on the LHS. The integrand is the divergence

$$\nabla \cdot \mathbf{F} = 2z - 2\sin y + 6z^2.$$

Every point within the cube has a divergence value (*e.g.* the origin has zero divergence). By inspection, we see that the maximum divergence in the cube is achieved at the surface $z = 1$ along the edge $y = -1$, with $(\nabla \cdot \mathbf{F})_{\max} = 4 + \sin 1 \approx 9.68$. Similarly, the minimum divergence occurs on the slice $z = 0$ (when the squared term vanish) and $y = 1$, giving $(\nabla \cdot \mathbf{F})_{\min} = -2\sin 1 \approx -1.68$

We can use Python to visualise the divergence on each $z = $ constant slice. Fig. 3.13 shows the heatmaps corresponding to the divergence of $\mathbf{F}$ at $z = -1, 0, 0.5$ and 1, varied using the slider below the figure. The arrows are the top-down views of the vector field $\mathbf{F}$ projected onto each $z = $ constant plane (this time we sample 400 points per slice to get a good view of the flow directions). The code `div.ipynb` was used to produce this figure.

The net flux is the sum of the divergence on all such slices. We can calculate this numerically by performing the triple integration:

$$\iiint_V \nabla \cdot \mathbf{F} \ dV = \iiint_V \left(2z - 2\sin y + 6z^2\right) dx \, dy \, dz$$

$$= \int_{-1}^{1} \int_{-1}^{1} \int_{-1}^{1} \left(2z - 2\sin y + 6z^2\right) dz \, dy \, dx$$

$$= 2 \int_{-1}^{1} \int_{-1}^{1} \left(2z - 2\sin y + 6z^2\right) dz \, dy$$

$$= 8 \int_{-1}^{1} \left(z + 3z^2\right) dz$$

$$= 16.$$

It remains to calculate the surface integral on the RHS of the Divergence Theorem. For our example, S comprises the 6 faces of the cube, where $x, y, z = \pm 1$.

Let's start by calculating the flux across the face $z = 1$. Here the outward-pointing normal is $\hat{\mathbf{n}} = (0, 0, 1)$, and $dS = dx \, dy$. Thus, the flux across this face is
```

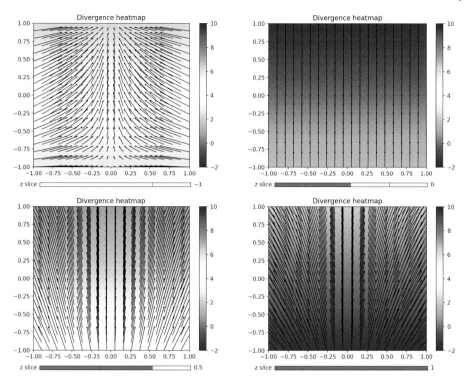

Fig. 3.13: Heatmaps corresponding to the divergence of $\mathbf{F}(x, y, z) = \left(2xz,\, z + 2\cos y,\, 2z^3\right)$ at $z = -1, 0, 0.5$ and 1 (varied using the slider). The arrows are the top-down views of the vector field $\mathbf{F}$ projected onto each $z = $ constant plane.

$$\iint_S \begin{pmatrix} 2xz \\ z + 2\cos y \\ 2z^3 \end{pmatrix} \cdot \begin{pmatrix} 0 \\ 0 \\ 1 \end{pmatrix} dS = \int_{-1}^1 \int_{-1}^1 2\, dx\, dy = 8.$$

Next, for the face $z = -1$, the outward-pointing normal is $\hat{\mathbf{n}} = (0, 0, -1)$, and $dS = dx\, dy$. The flux across this face also equals 8.

You should check that across the remaining 4 faces of the cube, the flux integrals all yield 0. Thus,

$$
\begin{aligned}
\text{Net flux } &= (\text{Flux across face with } x = -1) + (\text{face } x = 1) \\
&\quad + (\text{face } y = -1) + (\text{face } y = 1) \\
&\quad + (\text{face } z = -1) + (\text{face } z = 1) \\
&= 16.
\end{aligned}
$$

Hence, the Divergence Theorem is verified.

It is useful to check with the quiver plot (fig. 3.12) that these flux calculations make sense. Do this by rotating the figure to see if there are arrows piercing each of the 6 faces. You should see an abundance of upward arrows across $z = 1$, and similarly, many downward arrows across $z = -1$. Those are the faces that contribute to the net flux. On each of the remaining faces, you should see that every arrow has an 'anti-arrow' which cancels the total flux across that face down to zero.

| DISCUSSION |

- **The Fundamental Theorem of Calculus**. The Divergence Theorem expresses how a certain property inside a volume V is equivalent to a related property on the boundary surface S. In the simplest case, this idea can be demonstrated by a very simple calculus equation:

$$\int_a^b f'(x)\,dx = f(b) - f(a).$$

  The LHS sums up the contribution from all points in the interval $[a, b]$, whilst the RHS comprises contributions from the two boundary points $x = a$ and b. This equation is one form of what is known as the *Fundamental Theorem of Calculus*.

  Another example of this phenomenon of dimensional reduction is *Stokes' Theorem*, which we will discuss in the next section.

- **Divergence Theorem in physics**. Here are some important examples of physical laws that are manifestations of the Divergence Theorem.

  - *Gauss's law* in the theory of electromagnetism:

$$\Phi_E = \frac{Q}{\varepsilon_0},$$

    where Φ_E is the electric flux through the surface S, Q is the total charge contained in volume V, and ε_0 is the permittivity of free space (a constant).
  - *Poisson's equation* in the theory of gravitation:

$$\nabla^2 \phi = 4\pi G\rho,$$

    where ϕ is the gravitational potential, ρ is the density within the volume V, and G is Newton's constant. The operator $\nabla^2 = \nabla \cdot \nabla$ (div of grad) is also known as the *Laplacian* (sometimes written Δ).
  - The *continuity equation* in fluid mechanics:

$$\frac{\partial \rho}{\partial t} + \nabla \cdot (\rho \mathbf{v}) = 0,$$

    where ρ is fluid density, $\mathbf{v}$ is the flow velocity, and t is time.

  More about these important equations and their connections to the Divergence Theorem can be found in good undergraduate physics textbooks such as [184, 220].

div.ipynb (for plotting fig. 3.13)	
	``import numpy as np
import matplotlib.pyplot as plt
from matplotlib.widgets import Slider```
Interactive plot	```%matplotlib```
Create 400 grid points on each constant z slice	```x, y = np.meshgrid(np.linspace(-1, 1,20),
np.linspace(-1, 1,20))```
Initial slice	```z=0.5```
u, v are the x, y components of $\mathbf{F}$	```u = lambda x,y,z: 2*x*z```
We will use (u, v) to plot arrows on each slice	```v = lambda x,y,z: z+2*np.cos(y)```
div $\mathbf{F}$	```div = lambda x,y,z: 2*z-2*np.sin(y)+6*z**2```
Rough estimates of max and min of div $\mathbf{F}$	```dmax= 10```
(so the legend shows a sensible range)	```dmin= -2```
Initialise plot	```fig, ax = plt.subplots()```
Leave space for slider	```plt.subplots_adjust(bottom=0.15)
ax.axis([x.min(),x.max(),y.min(),y.max()])```
This function plots the heatmap at a given z	```def plotdiv(z):
D = div(x,y,z)
U = u(x,y,z)
V = v(x,y,z)```
Clear the canvas	``` ax.clear()
ax.set_title('Divergence heatmap')```
Heatmap of div $\mathbf{F}$	``` heat = ax.pcolormesh(x, y, D,
vmin = dmin, vmax=dmax,```
gouraud smooths the heatmap	``` shading='gouraud',cmap='RdBu')```
Plot the arrows ($\mathbf{F}$ projected onto each 2D slice)	``` arrow = ax.quiver(x, y, U, V,```
Use scale to adjust arrow size	``` units='xy', scale=5)
return arrow, heat```
Start with the heatmap at initial z slice	```arrow, heat= plotdiv(z)```
Add legend on the side	```fig.colorbar(heat, ax=ax)```
Dimensions and location of z slider	```axz = plt.axes([0.15, 0.05, 0.6, 0.02])```
Range and resolution of slider	```z_slide = Slider(axz, 'z slice', -1, 1,
valstep = 0.02, valinit = z)```
Update plot if slider is moved	```def update(val):```
Get new z value from slider	``` z = z_slide.val```
Replot	``` plotdiv(z)```
	``z_slide.on_changed(update)
plt.show()```

3.10 Stokes' theorem and the curl operator

Consider the vector field

$$\mathbf{F}(x, y, z) = (xz, \, yz, \, xy).$$

Calculate the net circulation of $\mathbf{F}$ on the surface of the semi-ellipsoid

$$x^2 + y^2 + \left(\frac{z}{a}\right)^2 = 1 \text{ and } z \geq 0,$$

where a is a positive constant.

Given a vector field $\mathbf{F}(x, y, z)$, the *circulation* at a point $P(x, y, z)$ on a surface S is given by the formula

$$\text{Pointwise circulation } = \nabla \times \mathbf{F} \cdot \hat{\mathbf{n}}, \qquad (3.36)$$

where $\hat{\mathbf{n}}$ is the unit normal to the surface at point P. The operator $\nabla \times$ is known as the *curl* of a vector field, sometimes written curl $\mathbf{F}$. It maps a vector to another vector. Let $\mathbf{F}(x, y, z) = (F_1, F_2, F_3)$ (where each component is a function of x, y, z), then[1]

$$
\begin{aligned}
\nabla \times \mathbf{F} &= \begin{vmatrix} \mathbf{i} & \mathbf{j} & \mathbf{k} \\ \frac{\partial}{\partial x} & \frac{\partial}{\partial y} & \frac{\partial}{\partial z} \\ F_1 & F_2 & F_3 \end{vmatrix} \\
&= \left(\frac{\partial F_3}{\partial y} - \frac{\partial F_2}{\partial z}, \; \frac{\partial F_1}{\partial z} - \frac{\partial F_3}{\partial x}, \; \frac{\partial F_2}{\partial x} - \frac{\partial F_1}{\partial y} \right).
\end{aligned}
\qquad (3.37)
$$

The pointwise circulation can be understood conceptually as follows. Think of $\mathbf{F}$ as the velocity field of a fluid. We would like to quantify the rotation of the fluid at P. This 'pointwise' rotation can be characterised by 2 properties: the speed of the rotation, and the orientation of the rotation plane. It turns out that the vector $\nabla \times \mathbf{F}$ completely captures these characteristics. Its length is proportional to the rotation speed, and its direction is normal to the rotation plane. For a mathematical justification of these statements, see for instance [174, 189, 192].

If P also lies on a surface S with normal $\hat{\mathbf{n}}$, we could ask, how much is the fluid circulating *around* $\hat{\mathbf{n}}$? This quantity is precisely the pointwise circulation (3.38): namely, curl $\mathbf{F}$ projected in the direction of $\hat{\mathbf{n}}$. The sign of this number tells us about the direction in which the fluid is locally rotating around $\hat{\mathbf{n}}$.

Make a thumbs-up gesture with your right hand, and align the thumb in the direction of $\hat{\mathbf{n}}$. If curl $\mathbf{F} \cdot \hat{\mathbf{n}}$ is positive, then the direction in which the fluid locally rotates around $\hat{\mathbf{n}}$ is the direction in which your other fingers curl. This is sometimes called the *right-hand rule*. This also means that it does not matter whether which one of the two possible directions for $\hat{\mathbf{n}}$ is chosen: the direction of local circulation remains consistent with either choice.

---

[1] The long vertical bars around the matrix in eq. 3.37 denote the *determinant* of the matrix. Don't worry if you are not familiar with matrices or the determinant at this point. We will discuss matrices in detail in chapter 5.

Now let's study the circulation of the vector field $\mathbf{F}(x, y, z)$ in our example. This vector field is shown in fig. 3.14. The flow looks fairly complicated, but try plotting this vector field (a template is given in the previous section) and rotating it to see the projection in the x-y plane. You will see that the vector field points radially outward away from the z-axis. Also note that the flow is upwards in the 1st and 3rd quadrant, and downwards otherwise.

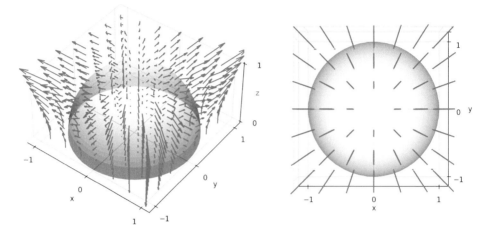

Fig. 3.14: *Left:* Sketch of the vector field $\mathbf{F}(x, y, z) = (xz, yz, xy)$ and the surface $z = \sqrt{1 - x^2 - y^2}$ (unit hemisphere). *Right:* View from the top down. In this projection, the vector field points radially outwards.

For this vector field, we find the curl:

$$\nabla \times \mathbf{F} = (x - y, \ x - y, \ 0).$$

Now, to find the normal $\hat{\mathbf{n}}$ to the ellipsoid, let's exploit the symmetry and change to cylindrical coordinates, (r, θ, z). The semi-ellipsoid can be parametrised by:

$$\mathbf{S}(r, \theta) = \left(r \cos \theta, \ r \sin \theta, \ a\sqrt{1 - r^2} \right),$$

where $r \in [0, 1]$ and $\theta \in [0, 2\pi]$.

The families of tangent vectors in the r and θ directions are

$$\frac{\partial \mathbf{S}}{\partial r} = \left(\cos \theta, \ \sin \theta, \ -ar/\sqrt{1 - r^2} \right), \quad \frac{\partial \mathbf{S}}{\partial \theta} = (-r \sin \theta, \ r \cos \theta, \ 0).$$

The cross product gives us the normal vector, $\mathbf{n}$, to the ellipsoid

$$\mathbf{n} = \frac{\partial \mathbf{S}}{\partial r} \times \frac{\partial \mathbf{S}}{\partial \theta} = \left(\frac{ar^2 \cos \theta}{\sqrt{1 - r^2}}, \ \frac{ar^2 \sin \theta}{\sqrt{1 - r^2}}, \ r \right).$$

We see from the vector components that the normal is outward (and upward) pointing.

Normalising $\mathbf{n}$ gives the outward-pointing unit normal, $\hat{\mathbf{n}}$.

$$\hat{\mathbf{n}} = \frac{1}{\sqrt{1 + r^2(a^2 - 1)}} \left(ar \cos\theta, \; ar^2 \sin\theta, \; \sqrt{1 - r^2} \right).$$

Therefore, the pointwise circulation is

$$\nabla \times \mathbf{F} \cdot \hat{\mathbf{n}} = \frac{ar^2 \cos 2\theta}{\sqrt{1 + r^2(a^2 - 1)}}. \tag{3.38}$$

In fig. 3.15, we plot the pointwise circulation projected onto the x-y plane for various values of a, corresponding to the following surfaces:

- $a = 0$ – a unit disc in the x-y plane
- $a = 0.5$: – an *oblate* (flat) ellipsoid
- $a = 1$: – the unit sphere
- $a = 2$: – a *prolate* (tall) ellipsoid

The code for producing these circular heatmaps (with a slider for a) is given in `curl.ipynb`.

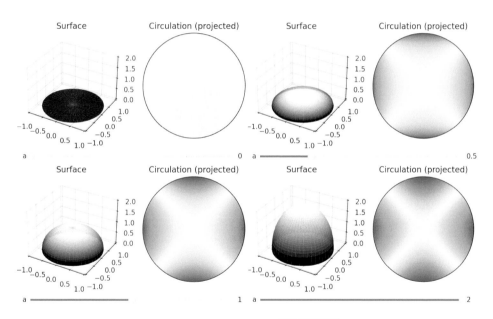

Fig. 3.15: Each panel shows a) the surface $z = a\sqrt{1 - x^2 - y^2}$, and b) a circular heatmap showing the projection in the x-y plane of the pointwise circulation patterns for $a = 0, 0.5, 1$ and 2 (varied using the slider). The net circulation on each heatmap is always zero regardless of the value of a.

Now let's sum up all the pointwise circulations on the surface. The *net circulation* on the surface is given by

$$\text{Net circulation} = \iint_S \nabla \times \mathbf{F} \cdot \hat{\mathbf{n}} \; dS. \tag{3.39}$$

Recall from eq. 3.24 that the area element on the surface $\mathbf{S}(u, v)$ is given by

$$dS = \left| \frac{\partial \mathbf{S}}{\partial r} \times \frac{\partial \mathbf{S}}{\partial \theta} \right| dr \, d\theta = |\mathbf{n}| \, dr \, d\theta.$$

Therefore, $\hat{\mathbf{n}} \, dS = \mathbf{n} \, dr \, d\theta$. Thus, the net circulation expressed entirely in r and θ is

$$\iint_S \nabla \times \mathbf{F} \cdot \hat{\mathbf{n}} \; dS = \int_0^{2\pi} \int_0^1 \frac{ar^3}{\sqrt{1 - r^2}} \cos 2\theta \, dr \, d\theta \qquad (3.40)$$

$$= 0.$$

The net circulation on the surface is zero. You may have already spotted this from the angular pattern in the heatmaps: we see that every point with positive pointwise circulation (red) has a partner with an equal but negative circulation (blue). You can find these partner points by folding each circular heatmap along the diagonal $\theta = 45°$.

It is also interesting to note that when a is large, the heatmap becomes insensitive to a. As the ellipsoid becomes very tall ($a \to \infty$), the heatmap pattern converges to $r \cos 2\theta$.

The fact that the net circulation on the surface is independent of the shape of the surface, in fact, points to a deeper result. Here is a very important result in vector calculus.

Theorem 3.4 *(Stokes' Theorem) Let $\mathbf{F}(x, y, z)$ be a vector field. Let S be a surface with unit normal $\hat{\mathbf{n}}$ and boundary curve C, oriented according to the right-hand rule. Then,*

$$\int_C \mathbf{F} \cdot d\mathbf{r} = \iint_S \nabla \times \mathbf{F} \cdot \hat{\mathbf{n}} \, dS. \qquad (3.41)$$

Like the Divergence Theorem, Stokes' Theorem reduces the dimensionality of the integral, in this case a 2-dimensional integral on the RHS (the net circulation) is reduced to a one-dimension one on the LHS. The amazing thing about Stokes' Theorem is that it holds for all smooth surfaces with boundary C. If we think of C as the rim of a butterfly net, then Stokes' Theorem holds regardless of the shape of the net.

Integrals of the form

$$\int_C \mathbf{F} \cdot d\mathbf{r}, \qquad (3.42)$$

are called *line integrals*. If $\mathbf{F}$ represents a force in moving an object, then the line integral represents the energy spent in moving it along the curve C (in physics, this is called the *work done* on the object). The direction in which the curve C is traversed is determined by the right-hand rule. This is often summed up by the phrase "*C is positively oriented*".

To evaluate the line integral, start by parametrising the curve C by a parameter t, and recast the integral in this single variable. Suppose C can be parametrised by $\mathbf{r}(t)$ where $t \in [a, b]$, then

$$\int_C \mathbf{F} \cdot d\mathbf{r} = \int_a^b \mathbf{F}(t) \cdot \mathbf{r}'(t) \, dt. \qquad (3.43)$$

Note that since $\mathbf{r}'(t)$ is the tangent vector along C, the line integral essentially quantifies the tendency of the vector field $\mathbf{F}$ to point in the same direction as C.

In our example, we parametrise the unit circle C as $\mathbf{r}(t) = (\cos t, \sin t, 0)$, $t \in [0, 2\pi)$. As t increases, C is traversed anti-clockwise, agreeing with the right-hand rule (the normals are all pointing upwards). Substituting x, y, z as the corresponding functions of $\mathbf{r}(t)$, we find

$$\int_C \mathbf{F} \cdot d\mathbf{r} = \int_0^{2\pi} \begin{pmatrix} 0 \\ 0 \\ \cos\theta\sin\theta \end{pmatrix} \cdot \begin{pmatrix} -\sin t \\ \cos t \\ 0 \end{pmatrix} dt = 0. \tag{3.44}$$

Hence Stokes' Theorem is verified.

George Stokes (1819–1903) was an Irish mathematician and physicist who made profound contributions particularly in fluid dynamics. Stokes' theorem, however, was not Stokes' own, but his name stuck because of his habit of setting it as an exam question at Cambridge. The theorem was probably first discovered by *George Green* (1793–1841).

Green was a remarkable English mathematician who, having taught himself mathematics at a library in Nottingham in his 30s, entered Cambridge University as an undergraduate at almost 40 years old. His name lives on today most notably in *Green's Theorem* (which we will shortly discuss) and *Green's function*, an indispensable tool in solving partial differential equations. See [108] for an in-depth historical account of Green's, Divergence and Stokes' Theorems.

DISCUSSION

- **Green's Theorem in the plane**. Apply Stokes' Theorem to the vector field $\mathbf{F} = (f(x, y), g(x, y), 0)$, with C a positively oriented curve in the x-y plane (so that $\hat{\mathbf{n}} = (0, 0, 1)$). This gives

$$\int_C (f\,dx + g\,dy) = \iint_R \left(\frac{\partial g}{\partial x} - \frac{\partial f}{\partial y} \right) dx\,dy, \tag{3.45}$$

  where the region R is bounded by C. This equation is called *Green's Theorem in the plane*. Here is one neat application: choose any f and g such that the integrand on the RHS is 1. The area of R will then be given by the double integral, and thus can be calculated by a 1D line integral thanks to Green's Theorem. This is the principle behind the *planimeter*, a device that, when used to trace the boundary of a region, tells us its area.

- **Stokes' Theorem in physics**. Here are some manifestations of Stokes' Theorem in physics.

  – *Conservative forces*. In mechanics, a force $\mathbf{F}$ is said to be *conservative* if the work done in moving an object from one point to another is independent of the path joining the two endpoints. If $\mathbf{F}$ is conservative, then

$$\nabla \times \mathbf{F} = 0.$$

  – *Faraday's Law* in electromagnetism:

$$\nabla \times \mathbf{E} = -\frac{\partial \mathbf{B}}{\partial t},$$

  where $\mathbf{E}$ is the electric field induced by the changing magnetic field $\mathbf{B}$.
  – *Ampère's Law*, also in electromagnetism:

$$\nabla \times \mathbf{B} = \mu_0 \mathbf{J},$$

where μ_0 is the permeability of free space, and $\mathbf{J}$ is the current density which generates the magnetic field.

Faraday's Law and Ampère's Law are part of *Maxwell's equations* in electromagnetism. See [83] for an excellent introduction to electromagnetism, and [184, 220] for a review of undergraduate physics.

- **Unifying Green's, Stokes' and Divergence Theorems**. In your senior undergraduate course (or beginning graduate course) in differential geometry, you would be pleasantly surprised to discover that Green's, Divergence and Stokes' Theorems can in fact be elegantly unified into a single equation. In this unified version, called *generalised Stokes' Theorem*, we have the equation

$$\int_\Omega d\omega = \int_{\partial\Omega} \omega,$$

where ω is a *differential form*, d is the *exterior derivative*, Ω is a *manifold* and $\partial\Omega$ its boundary. These technical terms are simply higher-dimensional generalisations of vector-calculus concepts such as partial derivatives, curves and surfaces. For a glimpse of these higher-dimensional objects, see textbooks on *differential geometry on manifolds* such as [42, 67, 148].

curl.ipynb (for producing fig. 3.15)

|  | ```python
import numpy as np
import matplotlib.pyplot as plt
from matplotlib.widgets import Slider
%matplotlib
``` |
| Create a grid of $50 \times 50$ pairs of $(r, \theta)$ | ```python
r, theta = np.meshgrid(np.linspace(0, 1),
                       np.linspace(0, 2*np.pi))
``` |
| Pointwise circulation (3.38)<br>This prevents zero division when $a = 0$<br>Return zeroes with the right dimension | ```python
def circulation(a):
    if (a==0): return 0*r
    else: return  a*r**2*np.cos(2*theta)/\
                  np.sqrt(1+r**2*(a**2-1))
``` |
| We'll plot 2 figures side by side<br>Left figure = 3D surface<br>Right figure = circular heatmap | ```python
fig = plt.figure(figsize=(6,4))
ax1 = fig.add_subplot(121, projection='3d')
ax2 = fig.add_subplot(122, projection='polar')
``` |
| Produce sample points on surface<br>Parametrise surface by $(r, \theta)$<br><br><br>Cylindrical coordinates | ```python
def data(a):
    Theta = np.linspace(0,2*np.pi)
    R = np.linspace(0, 1)
    rc, tc = np.meshgrid(R, Theta)
    X = rc*np.cos(tc)
    Y = rc*np.sin(tc)
    Z = a*np.sqrt(1-rc**2)
    return X, Y, Z
``` |
| **Left plot**: 3D surface given $a$<br><br><br><br><br><br>Get the coordinates<br>3D plot | ```python
def plotsurf(a):
    ax1.clear()
    ax1.set_title('Surface')
    ax1.set_xlim(-1,1)
    ax1.set_ylim(-1,1)
    ax1.set_zlim(0,2)
    X, Y, Z = data(a)
    P1 = ax1.plot_surface(X,Y,Z, cmap ='bone')
    return P1
``` |
| **Right plot**: Circular heatmap<br><br><br><br><br>Create the heatmap<br><br>Create filled 50 contour levels be-<br>tween $\pm 1$. With this cmap, positive=red,<br>negative=blue. | ```python
def plotcirc(a):
    ax2.clear()
    ax2.set_title('Circulation (projected)')
    ax2.set_xticklabels([])
    ax2.set_yticklabels([])
    C = circulation(a)
    heat = ax2.pcolormesh(theta, r, C,
            shading = 'gouraud')
    ctf = ax2.contourf(theta, r, C,
            levels= 50, cmap ='coolwarm',
            vmin = -1, vmax = 1)
    return heat, ctf
``` |
| Space for slider<br>Slider dimensions and location<br>Slider range and resolution | ```python
plt.subplots_adjust(bottom=0.15)
axa = plt.axes([0.16, 0.15, 0.7, 0.02])
a_slide = Slider(axa, 'a', 0, 2,
            valstep = 0.01, valinit = 0)
``` |
| Plot the initial surface and heatmap | ```python
plotsurf(0); plotcirc(0)
``` |
| Update plot if slider is moved<br>Get new $a$ value from slider<br>Replot | ```python
def update(val):
    a = a_slide.val
    plotsurf(a); plotcirc(a)
a_slide.on_changed(update)
plt.show()
``` |

3.11 Exercises

1 (*Cycloid revisited*) Derive the equation of the cycloid in the case when the point P is at a distance h from the centre of the unit circle. Show that its parametric equations are

$$x(t) = t - h \sin t,$$
$$y(t) = 1 - h \cos t.$$

By modifying `cycloid.ipynb`, plot the cycloid for $h = 0.8$ and 1.2. Use the initial condition $(x(0), y(0)) = (0, 1 - h)$. Are there cusps on the cycloid if $h \neq 1$?

2 (*Hypocycloid*) A circle radius $r < 1$ is rolling on the inside of a circle radius 1, centred at the origin O. The centre C of the smaller circle is initially at $(1 - r, 0)$. Let θ be the angle subtended by the line OC measured with respect to the positive x-axis. The point P, initially at $(1, 0)$, traces out a curve as the smaller circle rolls inside the unit circle. The setup is shown in the fig. 3.16.

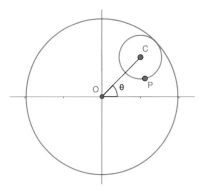

Fig. 3.16: The hypocycloid.

The curve traced out by P has the parametric equations:

$$x(\theta) = (1 - r) \cos \theta + r \cos\left(\frac{1 - r}{r} \theta\right),$$
$$y(\theta) = (1 - r) \sin \theta - r \sin\left(\frac{1 - r}{r} \theta\right).$$

This curve is called a *hypocycloid*.

  a. Plot the hypocycloids for $r = \frac{1}{2}, \frac{1}{4}, \frac{2}{3}$ and $\frac{1}{\sqrt{2}}$ with a slider for $\theta \in [0, 10\pi]$, showing the rolling action. Use `cycloid.ipynb` (§3.2) as a template.
  b. Conjecture the appearance of the hypocycloid for an arbitrary $r \in \mathbb{R}$. (Try $r > 0$ and $r < 0$ as well.)

3 (*Perimeter of the ellipse*) Investigate the fractional error of the following approximations for the perimeter of the ellipse with semi-major axis $a = 1$ and semi-minor axis b. Use `ellipse.ipynb` (§3.3) as a template. Be aware that the fractional error can be negative.

a. Euler's approximation (1773)

$$\pi\sqrt{2(a^2 + b^2)}.$$

b. Ramanujan's second approximation (1914)

$$\pi(a + b)\left(1 + \frac{3h}{10 + \sqrt{4 - 3h}}\right), \quad \text{where } h = \left(\frac{a - b}{a + b}\right)^2.$$

c. Padé approximation (one of many variations in the form of a rational function of h)

$$\pi(a + b)\left(\frac{64 - 3h^2}{64 - 16h}\right).$$

4 (*Witch of Agnesi*) Consider the *Witch of Agnesi* curve given by

$$y = \frac{1}{1 + x^2}.$$

The word *witch* was the result of an English mistranslation of the work of the Italian mathematician *Maria Gaetana Agnesi* (1718–1799), believed to be the first female professor of mathematics at a university.

  a. Parametrise the curve in the form $\mathbf{r}(t)$ where $t \in \mathbb{R}$.
  b. Calculate its curvature $\kappa(t)$.
  c. Plot the curve and its curvature on the domain $x \in [-2, 2]$. Use `curvature.ipynb` (§3.4) as a template.

5 (*Lemniscate of Bernoulli*) This question concerns the arc length and curvature of the Lemniscate of Bernoulli – see §3.4.

  a. Using eq. 3.11, plot the graph of the arc-length function $s(t)$ for the Lemniscate.
  b. By modifying `curvature.ipynb`, plot the curvature κ of the Lemniscate as a function of the arc length s.

6 Calculate the curvature $\kappa(t)$ for the curve given by eq. 3.12. Add its graph to fig. 3.7. Can you think of another parametric curve $\mathbf{r}(t)$ with the same curvature $\kappa(t)$ and torsion $\tau(t)$ as this curve?

7 (*Nielsen's spiral*) Consider the parametric curve defined by the equation $\mathbf{r}(t) = (\text{Ci}(t), \text{Si}(t))$, where $\text{Ci}(t)$ and $\text{Si}(t)$ are the cosine and sine integrals given by

$$\text{Si}(x) = \int_0^x \frac{\sin t}{t}\, dt, \qquad \text{Ci}(x) = \gamma + \ln x + \int_0^x \frac{\cos t - 1}{t}\, dt,$$

(γ is the Euler-Mascheroni constant).
The curve is known as a *Nielsen's spiral*. Plot it for $t \in [0, 100]$.
Suggestion: Use `scipy.special.sici` for the integrals.
Verify (using eq. 3.9) that the curvature κ grows linearly with t. Is this consistent with what you see in the plot?

8 (*Viviani's curve*) Consider the curve of intersection between the sphere $x^2 + y^2 + z^2 = 4$ and the cylinder $(x - 1)^2 + y^2 = 1$. This curve, called *Viviani's curve*, is named after *Vincenzo Viviani* (1622–1703), Italian mathematician and engineer who was also a student of Galileo. Viviani's curve can be expressed as

$$\mathbf{r}(t) = \left(1 + \cos t, \ \sin t, \ 2 \sin \frac{t}{2}\right), \quad t \in [0, 4\pi].$$

a. Find an expression for its arc length as an integral and evaluate it numerically. Suggestion: Express the arc length as an elliptic integral of the second kind (eq. 3.6) and use SciPy's `special.ellipeinc`. Answer ≈ 15.28.

b. The curvature and torsion of the curve are given by

$$\kappa(t) = \frac{\sqrt{13 + 3 \cos t}}{(3 + \cos t)^{3/2}}, \qquad \tau(t) = \frac{6 \cos \frac{t}{2}}{13 + 3 \cos t}.$$

You may like to verify these results using eqs. 3.9 and 3.13.
Create an interactive plot of the curve, its curvature and torsion. Use a slider to control the value of $t \in [0, 4\pi]$ on all three graphs simultaneously as in `curvature.ipynb`.

9 (*Osculating circle*) Consider of a curve $\mathbf{r}(t) = (f(t), g(t)) \in \mathbb{R}^2$. At point P on the curve, the *osculating circle* at P is a circle which touches the curve at P, where the circle and the curve share the same tangent line. It can be thought of as the 'best approximating circle' of the curve near P.
It can be shown that the centre and radius of the osculating circle are given by

$$\text{Radius} = \frac{1}{\kappa(t)},$$
$$\text{Centre} = (x_c(t), y_c(t)),$$
$$\text{where } x_c(t) = f - \frac{((f')^2 + (g')^2)g'}{f'g'' - f''g'}, \quad y_c(t) = g + \frac{((f')^2 + (g')^2)f'}{f'g'' - f''g'}.$$

Let $\mathbf{r}(t) = (t, t^3 - t)$ (a cubic curve) where $t \in [-2, 2]$ Create an interactive plot which plots the curve along with the osculating circle at a point which can be controlled by a slider. [Suggestion: Make sure to prevent zero division when $\kappa = 0$.]

10 (*Families of quadrics*) Let's explore more quadric surfaces. Use `quadrics.py` (§3.6) as a template for the following tasks.

a. Consider the family of surfaces given by the equation

$$x^2 + y^2 - z^2 = a,$$

for $-2 \leq a \leq 2$. Analyse the projections of the surface in the various coordinate planes and make a sketch (by hand) of the surfaces for $a = 0$, $a > 0$ and $a < 0$. Modify `quadrics.py` to create an interactive plot with a slider that controls the value of a.
As a increases from negative, to zero, to positive, you should see the quadric changing from a *hyperboloid of two sheets*, to a *double cone*, to a *hyperboloid of one sheet*.

b. Repeat part (a) for the family of surfaces given by the equation

$$z = x^2 + ay^2,$$

for $-2 \le a \le 2$. You should see the following quadrics as a increases: the *saddle* (also known as the *hyperbolic paraboloid*), the *parabolic cylinder*, and the *paraboloid*.

11 (*Surfaces of revolution*) Consider the curve parametrised by $\mathbf{r}(u) = (f(u), 0, g(u))$, where $f(u) \ge 0$ for all parameter values u. The curve lies on the x-z plane.
If the curve is then rotated anticlockwise about the z-axis, it can be shown that the resulting surface can be parametrised by

$$\mathbf{r}(u, v) = (f(u) \cos v, \; f(u) \sin v, \; g(u)),$$

where $v \in [0, 2\pi]$. The surface is called a *surface of revolution*.

a. Write a code that plots the surface of revolution for the generator curve $\mathbf{r}(u) = (\cosh u, 0, u)$, where $u \in [0, 1]$. Include contour lines on the surface, showing curves of constant u and v in two different colours. (This surface is called the *catenoid*.)

b. Use your code to plot some quadric surfaces that are surfaces of revolution. For example, a cone, a paraboloid, a hyperboloid of 1 and 2 sheets.

c. Plot the surface of revolution by rotating one arch of the cycloid $\mathbf{r}(u) = (1 - \cos u, 0, u - \sin u)$, where $u \in [0, 2\pi]$, about the z-axis (see §3.2). Suggestion: Use `ax.set_box_aspect((_,_,_))` to adjust the aspect ratio.
Use Theorem 3.1 to show that area of the surface of revolution equals $64\pi/3$.

12 (*Möbius strip*) The famous *Möbius strip* is formed by taking a long strip of paper, giving it a half twist, and closing it up in a loop. It can be parametrised as the surface

$$\mathbf{S}(u, v) = \left(\left(1 - u \sin \frac{v}{2}\right) \cos v, \; \left(1 - u \sin \frac{v}{2}\right) \sin v, \; u \cos \frac{v}{2}\right),$$

where $u \in [-1/2, 1/2]$ and $v \in [0, 2\pi)$.

a. Plot the Möbius strip.
b. Explain why the unit circle $\mathbf{C}(v) = (\cos v, \sin v, 0)$ lies along the centre of the strip. Overlay the unit circle on your plot.
c. The *standard unit normal* at the point P on the surface is defined as the vector

$$\hat{\mathbf{n}} = \frac{\mathbf{S}_u \times \mathbf{S}_v}{|\mathbf{S}_u \times \mathbf{S}_v|},$$

evaluated at P (where each subscript denotes a partial derivative).
  i. Show (by hand) that the unit normal $\hat{\mathbf{n}}$ along the centre of the strip is given by

$$\hat{\mathbf{n}} = \left(\cos \frac{v}{2} \cos v, \cos \frac{v}{2} \sin v, \sin \frac{v}{2}\right).$$

  Suggestion: Substitute $u = 0$ as soon as possible.
  ii. Create an interactive plot with a slider for $v \in [0, 2\pi]$ that controls the position of the normal $\hat{\mathbf{n}}$ along the unit circle $\mathbf{C}$ using a slider. Note the behaviour of

the normal as it traverses the Möbius strip. In particular, you should find that the normals as $v \to 0^+$ and $v \to 2\pi^-$ point in opposite directions.
Suggestion: Use `quiver` to create the arrow representing $\hat{\mathbf{n}}$.

Note: Since an 'outward' pointing normal can end up pointing 'inward', the Möbius strip has no well-defined notion of inside and outside - it only has one side! Such a surface is called a *non-orientable* surface.

13 (*Tangent plane and normal to a surface*) Use `grad.ipynb` (§3.8) to help you answer the following questions.
Consider the surface given by the equation

$$x^3 + y^2 + z = 1.$$

a. Plot the surface and its contour lines projected onto the x-y plane.
b. Use theorem 3.2 to show that the upward-pointing unit normal to the surface at $P(1, 0, 0)$ is $\hat{\mathbf{n}} = \frac{1}{\sqrt{10}}(3, 0, 1)$. Display this normal on your plot.
c. Show that the tangent plane to this surface at P is given by $3x + z = 3$. Plot the tangent plane on the same set of axes as the surface.

14 (*Torus*) Let $\mathbf{F}(x, y, z) = (x, 0, 0)$ and S be the surface of the *torus* with the parametrisation:

$$\mathbf{S}(u, v) = ((2 + \cos u) \cos v, \ (2 + \cos u) \sin v, \ \sin u), \quad \text{where } u, v \in [0, 2\pi].$$

a. Use Python to plot the torus, making sure that the aspect ratio is equal in all directions.
b. Overlay some lines of constant u and v on your plot. What do u and v measure?
c. The volume of the torus can be expressed in cylindrical coordinates as

$$4\pi \int_1^3 r\sqrt{1 - (r - 2)^2}\, dr.$$

Evaluate this integral with SciPy's `quad`.
d. Use the Divergence Theorem (theorem 3.3) to show that the volume of the torus agrees with part (c).

15 (*Stokes' Theorem*) Use `curl.ipynb` to help you with this question.
Consider the vector field $\mathbf{F}(x, y, z) = \left(y^2, x^2, z\right)$ and the surface S parametrised by

$$\mathbf{S}(r, \theta) = (r \cos \theta, \ r \sin \theta, \ 2 - r \sin \theta),$$

where $r \in [0, 1]$ and $\theta \in [0, 2\pi]$. We can think of surface S as the intersection between the plane $y + z = 2$ and the unit cylinder $x^2 + y^2 = 1$.

a. Plot the surface with Python.
b. Show that both sides of Stokes' Theorem (theorem 3.4) evaluate to zero.
c. Show that the local circulation on the surface is

$$\nabla \times F \cdot \hat{\mathbf{n}} = \sqrt{2}\, r \, (\cos \theta - \sin \theta).$$

Plot the heatmap of the pointwise circulation on the surface projected onto the unit circle on the x-y plane. Is the pattern consistent with the result in part (b)?

# CHAPTER FOUR

Differential Equations and Dynamical Systems

Fig. 4.1: *Leonhard Euler* (1707–1783), the Swiss mathematician many consider to be one of the greatest mathematicians who ever lived. By 1771 he became completely blind, yet continued to produce an enormous body of revolutionary mathematical work. *Euler's method* for solving a differential equation numerically will be discussed in this chapter (Image source: [137].)

Mathematical modelling of real world phenomena usually requires an understanding how a system evolves over time. If the change is continuous over time, then a *differential equation* can be used to model the system. If the change occurs in discrete time steps, one could model the system using a *recurrence relation*. This chapter shows how Python can be used to solve differential equations and recurrence relations, and help us visualise their solutions.

Along the way, we will show how a physical system can be modelled more abstractly as a *dynamical system*. In this approach, the evolution of the system is represented as a trajectory in an abstract *phase space* whose coordinates are the possible states of the system. More about this in §4.6. To learn more about dynamical systems, including the concepts of *chaos* and *fractals* (which we will touch on in this chapter), see [6, 96, 197] which offer excellent mathematical introductions to this fascinating subject. See also [135] which gives a very readable Python-led tour of dynamical systems.

Let us give an overview of the topics that we will study in this chapter.

4.1 *Basic concepts: ODEs, PDEs and recursions*

Ordinary differential equations

An ordinary differential equation (ODE) is one which contains derivatives. ODEs occur in essentially all areas of applied mathematics. At school, some students may have been taught how to solve simple ODEs such as

$$\frac{dy}{dt} = t + \frac{2y}{t+1} \tag{4.1}$$

$$\frac{d^2y}{dt^2} = \sin t \tag{4.2}$$

Eq. 4.1 is called a first-order ODE, whilst eq. 4.2 is a second-order ODE (the order being the highest derivative in the equation). Various ways to solve these equations numerically will be discussed in §4.3 and 4.4.

It is important to note that ODEs in real applications come with some prescribed initial condition(s). For instance, one could prescribe $y = 0$ when $t = 0$ for the ODE (4.1). Such an initial condition is required to solve the ODE numerically. This is why it is technically more appropriate to say that we are looking for the solution of the *initial-value problem* (IVP) rather than the solution "of the ODE". It is worth keeping in mind that different initial conditions can lead to very different solutions (or no solutions at all).

IVPs that arise in real-world systems cannot usually be solved by hand (or is too tedious to do so), and we have little choice but to tackle them numerically. It is therefore imperative that we know how Python can help us solve IVPs.

Incidentally, at university, the focus is not always on obtaining solutions to IVPs, but rather to prove whether a solution exists, and if so, whether it is unique. More about this later in §4.3.

Readers who are keen to delve more deeply into the mathematics of ODEs may like to take a look at introductory texts such as [97, 175]. More advanced analysis of ODEs can be found in [45, 96]. For an overview of how ODEs are used in real-world modelling, see [1]. For a strong focus on numerical approach to solving ODEs, see [13, 36]. Finally, for a classic compendium of ODEs and their solutions, see [168].

Partial differential equations

A partial differential equation (PDE) is one which contains partial derivatives.

Here is an example of a PDE. Let $y(x, t)$ be a function of x and t where x is defined on the domain $[0, 1]$ and $t \geq 0$. Both x and t are independent variables. The PDE:

$$\frac{\partial y}{\partial t} = \frac{\partial^2 y}{\partial x^2} \tag{4.3}$$

is called the *heat equation* (or *diffusion equation*) in one dimension.

Another example. Let $u(x, y, t)$ be a function of x, y and t, where $x, y \in [0, 1]$ and $t \geq 0$. The PDE:

$$\frac{\partial u^2}{\partial t^2} = \frac{\partial^2 u}{\partial x^2} + \frac{\partial^2 u}{\partial y^2} \tag{4.4}$$

is called the *wave equation* in two dimensions. These two PDEs will be solved numerically in §4.9 and 4.10.

Just as ODEs require initial conditions to solve numerically, PDEs also require additional data. For example, in the heat equation (4.3), the data required to solve for $y(x, t)$ could be

$$y(x, 0) = f(x),$$

for some function f. This is the *initial condition*. In addition, we may also be given that

$$y(0, t) = 0, \qquad y(1, t) = 1.$$

These are called *boundary conditions* since they prescribe values of y at the boundary of the domain $[0, 1]$ at all times. Thus, when we say we are "solving a PDE", we really mean that we are looking for the solution of the PDE given some initial and boundary conditions.

Like ODEs, PDEs also permeate much of applied mathematics. It is rare to be able to solve PDEs in real-world applications by hand (for instance, the simplest models of weather forecasting use 5-7 coupled PDEs). Therefore, numerical methods are indispensable in solving them.

At university, PDEs are normally only studied as a standalone topic in senior undergraduate (or beginning graduate) years. However, it is perfectly possible for beginning undergraduates to gain competency in solving them both by hand (using the 'separation of variable' technique) and numerically (using the 'finite-difference' method). These methods are the focus of §4.9 and 4.10.

For a deeper dive into the world of PDEs, see classic introductory texts such as [60, 196]. For an excellent introduction to numerical solutions of PDEs, see [124].

Recurrence relations

A *recurrence relation* (sometimes called *difference equation*) is an equation that expresses the nth term of a sequence in terms of previous terms. For example, the equation

$$x_{n+1} = x_n + x_{n-1},$$

is a recurrence relation which tells us how to generate a new term x_{n+1} given two initial terms x_n and x_{n-1}. Assuming the initial terms $x_0 = x_1 = 1$, this recurrence relation generates the famous *Fibonacci sequence*.

Another well-known recurrence relation is that for a complex sequence z_n:

$$z_{n+1} = z_n^2 + c, \quad z_0 = 0,$$

where $c \in \mathbb{C}$. This sequence gives rise to the famous *Mandelbrot set*, as we will see in §4.8.

Recurrence relations can be used to model a wide range of discrete-time systems, ranging from real-world applications in biology and economics, to theoretical applications in number theory and combinatorics. For good introductions to recurrence relations and their applications, see [58, 147, 180].

4.2 *Basics of Matplotlib animation*

In terms of Python visualisation, in this chapter we will move away from sliders and instead introduce the use of *Matplotlib*'s `animation` function to visualise "real-time" evolution of time-dependent systems. The result is a video animation which can be saved as an mpeg file or other suitable formats.

Matplotlib achieves this by stitching together a large number of plots and playing them back frame-by-frame. The impatient reader may like to jump ahead to our first animation in §4.4, where we explain in detail how to produce a simple animation of a swinging pendulum. The chapter closes with a more advanced 3D animation of a vibrating membrane (§4.10).

4.3 ODE I – first-order ODEs

> Plot $y(t)$ on the domain $[0, 3]$ given that y satisfies each of the following differential equations.
> a) $\dfrac{dy}{dt} = -y, \quad y(0) = 2,$
> b) $\dfrac{dy}{dt} - \left(\dfrac{2}{t+1}\right) y = t, \quad y(0) = 0,$
> c) $\dfrac{dy}{dt} = y(1-y), \quad y(0) = 1/2.$

Let's start by solving first-order ODEs, namely, solve for y as a function of t given the initial value problem (IVP):

$$\frac{dy}{dt} = f(t, y(t)), \quad y(t_0) = y_0, \tag{4.5}$$

where t is in some specified interval $[a, b]$. Higher-order ODEs can be recast as a system of first-order ODEs, as we will show in the next section.

All 3 ODEs in this example can be solved exactly with simple techniques that many students may have studied in school, but in any case let us walk through these techniques together.

a) $\boxed{\dfrac{dy}{dt} = -y.}$ This can be solved by rewriting the ODE "upside down" and integrating with respect to y.

$$\frac{dt}{dy} = -\frac{1}{y} \implies t = -\ln y + C.$$

Using the initial condition $y = 2$ when $t = 0$, we find $C = \ln 2$, yielding the exact solution

$$y = 2e^{-t}. \tag{4.6}$$

b) $\boxed{\dfrac{dy}{dt} - \left(\dfrac{2}{t+1}\right) y = t.}$ This can be solved by a technique called *integrating factor* which we summarise here. Let F and y be functions of t. The product rule gives $(yF)' = Fy' + F'y$. Now let

$$F(t) = e^{\int P(t)\,dt},$$ (4.7)

where we write the integral without limits to mean that the result is a function of t (ignoring the integration constant). Note that

$$F'(t) = P(t)F(t).$$

This means that if we were presented with an ODE of the form

$$y' + P(t)y = Q(t),$$ (4.8)

then multiplying the equation by the function $F(t)$ defined in eq. 4.7, we see that the LHS can be written as a product rule:

$$(yF)' = QF.$$

This can then be integrated with respect to t and a solution for y can be found. The function $F(t)$ defined in (4.7) is called the *integrating factor*.

The ODE in this example is of the form (4.8) so let's calculate the integrating factor.

$$F(t) = e^{-\int \frac{2}{t+1}\,dt} = (t+1)^{-2}.$$

Multiplying the ODE by this factor, we obtain

$$\left(\frac{y}{(t+1)^2}\right)' = \frac{t}{(t+1)^2}.$$

Now we can integrate both sides with respect to t. It helps to write the numerator on the RHS as $(t+1) - 1$ and split it into two terms. Then, integrating, we have

$$\frac{y}{(t+1)^2} = \ln(t+1) + \frac{1}{t+1} + C.$$

Now use the initial condition $y = 0$ when $t = 0$ to find that $C = -1$. Thus, we obtain the exact solution

$$y = (t+1)\left[(t+1)\ln(t+1) - t\right].$$ (4.9)

c) $\boxed{\dfrac{dy}{dt} = y(1-y).}$ Here we could repeat the "upside down" technique in part (a), or employ a very useful technique called *separation of variables*. This is useful for solving ODEs of the form

$$\frac{dy}{dt} = Y(y)T(t).$$ (4.10)

We can rewrite this ODE in the form

$$\frac{dy}{Y(y)} = \frac{dt}{T(t)},$$

(assuming Y and T are nonzero). In this form, the LHS only contains y, and the RHS only t (hence 'separation of variables'). We can then integrate both sides:

$$\int \frac{dy}{Y(y)} = \int \frac{dt}{T(t)},$$

giving us an expression in y and t.

The ODE in this example is in the form (4.10) with $Y = y(1 - y)$ and $T = 1$. Using separation of variables, we find

$$\int \frac{dy}{y(1 - y)} = \int dt,$$

where we assume $y \neq 0$ and $y \neq 1$ (we will revisit this point in the Discussion section). Then, the LHS can be tackled with partial fractions:

$$\text{LHS} = \int \left(\frac{1}{y} + \frac{1}{1 - y} \right) dy = \ln \frac{y}{1 - y}.$$

Thus we have

$$\frac{y}{1 - y} = Ce^t.$$

Using the initial condition $y(0) = 1/2$, we find $C = 1$. Finally, we make y the subject to find the exact solution

$$y = \frac{1}{1 + e^{-t}}. \tag{4.11}$$

Now that we have obtained the exact solutions (4.6), (4.9) and (4.11), let's study 3 methods to solve ODEs numerically, namely 1) the *forward-Euler method*, 2) *Heun's method* and 3) the *Runge-Kutta* method.

Forward-Euler (FE) method

The forward-Euler method (or sometimes simply the *Euler method*) is generally regarded as the simplest numerical method to solve ODEs. It is based on the Taylor expansion of $y(t + h)$, where h is a small parameter, to linear order.

$$y(t + h) \approx y(t) + hy'(t)$$
$$= y(t) + hf(t, y(t)), \tag{4.12}$$

where $f(t, y(t))$ denotes the RHS of the ODE (4.5). The parameter h is known as the *step size*. It plays a key role in determining the accuracy of the numerical solution, similar to its role in numerical differentiation in chapter 2. Choose h too large and the solution becomes inaccurate, but if h is too small the code might take too long. A very tiny h could even give hugely inaccurate solutions due to rounding error as we saw in §2.2.

Numerical solutions to ODEs are often written in terms of the step number i. For example, the FE method can be written as

$$y_{i+1} = y_i + hf(t_i, y_i), \tag{4.13}$$

where we use the shorthand y_i to mean $y(t_i)$. This formula says: "*the solution at step $i + 1$ is determined by data at step i.*" Thus, we see that numerical solutions to ODEs are calculated *step-by-step*, meaning that inaccuracies in each step can accumulate and grow with each additional step. A good numerical ODE-solving scheme would need to control the growth of these inaccuracies.

A pictorial description of the FE scheme (4.13) is shown in fig. 4.2. The black curve represents the exact solution $y(t)$ (which is unknown, except for one point, namely, the initial condition at $t = t_i$). The red line segment shown has gradient $(y_{i+1} - y_i)/h$, where h is the step size. The FE method approximates this gradient as $y'(t_i)$. In other words, at the point (t_i, y_i), we project forward a straight line in blue with gradient $y'(t_i)$, and claim that the solution at $t = t_{i+1}$ is the y value of this extrapolated line.

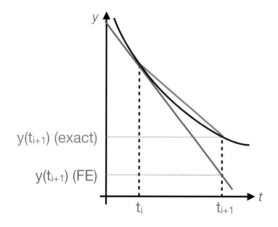

Fig. 4.2: The forward-Euler (FE) method. Knowing only $y(t_i)$, the FE method says that $y(t_{i+1})$ is the y value of the extrapolated blue line at $t = t_{i+1}$.

The code for solving the ODE (a) with the FE method is given in `odesolver.ipynb`. The result for 3 choices of the step size h are shown in fig. 4.3. We can see that in this case, decreasing the magnitude of h gives a more accurate solution. It is important to keep in mind that in real applications, we will not usually know what the exact solution is.

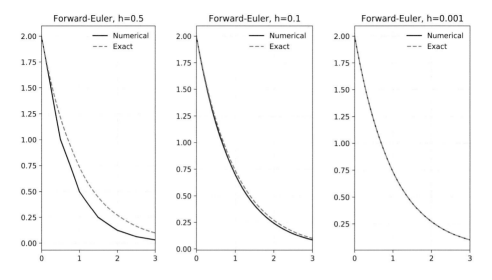

Fig. 4.3: The solution of $y' = -y$, with $y(0) = 2$, obtained using the forward-Euler method. The 3 panels correspond to 3 choices of the step size h, namely, 0.5, 0.1 and 0.001. The numerical solution (solid line) and exact solution (dashed line) agree better as h shrinks.

---

odesolver.ipynb (for plotting fig. 4.3)

```python
import numpy as np
import matplotlib.pyplot as plt
```

Forward-Euler routine
Number of steps (integer)
Initialise solution $y(t)$
Divide (a, b) into N subintervals
Initial condition

*** FE method (eq. 4.13)
Return t_i and y_i.

```python
def odeFE(f, y0, h, a , b):
    N = int(round((b-a)/h))
    y = np.zeros(N+1)
    t = np.linspace(a, b, N+1)
    y[0] = y0
    for i in np.arange(0,N):
        y[i+1]=y[i]+ h*f(t[i],y[i])
    return y, t
```

The RHS of the ODE (eq. 4.5)

```python
def f(t, y):
    return -y
```

Choose the step size h
Input the given initial condition
Choose interval (a, b) to plot
Solve the ODE!
t array for plotting exact solution

```python
h = 0.001
y0 = 2
a, b = 0, 3
y, t = odeFE(f, y0, h, a, b)
T = np.linspace(a, b)
```

Skinny figure
Plot numerical solution in black solid line,
exact solution in dashed line

```python
plt.figure(figsize=(3, 5))
plt.plot(t, y, 'k',
         T, 2*np.exp(-T), '--')
plt.xlim([0,3])
plt.legend(['Numerical', 'Exact'])
plt.title('Forward-Euler, dt=0.001')
plt.grid('on')
plt.show()
```

Let's briefly discuss the error, $E(h)$ defined in the usual way as a function of the step size h.

$$E(h) := |y_{\text{exact}} - y_{\text{numerical}}|.$$

Since the FE method is derived using a linear Taylor expansion in h, we might expect that the error scales like h^2. However, this is true for *each step* in the interval $[a, b]$. When we perform N steps (where $N = (b - a)/h$), the error accumulates to

$$h^2 N \propto h^2 \times \frac{1}{h} = h.$$

The conclusion is that for the FE method, $E(h)$ scales like h. We often write this as

$$\text{Forward-Euler method:} \qquad E(h) \sim h,$$

where we use the symbol $\sim$ to express the approximate scaling of error with step size. Alternatively, we say that forward Euler is an $O(h)$ approximation. You will verify this scaling in exercise 2.

In general, we want $E(h)$ to decrease as quickly as possible with decreasing h so that an accurate solution is obtained without taking too many steps. We present two more methods which improve on the FE method.

Heun's method

The forward-Euler formula (4.13) is asymmetric in the same way that the forward-difference formula for derivatives (Eq. 1.9) is. Just like the forward-difference formula, the forward-Euler method is not usually the method of choice for general practical use because small errors can quickly grow with each step.

One method to reduce the asymmetric effect and improve the stability of the FE method is known as the *modified Euler method*, or *Heun's method*, after the German mathematician *Karl Heun* (1859–1929). The idea is based on a more sophisticated estimate of the slope of the red line segment in fig. 4.2. Observe the following

- The slope of the solution curve at t_i is $f(t_i, y_i)$.
- The slope of the solution curve at t_{i+1}, according to the FE method, is $f(t_{i+1}, y_{i+1}^{\text{FE}})$, where

$$y_{i+1}^{\text{FE}} = y_i + h f(t_i, y_i).$$

- Heun's method states that the slope of the red line segment in fig. 4.2 is approximately the *average of the two slopes* above, *i.e.*

$$\frac{y_{i+1} - y_i}{h} = \frac{1}{2}\left(f(t_i, y_i) + f(t_{i+1}, y_{i+1}^{\text{FE}}) \right)$$

$$\implies y_{i+1} = y_i + \frac{h}{2}\left(f(t_i, y_i) + f(t_{i+1}, y_{i+1}^{\text{FE}}) \right). \tag{4.14}$$

It can be shown that this simple modification leads to one order of magnitude improvement in the convergence (see proof in [182]). In other words, the error for Heun's method satisfies:

$$\text{Heun's method:} \qquad E(h) \sim h^2.$$

You will verify this in exercise 2.

Fig. 4.4 shows the result of solving ODE (b) with Heun's method. A fairly terrible step-size $h = 0.1$ was purposely chosen to solve the ODE to make a point that the numerical solution still matches the exact solution well on this domain (visually at least).

To implement Heun's method in Python, simply replace the line annotated [*** FE method] in the code odesolver.ipynb by the snippet shown in fig. 4.4.

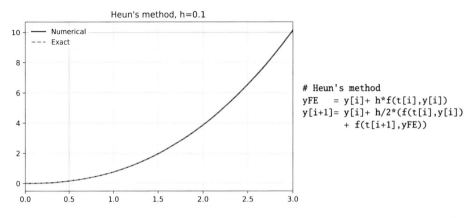

Heun's method, h=0.1

```
# Heun's method
yFE    = y[i]+ h*f(t[i],y[i])
y[i+1]= y[i]+ h/2*(f(t[i],y[i])
            + f(t[i+1],yFE))
```

Fig. 4.4: The solution of $y' - 2y/(t+1) = t$, with $y(0) = 0$, obtained using Heun's method (4.14). Even with $h = 0.1$, the numerical (solid line) and exact (dashed) solutions show good agreement on this domain. The code snippet on the right replaces the line annotated [*** FE method] in odesolver.ipynb.

Fourth-order Runge-Kutta (RK4) method

One of the most widely used methods of solving ODEs is the fourth-order *Runge-Kutta method* (RK4), named after two German scientists: the mathematical physicist *Carl Runge* (1856–1927) and the mathematician *Martin Kutta* (1867–1944). Here is the RK4 formula for solving $y' = f(x, y)$:

$$y_{n+1} = y_n + \frac{h}{6}(s_1 + 2s_2 + 2s_3 + s_4) \tag{4.15}$$

where
$$s_1 = f(x_n, y_n)$$
$$s_2 = f\left(x_n + \frac{h}{2}, \; y_n + \frac{h}{2}s_1\right)$$
$$s_3 = f\left(x_n + \frac{h}{2}, \; y_n + \frac{h}{2}s_2\right)$$
$$s_4 = f(x_n + h, \; y_n + hs_3).$$

The above is derived from an even more sophisticated estimate of the gradient of the red line in fig. 4.2 by a weighted average of four FE-like slopes, s_1, s_2, s_3, s_4, of straight lines drawn at $t = t_i$, t_{i+1} and the midpoint $t_i + h/2$. See, for example, [13, 36] for accessible derivations. The error $E(h)$ for the RK4 method satisfies

Runge-Kutta method: $E(h) \sim h^4$.

The order-4 scaling is the reason that RK4 is generally regarded as the go-to numerical method for solving ODEs accurately without too much coding effort.

Fig. 4.5 shows the result of solving ODE (c) with the RK4 method. Again we have chosen an impractically large step-size $h = 0.1$, yet we see that the numerical and exact solutions are almost visually indistinguishable on this domain.

To implement the RK4 method in Python, replace the line annotated [*** FE method] in the code odesolver.ipynb by the snippet shown in fig. 4.5.

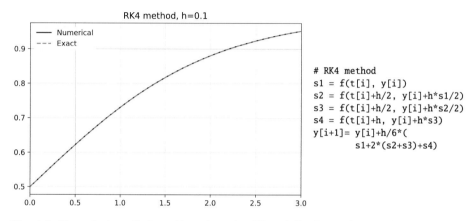

```
# RK4 method
s1 = f(t[i], y[i])
s2 = f(t[i]+h/2, y[i]+h*s1/2)
s3 = f(t[i]+h/2, y[i]+h*s2/2)
s4 = f(t[i]+h, y[i]+h*s3)
y[i+1]= y[i]+h/6*(
            s1+2*(s2+s3)+s4)
```

Fig. 4.5: The solution of $y' = y(1 - y)$, with $y(0) = 1/2$, obtained using the RK4 method (4.15), with $h = 0.1$, the numerical (solid line) and exact (dashed) solutions show excellent agreement on this domain. The code snippet on the right replaces the line annotated [*** FE method] in odesolver.ipynb.

DISCUSSION

• **Existence and uniqueness of solutions**. Before solving IVPs numerically, it would have been more logical to start by asking, do solutions exist? And if so, is the solution unique? Or do we expect the solver to yield just one of many solutions? Here's a theorem which sets out conditions for which there exists a unique solution to an IVP.

Theorem 4.1 *Consider the initial value problem*

$$y' = f(t, y(t)), \quad y(t_0) = y_0 \tag{4.16}$$

where $f : [t_0, t_1] \times \mathbb{R} \to \mathbb{R}$. The IVP has a unique solution if f is continuous in t and satisfies

$$|f(t, y_1) - f(t, y_2)| \le A|y_1 - y_2|, \tag{4.17}$$

where $A > 0$ is a fixed constant, for all pairs of points $(t, y_1), (t, y_2)$ in the domain of f.

See [36, 45] for proofs (which in fact depends on the observation in the next bullet point). The above is one variation of a fundamental theorem in the theory of ODEs called the *Picard-Lindelöf Theorem*, after the French mathematician *Émile Picard*

(1856–1941) and the Finnish mathematician *Ernst Lindelöf* (1870–1946), both of whom made contributions to a wide range of mathematics.

- **IVPs as integral equations**. The Fundamental Theorem of Calculus gives the following equivalence between an initial value problem and an integral equation:

$$y'(t) = f(t, y(t)), \quad y(t_0) = y_0 \iff y(t) = y(t_0) + \int_{t_0}^{t} f(s, y(s)) \, ds.$$

This equivalence suggests that that there is also a connection between numerical methods to solving ODEs and numerical integration. Indeed, we can regard $y(t)$ as the area under the graph $f(t, y(t))$, and thus $y(t) - y(t_0)$ is the area under the curve f on the interval $[t_0, t]$. In this sense, the numerical schemes presented in this section can be regarded as the form

$$y_{i+1} = y_i + \text{area under the curve } f \text{ over the strip } [t_i, t_{i+1}].$$

For example, in the FE formula (4.13), the area of the strip is $hf(t_i, y_i)$, which we recognise as the area of a rectangle width h, and height given by the function value at the left end point $t = t_i$. In other words, the FE method is equivalent to estimating the area under the curve as a sum of rectangular strips. Similarly, we can deduce from (4.14) that Heun's formula is equivalent to estimating area under the curve using trapezoidal strips.

- **n-th order Runge-Kutta**. The RK4 method is one in a family of infinitely many Runge-Kutta methods which can be iteratively refined to give arbitrarily high-order convergence (at the expense of an increased number of computations). It turns out that the FE and Heun's method can be subsumed into the RK family (they are RK1 and RK2 respectively). This is analogous to how the iterative Romberg integration encompasses the Trapezium and Simpson's Rules, as discussed in §2.7.

- **SciPy's `solve_ivp`**. SciPy has a built-in ODE solver `scipy.integrate.solve_ivp`. For example, the following short code solves the ODE (c) and plots the result over the interval $[0, 3]$. For clarity, the argument names for the routine `solve_ivp` are shown explicitly.

```
solving IVP with SciPy

import numpy as np
import matplotlib.pyplot as plt
from scipy.integrate import solve_ivp

def f(t, y):
    return y*(1-y)

T = np.linspace(0,3)
sol = solve_ivp(f, t_span = [0, 3], y0 = [0.5], t_eval = T)

plt.plot(sol.t, sol.y[0])
plt.show()
```

Note the following in the code:

- The argument `t_eval` specifies the values at which the numerical solution should be evaluated. Without this argument, the solver only picks a handful of points to evaluate.
- When defining the function `f`, the solver `solve_ivp` requires that the ordering of the arguments must be `f(t,y)`.
- By default, the solver uses variation of a fifth order RK method (dubbed 'RK45'). For details of this and other available methods, see SciPy's documentation on `solve_ivp`[1]

---

[1] https://docs.scipy.org/doc/scipy/reference/generated/scipy.integrate.solve_ivp.html

4.4 ODE II – the pendulum

A pendulum consists of a bob of mass m attached at the end of a taut string of length ℓ. The pendulum swings in the (x, y) plane about the pivot $(0, 0)$ without friction, making an angle $\theta(t)$ with respect to the vertical at time t. The equation of motion for θ is

$$\theta'' = -\frac{g}{\ell} \sin \theta, \tag{4.18}$$

where g is the acceleration due to gravity (9.8 m/s^2).
Assume that the pendulum starts from rest with $\theta(0) = \pi/4$ and that $\ell = 1$m. Plot $\theta(t)$ for from $t = 0$ to 10s.

This is a classic application of how a simple ODE can be used to describe a physical system. The ODE (4.18) can be derived from Newton's famous second law, $F = ma$ (see, for example, [17] for the derivation).

Unfortunately, even in this relatively simple form, the exact solution of the ODE involves elliptic integrals (see [23]). If, however, the pendulum were to swing at a sufficiently small angle such that $\sin \theta \approx \theta$, the ODE could then be approximated by:

$$\theta'' = -\omega^2 \theta, \quad \text{where } \omega = \sqrt{g/\ell}. \tag{4.19}$$

This second-order ODE can be solved by hand using the method of *auxiliary equation*, which we will briefly summarise below. For proof, see [97, 175].

Consider the second-order ODE

$$ay''(t) + by'(t) + cy(t) = 0,$$

where a, b, c are constant coefficients with $a \neq 0$. We associate it with a quadratic equation (the auxiliary equation):

$$am^2 + bm + c = 0.$$

Solving this quadratic equation yields two solutions

$$m_1, m_2 = \frac{-b \pm \sqrt{\Delta}}{2a}, \quad \Delta = b^2 - 4ac.$$

The solution of the ODE is then given by

$$y(t) = \begin{cases} Ae^{m_1 t} + Be^{m_2 t}, & \text{if } m_1, m_2 \text{ are real and distinct}, \\ e^{m_1 t}(A + Bt), & \text{if } m_1 = m_2, \\ e^{\alpha t}(A \cos \beta t + B \sin \beta t), & \text{if } m_1, m_2 = \alpha \pm i\beta \ (\beta \neq 0). \end{cases}$$

Note that there are two constants in the solution (A, B) which can be determined from two initial conditions. The above is equivalent to the following alternative form.

$$y(t) = \begin{cases} e^{-bt/2a} \left(A \cosh(t\sqrt{\Delta}/2a) + B \sinh(t\sqrt{\Delta}/2a) \right), & \text{if } \Delta > 0, \\ e^{-bt/2a}(A + Bt), & \text{if } \Delta = 0, \\ e^{-bt/2a} \left(A \cos(t\sqrt{|\Delta|}/2a) + B \sin(t\sqrt{|\Delta|}/2a) \right), & \text{if } \Delta < 0. \end{cases}$$

Returning to the ODE (4.19), the auxiliary equation is $m^2 = -\omega^2$, with $\Delta = -4\omega^2$, corresponding to the third case. Thus, $\theta(t) = A \cos \omega t + B \sin \omega t$. Finally, substituting the initial conditions gives $A = \pi/4$ and $B = 0$. We then have the approximation:

$$\theta(t) = \frac{\pi}{4} \cos \omega t. \tag{4.20}$$

The resulting periodic motion is an example of a *simple harmonic motion* (SHM), which occurs when the magnitude of the restoring force is proportional to the object's displacement.

Recall that our approximation was based on the small-angle approximation $\sin \theta \approx \theta$. But we also know that the given pendulum can swing by as much as $\theta = \pi/4 \approx 0.785$, whilst $\sin \theta = 1/\sqrt{2} = 0.707$. We should therefore keep in mind that our small-angle solution won't be an excellent approximation to the true solution. On the other hand, it won't deviate too wildly from the true solution either. We investigate this point next.

Numerical solution: solving a system of first-order ODEs

To solve the original second-order ODE (4.18) numerically, we first transform it into a system of two first-order ODEs. Let $Y_0 = \theta$ and $Y_1 = \theta' = Y_0'$. We see that (4.18) is equivalent to $Y_1' = -\omega^2 Y_0$. In other words, we have transformed a second-order ODE into a system of two first-order ODEs:

$$Y_0' = Y_1,$$
$$Y_1' = -\omega^2 Y_0, \tag{4.21}$$
with initial conditions $Y_0(0) = \pi/4$ and $Y_1(0) = 0$.

Let $\mathbf{Y} = \begin{pmatrix} Y_0 \\ Y_1 \end{pmatrix}$. We can write the above system in a matrix form as

$$\mathbf{Y}' = \begin{pmatrix} 0 & 1 \\ -\omega^2 & 0 \end{pmatrix} \mathbf{Y}, \quad \text{with } \mathbf{Y}(0) = \begin{pmatrix} \pi/4 \\ 0 \end{pmatrix}. \tag{4.22}$$

Since the above is in the form of an IVP, it can be solved with the methods described in the previous section. The only difference is that in each step in the `for` loop, two arrays are updated. The IVP can also be solved with SciPy's `solve_ivp` using the same syntax previously shown, except the derivative now returns an array with 2 elements. The code `pendulum.ipynb` shows how this is done (more about this code below).

You can see that if we were given an Nth-order ODE, we can similarly turn it into a system of N first-order ODE, and recast it as an IVP where the variable is an array with N elements.

Animating the pendulum

Let's visualise the numerical solution $\theta(t)$ as an animation of the pendulum. This is done in the code pendulum.ipynb using matplotlib.animation. The idea for the animation code is as follows.

1. We solve the IVP (with solve_ivp) once and store the solutions at a large number of time steps between $t = 0$ and 10s.
2. For each solution, we plot a straight line segment (representing the taut string) making an angle θ with the vertical, with two dots at the ends. One dot at the origin represents the pivot, and the other dot the swinging bob.
3. Stitch the plots (*frames*) together using FuncAnimation, yielding a video animation.

The code annotation explains how this is done in detail. Snapshots from the animation are shown in fig. 4.6.

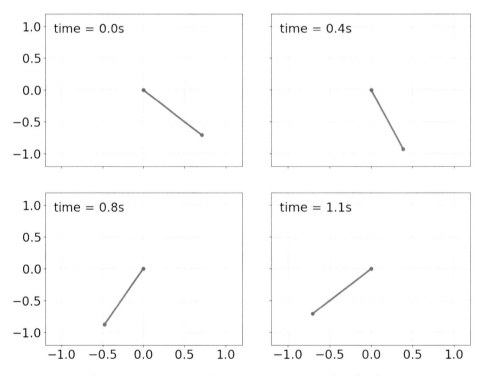

Fig. 4.6: Snapshots of the pendulum animation at $t = 0, 0.4, 0.8$ and 1.1s.

pendulum.ipynb (for plotting fig. 4.6)	
| | ```python
import numpy as np
import matplotlib.pyplot as plt
from scipy.integrate import solve_ivp
from matplotlib.animation import FuncAnimation
%matplotlib
``` |
| Increase font size | ```python
plt.rcParams.update({'font.size': 22})
``` |
| Acceleration due to gravity (m/s$^2$)<br>Length of pendulum (m)<br>(No mass dependence)<br>Derivatives (RHS of the ODEs (4.22)) | ```python
g = 9.8
l = 1.0

def derivs(t, Y):
    dYdt = np.zeros_like(Y)
    dYdt[0] = Y[1]
    dYdt[1] = -g/l*np.sin(Y[0])
    return dYdt
``` |
| Step size (s)<br>Time to end evolution (s)<br>Times at which to plot the pendulum | ```python
h = 0.025
Tmax = 10
T = np.arange(0, Tmax+h, h)
``` |
| Array of initial conditions [$\theta(0), \theta'(0)$] | ```python
Yinit = [np.pi/4, 0]
``` |
| Use SciPy to solve the IVP | ```python
sol = solve_ivp(derivs, t_span = [0, Tmax],
                y0= Yinit, t_eval=T)
``` |
| $\theta(t)$<br>$(x, y)$ coordinates of the bob | ```python
theta = sol.y[0]
xbob = l*np.sin(theta)
ybob = -l*np.cos(theta)
``` |
| | ```python
fig = plt.figure()
ax = fig.add_subplot(111, xlim=(-1.2, 1.2),
                     ylim=(-1.2, 1.2))
ax.grid('on')
``` |
| **Output 1**: The pendulum (red line joining 2 dots)<br>Display the time (to 1 dec. pl.)<br>**Output 2**: Text showing time elapsed (s) | ```python
line, = ax.plot([], [], 'ro-', lw=3)

text = 'time = %.1fs'
time = ax.text(-1.1, 0.9, '')
``` |
| Animation in the $i$th frame<br>$x$ coords of pivot and bob<br>$y$ coords of pivot and bob<br>Update **Output 1**<br>Update **Output 2** | ```python
def animate_frame(i):
    x_i = [0, xbob[i]]
    y_i = [0, ybob[i]]
    line.set_data(x_i, y_i)
    time.set_text(text % (i*h))
    return line, time
``` |
| Animate the pendulum<br>How many frames?<br>Delay between frames (s) | ```python
ani = FuncAnimation(fig, animate_frame,
        frames = len(T),
        interval = 10)

plt.show()
``` |
```

Finally, it is interesting to compare the numerical solution to the small-angle (SHM) approximation (4.4). Fig. 4.7 shows the angle θ as a function of time. We see that the oscillation period (time for one complete oscillation) actually increases slightly with time, whereas the small-angle oscillation period is a constant given by

$$T = \frac{2\pi}{\omega} \approx 2.007\text{s}.$$

You should experiment with changing the initial conditions, *e.g.* making $\theta(0)$ a larger or smaller angle, or giving it an initial kick $\theta'(0) < 0$. You should see that the larger θ can become, the larger the deviation between the numerical and SHM approximations. You should also try to reproduce fig. 4.7 (exercise 4).

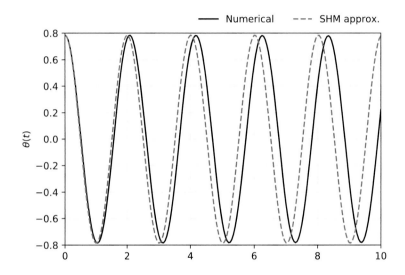

Fig. 4.7: Comparison between the numerical solution to the ODE (4.18) and the approximate solution (4.4) (simple harmonic motion).

DISCUSSION

- **Variations on the simple pendulum**. There are endless variations on the pendulum and its modelling using ODEs. For example, a more realistic pendulum may involve air resistance (exercise 6). The string could also be replaced by a spring or a solid rod. More dramatically, one could attach another pendulum at the bob of one pendulum, creating a double pendulum, which we will study in the next section.
  See [17] for a fascinating book specifically on the pendulum and its variations.
- **Saving an animation as a video file**. To save the animation produced in `pendulum.ipynb` as a video file (say, mp4), include the following lines at the end of the code

```
ani.save('pendulum.mp4')
```

where `ani` is the object of type `FuncAnimation`. You will first need to install the FFmpeg software[2].

---

[2] See https://www.ffmpeg.org for installation instructions.

4.5 ODE III – the double pendulum

A *double pendulum* consists of one pendulum attached at the end of another pendulum as shown in the figure below. The double pendulum swings in the (x, y) plane about the pivot $(0, 0)$. Let ℓ_1 and ℓ_2 be the length of the top and bottom pendulums respectively, and let $\theta_1(t)$ and $\theta_2(t)$ be their angular displacements with respect to the vertical at time t. Let m_1 and m_2 be the respective bob masses. The equations of motion for θ_1 and θ_2 are

$$\theta_1'' = \frac{1}{\ell_1 F} \left[g(\sin\theta_2 \cos\Theta - \mu\sin\theta_1) - \left((\theta_2')^2 \ell_2 + (\theta_1')^2 \ell_1 \cos\Theta\right) \sin\Theta \right],$$

(4.23)

$$\theta_2'' = \frac{1}{\ell_2 F} \left[g\mu(\sin\theta_1 \cos\Theta - \sin\theta_2) + \left(\mu(\theta_1')^2 \ell_1 + (\theta_2')^2 \ell_2 \cos\Theta\right) \sin\Theta \right],$$

(4.24)

where $\Theta := \theta_1 - \theta_2, \quad \mu := 1 + \dfrac{m_1}{m_2}, \quad F := \mu - \cos^2\Theta.$

where g is the acceleration due to gravity. Suppose that $m_1 = m_2 = \ell_1 = \ell_2 = 1$, and that the double pendulum starts from rest with $\theta_1(0) = \theta_2(0) = \pi/2$. Plot $\theta_1(t)$ and $\theta_2(t)$ for $t = 0$ to 30s.

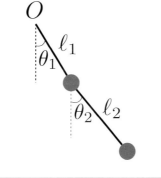

This modification gives the simple pendulum a surprising amount of mathematical richness. Firstly, the equations of motion are clearly a lot more complicated. It is possible to derive them by resolving forces and applying Newton's law, but this gets quite messy. Instead, it is more common to derive the equations of motion using *Lagrangian mechanics*, in which energy considerations lead to the equations of motion without having to draw a force diagram. Introduction to Lagrangian mechanics can be found in [88] and in good textbooks on classical mechanics (for example, [76, 112] are classic references). See [17] for a full derivation of the equations of motion of the double pendulum.

When working with very complicated equations, it is always a good idea to see if they reduce to known results in a simpler setup. In this case, setting $\theta_1 = \theta_2$ and $m_1 = 0$, we obtain the equation of motion (4.18) for the single pendulum, which reassures us of the equations' validity.

Take note of one key conceptual difference between the double and single pendulum: whilst the motion of the simple pendulum is independent of the bob mass, the motion of the double pendulum depends on the mass ratio m_1/m_2.

Numerical solution

To solve the ODEs numerically, we use the same technique as in the previous section: transform the ODEs into a system of first-order ODEs which we can solve with `solve_ivp`. In this case, the two second-order ODEs become four first-order ODEs. Let $\mathbf{Y}(t) = \begin{pmatrix} Y_0 \\ Y_1 \\ Y_2 \\ Y_3 \end{pmatrix} = \begin{pmatrix} \theta_1 \\ \theta_1' \\ \theta_2 \\ \theta_2' \end{pmatrix}$.

The equations of motion become:

$$Y_0' = Y_1,$$

$$Y_1' = \frac{1}{\ell_1 F}\left[g(\sin Y_2 \cos\Theta - \mu \sin Y_0) - \left(Y_3^2 \ell_2 + Y_1^2 \ell_1 \cos\Theta\right)\sin\Theta\right],$$

$$Y_2' = Y_3, \qquad\qquad\qquad\qquad\qquad\qquad\qquad\qquad\qquad\qquad\qquad\qquad (4.25)$$

$$Y_3' = \frac{1}{\ell_2 F}\left[g\mu(\sin Y_0 \cos\Theta - \sin Y_2) + \left(\mu Y_1^2 \ell_1 + Y_3^2 \ell_2 \cos\Theta\right)\sin\Theta\right],$$

$$\text{where } \Theta := Y_0 - Y_2, \quad \mu := 1 + \frac{m_1}{m_2}, \quad F := \mu - \cos^2\Theta,$$

with $\mathbf{Y}(0) = (\pi/2,\ 0,\ \pi/2,\ 0)$.

Animating the double pendulum

We could follow the steps outlined in the single pendulum animation and adjust the code by modifying the derivatives and adding a lower bob at position

$$(x, y)_{\text{lower bob}} = (x, y)_{\text{upper bob}} + (\ell_2 \sin\theta_2,\ -\ell_2 \cos\theta_2).$$

However, let's now add another feature to the visualisation. It would be interesting to display the trajectory of the lower bob as it swings around. We will call this trajectory the *trace* of the lower bob. In the ith frame, we want to plot the curve representing the history of motion so far from the first to the ith frame. Thus, in each frame, there are 3 outputs to display, namely:

- **Output 1**: The double pendulum (represented by two straight line segments joining 3 dots).
- **Output 2**: The trace of the lower bob so far.
- **Output 3**: Time elapsed.

The code `doublependulum.ipynb` is shown below. Snapshots from the animation are shown in fig. 4.8.

The technique we have used to animate the trace is to pass two function arguments to the animation function `animate_frame`, namely 1) the precalculated (x, y) coordinates of the lower bob (`xbob2, ybob2`), and 2) the trace, which gets longer in every frame. The key to plotting the trace so far is the line

```
trace.set_data(xbob2[:i], ybob2[:i])
```

which uses array slicing to extend the trace data in each new frame.

You should try experimenting with the code to study the effect of changing m_1, m_2, ℓ_1, ℓ_2, as well as the initial conditions. Again, it is worth checking first that with small angles $\theta_1(0)$ and $\theta_2(0)$ and large μ, the pendulum behaves effectively like a single pendulum. This gives us confidence in the validity of the numerical solution.

Once you are confident that the code is valid, you can go wild and explore how various initial angles and velocities result in hypnotic motions that are difficult to predict.

Chaotic behaviour

For the single pendulum, two very slightly different initial conditions lead to subsequent motion that are also very slightly different. However, this is not the case with the double pendulum. A tiny difference in the initial conditions can lead to huge differences in the subsequent evolution.

This is demonstrated in fig. 4.9, which shows a snapshot of the double pendulum with 3 sets of very slightly different initial conditions, namely,

$$(\theta_1, \theta_1', \theta_2, \theta_2') = (2 + \varepsilon, 0, 1, 0.1),$$

where $\varepsilon = 0, \ 10^{-3}$ and 10^{-5} (from left to right). We see that the final pendulum configurations after 30 seconds are all quite different, and so are the traces.

This sensitivity to initial conditions (the so-called 'butterfly effect') is a hallmark of *chaos*, which will be explored further in the next sections.

For an even more dramatic representation of the sensitivity to initial conditions, we can plot the dynamics of the double pendulum in the (θ_1, θ_2) phase space (exercise 8).

DISCUSSION

- **Building a double pendulum**. Building a real double pendulum and studying its dynamics is an ideal undergraduate research project, especially given that high-speed motion capture is now widely available and used in everything from gaming to sports science (see fig. 4.10 for an example). See [187] for a study using a real experimental setup.

- **Triple pendulum**. There is a lovely animation featuring 41 simultaneously released triple pendulums, demonstrating extreme sensitivity to initial condition in a most colourful and dramatic fashion[3], but beware that the equations of motion for the triple pendulum are extremely long, as you may expect. A systematic study of the equations of motion for a pendulum with N bobs is given in [219].

- **Lyapunov exponent**. We said that the double pendulum exhibits a chaotic behaviour, but surprisingly, there is no single universally accepted definition of what chaos constitutes. We loosely say that a chaotic system displays sensitivity to initial conditions. To measure this sensitivity, we could, say, take two systems with almost the same initial conditions, say, $\mathbf{Y}(0)$ and $\mathbf{Y}(0) + \Delta(0)$ (where $|\Delta(0)|$ is small). If the difference grows exponentially, *i.e.* $|\Delta(t)| \approx e^{\lambda t}|\Delta(0)|$, with $\lambda > 0$, then the system is said to be chaotic[4]. For instance, for the physical double pendulum constructed in [187], $\lambda \approx 7.5$ was measured, consistent with the expectation that double pendulums can display a chaotic behaviour.

---

[3] https://jakevdp.github.io/blog/2017/03/08/triple-pendulum-chaos/
[4] We are being slightly careless with this definition. There are actually n values of λ in an n-dimensional dynamical system. Here we are referring to the largest λ.

The exponent λ is called the *Lyapunov exponent*, named after *Aleksandr Mikhailovich Lyapunov* (1857–1918), a Russian mathematician who is best remembered today for his contributions to the study of dynamical systems and probability.

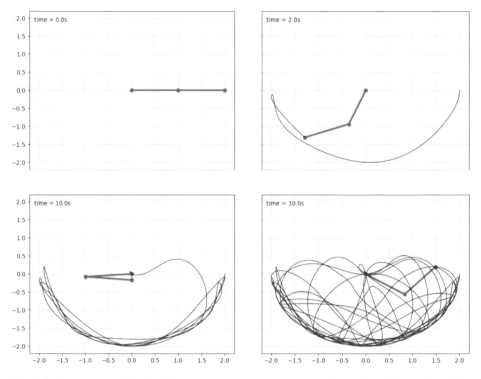

Fig. 4.8: Snapshots of the double pendulum animation at $t = 0, 2, 10$ and 30s, with initial conditions $(\theta_1, \theta_1', \theta_2, \theta_2') = (\pi/2, 0, \pi/2, 0)$.

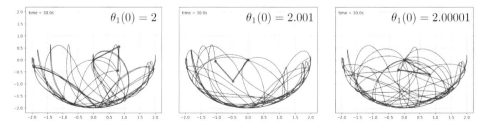

Fig. 4.9: Snapshots of the double pendulum animation after 30s of evolution, starting with very slightly different initial conditions, $(\theta_1, \theta_1', \theta_2, \theta_2') = (2 + \varepsilon, 0, 1, 0.1)$ where $\varepsilon = 0$ (left), 10^{-3} (centre) and 10^{-5} (right). This shows that the double pendulum is sensitive to initial conditions, *i.e.* it exhibits a *chaotic* behaviour.

doublependulum.ipynb (for plotting fig. 4.8)	
| Lots of these in the code | ```python
import numpy as np
import matplotlib.pyplot as plt
from numpy import sin, cos, pi
from scipy.integrate import solve_ivp
from matplotlib.animation import FuncAnimation
%matplotlib
``` |
| Acceleration due to gravity (m/s$^2$)<br>Pendulum lengths $\ell_1, \ell_2$ (m), masses $m_1, m_2$ (kg), and $\mu$. | ```python
g = 9.8
l1, l2, m1, m2 = 1, 1, 1, 1
mu = 1 + m1/m2
``` |
| Derivatives (RHS of the ODEs (4.25))<br><br><br><br><br>Eq. 4.23<br><br><br>Eq. 4.24 | ```python
def derivs(t, Y):
    dYdt = np.zeros_like(Y)
    Theta = Y[0]-Y[2]
    F = mu - cos(Theta)**2
    C, S = cos(Theta), sin(Theta)
    dYdt[0]= Y[1]
    dYdt[1]= (g*(sin(Y[2])*C-mu*sin(Y[0]))
        -(l2*Y[3]**2+l1*Y[1]**2*C)*S)/(l1*F)
    dYdt[2]= Y[3]
    dYdt[3]= (g*mu*(sin(Y[0])*C-sin(Y[2]))
        +(mu*l1*Y[1]**2+l2*Y[3]**2*C)*S)/(l2*F)
    return dYdt
``` |
| Step size (s)<br>Time to end evolution (s)<br>Times at which to plot the pendulum<br>I.C. $[\theta_1(0), \theta_1'(0), \theta_2(0), \theta_2'(0)]$ | ```python
h = 0.02
Tmax = 30
T = np.arange(0, Tmax+h, h)
Yinit = [pi/2, 0, pi/2, 0]
``` |
| Use SciPy to solve the IVP | ```python
sol = solve_ivp(derivs, t_span = [0, Tmax],
                y0= Yinit, t_eval=T)
``` |
| $\theta_1(t)$<br>$(x, y)$ coordinates of the upper bob | ```python
theta1 = sol.y[0]
xbob1 = l1*sin(theta1)
ybob1 = -l1*cos(theta1)
``` |
| $\theta_2(t)$<br>$(x, y)$ coordinates of the lower bob<br>(We will plot the trace of the lower bob) | ```python
theta2 = sol.y[2]
xbob2 = xbob1+l2*sin(theta2)
ybob2 = ybob1-l2*cos(theta2)
``` |
| <br>Adjust plot window if necessary<br><br><br>**Output 1:** Double pendulum (red)<br>**Output 2:** Trace of the lower bob (blue)<br>Display the time in s (to 1 dec. pl.)<br>**Output 3:** Time elapsed (s)<br>(specifying position of text) | ```python
fig = plt.figure()
ax = fig.add_subplot(111, xlim=(-2.2, 2.2),
                          ylim=(-2.2, 2.2))
ax.grid('on')
line, = ax.plot([], [], 'ro-', lw = 3)
trace,= ax.plot([], [], 'b-', lw = 1)
text = 'time = %.1fs'
time = ax.text(-2.1, 1.9, '')

# Code continues on the next page
``` |

doublependulum.ipynb (continued)	

Animation in the ith frame
Coordinates of the pivot, and 2 bobs
 	
Update **Output 1**
Update **Output 2** (see text)
Update **Output 3**
	``
	def animate_frame(i, xbob2, ybob2, trace):
	    x_i = [0, xbob1[i], xbob2[i]]
	    y_i = [0, ybob1[i], ybob2[i]]
	    line.set_data(x_i, y_i)
	    trace.set_data(xbob2[:i], ybob2[:i])
	    time.set_text(text % (i*h))
	    return line, trace, time
	``

Animate the pendulum
Passing two function arguments
How many frames?
Delay between frames (ms)
	``
	ani = FuncAnimation(fig, animate_frame,
	        fargs=(xbob2, ybob2, trace),
	        frames = len(T),
	        interval = 10)
	plt.show()
	``

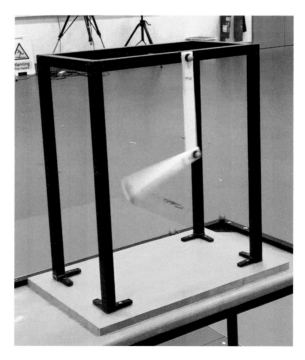

Fig. 4.10: A working double pendulum at the University of Hull (2019). Angle measurements were made using a high-speed motion-capture camera at the Sports Science Lab as part of an undergraduate research project. (Image credit: Mark Anguige.)

4.6 ODE IV – the Lorenz equations

Let x, y, z be functions of t, where σ, r, b are positive constants. The *Lorenz equations* are 3 coupled ODEs given by

$$\begin{aligned} x' &= \sigma(y - x) \\ y' &= x(r - z) - y \\ z' &= xy - bz \end{aligned} \qquad (4.26)$$

Study the trajectory in the (x, y, z) space for $\sigma = 10$, $b = 8/3$, $r = 28$ given the initial condition $(x, y, z)(0) = (1, 1, 1)$.

Edward Lorenz (1917–2008) was an American mathematician and meteorologist who serendipitously discovered that the above set of equations (which modelled fluid convection) have extreme sensitivity to initial conditions. In this model, x, y, z are time-dependent properties of the fluid (for details of these physical properties, see [6]). We are interested in the trajectory of the system in the (x, y, z) space, which is an example of a *phase space* of a dynamical system.

Fixed points

Whilst it is difficult to make analytical headway in solving this system, we can search for *fixed points* in the phase space.

Let $\mathbf{x}(t) = (x(t), y(t), z(t))$. A fixed point of a dynamical system is a point where all derivatives $\mathbf{x}'(t)$ vanish. (This is similar to the concept of stationary points.)

Fixed points tell us about the long-term behaviour of the system. A fixed point may be an *attractor* towards which the system converges, a *repellor* from which the system tends to migrate away, or a *saddle point* which behaves like an attractor or a repellor depending on the state of the system. We discuss more about this when we look at eigenvalues in chapter 5. All the textbooks on dynamical systems mentioned in the introduction give more complete discussions on fixed points of dynamical systems.

For now, let's find out where the fixed points for the Lorenz system are. Setting $x' = y' = z' = 0$ and solving the system of 3 simultaneous equations for (x, y, z), we find the following fixed points (you should verify this).

- The origin
- (If $r > 1$) A pair of fixed points

$$\mathbf{x}^{\pm} = \left(\pm\sqrt{b(r - 1)},\ \pm\sqrt{b(r - 1)},\ r - 1 \right).$$

In particular, for the given parameter values $(\sigma, b, r) = (10,\ 8/3,\ 28)$, we find the fixed points at

$$\mathbf{x}^{\pm} \approx (\pm 8.5,\ \pm 8.5,\ 27). \qquad (4.27)$$

Lorenz himself [133] showed that in this case $\mathbf{x}^{\pm}$ are points between which trajectories bounce around endlessly.

3D animation

Let's visualise the dynamics of the Lorenz equations by animating trajectories in the (x, y, z) space. The code `lorenz.ipynb` given here produces an animation of a trajectory, starting at $(x, y, z) = (1, 1, 1)$ for $0 \le t \le 60$.

The technique for plotting the trajectory is worth discussing. This is similar to plotting the trace in the double pendulum, but with a key difference: we use the command

```
traj.set_data_3d(data[0:3, :i])
```

to update the x, y, z coordinates of the trajectory so far (a quick reminder that the array slicing `0:3` excludes the index 3, so gives 3 rows of data).

The animation shows that indeed, the trajectory encircles one of the two fixed points $\mathbf{x}^{\pm}$, but switches between them in an apparently random manner. Over a long period, the trajectory traces out a 3D structure resembling a pair of butterfly wings. This structure is called a *strange attractor*. The geometric structure of the strange attractor is discussed further in the Discussion section.

Chaotic behaviour

Analogous to the double pendulum, the Lorenz equations also exhibit chaotic behaviour, *i.e.* sensitivity to initial conditions. In fig. 4.12, we plot the x coordinates of two trajectories whose initial conditions differ by one-billionth, namely

Trajectory A (red solid line): $(x, y, z) = (1, 1, 1)$,

Trajectory B (blue dashed line): $(x, y, z) = (1 + 10^{-9}, 1, 1)$.

We see that the trajectories evolve almost identically until $t \approx 37$, when they clearly separate into non-periodic orbits. The Lyapunov exponent in this case is known to be $\lambda \approx 0.9$. A positive λ indicates chaos as discussed at the end of §4.5.

It is interesting to note that whilst this holds for $r = 28$, changing the value of r reveals a fascinating range of behaviours, some chaotic, some not, and some only temporarily chaotic. You will investigate these behaviours in exercise 6.

DISCUSSION

- **Geometry of strange attractors**. The strange attractor in the Lorenz equations comprises trajectories (from different initial conditions) that trace out two 'butterfly wings'. Different trajectories tracing out the strange attractor cannot intersect (due to the uniqueness theorem discussed in§4.3). From this simple observation, one can deduce that the two wings do not merge (despite their appearance) and thus the strange attractor cannot be a single surface, but rather an "*an infinite complex of surfaces*," as Lorenz himself put it [133].

  Such a structure with infinite complexity is called a *fractal*, which we will explore further in §4.8.

- **Poincaré-Bendixson theorem**. Can a strange attractor be constructed using 2 differential equations rather than 3? The answer is no, thanks to a powerful result known as the *Poincaré-Bendixson theorem*, which roughly states that trajectories in a two-dimensional dynamical system either converges to a fixed points or a periodic orbit. A chaotic solution therefore cannot be found in a system with fewer than 3 differential equations. A more precise statement of this theorem and its proof can be found in [6, 96, 197].

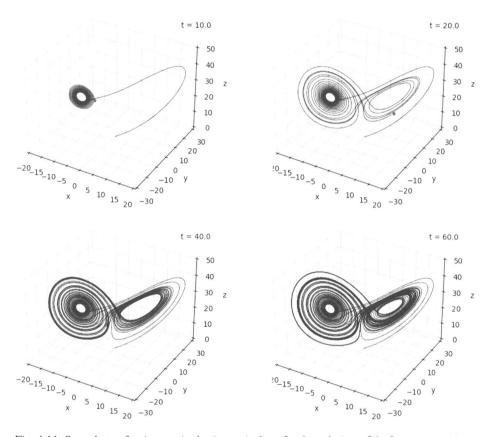

Fig. 4.11: Snapshots of trajectory in the (x, y, z) plane for the solution of the Lorenz equations at $t = 10, 20, 40$ and 60, with initial conditions $(x, y, z)(0) = (1, 1, 1)$ and parameter values $\sigma = 10$, $b = 8/3$, $r = 28$. The red dot in each panel indicates the instantaneous coordinates of the trajectory.

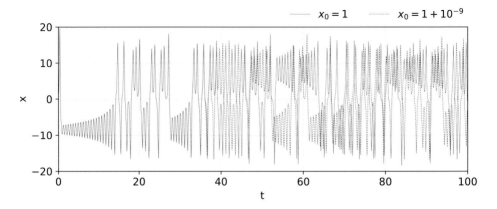

Fig. 4.12: A comparison between two trajectories with initial conditions $(x, y, z) = (1, 1, 1)$ (red solid line) and $(1 + 10^{-9}, 1, 1)$ (blue dashed line). Only the x-coordinates of the trajectories are shown. The parameter choices are the same as those in fig. 4.11. At $t \approx 37$, the two trajectories clearly separate into two different non-periodic orbits.

lorenz.ipynb (for plotting fig. 4.11)

| | ```python
import numpy as np
import matplotlib.pyplot as plt
from scipy.integrate import solve_ivp
from matplotlib.animation import FuncAnimation
%matplotlib
``` |
| Parameter choices for the Lorenz equations | ```python
sigma = 10
b = 8/3
r = 28
``` |
| Derivatives (RHS of the Lorenz system (4.26)) | ```python
def derivs(t, X):
    x,y,z = X
    dXdt = np.zeros_like(X)
    dXdt[0] = sigma *(y - x)
    dXdt[1] = x*(r-z) - y
    dXdt[2] = x*y - b*z
    return dXdt
``` |
| Step size<br>Time to end evolution<br>Times at which to plot the trajectory<br>I.C. [$x(0), y(0), z(0)$] | ```python
h = 0.01
Tmax = 60
T = np.arange(0, Tmax+h, h)
Xinit = [1, 1, 1]
``` |
| Use SciPy to solve the IVP | ```python
sol = solve_ivp(derivs, t_span=[0, Tmax],
                y0=Xinit, t_eval=T)
``` |
| Solutions $x(t)$, $y(t)$, $z(t)$<br>Create a data array - each column is $(x, y, z)^T$ at time $t$ | ```python
x, y, z = sol.y[0], sol.y[1], sol.y[2]
data = np.array([x, y, z])
``` |
| Create a 3D figure | ```python
fig = plt.figure()
ax = fig.add_subplot(111, projection='3d',
    xlim=(-20,20), ylim=(-30,30), zlim=(0,50))
ax.set_xlabel('x')
ax.set_ylabel('y')
ax.set_zlabel('z')
``` |
| **Output 1**: The trajectory in blue<br>**Output 2**: The current state as a red dot<br>Display $t$ to 1 dec. pl.<br>**Output 3**: Text displaying current $t$ value | ```python
traj,= ax.plot([],[],[],'b', lw=0.6)
pnt, = ax.plot([],[],[],'ro', markersize=3)
text = 't = %.1f'
time = ax.text(15,30,60, '')
``` |
| Animation in the $i$th frame<br>Update the $x$, $y$, $z$ values of **Output 1**<br>Update the $x$, $y$, $z$ values of **Output 2**<br>Update **Output 3** (time) | ```python
def animate_frame(i, data, traj):
    traj.set_data_3d(data[0:3, :i])
    pnt.set_data_3d(data[0:3, i:i+1])
    time.set_text(text % (i*h))
    return traj, pnt, time
``` |
| Animate trajectory evolution<br>Passing 2 function arguments<br>How many frames?<br>Delay between frames (ms) | ```python
ani = FuncAnimation(fig, animate_frame,
        fargs=(data, traj),
        frames= len(T),
        interval = 1)
``` |
| | ```python
plt.show()
``` |
```

4.7 The Logistic Map

> The *logistic map* is given by the relation
>
> $$x_{n+1} = f(x_n)$$
> $$\text{where} \quad f(x_n) := rx_n(1 - x_n), \tag{4.28}$$
>
> where $n \in \mathbb{N}$, $0 \leq x_n \leq 1$ and $r > 0$.
> Investigate the long-term behaviour of x_n for different choices of x_0 and r.

The dynamical systems we have studied so far are differential equations that evolve *continuously* with time t. This example, however, is a dynamical system with discrete time steps labelled by $n = 1, 2, 3 \ldots$. Equations like (4.28) are known as *recurrence relations* or *difference equations*.

Here is an example of a physical system which can be modelled by the logistic map: let N be the number of fish in a certain lake averaged over each month. Let $N(n)$ be the number of fish in month n (labelled by integer values, starting with $n = 0$). The fish naturally reproduce, but since the lake has finite resources, there is an absolute maximum number of fish, N_{max}, that the lake can sustain, *i.e.* if $N = N_{max}$ at any point, food runs out completely and the population declines due to starvation.

Let $x_n = N(n)/N_{max}$ (*i.e.* the fractional population measured with respect to the maximum capacity). It follows that x_n lies between 0 and 1. In addition, x_{n+1} (the fractional population next month) depends on x_n in a natural way: x_{n+1} increases with x_n due to reproduction up to a point, then decreases if x_n is too close to the maximum limit.

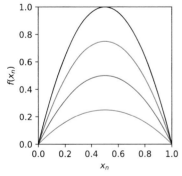

The logistic map models this behaviour as a one-parameter quadratic relation

$$x_{n+1} = rx_n(1 - x_n),$$

where r can be regarded as the growth rate of the population. Note that the limit $x_n \in [0, 1]$ restricts r to be between 0 and 4. The plot on the left shows x_{n+1} as a function of x_n for $r = 1, 2, 3, 4$ (bottom to top).

The logistic map was first studied by the Australian ecologist *Robert May* (1963–2020) in his hugely influential 1976 paper [140] which revealed surprisingly complicated dynamics for such a simple equation.

Fixed points and period

We look for fixed points in the same way we did in the previous section. These fixed points hold clues to the long-term behaviour of the system. Setting $f(x) = x$ (we drop the subscript n for clarity), we find different fixed points depending on the value of r.

- 0 is a fixed point for all r.
- $x^* = (r - 1)/r$ is a fixed point if $r > 1$.

These values of r give rise to x which remains constant from month to month. In addition, there may be values of x that return to the original value every 2 months, or 3 months, and so on. We say that x is a fixed point with *period k* if k is the smallest positive integer such that

$$f^k(x) = x. \tag{4.29}$$

Here f^k denotes function composition $f \circ f \circ f \ldots f$ (k times).
For example, to find fixed points with period 2, we set

$$f^2(x) = x \implies x \left(x - \frac{r-1}{r} \right) \left(r^3 x^2 - (r^2 + r^3)x + r^2 + r \right) = 0. \tag{4.30}$$

There are four solutions to this quartic equation: two of them are the fixed points (with period 1) found above. The other two are period-2 fixed points which exist if $r > 3$, namely:

$$x^\pm = \frac{r + 1 \pm \sqrt{(r - 3)(r + 1)}}{2r}.$$

For instance, with $r = 3.3$, we find the period-2 fixed points at $x^\pm \approx 0.479$ and 0.823. Another way to say this is that $x^\pm$ are a period-2 *orbit*, with x alternately bouncing between the two values. On the other hand, if $r = 3$, we have repeated roots at $x = 2/3$, which is precisely the period-1 fixed point x^*, so in this case the system does not have a period 2 orbit. The first two panels in fig. 4.13 show the period-1 orbit when $r = 3$ and the period-2 orbit when $r = 3.3$ (assuming some random initial condition $x_0 \in [0, 1]$).

It can be shown that one can always find values of r that give rise to orbits with periods 3, 4, 5 and so on. Some of these orbits are shown in fig. 4.13.

Stability

Let $p_1, p_2, \ldots p_k$ be a period-k orbit. The orbit is said to be *stable* if $|(f^k)'(p_1)| < 1$, and *unstable* if $|(f^k)'(p_1)| > 1$.

For example, $|f'(0)| = |r|$, so 0 is stable if $r < 1$. For the fixed point $x^* = \frac{r-1}{r}$, we find $|f'(x^*)| = |2 - r|$, which implies that x^* is stable if $1 < r < 3$. As r increases above 3, x^* loses its stability, but a stable period-2 orbit ($x^\pm$) appears. These points themselves lose their stability when r increases above $1 + \sqrt{6} \approx 3.449$ (you should verify this), and a stable period-4 orbit appears (as shown in the fourth panel of fig. 4.13). This phenomenon gives rise to all orbits which are powers of 2, and is known as a *period doubling cascade*.

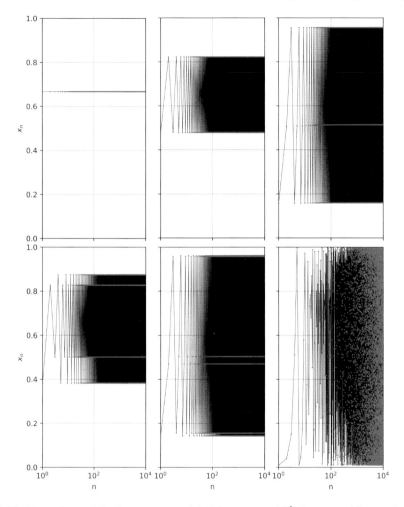

Fig. 4.13: The values of the logistic map (4.28) for n up to 10^4 (horizontal log scale) with parameter (from top left) $r = 3$ (period 1), $r = 3.3$ (period 2), $r = 3.8284$ (period 3), $r = 3.5$ (period 4), $r = 3.845$ (period 6) and $r = 3.99$ (chaotic).

Plotting the bifurcation diagram

As one dials up the value of r, we expect more period-k fixed points to appear, but where exactly? We can use Python to locate these points using the following strategy.

1. Fix a value of $r \in [0, 4]$.
2. Select a random initial condition $x_0 \in [0, 1]$.
3. Perform a large number of iterations to obtain, say, x_{500}. This should lock the trajectory into a steady state.
4. Record x_{500} up to, say, x_{550}. Plotting these values would reveal the periodicity of the orbit (or lack thereof).

The code `logistic.ipynb` executes the above strategy and produces a plot of x_n against r (fig. 4.14) This plot is known as a *bifurcation diagram* in which we can observe the following.

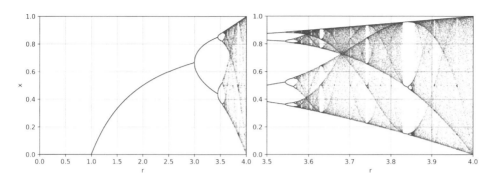

Fig. 4.14: Bifurcation diagram for the logistic map on the domain $r \in [0, 4]$ (left) and $[3.5, 4]$ (right).

- Creation of new stable period-k points occurs at $r = 1, 3$ and $1 + \sqrt{6} \approx 3.449$, in agreement with our stability analysis.
- Period-doubling continues up to $r \approx 3.57$ when the orbit appears blow up into a range of (apparently random) values. This means that very slightly different initial values of x_0 could potentially lead to very different x_n. Thus, we say that the logistic map gives rise to chaos.
- The chaotic behaviour is occasionally interrupted when, at certain values of r, the system settles into a period-k orbit (for example, a period-3 orbit at $r \approx 3.8284$, see the third panel of 4.13).
  As r increases a little further, the attractor becomes chaotic again, most dramatically at r close to 4 when x_n can take essentially any value between 0 and 1 (see the last panel of 4.13).
- The attractor behaviour observed above does not appear to depend on the value of the initial condition x_0.

It is truly astonishing that such a simple equation should give rise to such a rich spectrum of dynamics.

DISCUSSION

- **Feigenbaum constant.** Let b_i be the value of r where the period-doubling bifurcation occurs for the ith time. Our analysis showed that $b_1 = 3$, $b_2 \approx 3.449$, $b_3 \approx 3.544$, $b_4 \approx 3.564$ *etc.* We can quantify the rate at which b_i appears using the variable δ defined as

$$\delta := \lim_{n \to \infty} \frac{b_{n-1} - b_{n-2}}{b_n - b_{n-1}}.$$

It can be shown that $\delta \approx 4.669$. This constant is called the *Feigenbaum constant*, named after the American mathematical physicist *Mitchell Feigenbaum* (1944–2019). Remarkably, this behaviour holds for all one-dimensional maps with a single quadratic maximum, meaning that they all bifurcate at the same universal rate. See exercise 10d.

- **Li-Yorke theorem and Sharkovsky's theorem**. *Tien-Yien Li* and *James Yorke* showed in their legendary 1975 paper *'Period 3 implies chaos'* [129] that if a map $f : [a, b] \to [a, b]$ has a period-3 orbit (like the logistic map), then it has a period-n orbit for any positive integer n as well as chaotic orbits.
  It was later found that the Li-Yorke theorem was in fact a special case of an earlier result discovered in 1962 by the Ukrainian mathematician *Oleksandr Mikolaiovich Sharkovsky* (1936–2022).
- **Fractal**. Zooming in to the bifurcation diagram reveals a self-similar structure as shown in fig. 4.15 below. From left to right, each panel zooms into the bifurcation diagram on a scale $\Delta r \sim 10^{-1}, 10^{-2}, 10^{-4}$ respectively. Such a self-similar structure, known as a *fractal*, will be explored further in the next section.

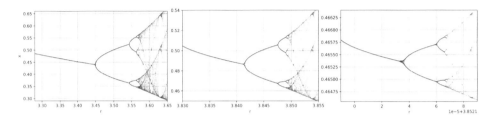

Fig. 4.15: Fractal structure of the bifurcation diagram for the logistic map. From left to right, each panel zooms into the bifurcation diagram on a scale $\Delta r \sim 10^{-1}, 10^{-2}, 10^{-4}$ respectively.

logistic.ipynb (for plotting fig. 4.14)	
	```python import numpy as np import matplotlib.pyplot as plt %matplotlib```
Range of r values to explore Initialise random number generator List for collecting x (steady states of x_n) List for collecting r (a single r may lead to many possible x values)	```python R = np.linspace(0, 4, 4000) rng = np.random.default_rng() xlist = [] rlist = []```

The logistic map	```python
	def f(x, r):
	    return r*x*(1-x)```

Random initial x_0 in (0, 1)
Perform $n = 500$ iterations towards a
steady state
Collect up to 50 steady states for each r
	``python
	for r in R:
	    x = rng.random()
	    for j in range(500):
	        x = f(x, r)
	    for k in range(50):
	        x = f(x, r)
	        xlist.append(x)
	        rlist.append(r)```

Plot steady states (tiny blue dots)	```python plt.plot(rlist, xlist, 'b.', markersize=0.05) plt.xlabel('r') plt.ylabel('x') plt.xlim([0,4]) plt.ylim([0,1]) plt.grid('on') plt.show()```

4.8 The Mandelbrot set

Consider the sequence of complex numbers

$$z_{n+1} = f_c(z_n)$$
$$\text{where} \quad f_c(z_n) = z_n^2 + c, \quad z_0 = 0, \tag{4.31}$$

where $c \in \mathbb{C}$. Locate values of c in the complex plane such that the set

$$\{z_n : n = 0, 1, 2, 3 \ldots\}$$

is bounded.

(This section requires some elementary knowledge of complex numbers. For a gentle introduction to complex numbers, see [9, 115].)

The question asks us to identify values of $c \in \mathbb{C}$ such the sequence of complex numbers

$$0 \to f_c(0) \to f_c(f_c(0)) \to f_c(f_c(f_c(0))) \ldots$$
$$\text{or, equivalently,} \quad 0 \to c \to c^2 + c \to (c^2 + c)^2 + c \ldots \tag{4.32}$$

remains in a bounded region in the complex plane even as the sequence continues forever. We will call this bounded sequence the *orbit* of 0 for a given c. The *Mandelbrot set*, $\mathcal{M}$, is defined as the set of all such permissible c values giving rise to bounded orbits.

The Mandelbrot set is named after the Polish-born mathematician *Benoit Mandelbrot* (1924–2010), who coined the term *fractal* and pioneered the use of computer to visualise complex geometry.

Analysing orbits in the complex plane

Let's work out the orbits of 0 for some values of c listed in the table below. Applying f_0 to these values repeatedly, we find the following orbits.

c	Orbit
0	$0 \to 0 \to 0 \ldots$
$-i$	$0 \to -i \to (-1 - i) \to i \to (-1 - i) \to i \ldots$
1	$0 \to 1 \to 2 \to 5 \to 26 \to 677 \to 458330 \to 210066388901 \ldots$
-2	$0 \to -2 \to 2 \to 2 \ldots$

It is clear that 0, $-i$ and -2 are in $\mathcal{M}$ since those corresponding sequences are bounded, but probably $1 \notin \mathcal{M}$, although we don't have a definite proof for the latter yet.

The code snippet below can be used to help calculate the orbits. Note that the syntax for the imaginary number i in Python is `1j`.

Calculating the orbit of 0 for the map $f_c(z) = z^2 + c$	
An example with $c = -i$	`c = -1j`
$f(c) = c$	`z = c`
Calculate $f^n(c)$ for $n = 2$ to 9 iteratively	`for n in range(2,9):`
	`    z = z**2 + c`
Print using f flag	`    print(f"f^{n}(z) = {z}")`

Our calculations so far have yielded three points in $\mathcal{M}$, but the goal is to determine *all* such $c \in \mathcal{M}$. At this point, one could use Python to do a brute search by calculating orbits for a large number of points in the complex plane, but there are analytic properties that can help us reduce the search time.

Lemma 1 $\mathcal{M}$ *is symmetric about the real axis.*

In other words, if $c \in \mathcal{M}$ then its complex conjugate $\bar{c}$ also belongs to $\mathcal{M}$. One can deduce this by examining the orbit 4.32: the nth term in the sequence is clearly a polynomial, say $P_n(c)$, with real coefficients. Taking the complex conjugate of $P_n(c)$ shows that $\bar{c}$ satisfies the same polynomial, *i.e.* $\overline{P_n(c)} = P_n(\bar{c})$. Since conjugation does not change the magnitude of a complex number, both $P_n(c)$ and $P_n(\bar{c})$ are either both bounded or both unbounded orbits. Hence the lemma follows.

Lemma 1 implies that it is sufficient to analyse values of $c = x + iy$ for $y \geq 0$ only, hence halving our work.

Lemma 2 *All orbits containing* $c \in \mathcal{M}$ *remain within the circle radius 2, centred at the origin.*

The proof requires a little manipulation of inequalities for complex numbers and can be found in [162] for example (we will not reproduce it here). This lemma means that if an orbit ever escapes outside the circle $|z| = 2$, then $c \notin \mathcal{M}$. We will use this observation in our code as a criterion to decide whether $c \in \mathcal{M}$. This criterion is sometimes called the *escape criterion*.

Plotting the Mandelbrot set

Our first attempt to plot the Mandelbrot set might be to pick a point c in the complex plane, and use the escape criterion to determine whether it lies in $\mathcal{M}$ or not. In this case, Lemmas 1-2 imply that it is sufficient to perform this membership test for points within and on the boundary in the upper semicircle $\{c \in \mathbb{C} : |c| \leq 2 \text{ and } \text{Im}(c) \geq 0.\}$

However, the image resulting from this yes/no test is not ideal. Unless we use a sufficiently high resolution (and spend a lot of computing time), we will see an image comprising dots that are sometimes isolated from the main bulbous structure (we will try this method later when plotting fig. 4.18). In fact, a famous result concerning the Mandelbrot set is that it is a single *connected* region with no gaps [55].

You may be more familiar with pictures of the Mandelbrot set that look more like fig. 4.16, with glowing bulbs and root-like substructures. You will notice that the picture contains more than just black and white pixels. Indeed, we have added an additional piece of information to produce shades and tones that help outline $\mathcal{M}$ as a connected structure. The additional information is this: *how many iterations does it take for the orbit to escape the circle radius 2?*. The fewer the iterations required, the lighter the colour. If a point does not ever escape (within a fixed maximum number of iterations), it is shaded the darkest colour.

Here is a summary of the algorithm which is coded in `mandelbrot.ipynb`.

1. Pick $c = x + iy$ in a rectangular grid above the real axis. For instance, $x \in [-2, 0.5]$ and $y \in [0, 1.2]$.
2. Calculate the orbit of 0 by iterating the function f_c k times.
3. If at some point the result satisfies $|z_k| > 2$ (or, to avoid square-rooting, $|z_k|^2 = z_k \bar{z}_k > 4$), we record k (the number of iterations needed to escape).

4. If the orbit never escapes even after k_{max} iterations (say 100), then we record k_{max} and conclude that $c \in \mathcal{M}$.
5. Collect the k value associated with each point in the grid and form a matrix. Display the colour around each point according to the value of k (the larger the k value, the darker the cell).

The final image is the result of the above algorithm together with its reflection across the x-axis.

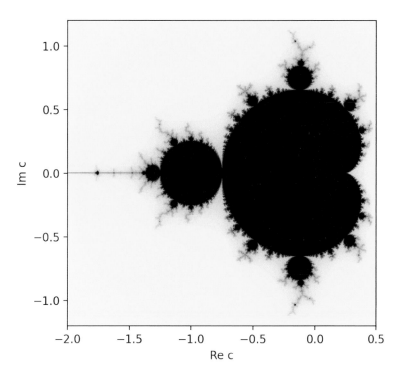

Fig. 4.16: The Mandelbrot set.

The Mandelbrot set is of course the poster child for the concept of *fractal*, or self-similarity on infinitely decreasing scales. Using our code, the plot produced in interactive mode allows zooming into different parts of $\mathcal{M}$, revealing the fractal structure.

In fig. 4.17, we show 3 similar structures on different scales (analogous to fig. 4.15 for the logistic map). We obtain these pictures by repeatedly zooming in on the small bulb on the immediate left of the main bulb. (You may need to decrease eps in the code to see structures on smaller scales and reduce colour bleed.)

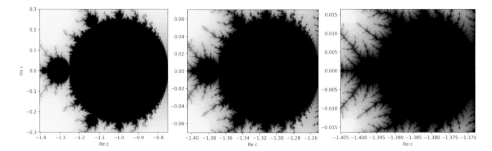

Fig. 4.17: The fractal structure of the Mandelbrot set. The same nodular structure is seen as we zoom into smaller and smaller scales.

mandelbrot.ipynb (for plotting fig. 4.16)	
| | ```python
import numpy as np
import matplotlib.pyplot as plt
%matplotlib
``` |
| Grid resolution (smaller eps = finer)<br>Range of $x$ (where $c = x + iy$)<br>Range of $y$ is [0, ymax] | ```python
eps = 3e-3
xmin, xmax = -2, 0.5
ymax = 1.2
X = np.arange(xmin, xmax, eps)
Y = np.arange(0, ymax, eps)
``` |
| Max no. of iterations<br>How many $x$ and $y$ values are there? | ```python
kmax = 100
nx, ny = len(X), len(Y)
``` |
| Initialise an ny by nx matrix | ```python
kmatrix = np.zeros((ny, nx))
``` |
| The function $g(z) = \|z\|^2 = z\bar{z}$ | ```python
modsq= lambda z: z*np.conj(z)
``` |
| Looping over all $c = x + iy$ in the grid<br><br>$c = x + iy$<br>k counts how many iterations<br>Start the orbit with $z_0 = 0$<br>Keep iterating as long as orbit does not escape<br>and $k \leq k_{max}$<br><br>Record k for each $c$ | ```python
for row in np.arange(ny):
    for col in np.arange(nx):
        c = X[col] + Y[row]*1j
        k = 0
        z = 0
        while (modsq(z)<=4)and(k<=kmax):
            z = z**2 + c
            k = k+1
        kmatrix[row,col]=k-1
``` |
| Lemma 1 $\Longrightarrow$ flip kmatrix upside down to get<br>the image for $y < 0$.<br>Stitch the two matrices together to form K (avoid<br>recording the axis of symmetry twice) | ```python
kflip = np.flipud(kmatrix)
K = np.concatenate((kflip, kmatrix[1:]))
``` |
| Display the colour-coded elements of K | ```python
fig, ax= plt.subplots()
ax.imshow(K, cmap = "magma_r",
             origin = 'lower',
             extent = [xmin,xmax,-ymax,ymax])
``` |
| Relabel the ticks on the axes | ```python
ax.set_xlabel("Re c")
ax.set_ylabel("Im c")
plt.show()
``` |
```

3D view and connection to the logistic map

Let's explore a surprising connection between the Mandelbrot set and the logistic map.
Recall that the Mandelbrot set $\mathcal{M}$ is generated using the mapping

$$z_{n+1} = z_n^2 + c, \tag{4.33}$$

where we studied the orbits of $z_0 = 0$ for different values of $c \in \mathbb{C}$.

To connect $\mathcal{M}$ to the logistic map, let r be a real number such that $0 < r < 4$ and let
$x_n \in [0, 1]$. Let's make the substitutions

$$z_n = r(\frac{1}{2} - x_n) \text{ and } c = \frac{r}{2}\left(1 - \frac{r}{2}\right) \tag{4.34}$$

In other words, we are restricting ourselves to real orbits z_n and real c, but parametrised by
x_n and r respectively.

It is easy to check that $x_n \in [0, 1] \implies z_n \in [-2, 2]$, consistent with Lemma 2. We
can also check that $r \in (0, 4] \implies c \in [-2, \frac{1}{4}]$. You can verify this in the plot for $\mathcal{M}$: its
intersection with the real axis is indeed the interval $[-2, \frac{1}{4}]$.

In terms of x_n and r, eq. 4.33 becomes:

$$r\left(\frac{1}{2} - x_{n+1}\right) = r^2\left(\frac{1}{2} - x_n\right)^2 + \frac{r}{2}\left(1 - \frac{r}{2}\right)$$

Making x_{n+1} the subject, we obtain

$$x_{n+1} = rx_n(1 - x_n), \tag{4.35}$$

which is precisely the logistic map!

Let's go further and see if we can locate the bifurcation diagram (fig. 4.14) somewhere
in the Mandelbrot set. Recall that for the bifurcation diagram, we start the sequence with
a random initial $x_0 \in [0, 1]$ for each fixed $r \in [0, 4]$. This means that we should see the
same bifurcation if, for each fixed $c \in [-2, \frac{1}{4}]$, we start the orbit off with a random initial
$z_0 \in [-2, 2]$.

Over to Python. Firstly, for clarity, let's isolate the points in $\mathcal{M}$ using the binary (yes/no)
criterion that we discussed earlier, *i.e.*, for each $c \in \mathbb{C}$, we test its membership of $\mathcal{M}$ using
the escape criterion, and plot it as a dot if $c \in \mathcal{M}$.

For each dot representing $c \in \mathcal{M}$, let's also add a third dimension: the real part of the
complex number z which each orbit converges to after 100 iterations.

The resulting 3D plot in the $(\text{Re}(c), \text{Im}(c), \text{Re}(z))$ plane is shown in fig. 4.18. Here we
only focus on the region $\text{Im}(c) > 0$ (thanks to the symmetry discussed in Lemma 1). We
see (a dotty version of) the Mandelbrot set in shades of blue, but note that the main bulb
bifurcates into two secondary nodes (corresponding to two possible values of $\text{Re}(z)$), and
so on.

If we chop the 3D Mandelbrot set above along the plane $\text{Im}(c) = 0$ (if we are only
interested in real values of c), we do indeed see the bifurcation diagram for the logistic map
(shown in red). It is upside down compared to fig. 4.14 because of the minus sign in the
substitution (4.34).

Finally, fig. 4.19 shows how the various nodes in the Mandelbrot set correspond to features in the same bifurcation diagram in red. Period-1 points (along the main branch) are precisely those along the cross-section of the largest bulb. Period-2 points are those of the secondary bulb and so on. This also means that the Feigenbaum constant also appears in the Mandelbrot set (see Discussion section in §4.7).

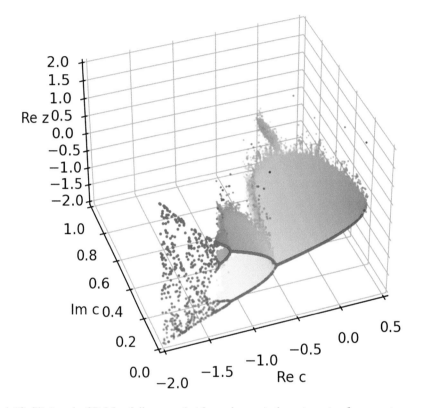

Fig. 4.18: Slicing the 3D Mandelbrot set (with random z_0) along its axis of symmetry reveals a surprising cross section, namely, the bifurcation diagram of the logistic map (shown in red).

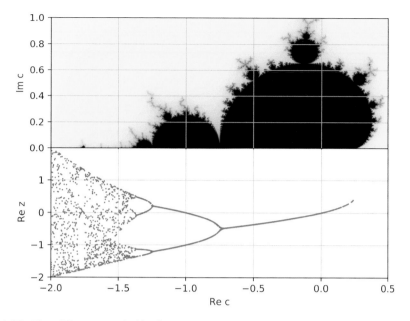

Fig. 4.19: *Top:* The upper half of the Mandelbrot set. *Bottom:* The subset of points with Im(c)=0 shown in the Re(c)-Re(z) plane. These points are precisely the bifurcation diagram of the logistic map. Note the correspondence between the period-n points and the substructures of the Mandelbrot set.

$\boxed{\textsc{Discussion}}$

- **The Cardioid**. The boundary of the main bulb might look like a familiar polar curve. To investigate further, we recall a result on orbit stability from §4.7: a fixed point p of period k is stable if $|(f^k)'(p)| < 1$. We leave it as an exercise to show that for the Mandelbrot set, the fixed points of period 1 (*i.e.* the solution of $f_c(z) = z$) are $z_\pm = (1 \pm \sqrt{1 - 4c})/2$. To look for the boundary between stability and instability, we solve $|f'(z_\pm)| = 1$ for $c = x + iy$. This yields the parametric polar equations of the *Cardioid*:

$$x = \frac{1}{2}\cos\theta\,(1 - \cos\theta) + \frac{1}{4}, \qquad y = \frac{1}{2}\sin\theta\,(1 - \cos\theta). \qquad (4.36)$$

  Similarly, the boundary of the secondary bulb (to the left of the main bulb) is formed by fixed points of period 2. The same technique shows that the boundary is a circle centered at $c = -1$, radius $\frac{1}{4}$. See [135] for calculation details.

- **Mathematical art**. If a picture paints a thousand words, then the Mandelbrot set paints an infinite volume on mathematical art. Since the popularisation of computer-generated art in the 1980s, today we can find many ultra high-resolution zoom-ins of the Mandelbrot set on YouTube [5]. They reveal a whole universe of hidden structures and psychedelic fractals, all stemming from a single quadratic equation.

---

[5] https://www.youtube.com/c/MathsTown

- **The Mandelbrot set is universal**. You may be wondering if the rich structure of the Mandelbrot set is unique to the function $f_c(z) = z^2 + c$. Try generalising the function f_c by changing to other polynomials. Other generalisations are explored in exercise 12. Furthermore, it appears that the characteristic shape of the Mandelbrot set occurs in the fractal patterns of associated with a wide range of complex functions. Like the Feigenbaum's constant, the Mandelbrot set appears to be a universal phenomenon associated with fractals. See [54, 144] for in-depth studies of this phenomenon. One particularly intriguing appearance is that in *Newton's fractal* arising from solutions obtained by the Newton-Raphson root finding method (Chapter 1, exercise 11). This connection is clearly visualised and explained in this video[6].

---

[6] https://www.youtube.com/watch?v=LqbZpur38nw

mandelbrot3D.ipynb (for plotting fig. 4.18)	
	```import numpy as np```
	```import matplotlib.pyplot as plt```
	```%matplotlib```
Collect real and imaginary parts of $c \in \mathcal{M}$...	```creallist = []```
	```cimaglist = []```
and the real part of $z = \lim z_n$	```zreallist = []```
Collect the subset of the above 3 quantities when	```crealred = []```
c is real (we will colour them red)	```cimagred = []```
	```zrealred = []```
Initialise random number generator	```rng = np.random.default_rng()```
Max no. of iterations	```kmax = 100```
Grid resolution (smaller eps = finer)	```eps = 3e-3```
On the domain $x \in [-2, 0.4]$, loop over	```for x in np.arange(-2, 0.4, eps):```
all points (x, y) within and on the upper	``` ymax = np.sqrt(4-x*x)```
semicircle radius 2	``` for y in np.arange(0, ymax+eps, eps):```
$c = x + iy$	``` c = x + y*1j```
Start the orbit with a random $z_0 \in [-2, 2]$	``` z = 4*rng.random() - 2```
Initialise Boolean value for membership of $\mathcal{M}$	``` test = True```
Applying f_c iteratively up to kmax times	``` for k in range(1,kmax+1):```
$f_c(z)$	``` z = z**2 + c```
Calculate $|z|^2 = z\bar{z}$	``` modsq = z*np.conj(z)```
Escape criterion	``` test = (modsq<=4)```
If orbit escapes, stop applying f_c	``` if (not test):```
	``` break```
Otherwise, $c \in \mathcal{M}$	``` if (test):```
and we record the 3 quantities	``` creallist.append(c.real)```
	``` cimaglist.append(c.imag)```
	``` zreallist.append(z.real)```
If $\mathrm{Im}(c) = 0$ (mind the machine epsilon)	``` if (np.abs(c.imag)<1e-16):```
record them as special red points	``` crealred.append(c.real)```
	``` cimagred.append(c.imag)```
	``` zrealred.append(z.real)```
Turn lists into arrays for easier handling	```creal = np.array(creallist)```
	```cimag = np.array(cimaglist)```
	```zreal = np.array(zreallist)```
	```crealR = np.array(crealred)```
	```cimagR = np.array(cimagred)```
	```zrealR = np.array(zrealred)```
	```fig = plt.figure()```
3D plot	```ax = fig.add_subplot(111, projection='3d')```
	```ax.set_xlabel('Re c')```
	```ax.set_ylabel('Im c')```
	```ax.set_zlabel('Re z')```
	```ax.set_xlim(-2,0.5)```
	```ax.set_ylim(0,1)```
	```ax.set_zlim(-2,2)```
Plot $c \in \mathcal{M}$ as blue points, size 0.4	```ax.scatter(creal, cimag, zreal,```
and shaded according to the magnitude of $\mathrm{Re}(z)$	``` s=0.4, c=zreal, cmap="Blues")```
Plot the subset of the above points with real c	```ax.scatter(crealR, cimagR, zrealR,```
as bigger red points	``` s=1, color='r')```
	```plt.show()```

4.9 PDE I – the heat equation

Let $u(x,t)$ denote the temperature (in $°C$) of an insulated rod of unit length at position x (where $x \in [0,1]$) at time t (in seconds). The rod is heated so that its temperature is initially given by $u(x,0) = \sin(5\pi x/2)$.

The heating suddenly stops. Subsequently, at $x = 0$, the rod is maintained at $0°C$, whilst at $x = 1$ the rod is maintained at $1°C$. The subsequent temperature distribution for $t > 0$ is governed by the *heat equation*

$$\frac{\partial u}{\partial t} = \frac{\partial^2 u}{\partial x^2}, \tag{4.37}$$

$$\text{with initial condition} \quad u(x,0) = \sin(5\pi x/2), \tag{4.38}$$

$$\text{and boundary conditions} \quad u(0,t) = 0, \quad u(1,t) = 1. \tag{4.39}$$

Determine the temperature distribution $u(x,t)$. What happens as $t \to \infty$?

The heat equation (also called *diffusion equation*) was discovered and studied by *Joseph Fourier* (whom we have met in Chapter 2) around 1822. The derivation of the heat equation (and solution methods) can be found in classic textbooks on PDEs such as [60, 196].

Finite-difference method

Here is a method called *finite-difference method* which demonstrates the general principal of solving PDEs numerically. It is important to keep in mind solving the PDE requires not only the differential equation itself, but also the boundary and initial conditions. Changing the boundary conditions, for instance, could give rise to an entirely different solution (or no solutions at all).

The method starts by discretising the rod, *i.e.* dividing into N equal subintervals $[x_i, x_{i+1}]$ where $i = 0, 1, 2 \ldots N$. Note that there are $N + 1$ grid points x_i, where

$$x_i = \frac{i}{N}.$$

Let $\Delta x := x_{i+1} - x_i = \frac{1}{N}$ be the length of each subinterval. We will use Δx in the approximation of the partial derivative $\frac{\partial^2 u}{\partial x^2}$ on the RHS of the heat equation, so we will need to keep Δx small.

To do this, we express the function $f(x \pm \Delta x)$ as a Taylor series in Δx for small Δx.

$$f(x + \Delta x) = f(x) + f'(x)\Delta x + \frac{f''(x)}{2}(\Delta x)^2 + \frac{f'''(x)}{3!}(\Delta x)^3 + O\left((\Delta x)^4\right),$$

$$f(x - \Delta x) = f(x) - f'(x)\Delta x + \frac{f''(x)}{2}(\Delta x)^2 - \frac{f'''(x)}{3!}(\Delta x)^3 + O\left((\Delta x)^4\right),$$

(where O expresses the next leading order in the tail of the series). Adding the above equations and making $f''(x)$ the subject gives

$$f''(x) = \frac{f(x + \Delta x) - 2f(x) + f(x - \Delta x)}{(\Delta x)^2} + O\left((\Delta x)^2\right). \tag{4.40}$$

In other words, we have obtained a *symmetric-difference* approximation for the second derivative. This means that the second-derivative term in the heat equation can be expressed in terms of u evaluated at grid points as

$$\frac{\partial^2 u(x_i, t)}{\partial x^2} \approx \frac{u(x_{i+1}, t) - 2u(x_i, t) + u(x_{i-1}, t)}{(\Delta x)^2}. \tag{4.41}$$

This holds for all internal grid points ($i = 1, 2, \ldots N - 1$). We exclude the boundary grid points since we already know that $u(x_0, t) = 0$ and $u(x_N, t) = 1$ at all times.

In summary, a single PDE (4.37) is converted into a system of $(N - 1)$ ODEs. Writing $u(x_i, t)$ as $u_i(t)$, we have

For $i = 1, 2, \ldots N - 1$,
$$\frac{du_i(t)}{dt} = \frac{u_{i+1}(t) - 2u_i(t) + u_{i-1}(t)}{(\Delta x)^2}, \tag{4.42}$$

with initial conditions $u_i(0) = \sin(5\pi x_i/2)$, $\tag{4.43}$

and boundary conditions $u_0(t) = 0, \quad u_N(t) = 1.$ $\tag{4.44}$

We can then use one of the methods discussed in §4.3 to solve the ODE for the $(N - 1)$ variables u_i as a function of t. The code `heat.ipynb` below uses SciPy's `solve_ivp` to solve the ODE. We then use *Matplotlib* to create an animation of the temperature evolution from $t = 0$ to 0.5s. Snapshots from the animation are shown in fig. 4.20.

We see that as time evolves, the temperature pattern flattens towards a linear variation $u(x, t) = x$. This makes sense since the rod is insulated and is maintained at a constant temperature at each end, so in the long run the temperature will vary linearly along the rod.

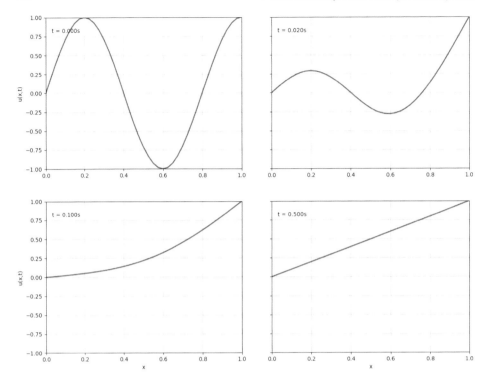

Fig. 4.20: Snapshots from the animation of the temperature $u(x, t)$ at $t = 0, 0.02, 0.1$ and 0.5s. The x-axis shows the position along the rod.

heat.ipynb (for plotting fig. 4.20)

```python
import numpy as np
import matplotlib.pyplot as plt
from scipy.integrate import solve_ivp
from matplotlib.animation import FuncAnimation
%matplotlib
```

Divide the rod into N subintervals
The grid points along the rod
The length of each subinterval Δx

```python
N = 40
x = np.linspace(0,1,N+1)
dx = x[1]-x[0]
```

Derivatives (RHS of ODEs (4.42))

Vectorised form of (4.42)

```python
def derivs(t, U):
    dUdt = np.zeros_like(U)
    dUdt[1:-1]=(U[2:]-2*U[1:-1]+U[:-2])/dx**2
    return dUdt
```

Step size (in time)
Time to end evolution (s)
Times to record temperature snapshots
The initial temperature distribution (4.43)

```python
h = 5e-4
Tmax = 0.5
T = np.arange(0, Tmax+h, h)
Uinit = np.sin(np.pi*5*x/2)
```

Use SciPy to solve the IVP for $u_i(t)$

```python
sol = solve_ivp(derivs, t_span=[0, Tmax],
                y0=Uinit, t_eval=T)
```

Create a 2D array U. Each row = temp. at each time step. Each column = temp. at a fixed point on the rod.
Fill in the columns with solutions from solve_ivp

```python
U = np.zeros((len(T),len(x)))

for i in np.arange(1,N):
        U[:,i] = sol.y[i]
```

The last column is the boundary condition $u_N(t) = 1$ (4.44).

```python
U[:,-1] = np.ones(len(T))

fig = plt.figure()
ax = fig.add_subplot(111, xlim=(0, 1),
                          ylim=(-1, 1))
ax.grid('on')
ax.set_xlabel('x')
ax.set_ylabel('u(x,t)')
```

Template for plotting each frame
Display time t
Position of the time display

```python
line, = ax.plot([], [], 'r', lw=2)
text = 't = %.3fs'
time = ax.text(0.03, 0.8, '')
```

Animate each frame
Plot each row of U in each frame
Update time display

```python
def animate_frame(i):
    line.set_data(x, U[i,:])
    time.set_text(text % (i*h))
    return line, time

ani = FuncAnimation(fig, animate_frame,
        frames = len(T),
```

Set delay between frames

```python
        interval = 20)

plt.show()
```

Separation of variables

Let's try to solve the heat equation by hand and see if the solution is consistent with our numerical solution.

Here is a very useful method which can help us solve certain types of PDEs exactly. This is the method called *separation of variables*, which we will outline below step-by-step (you may want to fill in the details). This is a multi-step process with nuances in every step. Whilst the calculations involved can be overwhelming at first reading, your introductory PDE course will provide plenty of practice. For those who want to read ahead, see [33] for plenty of solved problems.

1. **Simplify the system** Since the numerical investigation showed that $u(x,t) \approx x$ eventually, let's define

$$v(x,t) := u(x,t) - x,$$

   and solve for $v(x,t)$ instead of $u(x,t)$. A quick calculation shows that v satisfies the same PDE, whilst the boundary conditions become simpler (and the initial condition gains an extra term).

$$\frac{\partial v}{\partial t} = \frac{\partial^2 v}{\partial x^2}, \tag{4.45}$$

$$\text{with initial condition} \quad v(x,0) = \sin(5\pi x/2) - x, \tag{4.46}$$

$$\text{and boundary conditions} \quad v(0,t) = 0, \quad v(1,t) = 0. \tag{4.47}$$

   The fact that the boundary conditions are now zero at both endpoints is the key to the success of this method.

2. **Separate the variables** Let's try looking for a solution of the form

$$v(x,t) = X(x)T(t), \tag{4.48}$$

   where X is a function of x alone and T is a function of t alone. Substituting this into the heat equation (4.45) and dividing by XT, we obtain

$$\frac{T'}{T} = \frac{X''}{X}. \tag{4.49}$$

   (This is the *separation of variables*.) Note that the LHS of (4.49) is a function of t alone, whilst the RHS is a function of x alone. What function is both a function of t alone and x alone? The answer is of course, a constant, say k, i.e.

$$\frac{T'}{T} = \frac{X''}{X} = k. \tag{4.50}$$

3. **Solve for $X(x)$** We will now show that the constant k must be strictly negative. To do this, we will rule out the case $k = 0$ and $k > 0$.

   - Can $k = 0$? If this is the case, eq. 4.50 tells us that:

$$\frac{X''}{X} = 0 \implies X = \alpha x + \beta,$$

for some constants α and β to be determined from the boundary conditions. Since we require $X(0) = X(1) = 0$, the only solution is $X(x) \equiv 0$, which is invalid since this would mean that the temperature is 0 at all times. So $k \neq 0$.

• Can $k > 0$? Let $k = \lambda^2$ for some $\lambda \neq 0$. If this is the case, we have

$$\frac{X''}{X} = \lambda^2 \implies X = \alpha \cosh x + \beta \sinh x.$$

Using $X(0) = X(1) = 0$ again gives $X(x) \equiv 0$, which we rule out.

• Finally, let's try $k < 0$. Write $k = -\lambda^2$ ($\lambda \neq 0$). In this case, the solution of $X'' = 0$ with boundary conditions $X(0) = X(1) = 0$ is nontrivial:

$$X(x) = \alpha \sin \lambda x, \quad \text{where} \quad \lambda = n\pi \quad (n = 1, 2, 3 \ldots). \tag{4.51}$$

These values of λ are called the *eigenvalues* of the boundary-value problem.

4. **Solve for** $T(t)$ Equating the LHS of (4.50) to $-\lambda^2$ and solving the ODE gives

$$\frac{T'}{T} = -\lambda^2 \implies T = \beta e^{-n^2\pi^2 t}.$$

Combining this with $X(x)$ (4.51), we have so far found solutions of the form

$$v_n(x, t) = be^{-n^2\pi^2 t} \sin n\pi x, \quad (n \in \mathbb{N}),$$

where b is a constant.

5. **Superposition** An important observation is that if v_1 and v_2 are solutions to the PDE, then $v_1 + v_2$ is another solution. This property (called the *superposition principle*) simply follows from the linearity of the partial derivatives. This means that the following infinite sum (if convergent) is also a solution:

$$v(x, t) = \sum_{n=1}^{\infty} b_n e^{-n^2\pi^2 t} \sin n\pi x.$$

6. **Find the Fourier series** It remains to use the initial condition (4.46), which requires

$$\sin(5\pi x/2) - x = \sum_{n=1}^{\infty} b_n \sin n\pi x,$$

where $x \in [0, 1]$, *i.e.* we want to write the LHS as a linear combination of sine waves. Recall our earlier encounter with Fourier series in §2.9, where we saw how a function can be expressed a sum of sines and cosines. It is also possible to express the function using only the sine terms. To do this, we extend the rod (defined on $[0, 1]$) to $[-1, 0]$ in such a way that $v(x, t)$ is an *odd function* defined on the interval $[-1, 1]$. When we calculate the Fourier series for this odd-extended function, we find that only the sine terms are needed, and that the coefficient b_n is given by

$$b_n = 2 \int_0^1 (\sin(5\pi x/2) - x) \sin n\pi x \, dx$$

$$= \frac{2(-1)^n}{\pi} \left(\frac{4n}{25 - 4n^2} + \frac{1}{n} \right). \tag{4.52}$$

The integration above is fiddly but definitely worth trying!

In conclusion, we have found the exact solution for the temperature

$$u(x, t) = x + \sum_{n=1}^{\infty} b_n e^{-n^2 \pi^2 t} \sin n\pi x, \tag{4.53}$$

where b_n is given by eq. 4.52.

We can now easily see the asymptotic behaviour of the solution from the exact solution (4.53). The sinusoidal terms are exponentially damped when t is large, leaving us with $u(x, t) \approx x$.

In exercise 13c, you will investigate how the exact solution compares with the numerical solution.

$\boxed{\text{DISCUSSION}}$

- **Uniqueness.** In the search for the exact solution, we started by assuming that the solution is in the separable form 4.48. Are there other solutions? The answer is no. The uniqueness of the solution can be shown using an elegant trick known as the *energy method*. Details in [196], for example.

  Uniqueness also holds for the solution to the wave equation, which we will study in the next section.

- **The heat equation in** $\mathbb{R}^3$. Suppose that heat can move in all directions in $\mathbb{R}^3$. The heat equation then becomes

$$\frac{1}{\alpha} \frac{\partial u}{\partial t} = \frac{\partial^2 u}{\partial x^2} + \frac{\partial^2 u}{\partial y^2} + \frac{\partial^2 u}{\partial z^2} = \nabla^2 u,$$

  where $\nabla^2 = \nabla \cdot \nabla$ (called the *Laplacian* operator). We have also introduced the constant α, the *thermal diffusivity* of the medium in which the heat is being conducted. Better heat conductors have higher α values. For example, $\alpha_{\text{silver}} \approx 166$ mm^2/s whereas $\alpha_{\text{rubber}} \approx 0.1$ mm^2/s.

- **Stability** The timestep Δt and grid size Δx need to be chosen carefully so that the numerical solution is *stable*, *i.e.* numerical errors do not grow and dominate the solution. For the heat equation, for a given Δx, stability is guaranteed if

$$\Delta t \leq C \frac{(\Delta x)^2}{\alpha}, \tag{4.54}$$

  where α is the diffusivity and C is a constant which depends on the numerical method used to solve the differential equation (4.42). For example, $C = 0.5$ if the forward Euler method is used.

  Nevertheless, it is possible to avoid having to choose a small Δt dictated by the stability criterion. A famous trick known as the *Crank-Nicolson method* uses a small modification of the discretised scheme (4.42) to guarantee stability for *any* Δt and Δx.

  More details on the analysis of numerical stability, and the Crank-Nicolson method, can be found in [59, 124, 170]. The open-access reference [124] in particular contains excellent expositions on the finite-difference method with Python implementations.

4.10 PDE II – the wave equation

A thin elastic membrane is stretched over a square frame $S = \{(x, y) \in (0, 1) \times (0, 1)\} \subset \mathbb{R}^2$. The edges of the membrane are kept fixed to the frame but the membrane itself is otherwise free to vibrate vertically within the frame. Let $u(x, y, t)$ be the amplitude of the vibration at position (x, y) at time t. It is known that u satisfies the two-dimensional *wave equation*

$$\frac{\partial^2 u}{\partial t^2} = \frac{\partial^2 u}{\partial x^2} + \frac{\partial^2 u}{\partial y^2}, \quad (x, y) \in S, \quad (4.55)$$

with initial conditions $u(x, y, 0) = \sin(2\pi x)\sin(2\pi y),$ (IC1)

$$\frac{\partial u}{\partial t}(x, y, 0) = 0, \quad \text{(IC2)}$$

and boundary condition $u(x, y, t) = 0, \quad (x, y) \in \partial S,$ (BC)

where ∂S denotes the boundary of S. Determine the subsequent amplitude $u(x, y, t)$ where $t > 0$.

As we will be discussing a variety of partial derivatives, let us use the notation $\partial_{xx} u$ to mean $\frac{\partial^2 u}{\partial x^2}$. In this notation, the wave equation in 2D reads

$$\partial_{tt} u = \partial_{xx} u + \partial_{yy} u.$$

You can probably guess that the wave equation in 1D simply reads $\partial_{tt} u = \partial_{xx} u$. You will solve in this in exercise 14a. The wave equation in 3D is equally obvious to generalise.

The 1D wave equation was discovered and solved in 1747 by the French mathematician *Jean d'Alembert* (1717–1783) who made important contributions to mechanics. You may recall that the Ratio Test in §2.3 was also named after d'Alembert.

Finite-difference method

First, let's see how we can discretise the problem and apply a similar finite-difference method as that for the heat equation.

Let's divide the unit square into $N \times N$ square cells with grid points

$$(x_i, y_j) = \left(\frac{i}{N}, \frac{j}{N}\right), \quad i, j = 0, 1, 2 \dots N.$$

Of course one may instead divide the square into $N_x \times N_y$ rectangular cells where N_x and N_y may not necessarily be equal. Let $\Delta x := x_{i+1} - x_i$ and $\Delta y := y_{i+1} - y_i$ be the width and height of each cell.

The RHS of the wave equation 4.55 can be discretised using eq. 4.41. Let $u_{i,j}$ denote u evaluated at grid point (x_i, y_j). The second-order partial derivatives become:

$$(\partial_{xx}u)_{i,j} \approx \frac{u_{i+1,j} - 2u_{i,j} + u_{i-1,j}}{(\Delta x)^2},$$ (4.56)

$$\left(\partial_{yy}u\right)_{i,j} \approx \frac{u_{i,j+1} - 2u_{i,j} + u_{i,j-1}}{(\Delta y)^2}.$$ (4.57)

This holds for all internal grid points $(i, j = 1, 2, \ldots N - 1)$ at all times. We exclude the boundary grid points since we already know from eq. BC that $u_{i,j}(t) = 0$ at all times whenever i or j equals 0 or N.

The discretisation so far has reduced the PDE to a second-order ODE in terms of t, namely

$$\left(\frac{d^2u}{dt^2}\right)_{i,j} = (4.56) + (4.57).$$ (4.58)

One way to solve this ODE numerically is to transform it into a system of two first-order ODEs (as we did in §4.4) and then use SciPy's `solve_ivp` to solve for $u(x_i, y_j, t)$. The difficulty with this method, however, is that `solve_ivp` can only handle vectors (1D arrays), so our data on the 2D plane must first be reshaped into a 1D array, then the output from `solve_ivp` must then be again reshaped into a 2D array for plotting. Whilst the method is viable, there is a lot of fiddly indexing and array reshaping to keep track of. You can imagine that this reshaping gets even more complicated if we were dealing with a 3D problem.

An alternative method is to forgo `solve_ivp` and instead discretise the *time* derivative as well (using eq. 4.41). Let each time step be of size h and let the nth time step be

$$t_n = nh, \quad n = 0, 1, 2 \ldots$$

We then find that for all interior grid points (x_i, y_j) at time step t_n $(n \geq 1)$, we have

$$u_{i,j}(t_{n+1}) = -u_{i,j}(t_{n-1}) + 2u_{i,j}(t_n) + h^2 (\partial_{xx}u)_{i,j}(t_n) + h^2 \left(\partial_{yy}u\right)_{i,j}(t_n)$$ (4.59)

which tells us how to advance u to the next time step using data from previous time steps.

However, we can't yet use eq. 4.59 to advance from t_0 to t_1 since we do not know the data at t_{-1}. To this end, note that the symmetric-difference formula for the time derivative $\frac{\partial}{\partial t}$ at time step $t_0 = 0$ reads

$$(\partial_t u)_{i,j}(t_0) = \frac{u_{i,j}(t_1) - u_{i,j}(t_{-1})}{2h} \implies u_{i,j}(t_{-1}) = \underbrace{u_{i,j}(t_1) - 2h (\partial_t u)_{i,j}(t_0)}_{(\text{IC2})}.$$

Substituting the above into (4.59) (with $n = 0$) and rearranging gives

$$u_{i,j}(t_1) = \underbrace{u_{i,j}(t_0)}_{(\text{IC1})} + \underbrace{h (\partial_t u)_{i,j}(t_0)}_{(\text{IC2})} + \underbrace{\frac{h^2}{2} (\partial_{xx}u)_{i,j}(t_0)}_{\text{Eq. 4.56}} + \underbrace{\frac{h^2}{2} \left(\partial_{yy}u\right)_{i,j}(t_0)}_{\text{Eq. 4.57}}.$$ (4.60)

In summary, we use (4.60) to advance from t_0 to t_1, then (4.59) for the subsequent time steps. This gives us the amplitude $u(x_i, y_j, t_n)$ at all interior grid points at all times.

The code `wave.ipynb` implements the above algorithm and creates an animation for $t \geq 0$. Some snapshots from the animation are shown in fig. 4.21.

Some points of interest in the code.

- There is a lot of vectorisation happening in the functions `first_step` and `next_step`. Make sure you understand why these are equivalent to eqs. 4.60 and 4.59.
- In defining the initial condition (IC2) as a lambda function, we write `0*x` to make the output a 2D array. What do you think would happen if we make the function return `0` instead of `0*x`? (We used the same technique in the code `curl.ipynb` in §3.10).
- To update the 3D surface in each frame, we clear the deck and replot the surface from scratch (using `ax.clear()`). This is in contrast with how we updated the 3D curve in `lorenz.ipynb` (§4.6), where we kept the axes and just updated the data. At the time of writing, it is not possible to do this kind of data update for 3D surfaces without using some unwieldy workaround.

Separation of variables

Let us investigate if we can solve for $u(x, y, t)$ exactly by following the steps outlined in the previous section. We begin by searching for a separable solution of the form

$$u(x, y, t) = X(x)Y(y)T(t).$$

Using the boundary condition and the superposition principle gives us

$$u(x, y, t) = \sum_{n=1}^{\infty} \sum_{m=1}^{\infty} \sin(m\pi x) \sin(n\pi y) (a_{mn} \cos \lambda t + b_{mn} \sin \lambda t), \tag{4.61}$$

where the eigenvalues $\lambda = \pi\sqrt{m^2 + n^2}$, and a_{mn}, b_{mn} are functions of m and n to be determined.

Next, using the initial condition (IC1), we can determine a_{mn} as follows.

$$\sin(2\pi x) \sin(2\pi y) = \sum_{n=1}^{\infty} \sum_{m=1}^{\infty} a_{mn} \sin(m\pi x) \sin(n\pi y) \implies a_{mn} = \begin{cases} 1 & \text{if } (m, n) = (2, 2), \\ 0 & \text{otherwise.} \end{cases}$$

Using (IC2), we determine b_{mn} as follows.

$$0 = \sum_{n=1}^{\infty} \sum_{m=1}^{\infty} \lambda b_{mn} \sin(m\pi x) \sin(n\pi y) \implies b_{mn} = 0.$$

Putting everything together, we find the solution

$$u(x, y, t) = \sin(2\pi x) \sin(2\pi y) \cos(2\sqrt{2}\pi t). \tag{4.62}$$

This can be used to verify the validity of the numerical solution – see exercise 14d. As a rough check, we see from (4.62) that the cosine term first vanishes at time $t = \sqrt{2}/8 \approx 0.177$ s, and the amplitude of the membrane should also be zero. The middle panel of fig. 4.21 corroborates this calculation.

$\boxed{\textbf{DISCUSSION}}$

- **Normal modes**. We were quite lucky that the initial condition and boundary conditions of the problem allowed us to solve the wave equation by hand exactly. The relatively simple solution (4.62) is called a *normal mode* of the system, meaning that all points on the membrane oscillates at the same frequency. This is apparent in the animation which shows that the solution behaves like a 'standing wave'.

  In general, a different set of initial and boundary conditions will yield a more complicated solution which, as can be seen in eq. 4.61, can be decomposed into a combination of normal modes (this is precisely the idea behind Fourier series).

- **The wave equation in** $\mathbb{R}^3$. Suppose that the wave can propagate in all directions in $\mathbb{R}^3$, then the amplitude $u(x, y, z, t)$ satisfies the 3D wave equation

$$\frac{1}{c^2}\frac{\partial^2 u}{\partial t^2} = \nabla^2 u,$$

  where the constant c is the speed of the wave in the medium. The above can also be written as

$$\Box u = 0,$$

  where the *d'Alembertian* operator (or *box operator*) is $\Box := \frac{1}{c^2}\frac{\partial^2}{\partial t^2} - \nabla^2$.

- **Wave equation in physics**. Here are some examples.

  - The *electric field* **E** in a vacuum obeys the wave equation

$$\Box \mathbf{E} = 0.$$

    The above equation is in fact 3 equations written in vectorial notation (*i.e.* each component of **E** satisfies the same PDE). The magnetic field **B** obeys the same equation. The wave speed c is the speed of light.

  - *Gravitational waves*, predicted by Einstein in 1916 and first detected in 2016, can be described by a *tensor*, $\bar{h}_{\mu\nu}$, which, in this case, can be thought of as a 4×4 matrix and $\mu, \nu = 0, 1, 2, 3$ are the indices for its rows and columns. The amplitude $|\bar{h}_{\mu\nu}|$ can be thought as the magnitude of the ripples propagating in spacetime (3 space dimensions + time). General relativity predicts that in vacuum, $\bar{h}_{\mu\nu}$ satisfies the wave equation

$$\Box h_{\mu\nu} = 0.$$

    This means that gravitational waves travel at the speed of light.

  - In quantum mechanics, a particle is described by a *wave function* $\psi(\mathbf{x}, t)$ which describes the physical state of the particle probabilistically. If a particle has mass m, it obeys the equation

$$\Box \psi = -\frac{mc^2}{\hbar}\psi,$$

    where c is the speed of light and $\hbar$ is Planck's constant. This can be regarded as a wave equation with a *source term* on the RHS.

These examples tell us that the wave phenomenon is ubiquitous on all physical scales, from quantum to cosmological.

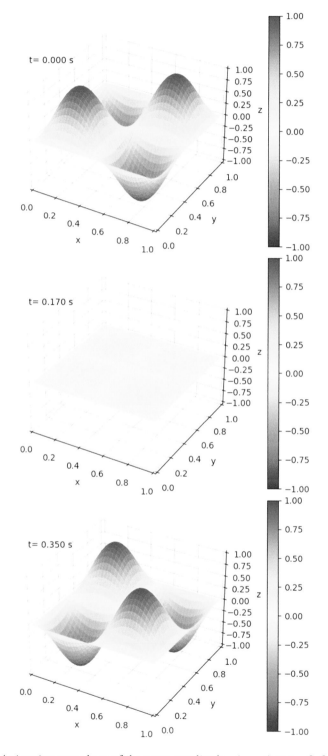

Fig. 4.21: Animation snapshots of the wave amplitude $u(x, y, t)$ at $t = 0, 0.17, 0.35$s.

wave.ipynb (for creating the wave animation (fig. 4.21))	

```
import numpy as np
import matplotlib.pyplot as plt
from matplotlib.animation import FuncAnimation
%matplotlib
```

Divide the domain into $N_y \times N_x$ cells
The x and y coordinates of the grid points

```
Nx, Ny = 40, 40
x = np.linspace(0,1,Nx+1)
y = np.linspace(0,1,Ny+1)
```

Cell dimension $\Delta x, \Delta y$
Step size (in time)
Time to end evolution
Times to record snapshots
Constants $(h/\Delta x)^2$ and $(h/\Delta y)^2$

```
dx, dy  = x[1]-x[0], y[1]-y[0]
h = 1e-3
Tmax = 3
T = np.arange(0, Tmax+h, h)
Cx2, Cy2 = (h/dx)**2, (h/dy)**2
```

Initial condition (IC1)

```
IC1 = lambda x, y : np.sin(2*np.pi*x)*\
                    np.sin(2*np.pi*y)
```

Initial condition (IC2)

```
IC2 = lambda x, y: 0*x
```

Grid points (x_i, y_j) (2D arrays)
Initialising (IC1) and (IC2) at grid points

```
X, Y = np.meshgrid(x, y)
U = IC1(X, Y)
V = IC2(X, Y)
```

$u_{i,j}(t_{n-1})$
$u_{i,j}$ at snapshot times T

```
Unm1 = np.zeros((len(x),len(y)))
Usnap = np.zeros((len(x),len(y),len(T)))
```

Eq. 4.60 – to go from t_0 to t_1
Setting boundary condition (BC)
Numerator of (4.56)
Numerator of (4.57)
Set $u_{i,j}$ at interior grid points

```
def first_step(U):
    Ustep = np.zeros((len(x),len(y)))
    Uxx=U[:-2, 1:-1]-2*U[1:-1,1:-1]+U[2:,1:-1]
    Uyy=U[1:-1,:-2] -2*U[1:-1,1:-1]+U[1:-1,2:]
    Ustep[1:-1,1:-1]= U[1:-1,1:-1] + \
        h*V[1:-1,1:-1]+ 0.5*(Cx2*Uxx + Cy2*Uyy)
    return Ustep
```

Eq. 4.59 – for subsequent time steps

```
def next_step(U, Unm1):
    Ustep = np.zeros((len(x),len(y)))
    Uxx=U[:-2, 1:-1]-2*U[1:-1,1:-1]+U[2:,1:-1]
    Uyy=U[1:-1,:-2] -2*U[1:-1,1:-1]+U[1:-1,2:]
    Ustep[1:-1,1:-1]=-Unm1[1:-1, 1:-1]+\
        2*U[1:-1,1:-1]+ Cx2*Uxx + Cy2*Uyy
    return Ustep
```

Record $u_{i,j}(t_0)$
Record $u_{i,j}(t_1)$

```
Usnap[:,:,0] = U
Usnap[:,:,1] = first_step(U)
```

Record $u_{i,j}(t_n)$ for $n \geq 2$

```
for i in np.arange(2, len(T)):
    Usnap[:,:,i] = next_step(Usnap[:,:,i-1],
                             Usnap[:,:,i-2])
```

Display every 10th time step in each frame
for the sake of speed (reduce fskip for a
smoother animation)

```
fskip = 10
frames = Usnap[:,:, 0::fskip]

# Code continues on the next page
```

wave.ipynb (continued)	
Plotting commands for the ith frame	```def plot_ith_frame(i):```
	``` ax.set_xlim([0,1])```
	``` ax.set_ylim([0,1])```
	``` ax.set_zlim([-1,1])```
	``` ax.set_xlabel('x')```
	``` ax.set_ylabel('y')```
	``` ax.set_zlabel('z')```
Plot 3D surface in the ith frame	``` surf = (ax.plot_surface(X,Y,frames[:,:,i],```
	``` cmap='Spectral', vmax = 1, vmin = -1))```
Display t value on the plot	``` ax.text(0,0,1.6, f't= {i*fskip*h:.3f} s')```
	``` return surf```
	```fig = plt.figure()```
	```ax = fig.add_subplot(111, projection = '3d')```
Plot the surface at $t = 0$	```surf = plot_ith_frame(0)```
Add colour bar on the right	```fig.colorbar(surf, pad=0.1)```
Animate each frame	```def animate(i):```
	``` ax.clear()```
	``` surf = plot_ith_frame(i)```
	``` return surf```
	```ani = FuncAnimation(fig, animate,```
How many frames?	``` frames = frames.shape[2])```
	```plt.show()```

4.11 Exercises

1 (*Solving ODEs numerically*)

   a. Use separation of variables to show that IVP

$$y'(t) = \frac{y}{t(t+1)}, \quad y(1) = 2, \tag{4.63}$$

    has exact solution $y = 4t/(t+1)$.
    Now solve the IVP numerically using the forward-Euler, Heun's and RK4 methods
    with $h = 0.01$. Obtain the values of the absolute error $|y_{\text{exact}} - y_{\text{numerical}}|$ at $t = 2$.
    Which method gives the most accurate answer?

   b. Use integrating factor to show that the IVP

$$w'(t) = w \tan t - \sec t, \quad w(0) = 1,$$

    has exact solution $w = (1-t) \sec t$.
    Solve the IVP numerically using 3 methods in part (a) with $h = 0.01$. Obtain
    the absolute error of the solution at $t = 1.57$. You should find that the numerical
    solutions are much less accurate than in part (a). Why?

2 (*Scaling of $E(h)$*) Consider the IVP

$$y'(t) = -y, \quad \text{with } y(0) = 2$$

(see §4.3a).

   a. Solve the IVP numerically using the forward-Euler, Heun's and RK4 methods using
    step size $h = 0.01$ to obtain the solution y at $t = 3$.

   b. Vary h and plot the absolute error

$$E(h) = |y_{\text{exact}}(3) - y_{\text{numerical}}(3)|.$$

    You should obtain something similar to fig. 4.22.
    Calculate the gradient of the lines and hence verify that $E(h)$ scales like h, h^2 and
    h^4 for the three methods.
    Suggestion: Combine `Eh.ipynb` (§2.2) and `odesolver.ipynb` (§4.3). Vary the
    number of subintervals N in $[0, 3]$, then calculate the corresponding step size.
    Why do we see the wiggles in the graph when using the RK4 method?

3 (*Third-order ODE*) Consider the following IVP

$$y''' + y'' - y' - y = 0, \quad \text{with } y(0) = 7, \ y'(0) = -3, \ y''(0) = C.$$

Solve the IVP with SciPy's `solve_ivp` (see code at the end of §4.3). Hence, produce a
plot of $y(t)$ for $t \in [0, 5]$ for a few values of C on the same set of axes.
Suggestion: Start by writing the third-order ODE a system of three first-order ODEs.
Find C such that $y(t) \to 0$ as $t \to \infty$.
(Note: This problem could also be solved analytically using the auxiliary-equation
method.)

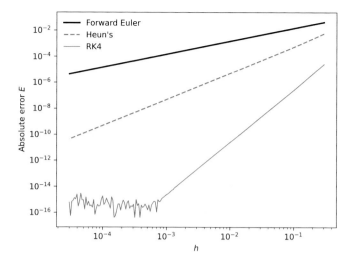

Fig. 4.22: Scaling of the absolute error $E(h)$ as a function of step size h for three ODE solving schemes, namely, forward Euler, Heun's and RK4.

4 Reproduce figure 4.7.

Suggestion: Use the command `bbox_to_anchor(x,y)` to place the legend outside the plot by adjusting x and y.

5 (*Satellite orbits*) Let $(x(t), y(t))$ describe the 2-dimensional trajectory of a satellite orbiting a planet. Assume the planet is stationary at the origin $(0, 0)$. Newtonian mechanics tells us that, if the mass of the satellite is negligible compared to that of the planet, then the orbit satisfies the following coupled second-order ODEs:

$$\frac{d^2x}{dt^2} = -\frac{Mgx}{(x^2 + y^2)^{3/2}},$$

$$\frac{d^2y}{dt^2} = -\frac{Mgy}{(x^2 + y^2)^{3/2}},$$

where M is the mass of the planet and g its gravitational acceleration. Note that the mass of the satellite does not appear in these equations.

a. Set $Mg = 3$ and set the initial conditions $(x, x', y, y') = (2, 0, 0, -1)$. Produce an animation of the orbit up to $t = 100$ by solving the ODEs using the RK4 method with $h = 0.1$.

b. From Newton's theory of gravitation, it can be shown that all possible satellite orbits are *conic sections* (*i.e.* ellipses, hyperbolae or straight lines). Which shape is your orbit in part (a)? Try changing the initial conditions to produce the other kinds of orbit.

c. Re-calculate the orbit in part (a) using the forward Euler and Heun's methods with $h = 0.1$. Are the resulting orbits physically viable?

6 (*Damped pendulum*) A pendulum is set up as in §4.4, with an addition of friction (*e.g.* due to air resistance). Suppose that the friction is proportional to the angular velocity of the pendulum, the ODE for the angular displacement $\theta(t)$ is

$$\theta'' + \frac{b}{m}\theta' + \frac{g}{\ell}\sin\theta = 0, \tag{4.64}$$

where m is the mass of the bob (in kg) and the constant b is the *damping coefficient*.

a. Let $\Omega := \sqrt{\left|\frac{b^2}{4m^2} - \frac{g}{\ell}\right|}$.
   Using the small-angle approximation ($\sin\theta \approx \theta$) and the method of auxiliary equation, show that the approximate solution is given by

$$\theta_{\text{approx}}(t) = \begin{cases} e^{-bt/2m}(A\cosh\Omega t + B\sinh\Omega t), & \text{if } \Omega^2 > 0, \\ e^{-bt/2m}(A + Bt), & \text{if } \Omega^2 = 0, \\ e^{-bt/2m}(A\cos\Omega t + B\sin\Omega t), & \text{if } \Omega^2 < 0. \end{cases}$$

   where A and B are constants to be determined.

b. Suppose that $\theta(0) = 0$ rad, $\theta'(0) = 2$ rad/s and $\ell = 1$m.
   Using the code `pendulum.ipynb` as a starting point, plot $\theta(t)$ from $t = 0$ to 5 s assuming that

$$\frac{b}{m} = 0.5\sqrt{g}, \ 2\sqrt{g} \text{ and } 3\sqrt{g}.$$

   Plot these 3 curves on the same set of axes.

c. For the given the initial conditions in part (b), obtain $\theta_{\text{approx}}(t)$ by showing that $A = 0$, and that

$$B = \begin{cases} \theta'(0) & \text{if } \Omega = 0, \\ \theta'(0)/\Omega, & \text{otherwise.} \end{cases}$$

d. For each of the 3 values of b/m in part (b), plot the approximation and the numerical solution on the same set of axes. Do the approximations appear to be accurate? (You may like to quantify the accuracy.)
   Verify that as the initial angular velocity $\theta'(0)$ increases, the small-angle approximation becomes less accurate.

e. The 3 values of b/m in part (b) give rise to an *underdamped, overdamped* or *critically damped* pendulum. Explain which case is which, and suggest why they are given these names.

7 (*The upside-down pendulum*) Use the code `pendulum.ipynb` as a starting point for the following questions.
   Watch this video[7].
   The video shows a pendulum made from a rod of length ℓ attached to a pivot that moves up and down. Let θ be the angle that the rod makes with the vertical (*i.e.* $\theta = 0$ when the rod points downward). The goal of this question is to investigate the situation in which the upside-down pendulum ('inverted') pendulum becomes stable.
   Suppose that at time t, the pivot's position is at

$$y_{\text{pivot}} = A\cos(\omega t),$$

---

[7] https://www.youtube.com/watch?v=5oGYCxkgnHQ

where A and ω are constants that determine the amplitude and frequency of the pivot's oscillation. Intuitively, it is clear that if $A = 0$ (when the pivot does not move), the downward pendulum $\theta = 0$ is a stable equilibrium, and the inverted pendulum $\theta = \pi$ is an unstable equilibrium.

When the pivot starts moving up and down, it can be shown that $\theta(t)$ satisfies the ODE:

$$\theta'' + \left(\frac{g}{\ell} - \frac{\omega^2 A}{\ell} \cos \omega t \right) \sin \theta = 0.$$

In this question, let $\ell = 0.5$ m, $g = 9.8$ m/s^2. Use the initial condition $\theta(0) = 3.1$.

    a. Produce an animation of the upside-down pendulum.

    b. Set $A = 0$ to make sure that the pendulum behaves as expected.

    c. Adjust A, ω and $\theta'(0)$ so that the pendulum is stable in the upside-down position.

8 (*Exploring the double pendulum further*) Use the code `doublependulum.ipynb` as a starting point for each of these questions.

    a. Set the double pendulum to be initially perfectly upside down, *i.e.* set $\theta_1(0) = \theta_2(0) = \pi$ and $\theta_1'(0) = \theta_2'(0) = 0$ (assume the bobs are connected by massless rods). This is an unstable equilibrium. Nonetheless, in theory we should expect no movement at all. Does the animation support this prediction? If not, explain why.

    b. Visually demonstrate the sensitivity of the double pendulum to initial conditions by producing an animation of 2 double pendulums (on the same set of axes) using almost the same initial conditions.
       The animation should show that the pendulums evolve almost identically up to a certain time when they suddenly become wildly different.

    c. Here is another way to visualise the sensitivity of the double pendulum to initial conditions. Using the initial conditions $(\theta_1, \theta_1', \theta_2, \theta_2') = (2 + \varepsilon, 0, 1, 0.1)$ where $\varepsilon = 0$, 10^{-3} and 10^{-5} (as in fig. 4.9), plot the trajectories in the (θ_1, θ_2) phase space from $t = 0$ to 30s.
       The plot for $\varepsilon = 10^{-3}$ is shown in fig. 4.23. Your other plots should look very different. The dramatic changes in these plots as a result of tiny changes in ε demonstrates the chaotic nature of the double pendulum.

    d. How do we make the double pendulum *more chaotic*? For example, how does increasing the ratio m_1/m_2 affect the behaviour of the double pendulum? Does it become 'more' or 'less' chaotic? Make a conjecture and verify it with the animation. (One could actually measure how chaotic a system is using the Lyapunov exponent λ, but you are not asked to calculate λ in this question.)

    e. Investigate the sensitivity of the evolution of the double pendulum to the choice of step size h. You may like to plot the trajectory in the (θ_1, θ_2) plane for various choices of h.

9 (*Exploring the Lorenz equations further*) Use the code `lorenz.ipynb` as a starting point for each of these questions.

    a. Use the animation to verify the behaviour of trajectories given the values of r listed in the following table, keeping $\sigma = 10$ and $b = 8/3$. (Table adapted from [6].) Use the animation to explain what the phrase '*transient chaos*' means.

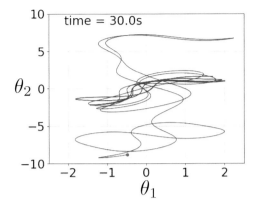

Fig. 4.23: A trajectory of the double pendulum in the θ_1-θ_2 phase space.

$0 < r < 1$	The origin is an attractor.
$1 < r < 13.93$	$\mathbf{x}^{\pm}$ are attractors.
$13.93 < r < 24.06$	'Transient chaos'.
$r > 24.06$	A strange attractor appears.
$r > 313$	Periodic orbits exist.

b. Plot fig. 4.12, which shows $x(t)$ for two trajectories with almost identical initial conditions.

c. Create a 3D animation of the two trajectories in fig. 4.12 evolving simultaneously on the same set of (x, y, z) axes.

d. Does the numerical method used to solve the Lorenz equations matter?
Try solving the Lorenz equations with two different methods in `scipy.solve_ivp` (say, RK45 and DOP853 – type `help(solve_ivp)` to see other methods). Plot $x(t)$ (say) arising from these methods on the same diagram.

10 (*Exploring the logistic map further*) Use the code `logistic.ipynb` as a starting point for each of these questions.

a. Plot the bottom right panel of fig. 4.13. Verify that chaotic behaviour occurs regardless of the initial value $x_0 \in [0, 1]$.

b. Produce a fly-through animation which zooms into the bifurcation diagram of the logistic equation, showing its fractal structure.
Suggestion: start with the window $[3.84, 3.855] \times [0.45, 0.54]$ and end with the window $[3.85213, 3.85219] \times [0.4646, 0.4664]$. Do linear interpolations of the endpoints and plot the bifurcation diagram in each frame. Use fewer frames to start off with and increase the number of frames when you are confident that the code works. Save your animation as a *.mp4* file.

c. By examining the bifurcation diagram, or otherwise, give an estimate of r which produces a period-5 orbit.

d. Consider the map $x_{n+1} = r - x_n^2$ where x_0 is drawn randomly from $[-2, 2]$ and $r \in [0, 2]$ (this is a variation on the logistic map).

Plot the bifurcation diagram and locate the first 4 bifurcation points b_i. Calculate the ratio

$$\frac{b_{n-1} - b_{n-2}}{b_n - b_{n-1}}.$$

(You should obtain a number close to the Feigenbaum constant.)

11 (*SIR model in epidemiology*)

The spread of a disease in a population can be modelled using the famous *SIR* model, where S, I, and R are all functions of time t (in days) and take values in $[0, 1]$.

- $S(t)$ is the fraction of the population who are *susceptible* to the disease (but not yet infected),
- $I(t)$ is the fraction of the population who are *infectious* to others,
- $R(t)$ is the fraction of the population who have *recovered* and are now immune to the disease.

The ODEs for the SIR model are:

$$\frac{dS}{dt} = -\beta SI,$$

$$\frac{dI}{dt} = \beta SI - \gamma I,$$

$$\frac{dR}{dt} = \gamma I,$$

where the constant β is called the *contact rate* (the transition rate from susceptible to infectious), and γ is the *recovery rate* (the transition rate from infectious to recovered). The following simple diagram summarises this model.

$$S \xrightarrow{\beta} I \xrightarrow{\gamma} R.$$

Summing the ODEs tells us that $(S + I + R)' = 0$, meaning that $S + I + R$ is constant at all times. We shall set $S + I + R = 1$. We say that the population is *compartmentalised* into these 3 categories

In this question, let $\gamma = 0.1$, $I(0) = 10^{-3}$, $R(0) = 0$. We take $\beta = 0.2$ for now, but we will vary it in part (c).

a. Use `solve_ivp` to solve the system. Hence, on the same set of axes, plot S, I, R as functions of $t \in [0, 100]$.

b. Plot the trajectory showing how the system evolves in the (I, S) phase space. In other words, plot S against I. Furthermore,

i. Determine whether the trajectory travel up or down in this plot.

ii. From the graph, write down the value of S where the trajectory has a vertical tangent. Explain why this value corresponds to γ/β.

c. Add a slider for β (ranging from 0 to 1) to part (a). Hence, answer the following questions.

i. Describe the long-term behaviours of S, I and R when the contact rate is very high (≈ 1) or very low (≈ 0). Do these behaviours reflect what might happen in the real world?

ii. When $\beta = 1$, what fraction of the population was infected at the peak of the disease outbreak? When did this peak occur? (Answer: 67%, occurring on day 10.)

For further reading on mathematical epidemiology see [32] and [120] (the latter is an accessible introduction to the mathematical modelling of the spread of COVID-19).

12 (*Exploring fractals further*)

a. Use the code `mandelbrot.ipynb` as a starting point for this question.
Replace the map $f_c(z)$ in eq. 4.31 by the following maps and plot the resulting fractals. You may assume that the escape criterion (Lemma 2) still applies.
   i. ('*Burning ship*' [145])

$$f_c(z) = (|\mathrm{Re}(z)| + i|\mathrm{Im}(z)|)^2 + c, \quad z_0 = 0.$$

   ii. (*Tricorn* and *Multicorns* [51]):

$$f_c(z) = \bar{z}^k + c, \quad z_0 = 0,$$

   where $k = 2, 3, 4 \ldots$
   iii. (*The 'Multibrot' set*)

$$f_c(z) = z^k + c \quad z_0 = 0,$$

   where $k = 3, 4, 5 \ldots$
   In each case, zoom into the image and explore its fractal structure.
   Suggestion: You may need to do more than just changing the iterative map in the code. In particular, some of these fractals may not have the same symmetries as the Mandelbrot set. Also, to speed up the investigation, start with a relatively poor resolution to figure out the best window to plot.

b. By modifying `mandelbrot3D.ipynb`, plot figure 4.19.
Suggestion: Use the code snippet below to stack the two figures on top of each other without any space between.

```
fig = plt.figure()
ax1 = fig.add_subplot(211)
...

ax2 = fig.add_subplot(212)
...

plt.subplots_adjust(hspace=0)
plt.setp(ax1.get_xticklabels(), visible=False)
yticks = ax2.yaxis.get_major_ticks()
yticks[-1].label1.set_visible(False)
```

c. (*Julia set*) In the complex map f_c (eq. 4.31), let's take an alternative view: instead of fixing z_0 and vary c, let's fix c and vary z_0.
More precisely, we want to investigate orbits of $z_0 \in \mathbb{C}$ under the iterated map $f_c(z)$. An orbit either escapes to infinity or remains bounded. The boundary between these two behaviours is called the *Julia set* of f_c. Gaston Julia (1893–1978) along with fellow French mathematicians *Pierre Fatou* (1878–1929) are regarded pioneers in the field of dynamical systems.

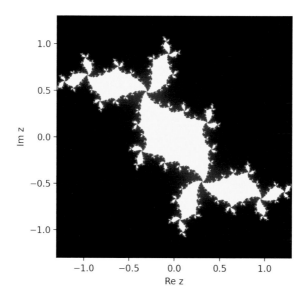

Fig. 4.24: *Douady's Rabbit*. The boundary between the light and dark regions is the Julia set of $f_c(z) = z^2 + c$ with $c = -0.17 + 0.78i$.

The Julia set for $c = -0.17 + 0.78i$ is shown in fig. 4.24.
  i. Reproduce this figure and investigate the Julia set for other values of c.

  To plot Julia sets, you only need to perform very few modifications to `mendelbrot.ipynb`. Here are some pointers.
    • Fix c at the start of the code.
    • For each z_0, iterate the map f_c while checking that $|z_n|^2 < 1000$ (say).
    • Exploit the symmetry $z \to -z$ under the map f_c. You only need to check points in the region $\operatorname{Im} z \geq 0$. To obtain the image for $\operatorname{Im} z \leq 0$, flip the image left-to-right and upside down.

  ii. Produce the Julia sets for other values of c. What can you conclude about the Julia sets for $c \in \mathcal{M}$ and $c \notin \mathcal{M}$? (*Hint:* think in terms of *connectedness*.)

13 (*Exploring the heat equation further*) Use the code `heat.ipynb` as a starting point for each of these questions.

    a. Solve the 1D heat equation 4.3 using the initial condition $u(x, 0) = \sin(5\pi x/2)$ with the following boundary conditions. Produce an animation of $u(x, t)$ where $t \in [0, 0.5]$ and $x \in [0, 1]$.

        i. (*variable left endpoint*) $u(0, t) = \sin(20t)$, $u(1, t) = 1$.

        ii. (*variable endpoints*) $u(0, t) = \sin(20t)$, $u(1, t) = \cos(10t)$.

        iii. (*Neumann boundary condition*) $\frac{\partial u}{\partial x}(0, t) = \frac{\partial u}{\partial x}(1, t) = 0$.

    Suggestion: Only a few modifications to the function `derivs` in `heat.ipynb` are required to include data at the endpoints. Symmetric-difference formulae for x derivatives cannot be applied at the endpoints, so use forward or backward-difference formulae instead. Your animation of $u(x, t)$ should not contain any kinks or discontinuities.

    b. (*Heat equation with a source term*) If at $t > 0$, the entire rod continues to be heated (or cooled), the heat equation becomes

$$\frac{\partial u}{\partial t} = \frac{\partial^2 u}{\partial x^2} + q(x),$$

    where $q(x)$ is called the *source term*, where $q > 0$ corresponds to heating and $q < 0$ cooling. Suppose that for $x \in [0, 1]$, the source term is given by:

$$q(x) = x(1 - x).$$

    Using the initial condition $u(x, 0) = \sin(5\pi x/2)$ and boundary conditions $u(0, t) = 0$ and $u(1, t) = 1$, produce an animation of $u(x, t)$ up to $t = 1$.

    Obtain an expression for the steady-state temperature distribution (by solving the PDE and ignoring the time derivative). Plot the curve representing the steady state temperature on your animation. You should see the time-dependent solution approaching this curve.

    c. For the original heat-equation problem posed in §4.9, compare the series (exact) solution 4.53 to the numerical solution at some fixed time (say, $t = 0.1$). Plot them on the same set of axes, noting the number terms needed for the series solution to resemble the numerical solution. Obviously this will depend on your choices of Δx and h.

    Quantify the accuracy of the numerical solution by plotting the fractional difference.

    d. Solve the discretised heat equation (4.42)-(4.44) again but, instead of using `solve_ivp`, use the forward-difference method.

    Explore different choices of Δx and Δt. What happens when the stability condition (4.54) is not satisfied?

14 (*Exploring the wave equation further*) Use the code `wave.ipynb` as a starting point for each of these questions.

    a. Reduce the dimensionality of code `wave.ipynb` so that it solves the following one-dimensional wave equation:

$$\frac{\partial^2 u}{\partial t^2} = \frac{\partial^2 u}{\partial x^2}, \quad x \in (0, 1),$$

with initial conditions $u(x, 0) = \sin(2\pi x),$

$$\frac{\partial u}{\partial t}(x, 0) = 0,$$

and boundary condition $u(1, t) = u(0, t) = 0.$

This involves going through the code and deleting the y dimension. When done correctly, your code should produce a 2D animation of a string vibrating in a normal mode.

Repeat the animation with the initial triangular waveform

$$u(x, 0) = 1 - |2x - 1|.$$

    b. (*Non-trivial boundary condition*) Suppose the vibrating membrane modelled by the 2D wave equation (4.55) is forced to move on one side, so that the boundary condition (BC) becomes:

$$u(x, y, t) = \begin{cases} \sin(2t)\sin(2\pi x) & \text{if } y = 0, \\ 0 & \text{if } y \neq 0 \text{ and } (x, y) \in \partial S. \end{cases}$$

Produce an animation of the vibrating membrane for $t \in [0, 3]$. Suggestion: This involves a minor modification of the function `next_step` in the code.

    c. Modify the animation to produce an evolving heatmap (see §3.9) instead of a vibrating 3D surface.

    d. For the original wave-equation problem posed in §4.10, compare the numerical solution of the wave equation against the separable solution (4.62). Do this by plotting the difference between them and observing the maximum values. (This depends on your choices of Δx, Δy and h – watch out for potential instability.)

# CHAPTER FIVE

Linear Algebra

Fig. 5.1: *Seki Takakazu* (ca.1640–1708), a Japanese scholar whose many mathematical discoveries preceded those in the western world. He was the first person to study matrix determinants and use them to solve linear systems ten years before the same discoveries were made by Leibniz. (Image source: [214].)

Linear algebra introduces students to abstract concepts associated with vectors and matrices. Some students may have already worked with matrices at school (some may even know how to calculate a 3×3 determinant), but it is only at university that matrices are viewed as objects in a *vector space*. Students will discover that a matrix is not just a grid of numbers, but has associated with it structures such as *row space*, *column space* and *nullspace*. Each matrix can also be viewed as a representation of a *linear transformation* which has a geometric interpretation (*e.g.* rotation, magnification, shear or a combination of transformations). In this chapter, we will see how these abstract structures and transformations can be understood and visualised with the help of Python.

Linear algebra has been studied for millennia, mostly in the context of solving systems of linear equations with many unknowns (the first matrix appeared in China as early as 200BC – see fig. 5.2). The subject was given a modern synthesis in the seminal book by Birkhoff and Mac Lane [26] (and its subsequent editions [27]). Today, linear algebra plays a key role

S. Chongchitnan, *Exploring University Mathematics with Python*,
215

in the development of machine learning and data science – see [2, 195]. An application to image reduction will be explored in this chapter (§5.8).

There are many good introductory textbooks on linear algebra. Some notable modern references are [14, 104, 125] and, in particular, [194].

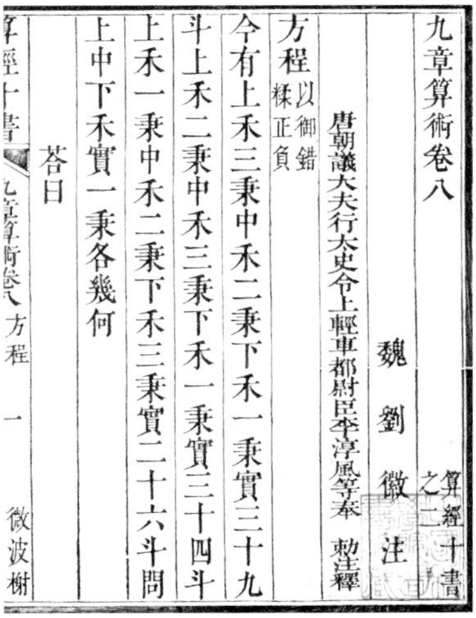
九章算術卷八
算經十書之二
魏 劉 徽 注
唐朝議大夫行太史令上輕車都尉臣李淳風等奉 勅注釋
方程 以御錯糅正負
今有上禾三秉中禾二秉下禾一秉實三十九
斗上禾二秉中禾三秉下禾一秉實三十四斗
上禾一秉中禾二秉下禾三秉實二十六斗問
上中下禾實一秉各幾何
荅曰
算經十書 九章算術卷八 方程 一 微波榭

Fig. 5.2: A page from a reprint of the *Nine Chapters of Mathematical Art*, an anonymous Chinese text originating from 200-100BC. This page describes a problem related to grain production involving 3 unknowns. The translation and solution are discussed in this chapter – see eq. 5.2. (Image sources: [221])

5.1 *Basics of SymPy*

One of the main tools we will use in this chapter is *SymPy*, a Python library for symbolic computing. It can help us solve fiddly algebraic equations and simplify complicated expressions, although, as we will see later, the user must be vigilant when interpreting SymPy's results. The SymPy website has an excellent introductory guide[1], including common mistakes and pitfalls (essential reading for all new users). SymPy has a large number of linear-algebra functions, some of which will be used in this chapter when we require symbolic results or exact expressions.

Another tool for computational linear algebra is SciPy's `linalg` library (`scipy.linalg`). We will use this when numerical (rather than symbolic) answers are required (especially when numpy arrays are needed for plotting). NumPy also has a *linalg* library (`numpy.linalg`). However, *scipy.linalg* contains all the functionality of *numpy.linalg*, so it is recommended that you always use *scipy.linalg*.

It is worth mentioning other popular alternatives for numerical linear algebra that you can use for free. (Their uses are not only limited to linear algebra.)

[1] `https://docs.sympy.org/latest/tutorials/intro-tutorial/index.html`

- *Julia* (https://julialang.org)
- *Sage* (https://www.sagemath.org)
- *GNU Octave* (https://octave.org)

In the next section, we will give a quick summary of essential linear algebra and show how both the numerical method (using NumPy and SciPy) and symbolic method (using SymPy) can be used to demonstrate those concepts.

5.2 Basic concepts in linear algebra

All code snippets in this section are to begin with the following lines:

```
import numpy as np
import scipy.linalg as LA
import sympy as sp
```

Matrices

Matrices and basic operations	
Defining a matrix Numerical : $A = \begin{pmatrix} 1 & 2 \\ 3 & 4 \end{pmatrix}$	`A = np.array([[1,2], [3,4]])`
Symbolic : $B = \begin{pmatrix} a & b \\ c & d \end{pmatrix}$	`a,b,c,d = sp.symbols("a,b,c,d")` `B = sp.Matrix([[a,b], [c,d]])`
Identity and zero matrices $I = \begin{pmatrix} 1 & 0 \\ 0 & 1 \end{pmatrix}$ and $Z = \begin{pmatrix} 0 & 0 & 0 \\ 0 & 0 & 0 \end{pmatrix}$	
Numerical	`I = np.eye(2)` `Z = np.zeros((2,3))`
Symbolic	`I = sp.eye(2)` `Z = sp.zeros(2,3)`
Multiplying two matrices Numerical: A^2 (Be careful! Both A*A and A**2 mean squaring elementwise)	`A@A`
Symbolic: B^2	`B*B    # or B**2 or B@B`
Transposing Numerical: A^T	`A.T    # or np.transpose(A)`
Symbolic: B^T	`B.T`
Finding the determinant Numerical: $\det A = -2$	`LA.det(A)`
Symbolic: $\det B = ad - bc$	`B.det()`
Finding the inverse Numerical: A^{-1}	`LA.inv(A)`
Symbolic: B^{-1}	`B**-1`

Here are a few important pointers for the above concepts.

- **Matrix multiplication** is associative, meaning that

$$(AB)C = A(BC).$$

But matrix multiplication is not commutative, *i.e.* AB is not always equal to BA.

- The **transpose** of A (denoted A^T) is a matrix in which the rows/columns of A become the columns/rows of A^T (*i.e.* rows and columns are switched). The transpose of AB is not $A^T B^T$. In fact, the order is reversed:

$$(AB)^T = B^T A^T.$$

- The **determinant** of a 2×2 matrix is defined as

$$\begin{vmatrix} a & b \\ c & d \end{vmatrix} = ad - bc.$$

This comes from the expression of the area of a parallelogram – see the Discussion in §5.5.

The determinant of a 3×3 matrix is

$$\begin{vmatrix} a & b & c \\ d & e & f \\ g & h & i \end{vmatrix} = a \begin{vmatrix} e & f \\ h & i \end{vmatrix} - b \begin{vmatrix} d & f \\ g & i \end{vmatrix} + c \begin{vmatrix} d & e \\ g & h \end{vmatrix}$$

$$= a(ei - fh) - b(di - fg) + c(dh - eg)$$

when expanded by the first row, although any row or column can also be used. The determinant of a matrix A may be written as $|A|$ or $\det A$.

- The **inverse** of A is the matrix A^{-1} such that $AA^{-1} = A^{-1}A = I$. To find the inverse of AB, the same order-reversal rule as the transpose applies:

$$(AB)^{-1} = B^{-1}A^{-1}.$$

The inverse of a matrix exists if and only if its determinant is nonzero, in which case the matrix is said to be *invertible*. For most numerical purposes, finding an explicit expression for the inverse of a matrix is usually avoidable. We explain how and why in §5.4.

Row reduction and the RREF

Given a matrix, one can perform row operations to reduce it to a simpler form. These row operations (also called *Gaussian elimination* despite its use long before Gauss) are:

- Multiplying row i by a constant c (denoted $R_i : cR_i$)
- Swapping row i with row j (denoted $R_i \leftrightarrow R_j$)
- Adding c times row i to row j (denoted $R_j : R_j + cR_i$)

These are normally performed to reduce a matrix to *row-echelon form* (REF), meaning that:

1. Rows that are entirely zeros are the last rows of the matrix.
2. The leading entry (*i.e.* the first nonzero entry) in each row is to the right of the leading entry of the row above.

In addition, if all the leading entries are 1, and each leading 1 is the only nonzero entry in its column, then the matrix is said to be in *reduced row-echelon form* (RREF). ,

The following example shows how a 3×3 matrix A can be reduced to row-echelon form.

$$A = \begin{pmatrix} 1 & 2 & -1 \\ 2 & 1 & -2 \\ -3 & 1 & 1 \end{pmatrix} \xrightarrow[R_3:R_3+3R_1]{R_2:R_2-2R_1} \begin{pmatrix} 1 & 2 & -1 \\ 0 & -3 & 0 \\ 0 & 7 & -2 \end{pmatrix} \xrightarrow{R_3:R_3+\frac{7}{3}R_2} \begin{pmatrix} 1 & 2 & -1 \\ 0 & -3 & 0 \\ 0 & 0 & -2 \end{pmatrix}. \quad \text{(REF)}$$

One can go further and reduce it to reduced row-echelon form

$$\begin{pmatrix} 1 & 2 & -1 \\ 0 & -3 & 0 \\ 0 & 0 & -2 \end{pmatrix} \xrightarrow[R_3:-\frac{1}{2}R_3]{R_2:-\frac{1}{3}R_2} \begin{pmatrix} 1 & 2 & -1 \\ 0 & 1 & 0 \\ 0 & 0 & 1 \end{pmatrix} \xrightarrow[R_1:R_1+R_3]{R_1:R_1-2R_2} \begin{pmatrix} 1 & 0 & 0 \\ 0 & 1 & 0 \\ 0 & 0 & 1 \end{pmatrix} \quad \text{(RREF)}$$

We see that the RREF of the original matrix is simply the 3×3 identity matrix. (In fact, the RREF of any invertible matrix is the identity matrix.)

To obtain the RREF of the above matrix in Python, we use SymPy.

Finding the reduced row-echelon form of a matrix with SymPy

```
A = sp.Matrix([[1, 2, -1], [2, 1, 2], [-3, 1, 1]])
A.rref()[0]
```

The function `rref` gives two outputs: the first is the rref of the matrix, the second is a tuple containing the indices of columns in which the leading entries appear.

Reducing a matrix to its REF or RREF can be expressed as a sequence of matrix multiplications on A. For example, swapping the first two rows of a 3×3 matrix A can be written as $L_1 A$ where

$$L_1 = \begin{pmatrix} 0 & 1 & 0 \\ 1 & 0 & 0 \\ 0 & 0 & 1 \end{pmatrix}.$$

Similarly, the row operation $R_3 : R_3 - 7R_2$ performed on A can be expressed as $L_2 A$ where

$$L_2 = \begin{pmatrix} 1 & 0 & 0 \\ 0 & 1 & 0 \\ 0 & -7 & 1 \end{pmatrix}.$$

We will see why these matrices are important when we discuss LU decomposition in §5.4.

Solving linear systems

One of the most important topics in introductory linear algebra is the understanding of the solution(s) to the linear system

$$A\mathbf{x} = \mathbf{b}. \quad (5.1)$$

Here A is an $m \times n$ matrix, $\mathbf{x} = (x_1, x_2, \ldots x_n)^T$ and $\mathbf{b} \in \mathbb{R}^n$. Eq. 5.1 represents a system of m linear equations with n unknowns. We want to understand when the system permits a unique solution, more than one solution (how many?), or no solutions at all. This will be discussed in §5.3.

Four methods of solving $A\mathbf{x} = \mathbf{b}$ (by hand) will be discussed in §5.4. For now, let's see how Python can help us solve a linear system that appeared in the legendary Chinese text *Nine Chapters of Mathematical Art* (*jiǔ zhāng suàn shù*) written anonymously in circa 200–100 BC. Below is the first problem posed in Chapter 8. The translation is adapted from [122].

> "3 bundles of top grade cereal, 2 bundles of medium grade cereal and 1 bundle of low grade cereal yield 39 *dŏu* of grains.
>
> 2 bundles of top grade cereal, 3 bundles of medium grade cereal and 1 bundle of low grade cereal yield 34 *dŏu*.
>
> 1 bundle of top grade cereal, 2 bundles of medium grade cereal and 3 bundles of low grade cereal yield 26 *dŏu*.
>
> How many *dŏu* of grains can be obtained from each bundle of the top, medium and low grade cereal?"

(A *dŏu* is approximately 10 litres.)

Let x, y, z be the number of *dŏu* of grains that can be obtained from a bundle of top, medium and low grade cereal respectively. The question then translates to the system:

$$
\begin{aligned}
3x + 2y + z &= 39 \\
2x + 3y + z &= 34 \\
x + 2y + 3z &= 26
\end{aligned}
\implies
\begin{pmatrix} 3 & 2 & 1 \\ 2 & 3 & 1 \\ 1 & 2 & 3 \end{pmatrix}
\begin{pmatrix} x \\ y \\ z \end{pmatrix}
=
\begin{pmatrix} 39 \\ 34 \\ 26 \end{pmatrix}.
\tag{5.2}
$$

Solving the linear system (5.2)	

Numerical
Answer: $(9.25, 4.25, 2.75)$
	```A = np.array([[3,2,1],[2,3,1],[1,2,3]])```
	```b = np.array([39, 34, 26])```
	```LA.solve(A,b)```

Symbolic
Answer: $\left(\dfrac{37}{4}, \dfrac{17}{4}, \dfrac{11}{4} \right)$
	```x,y,z = sp.symbols("x,y,z")```
	```A = sp.Matrix([[3,2,1],[2,3,1],[1,2,3]])```
	```b = sp.Matrix([39, 34, 26])```
	```sp.linsolve((A, b), x, y, z)```

In the *Nine Chapters*, the solution to this problem is obtained essentially by row reduction (although columns are used instead). It is widely accepted that this was the earliest explicit use of a matrix to solve a mathematical problem.

More advanced concepts and definitions

Here we collect a non-exhaustive list of definitions that are useful in introductory linear algebra.

Definition of a vector space

A **vector space** V is a set containing objects (called *vectors*) that can be added together or multiplied by real numbers. Take any $\mathbf{u}, \mathbf{v}, \mathbf{w} \in V$ and $a, b \in \mathbb{R}$. The following are axioms must be satisfied.

- (Closure under addition and scalar multiplication) $\mathbf{u} + \mathbf{v} \in V$ and $a\mathbf{u} \in V$.
- (Commutativity of addition) $\mathbf{u} + \mathbf{v} = \mathbf{v} + \mathbf{u}$.
- (Associativity of addition) $(\mathbf{u} + \mathbf{v}) + \mathbf{w} = \mathbf{u} + (\mathbf{v} + \mathbf{w})$.
- (Associativity of scalar multiplication) $a(b\mathbf{u}) = (ab)\mathbf{u}$.
- (Existence of identity) $\exists \mathbf{0} \in V$ such that $\forall \mathbf{u} \in V, \mathbf{u} + \mathbf{0} = \mathbf{u}$.
- (Existence of additive inverse) $\forall \mathbf{u} \in V, \exists (-\mathbf{u}) \in V$ such that $\mathbf{u} + (-\mathbf{u}) = \mathbf{0}$.
- (Distributivity over vector addition) $a(\mathbf{u} + \mathbf{v}) = a\mathbf{u} + a\mathbf{v}$.
- (Distributivity over scalar addition) $(a + b)\mathbf{u} = a\mathbf{u} + b\mathbf{u}$.
- (Scalar identity) $1\mathbf{u} = \mathbf{u}$.

Linear algebra can be regarded, in the broadest terms, as the study of vector spaces. You are by now familiar with with vectors in $\mathbb{R}^n$, which is a vector space containing vectors which are n-tuples $\mathbf{x} = (x_1, x_2, \ldots x_n)$. There are less obvious examples of vector spaces, such as $\mathcal{P}_n$ – the set of all polynomials with degree at most n – which we will study in §5.10 (the Gram-Schmidt process).

Definitions related to vectors and vector spaces

- **Linear independence** The set $\{\mathbf{v}_1, \mathbf{v}_2, \ldots \mathbf{v}_n\}$ is said to be linearly independent if the equation
$$c_1\mathbf{v}_1 + c_2\mathbf{v}_2 + \cdots + c_n\mathbf{v}_n = \mathbf{0}$$
only has the trivial solution $c_1 = c_2 = \ldots = c_n = 0$.
- **Linear combination** A linear combination of vectors $\mathbf{v}_1, \mathbf{v}_2, \ldots \mathbf{v}_n$ is the vector
$$c_1\mathbf{v}_1 + c_2\mathbf{v}_2 + \cdots + c_n\mathbf{v}_n$$
for some given constants $c_1, c_2, \ldots c_n$.
- **Span** The span of a set of vectors $\{\mathbf{v}_1, \mathbf{v}_2, \ldots \mathbf{v}_n\}$ is the set of all possible linear combinations of $\mathbf{v}_i$.
- **Basis** A basis of a vector space V is a linearly independent set of vectors $\{\mathbf{e}_i\}_{i=1}^n$ whose span is V.
- **Dimension** The dimension of a vector space V is the number of vectors in its basis. For example, if V has basis $\{\mathbf{e}_i\}_{i=1}^n$, then the dimension of V is n. We write $\dim V = n$.

Definitions related to matrices

Let A be an $m \times n$ matrix. Denote its rows as $\mathbf{r}_1, \mathbf{r}_2, \ldots \mathbf{r}_m$. Denote its columns as $\mathbf{c}_1, \mathbf{c}_2, \ldots \mathbf{c}_n$.

- **Row and column spaces** The row space of A is $\mathrm{span}(\mathbf{r}_1, \mathbf{r}_2, \ldots \mathbf{r}_m)$. The column space is $\mathrm{span}(\mathbf{c}_1, \mathbf{c}_2, \ldots \mathbf{c}_n)$.
- **Rank** The rank of A is the dimension of its row space (or, equivalently, its column space).

- **Nullspace** The nullspace (sometimes written *null space*) of A is the set of all solutions **x** to the equation $A\mathbf{x} = \mathbf{0}$.
- **Nullity** The nullity of A is the dimension of its nullspace.

Definitions related to linear transformations

Let $T : V \rightarrow W$ be a mapping from V to W (both are vector spaces). Suppose that for all $\mathbf{u}, \mathbf{v} \in V$ and for all $\alpha \in \mathbb{R}$, we have

$$T(\mathbf{u} + \alpha\mathbf{v}) = T(\mathbf{u}) + \alpha T(\mathbf{v}). \tag{5.3}$$

Then, T is said to be a *linear transformation*.

- **Image** The image of T is defined by

$$\mathrm{im}\, T = \{T(\mathbf{v}) : \mathbf{v} \in V\}.$$

- **Kernel** The kernel of T is defined by

$$\ker T = \{\mathbf{v} : T(\mathbf{v}) = \mathbf{0}\}.$$

In the case that $V = \mathbb{R}^n$ and $W = \mathbb{R}^m$, the linear transformation T can be represented by an $m \times n$ matrix A. In §5.5, we will visualise the geometric action of the linear transformation associated with a matrix.

In particular, $\mathrm{im}\, T$ is precisely the column space of A, and $\ker T$ is the nullspace of A. These relations will be discussed in §5.9 (the rank-nullity theorem).

Some advanced linear-algebra concepts in Python

Let $A = \begin{pmatrix} 1 & -1 & -1 \\ 0 & 0 & 1 \\ 0 & 0 & 1 \end{pmatrix}$	`A=sp.Matrix([[1,-1,-1],[0,0,1],[0,0,1]])`
Rank	`A.rank()`
Row and column spaces The results are the basis vector(s) for the row/column spaces	`A.rowspace()` `A.columnspace()`
Nullspace and nullity with SymPy Nullspace Nullity	`NullSpace=A.nullspace()` `len(NullSpace)`
Nullspace and nullity with `scipy.linalg` Nullspace Nullity	`A=np.array([[1,-1,-1],[0,0,1],[0,0,1]])` `NullSpace = LA.null_space(A)` `NullSpace.shape[1]`

The output for SymPy's `nullspace` command or SciPy's `null_space` command is the set of basis vector(s) for the nullspace.

$$\text{SymPy}: \begin{pmatrix} 1 \\ 1 \\ 0 \end{pmatrix} \qquad \text{scipy.linalg}: \begin{pmatrix} -7.07106781 \times 10^{-1} \\ -7.07106781 \times 10^{-1} \\ 8.86511593 \times 10^{-17} \end{pmatrix}.$$

In the second vector, the last entry is a sub-machine-epsilon quantity (see §2.2). Thus, the two answers agree. Also note that $-7.07106781 \times 10^{-1}$ is meant to represent $-\frac{1}{\sqrt{2}}$. This comes from the fact that the SciPy method always gives normalised basis vectors.

Invertible Matrix Theorem

A huge chunk of introductory linear algebra can be summarised by the following collection of results which link many concepts and definitions introduced in this section.

Theorem 5.1 (*Invertible Matrix Theorem*) *Let A be an $n \times n$ matrix. Let $T : \mathbb{R}^n \to \mathbb{R}^n$ be the linear transformation represented by A. The following are equivalent:*

- *A is invertible*
- *$\det A \neq 0$*
- *The RREF of A is I*
- *The columns of A are linearly independent*
- *The column space of A is $\mathbb{R}^n$*
- *The row of A are linearly independent*
- *The row space of A is $\mathbb{R}^n$*

- *rank $(A) = n$*
- *The nullspace of A is $\{\mathbf{0}\}$*
- *The nullity of A is 0*
- *$\ker T = \{\mathbf{0}\}$*
- *$\operatorname{im} T = \mathbb{R}^n$*
- *The eigenvalues of A are nonzero*
- *A has n positive singular values*

Eigenvalues and eigenvectors will be discussed in §5.6. Singular values will be discussed in §5.8.

5.3 Linear systems in $\mathbb{R}^3$

Consider the system of equations

$$x - 2z = 1$$
$$x - y + z = 1 \qquad\qquad (5.4)$$
$$2x + \alpha y - z = \beta$$

where α and β are real constants. Investigate the number of solutions for different values of α and β.

Algebraic solutions

Let's solve the system by hand. We start by writing it in matrix form:

$$\begin{pmatrix} 1 & 0 & -2 \\ 1 & -1 & 1 \\ 2 & \alpha & -1 \end{pmatrix} \begin{pmatrix} x \\ y \\ z \end{pmatrix} = \begin{pmatrix} 1 \\ 1 \\ \beta \end{pmatrix}. \qquad (5.5)$$

Next, we row-reduce the *augmented matrix* to find the RREF:

$$\left(\begin{array}{ccc|c} 1 & 0 & -2 & 1 \\ 1 & -1 & 1 & 1 \\ 2 & \alpha & -1 & \beta \end{array}\right) \xrightarrow{\text{RREF } (\alpha \neq -1)} \left(\begin{array}{ccc|c} 1 & 0 & 0 & (3\alpha + 2\beta - 1)/(3\alpha + 3) \\ 0 & 1 & 0 & (\beta - 2)/(\alpha + 1) \\ 0 & 0 & 1 & (\beta - 2)/(3\alpha + 3) \end{array}\right), \qquad (5.6)$$

where we have assumed that $\alpha \neq -1$. If $\alpha = -1$, we instead have

$$\left(\begin{array}{ccc|c} 1 & 0 & -2 & 1 \\ 1 & -1 & 1 & 1 \\ 2 & -1 & -1 & \beta \end{array}\right) \xrightarrow{\text{RREF}} \left(\begin{array}{ccc|c} 1 & 0 & -2 & 1 \\ 0 & 1 & -3 & 0 \\ 0 & 0 & 0 & \beta - 2 \end{array}\right). \qquad (5.7)$$

Do try to obtain these RREFs by hand, and, if you like, use SymPy to verify your results (more about this in the Discussion section). Let's now investigate the possibilities.

Case 1 $\alpha \neq -1$. Reading off eq. 5.6, we have the following unique solution for (x, y, z):

$$x = \frac{3\alpha + 2\beta - 1}{3\alpha + 3}, \qquad y = \frac{\beta - 2}{\alpha + 1}, \qquad z = \frac{\beta - 2}{3\alpha + 3}. \qquad (5.8)$$

Case 2 $\alpha = -1, \beta \neq 2$. The last row of the RREF matrix in eq. 5.7 reads

$$0 = \beta - 2.$$

Therefore, if $\beta \neq 2$, the RHS of the above equation is nonzero. The system has no solutions and is said to be *inconsistent*.

Case 3 $\alpha = -1, \beta = 2$. The last row of eq. 5.7 is now consistent, but also redundant ($0 = 0$). We have 3 unknowns with only 2 equations, and so one unknown can be regarded as a free variable. This means that there are infinitely many solutions to this system. We say that the system is *underdetermined*. (*Overdetermined* systems will be explored in exercise 1.)

Let's say we regard z as our free variable. Writing $z = t$, eq. 5.7 then gives us the solution (x, y, z) in terms of the free variable t as:

$$\begin{pmatrix} x \\ y \\ z \end{pmatrix} = \begin{pmatrix} 1 + 2t \\ 3t \\ t \end{pmatrix} = \begin{pmatrix} 1 \\ 0 \\ 0 \end{pmatrix} + t \begin{pmatrix} 2 \\ 3 \\ 1 \end{pmatrix}, \quad t \in \mathbb{R}. \tag{5.9}$$

Geometric interpretations

The 3 linear equations (5.4) represent 3 planes in $\mathbb{R}^3$. Different values of α and β will determine the configuration of the planes, which can intersect at a single point, a line, a plane, or at no points at all.

Case 1 $\alpha \neq -1$. The 3 planes intersect at a point, corresponding to the unique solution (5.8). A visualisation of this situation is shown in the top panel of fig. 5.3, where $\alpha = 1$, $\beta = 0$. Substituting these values into eq. 5.8 gives the coordinates of the intersection point.

$$(x, y, z) = \left(\frac{1}{3}, -\frac{2}{3}, -\frac{1}{3} \right).$$

Case 2 $\alpha = -1, \beta \neq 2$. The 3 planes do not have a common intersection. Perhaps two of the planes are parallel, or perhaps the 3 planes are 3 sides of a triangular prism. The latter case is shown in the middle panel of fig. 5.3, which shows the configuration with $\alpha = -1$ and $\beta = 0$.

Case 3 $\alpha = -1, \beta = 2$. The solution (5.9) represents a line through the point $(1, 0, 0)$ parallel to the vector $(2, 3, 1)$. All solutions to the system lie along this line. This configuration is shown in the bottom panel of fig. 5.3, which shows the 3 planes intersecting along the line (5.9).

Plotly

The code `planes.ipynb` produces fig. 5.3. This code looks quite different from our previous programs. Rather than using Matplotlib, we use the `Plotly`[2] package to plot the planes. Whilst Matplotlib can certainly plot 3 planes on the same set of axes, unfortunately, it looks *bad*. To save you the trouble of trying this, the Matplotlib plot is shown in fig. 5.4. You can see that Matplotlib does a poor job at rendering the intersection of planes, and the confusing figure gives us little sense of the positions of the planes relative to one another.

Fig. 5.3, produced by Plotly, is a 3D interactive figure showing 3 planes in the cubic plot window $(x, y, z) \in [-5, 5]^3$. The planes are in different colours, and a slider is added for the value of α (say, $\alpha \in [-2, 2]$ in steps of 0.25). When we move the slider, the corresponding plane is replotted accordingly. To keep things relatively simple with only one slider, we specify the value of β manually in the code.

---

[2] For installation, see `https://plotly.com/python/getting-started/`

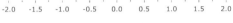

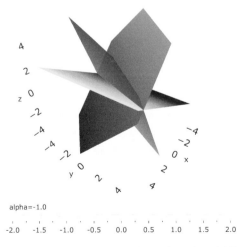

Fig. 5.3: Three planes representing the system (5.4), plotted with Plotly. *Top*: $\alpha = 1, \beta = 0$ (unique solution). *Middle*: $\alpha = -1, \beta = 0$ (no solution). *Bottom*: $\alpha = -1, \beta = 2$ (infinitely many solutions).

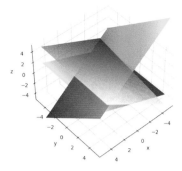

Fig. 5.4: Matplotlib produces a confusing figure.

The syntax for creating a 3D plot with a slider in Plotly is quite involved, but here are some pointers.

- The 3 planes are called `plane1`, `plane2` and `plane3`. Only the third plane is changed by the α value determined by the slider.
- The `for` loop marked **Loop 1** plots all possible positions of `plane3` for all α values. However, we make them all invisible for now.
- The `for` loop marked **Loop 2** switches the visibility of the planes on and off according to the position of the slider.
- There are 17 possible values of α. For each α, there are 3 planes to display. Altogether, there are $17 \times 3 = 51$ objects whose visibility must be controlled. These 51 objects are stored in `fig.data`. (Try checking `len(fig.data)`.)
- `plane1` is stored as objects number 0, 3, 6...48, `planes2` as objects 1, 4, 7...49 and `plane3` as objects 2, 5, 8...50.
- **Loop 2** starts by making everything in `fig.data` invisible. If α takes the ith value on the slider ($\alpha = -2 + i/4$, $i = 0, 1, \ldots 16$), then **Loop 2** turns on the visibility for objects $3i$, $3i + 1$ and $3i + 2$.

Throughout the code, properties of the objects in each frame are stored in a *dictionary*. A dictionary is a Python object containing an ordered list of properties. See Appendix A for more details.

In exercise 1, you will use the code to explore a scenario where the slider changes two planes at the same time.

DISCUSSION

- **SymPy's naivety**. Using SymPy to find the RREF in eq. 5.7, you will be shown the augmented matrix

$$\begin{pmatrix} 1 & 0 & -2 & 0 \\ 0 & 1 & -3 & 0 \\ 0 & 0 & 0 & 1 \end{pmatrix}.$$

You ought to be suspicious of this answer: somehow β has vanished from the matrix! There are in fact two errors in this augmented matrix, stemming from SymPy's naivety in performing algebraic simplifications. Since a RREF has leading 1s, SymPy simply performs the division $\frac{\beta-2}{\beta-2}$ to obtain the 1 on the lower right entry without considering the possibility of zero division. It goes on to use this 1 to create a 0 on the top right entry. Both numbers are wrong (the correct answer is on the RHS of eq. 5.7).

In simple terms, if x is a SymPy symbol, then $x/x = 1$. You must take care of the possibility that $x = 0$ separately.

It is far too easy to delegate all fiddly algebraic manipulations to computer algebra systems, but the lesson here is that you must also be vigilant of the results, and always ask yourself if the computer's answers are sensible. There are many other gotchas and pitfalls worth reading on SymPy documentation[3].

- **Determinant and rank**. The following theorem allows us to rephrase the uniqueness of the solutions for $A\mathbf{x} = \mathbf{b}$ in terms of the rank or determinant of the matrix A.

Theorem 5.2 *Consider the linear system* $A\mathbf{x} = \mathbf{b}$ *where* A *is an* $n \times n$ *matrix,* $\mathbf{x} = (x_1, x_2, \ldots x_n)$ *and* $\mathbf{b}$ *is a constant vector in* $\mathbb{R}^n$. *The following are equivalent:*

1. The system has a unique solution
2. rank $A = n$ (we say that A is 'full rank')
3. $\det A \neq 0$ (i.e. A is invertible)

Returning to the matrix in the question, let $A = \begin{pmatrix} 1 & 0 & -2 \\ 1 & -1 & 1 \\ 2 & \alpha & -1 \end{pmatrix}$. If $\alpha \neq -1$, we see from eq. 5.6 that the RREF of A is the identity matrix with rank 3.

You may like to check the rank with SymPy, but be careful. SymPy will tell you (falsely) that rank $A = 3$ regardless of α. The previous bullet point explains why this isn't true. More helpfully, SymPy's expression for the determinant is $\det A = -3\alpha - 3$, which is correct. Clearly the determinant is nonzero if and only if $\alpha \neq -1$, giving us the unique solution $\mathbf{x} = A^{-1}\mathbf{b}$.

---

[3] https://docs.sympy.org/latest/explanation/gotchas.html

planes.ipynb (for plotting fig. 5.3)

```
import plotly.graph_objects as go
import numpy as np
```

Define matrix A and vector $\mathbf{b}$
where $A\mathbf{x} = \mathbf{b}$, from eq. 5.5

```
A = np.array([[1,0,-2], [1,-1,1], [2,1,-1]])
b = np.array([1,1,0])
```

The (x, y) domain to plot the planes

```
x = np.linspace(-5,5)
y = np.linspace(-5,5)
```

Create 50×50 grid points
Express z in terms of x and y...
for the first two planes

```
X,Y= np.meshgrid(x,y)
Z1 = (A[0,0]*X + A[0,1]*Y - b[0])/-A[0,2]
Z2 = (A[1,0]*X + A[1,1]*Y - b[1])/-A[1,2]

fig = go.Figure()
```

Plot the first plane
Suppress the legend for z values
Plot the second plane

If not specified, the default is 'electric'

```
plane1 = go.Surface(x=x, y=y, z=Z1,
                    showscale=False)
plane2 = go.Surface(x=x, y=y, z=Z2,
                    showscale=False,
                    colorscale='viridis')
```

Loop 1: Prepare plots for all α

Plot all possible positions of the third plane

Make them all translucent blue planes

Make them all invisible for now
Add the 3 planes to data

```
alpha = np.arange(-2, 2.1, 0.25)
for a in alpha:
    Z3= (A[2,0]*X + a*Y - b[2])/-A[2,2]
    plane3 = go.Surface(x=x, y=y, z=Z3,
                        showscale=False,
                        colorscale='blues',
                        opacity = 0.9,
                        visible=False)
    fig.add_traces([plane1,plane2,plane3])
```

Make the third plane visible initially

```
fig.data[2].visible = True
```

This will be filled with 51 plots

```
steps = []
```

Loop 2: Selectively display plot

Make everything invisible
Display the current α value on the slider
Turn on the visibility of the 3 planes

```
for i, a in enumerate(alpha):
    step = dict(
      method="restyle",
      args=[{"visible":[False]*len(fig.data)}],
      label=str(a))
    step["args"][0]["visible"][3*i] = True
    step["args"][0]["visible"][3*i+1] = True
    step["args"][0]["visible"][3*i+2] = True
    steps.append(step)
```

Add a slider
What to display on the slider

```
sliders = [dict(steps=steps,
                currentvalue={"prefix": 'alpha='},
                font=dict(size=20))]
```

Update the plot as the slider is changed
Make the plotting area bigger
Specify the z range in the plot
Set the aspect ratio to be equal

```
fig.update_layout(sliders=sliders, width=700,
    margin=dict(r=20, l=10, b=5, t=5),
    scene = dict(zaxis = dict(range=[-5,5])),
    scene_aspectmode='cube',
    font=dict(size=18))
```

Display in a new browser tab

```
fig.show(renderer='browser')
```

5.4 Four methods for solving $A\mathbf{x} = \mathbf{b}$

Solve the system

$$
\begin{aligned}
x_1 + 2x_2 - x_3 &= 3 \\
2x_1 + x_2 - 2x_3 &= 3 \\
-3x_1 + x_2 + x_3 &= -6
\end{aligned}
\tag{5.10}
$$

using each of the following methods.
1) Row reduction and back substitution
2) LU decomposition
3) Inverse matrix
4) Cramer's rule
Which method is the quickest?

We wish to solve $A\mathbf{x} = \mathbf{b}$ for $\mathbf{x} = \begin{pmatrix} x_1 & x_2 & x_3 \end{pmatrix}^T$ where

$$
A = \begin{pmatrix} 1 & 2 & -1 \\ 2 & 1 & -2 \\ -3 & 1 & 1 \end{pmatrix}, \qquad \mathbf{b} = \begin{pmatrix} 3 \\ 3 \\ -6 \end{pmatrix}.
$$

Method 1: Row reduction and back substitution

Let's obtain the row-echelon form of the augmented matrix $(A|\mathbf{b})$ whilst keeping track of the row operations along the way (this will be important for the next method).

$$
\left(\begin{array}{ccc|c} 1 & 2 & -1 & 3 \\ 2 & 1 & -2 & 3 \\ -3 & 1 & 1 & -6 \end{array}\right)
\xrightarrow[R_3:R_3+3R_1]{R_2:R_2-2R_1}
\left(\begin{array}{ccc|c} 1 & 2 & -1 & 3 \\ 0 & -3 & 0 & -3 \\ 0 & 7 & -2 & 3 \end{array}\right)
\xrightarrow{R_3:R_3-\frac{7}{3}R_2}
\left(\begin{array}{ccc|c} 1 & 2 & -1 & 3 \\ 0 & -3 & 0 & -3 \\ 0 & 0 & -2 & -4 \end{array}\right)
\tag{5.11}
$$

Next to solve for $\mathbf{x}$, we perform *back substitution*, meaning that we start from the last row, solve for x_3, and make our way up the rows.

$$
\begin{aligned}
\text{From } R_3: \qquad & -2x_3 = -4 \implies x_3 = 2, \\
\text{From } R_2: \qquad & -2x_2 = -3 \implies x_2 = 1, \\
\text{From } R_1: \quad x_1 + 2x_2 - x_3 = 3 \quad & \implies x_1 = 3.
\end{aligned}
$$

Method 2: LU decomposition

The row operations in the previous method are, in order,

$$
R_2 : R_2 - 2R_1, \quad R_3 : R_3 + 3R_1, \quad R_3 : R_3 - \frac{7}{3}R_2.
$$

These operations can be expressed as left-multiplication by the following *elementary matrices*:

$$L_1 = \begin{pmatrix} 1 & 0 & 0 \\ -2 & 1 & 0 \\ 0 & 0 & 1 \end{pmatrix}, \quad L_2 = \begin{pmatrix} 1 & 0 & 0 \\ 0 & 1 & 0 \\ 3 & 0 & 1 \end{pmatrix}, \quad L_3 = \begin{pmatrix} 1 & 0 & 0 \\ 0 & 1 & 0 \\ 0 & \frac{7}{3} & 1 \end{pmatrix}.$$

Their inverses are simple to write down (think about undoing the row operations).

$$L_1^{-1} = \begin{pmatrix} 1 & 0 & 0 \\ 2 & 1 & 0 \\ 0 & 0 & 1 \end{pmatrix}, \quad L_2^{-1} = \begin{pmatrix} 1 & 0 & 0 \\ 0 & 1 & 0 \\ -3 & 0 & 1 \end{pmatrix}, \quad L_3^{-1} = \begin{pmatrix} 1 & 0 & 0 \\ 0 & 1 & 0 \\ 0 & -\frac{7}{3} & 1 \end{pmatrix}.$$

Hence, we have the following relation between A and its REF.

$$L_3 L_2 L_1 A = \begin{pmatrix} 1 & 2 & -1 \\ 0 & -3 & 0 \\ 0 & 0 & -2 \end{pmatrix} \implies A = L_1^{-1} L_2^{-1} L_3^{-1} \begin{pmatrix} 1 & 2 & -1 \\ 0 & -3 & 0 \\ 0 & 0 & -2 \end{pmatrix} = \underbrace{\begin{pmatrix} 1 & 0 & 0 \\ 2 & 1 & 0 \\ -3 & -\frac{7}{3} & 1 \end{pmatrix}}_{L} \underbrace{\begin{pmatrix} 1 & 2 & -1 \\ 0 & -3 & 0 \\ 0 & 0 & -2 \end{pmatrix}}_{U}$$

$$(5.12)$$

In summary, we have decomposed A into two matrices, L (a lower triangular matrix) and U (an upper triangular matrix). A matrix L is said to be *lower triangular* if $L_{ij} = 0$ whenever $i < j$. A matrix U is said to be *upper triangular* if $U_{ij} = 0$ for $i > j$.

The decomposition $A = LU$ is known as *LU decomposition* (or *LU factorisation*). How does this help us solve $A\mathbf{x} = \mathbf{b}$? Well, now that we have $LU\mathbf{x} = \mathbf{b}$, let $\mathbf{y} = U\mathbf{x}$ and write $\mathbf{y} = \begin{pmatrix} y_1 & y_2 & y_3 \end{pmatrix}$. Let's first solve $L\mathbf{y} = \mathbf{b}$. The triangular structure of L allows us to easily obtain $\mathbf{y}$ by substitution from the top row down.

$$\begin{pmatrix} 1 & 0 & 0 \\ 2 & 1 & 0 \\ -3 & -\frac{7}{3} & 1 \end{pmatrix} \begin{pmatrix} y_1 \\ y_2 \\ y_3 \end{pmatrix} = \begin{pmatrix} 3 \\ 3 \\ -6 \end{pmatrix} \implies \begin{pmatrix} y_1 \\ y_2 \\ y_3 \end{pmatrix} = \begin{pmatrix} 3 \\ -3 \\ -4 \end{pmatrix}.$$

Finally, we solve $U\mathbf{x} = \mathbf{y}$. This is also easy – just do back substitution (bottom row up).

$$\begin{pmatrix} 1 & 2 & -1 \\ 0 & -3 & 0 \\ 0 & 0 & -2 \end{pmatrix} \begin{pmatrix} x_1 \\ x_2 \\ x_3 \end{pmatrix} = \begin{pmatrix} 3 \\ -3 \\ -4 \end{pmatrix} \implies \begin{pmatrix} x_1 \\ x_2 \\ x_3 \end{pmatrix} = \begin{pmatrix} 3 \\ 1 \\ 2 \end{pmatrix}.$$

Of course, this agrees with the solution from the previous method.

But *why do LU*? It *feels* like we're doing extra work with LU decomposition compared with row reduction. Why do we bother?

Firstly, LU decomposition keeps a record of the row operations, which remains the same even if $\mathbf{b}$ is changed. There is no need to work with $\mathbf{b}$ until after the LU decomposition is complete. On the other hand, with row reduction, changing $\mathbf{b}$ means the augmented matrix is changed, and so the row reduction would have to be redone.

Secondly, there is really no extra computational work compared with row reduction. It feels as though there are extra steps in LU decomposition because we only dealt with $\mathbf{b}$ separately in the final stage, whereas in row reduction, we incorporated $\mathbf{b}$ into our calculation from the beginning in the augmented matrix.

For problems involving multiple $\mathbf{b}$'s, the LU approach is *significantly* faster than row reduction. More precisely, for a given $n \times n$ matrix A, the number of operations needed to obtain its row-echelon form by row reduction (or to obtain its LU decomposition) is approximately $\frac{2}{3}n^3$, *i.e.* the complexity is $O(n^3)$. On the other hand, when solving $LU\mathbf{x} = \mathbf{b}$ by back substitution, only $O(n^2)$ operations are required. (See [182] for details.)

In Python, the syntax

```
scipy.linalg.solve(A, b)
```

performs (essentially) an LU decomposition on A, then uses back substitution to obtain the solution. The actual decomposition is a slightly more sophisticated algorithm (called $PA = LU$ decomposition, where the matrix P encodes additional row swaps performed on A to minimise numerical errors).

Method 3: Inverse matrix

In this method we will calculate A^{-1} and obtain the solution $\mathbf{x} = A^{-1}\mathbf{b}$. One way to find A^{-1} is to use the formula

$$A^{-1} = \frac{\mathrm{adj}\, A}{\det A},$$

where adj A is the *adjoint* of A. It's logical to first calculate $\det A$ – if that's zero, then the matrix is not invertible and this method won't work. For the matrix in the question, we find that the determinant is nonzero, with $\det A = 6$. Let's now try using the adjoint formula.

The recipe for obtaining the adjoint of a 3×3 matrix A is as follows. First, create the following skeleton in which there are alternating signs.

$$\begin{pmatrix} \square & -\square & \square \\ -\square & \square & -\square \\ \square & -\square & \square \end{pmatrix}^T$$

To calculate the number in the box in row i and column j, go to matrix A, delete row i and column j. What remains is a 2×2 matrix. Find the determinant of that matrix, and that's the number in the box. (See any introductory linear algebra textbooks for more details and examples.)

Following this recipe (and don't forget to take the transpose in the end), we find:

$$\mathrm{adj}\, A = \begin{pmatrix} 3 & -3 & -3 \\ 4 & -2 & 0 \\ 5 & -7 & -3 \end{pmatrix}.$$

Dividing this by the determinant gives:

$$A^{-1} = \begin{pmatrix} \frac{1}{2} & -\frac{1}{2} & -\frac{1}{2} \\ \frac{2}{3} & -\frac{1}{3} & 0 \\ \frac{5}{6} & -\frac{7}{6} & -\frac{1}{2} \end{pmatrix}.$$

The solution $\mathbf{x}$ is therefore

$$\mathbf{x} = A^{-1}\mathbf{b} = \begin{pmatrix} \frac{1}{2} & -\frac{1}{2} & -\frac{1}{2} \\ \frac{2}{3} & -\frac{1}{3} & 0 \\ \frac{5}{6} & -\frac{7}{6} & -\frac{1}{2} \end{pmatrix} \begin{pmatrix} 3 \\ 3 \\ -6 \end{pmatrix} = \begin{pmatrix} 3 \\ 1 \\ 2 \end{pmatrix}.$$

It's worth noting that in Python, the command `scipy.linalg.inv(A)` does not use the adjoint formula to calculate A^{-1} because determinants are computationally expensive. Instead, an LU decomposition is performed on A, and the result is used to solve 3 systems:

$$A\mathbf{c_1} = \begin{pmatrix} 1 \\ 0 \\ 0 \end{pmatrix}, \quad A\mathbf{c_2} = \begin{pmatrix} 0 \\ 1 \\ 0 \end{pmatrix}, \quad A\mathbf{c_3} = \begin{pmatrix} 0 \\ 0 \\ 1 \end{pmatrix}.$$

The solutions are columns of the inverse matrix $A^{-1} = \begin{pmatrix} \mathbf{c_1} & \mathbf{c_2} & \mathbf{c_3} \end{pmatrix}$, which is then post-multiplied by $\mathbf{b}$ to obtain the solution $\mathbf{x}$.

Compared with the previous method, we see that there are two extra sets of back substitutions to perform, plus the final matrix multiplication $A^{-1}\mathbf{b}$. If A is an $n \times n$ matrix, the complexity of both methods 2 and 3 is still dominated by LU decomposition, costing $O(n^3)$. However, when n is large ($\gtrsim 1000$), the extra calculations in method 3 can make a difference.

The code `solvetimes.ipynb` produces the top panel of fig. 5.5. Warning: this code will probably take a few minutes to run. If your computer is struggling, try considering smaller matrices or use fewer points in the plot.

The figure shows the time taken (in seconds) to solve the system $A\mathbf{x} = \mathbf{b}$ where A is a large matrix filled with random numbers between 0 and 1, and $\mathbf{b} = (1, 1, 1, \ldots 1)^T$. The graphs are plotted against increasing size of the matrix A. The comparison between SciPy's `linalg.solve(A,b)` method and the `linalg.inv(A)*b` method corroborates our observations in the previous paragraph. When $n \approx 5000$, the inverse method takes around 50% longer in this particular run.

Saving fractions of a second on computation seems hardly impressive. However, remember that with more computations come greater numerical errors, so keeping the number of operations low not only means that your code will run faster, but the output will probably be more accurate too.

The main message here is that when it comes to linear algebra with Python, you should *avoid inverting matrices* unless A^{-1} is explicitly required (which is seldom the case). In any case, you should never solve $A\mathbf{x} = \mathbf{b}$ by the inverse-matrix method.

Method 4: Cramer's rule

Cramer's rule is a method for solving $A\mathbf{x} = \mathbf{b}$ using nothing but determinants. The rule is named after *Gabriel Cramer* (1704–1752), a mathematician born in the former Republic of Geneva. His 'rule' was given in an appendix which accompanied his book on algebraic curves (see [116] for an interesting history of Cramer's rule).

Cramer's rule states that the solution for the unknown x_i is given by

$$x_i = \frac{\det B_i}{\det A},$$

where B_i is the matrix A with the ith column replaced by the vector $\mathbf{b}$ (see [194] for proof).

In our example, we have $\det A = 6$, and thus the solutions by Cramer's rule are

$$x_1 = \frac{1}{\det A} \begin{vmatrix} 3 & 2 & -1 \\ 3 & 1 & -2 \\ -6 & 1 & 1 \end{vmatrix} = 3, \quad x_2 = \frac{1}{\det A} \begin{vmatrix} 1 & 3 & -1 \\ 2 & 3 & -2 \\ -3 & -6 & 1 \end{vmatrix} = 1, \quad x_3 = \frac{1}{\det A} \begin{vmatrix} 1 & 2 & 3 \\ 2 & 1 & 3 \\ -3 & 1 & -6 \end{vmatrix} = 2.$$

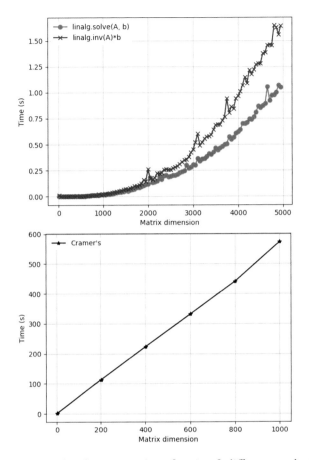

Fig. 5.5: Time taken to solve the system $A\mathbf{x} = \mathbf{b}$ using 3 different methods, plotted as a function of the size of matrix A. We take A to be an $n \times n$ matrix of random small numbers between 0 and 1, and $\mathbf{b} = (1, 1, 1 \dots 1)^T$. The top panel (produced by `solvetimes.ipynb`) shows that SciPy's `linalg.solve` is the quickest, followed by the matrix inversion method $(\mathbf{x} = A^{-1}\mathbf{b})$. Cramer's rule (bottom) is the slowest by far. We only considered n up to 1000 in the bottom panel.

One useful thing about Cramer's rule is that it allows us to solve for a special subset of unknowns (say, if you only want x_{17} out of 200 unknowns $x_1, x_2, \dots x_{200}$). Otherwise, it is a computationally inefficient way to solve a linear system, due to the need to compute many determinants.

In SciPy, the command `linalg.det(A)` first performs LU factorisation, then calculates the det A by multiplying the diagonal elements of U (you can verify this property for our example by inspecting the diagonal elements of U in eq. 5.12). Solving for n unknowns requires $(n + 1)$ LU decompositions, each costing $\sim n^3$ operations. This means that the complexity of Cramer's rule in Python is $O(n^4)$, which is an order of magnitude larger than the previous two methods. It goes without saying that when it comes to solving linear systems numerically, Cramer's rule should not be used.

To illustrate this point, let's repeat the same large-matrix analysis as in the previous method. The result is shown on the lower panel of fig. 5.5, which was calculated for random matrices up to $n = 1000$ (at which point it took my poor computer almost 10 minutes to solve – whilst the previous methods would only have taken a fraction of a second).

DISCUSSION

• **Other matrix decompositions**. Besides the LU decomposition discussed in this section (where we obtained $A = LU$), here are other famous decompositions for a given matrix A.

  – *Eigendecomposition* $A = PDP^{-1}$, where D is a diagonal matrix. We will discuss this in §5.6.
  – *QR decomposition* $A = QR$, where R is an upper triangular matrix and Q satisfies $Q^T Q = I$ (Q is said to be an *orthogonal matrix*).
  – *Cholesky decomposition* $A = R^T R$, where R is an upper triangular matrix. (This can be thought of as taking the square root of A.)
  – *Singular-value decomposition* (SVD) $A = U\Sigma V^T$, where U and V are orthogonal matrices and Σ is a diagonal matrix of nonnegative real numbers. We will discuss this in §5.8.

Matrix decompositions are generally deployed to speed up computations by reducing the complexity of the problems at hand. Certain decompositions are only applicable when the matrix A satisfies certain properties. For example, the Cholesky decomposition can help us solve $A\mathbf{x} = \mathbf{b}$ twice as fast as LU decomposition, but can only be performed if A is symmetric ($A = A^T$) and positive definite ($\mathbf{x} \cdot A\mathbf{x} > 0$ for all nonzero vectors $\mathbf{x}$). The QR decomposition can help solve the *least-square* problem (*i.e.* find the best solution to $A\mathbf{x} = \mathbf{b}$ when there are no exact solutions). The SVD is more general and can be performed on any matrix. It is used frequently in data analysis especially in reducing big data, as we will see later.

These matrix decompositions are typically taught in a second course in linear algebra (or numerical analysis). Mathematical details can be found in more advanced linear algebra books such as [103, 134, 195].

solvetimes.ipynb (for plotting the top panel of fig. 5.5)	
Load the timer	```import numpy as np
import matplotlib.pyplot as plt
from scipy import linalg as LA
from time import perf_counter as timer
%matplotlib```
Range of sizes of matrix A to solve Initialise random number generator To be filled with computational times in seconds	```N=np.arange(1,5001,50)
rng = np.random.default_rng()
t1 = []
t2 = []```
n = matrix size A = matrix of random numbers between 0 and 1 **b** = column vector of 1's	```for n in N:
A = rng.random((n,n))
b = np.ones((n,1))```
Start the clock! *LU* method *Stop the clock!* Note the time taken	``` tic = timer()
x1 = LA.solve(A, b)
toc = timer()
t1.append(toc-tic)```
Start the clock! inverse-matrix method *Stop the clock!* Note the time taken	``` tic = timer()
x2 = LA.inv(A)*b
toc = timer()
t2.append(toc-tic)```
Join up the data points (red circles vs blue crosses)	```plt.plot(N,t1, '-ro',N,t2, '-bx')
plt.xlabel('Matrix dimension')
plt.ylabel('Time (s)')
plt.legend(['linalg.solve(A, b)',
'linalg.inv(A)*b'])
plt.grid('on')
plt.show()```

5.5 Matrices as linear transformations

Let S be the unit square $[0, 1] \times [0, 1]$. Consider the linear transformation $T : S \to \mathbb{R}^2$ defined by $T(\mathbf{x}) = M_i \mathbf{x}$, where

a) $M_1 = \begin{pmatrix} \cos t & -\sin t \\ \sin t & \cos t \end{pmatrix}$ b) $M_2 = \begin{pmatrix} 1 & t \\ 0 & 1 \end{pmatrix}$ c) $M_3 = \begin{pmatrix} t & 0 \\ 0 & 1 \end{pmatrix}$.

Find the image of S under each transformation. Describe each transformation geometrically.

Recall the definition of a linear transformation $T : V \to W$. It satisfies

$$T(\mathbf{x_1} + \alpha \mathbf{x_2}) = T(\mathbf{x_1}) + \alpha T(\mathbf{x_2}), \tag{5.13}$$

for all $\mathbf{x_1}, \mathbf{x_2} \in V$ and $\alpha \in \mathbb{R}$. The equation says that a linear expression is mapped to another linear expression. In particular, if V and W are $\mathbb{R}^2$ or $\mathbb{R}^3$, it follows that straight lines in V are mapped onto straight lines in W. Also note that a linear transformation maps the zero vector onto itself (just set $\mathbf{x_1} = \mathbf{x_2}$ and $\alpha = -1$).

It is easy to check that the transformation $T(\mathbf{x}) = M\mathbf{x}$ satisfies the linearity condition (since matrix multiplication is distributive over vector addition). Conversely, any linear transformation $T : \mathbb{R}^n \to \mathbb{R}^m$ can be represented by a matrix multiplication $T(\mathbf{x}) = M\mathbf{x}$. Let's see why this holds in $\mathbb{R}^2$. Write $\mathbf{x} = x\mathbf{e_1} + y\mathbf{e_2}$ where $\mathbf{e_1} = \mathbf{i} = \begin{pmatrix} 1 \\ 0 \end{pmatrix}$, and $\mathbf{e_2} = \mathbf{j} = \begin{pmatrix} 0 \\ 1 \end{pmatrix}$ are the standard basis vectors. Linearity of T means that

$$T(x\mathbf{e_1} + y\mathbf{e_2}) = xT(\mathbf{e_1}) + yT(\mathbf{e_2}) = \underbrace{\left(T(\mathbf{e_1}) \mid T(\mathbf{e_2})\right)}_{M} \begin{pmatrix} x \\ y \end{pmatrix}. \tag{5.14}$$

This shows that the transformation $T(\mathbf{x})$ can be represented by the matrix multiplication $M\mathbf{x}$ where the first column of M is $T(\mathbf{e_1})$ and the second column is $T(\mathbf{e_2})$. In other words, the action of T on the basis vectors tells us about its action on *any* vectors.

It's helpful to distinguish the two concepts: the transformation is an action (which is independent of coordinate choices), but the matrix is its representation in a certain coordinate choice. Knowing this subtlety, you can see that one should not say "*the transformation M_1*", but rather "*the transformation represented by the matrix M_1*".

Here are some examples of linear transformations in $\mathbb{R}^2$.

Rotation

In $\mathbb{R}^2$, consider a rotation by angle t about the origin. Clearly straight lines are mapped onto straight lines, and the origin stays put, so we suspect that it is a linear transformation that can be represented by a matrix. What would such a matrix look like? As we discovered earlier, we only need to find the images of $\mathbf{e_1}$ and $\mathbf{e_2}$ to obtain the matrix.

The diagram below tells us that

$$T(\mathbf{e_1}) = \cos t\mathbf{e_1} + \sin t\mathbf{e_2}, \qquad T(\mathbf{e_2}) = -\sin t\mathbf{e_1} + \cos t\mathbf{e_2}.$$

Putting them into columns, we obtain the matrix

$$M_1 = \begin{pmatrix} \cos t & -\sin t \\ \sin t & \cos t \end{pmatrix}.$$

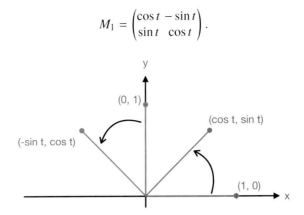

The image of the square S under this matrix can be visualised with Python as shown in fig. 5.6, created by the code `transformation.ipynb`. The code finds the images of 3 vertices (corners) of the square by matrix multiplication, and join the image points again to find the transformed square (the vertex at the origin stays put due to linearity). We have included a slider for the variable $t \in [-2\pi, 2\pi]$. The original square is plotted with reduced opacity `alpha=0.4`. The red and green edges are the images of $\mathbf{e}_1$ and $\mathbf{e}_2$ respectively. Indeed the image is a rotated square.

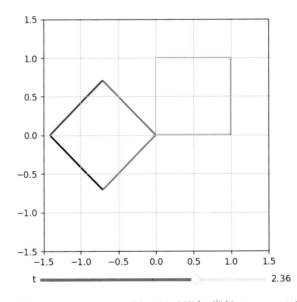

Fig. 5.6: The transformation represented by M_1 ($\begin{smallmatrix} \cos t & -\sin t \\ \sin t & \cos t \end{smallmatrix}$) is an anticlockwise rotation by angle t (adjustable using the slider) about the origin. In the figure, M_1 rotates the unit square by an angle $t \approx 3\pi/4$.

Shear

The code allows us to find the image of the square under any 2×2 matrix. The image of the square under $M_2 = \begin{pmatrix} 1 & t \\ 0 & 1 \end{pmatrix}$ is shown in the top panel of fig. 5.7.

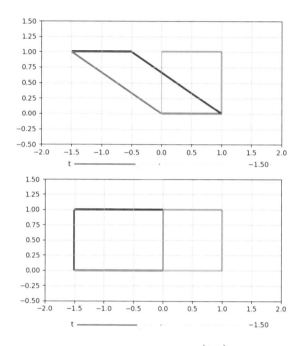

Fig. 5.7: *Top*: The transformation represented by $M_2 = \begin{pmatrix} 1 & t \\ 0 & 1 \end{pmatrix}$ is a shear along the x direction. *Bottom*: The transformation represented by $M_3 = \begin{pmatrix} t & 0 \\ 0 & 1 \end{pmatrix}$ is a scaling along the x direction. Both transformations are shown here with $t = -1.5$.

In the figure, we can see that the vector e_1 stays put under this transformation, and this is confirmed by the first column of M_2, which tells us that $T(e_1) = e_1$. However, $T(e_2) = \begin{pmatrix} t \\ 1 \end{pmatrix}$. If $t > 0$, the green side of the square is pushed, or *sheared*, to the right, whilst maintaining the same height. The amount of shear that a point undergoes is directly proportional to its y coordinates. If $t < 0$, the green side is sheared to the left, as shown in the figure (with $t = -1.5$).

We conclude that the transformation represented by M_2 maps the square S to a parallelogram of the same height but sheared left or right. This transformation is said to be a horizontal shear and t is called the shear factor. It is worth noting that a sheared straight line is another straight line, and the origin remains unchanged, as expected.

Scaling and reflection

Repeating the same analysis, we conclude that $M_3 = \begin{pmatrix} t & 0 \\ 0 & 1 \end{pmatrix}$ represents *scaling (i.e.* stretching) along the x direction. If $t > 0$, the square is stretched to the right, mapping it to a rectangle. If $t < 0$, the stretch is to the left (as shown in fig. 5.7 with $t = -1.5$). In particular, when $t = -1$, the transformation is a *reflection* along the y-axis.

> **DISCUSSION**

- **Orthogonal matrices**. Let $R(t)$ be the rotation matrix $\begin{pmatrix} \cos t & -\sin t \\ \sin t & \cos t \end{pmatrix}$. Note that the inverse matrix is precisely its transpose, *i.e.*

$$R^{-1}(t) = \begin{pmatrix} \cos t & \sin t \\ -\sin t & \cos t \end{pmatrix} = R^T(t).$$

Matrices satisfying the condition $R^T = R^{-1}$ are called *orthogonal matrices*. These matrices occur frequently in mathematics and physics, and have a rich mathematical structure.
For our rotation matrix $R(t)$, note the following:

  - $R(0)$ is the identity matrix I.
  - $R^{-1}(t) = R(-t)$, meaning that the inverse of a rotation matrix is another rotation matrix (which does the rotation in the opposite direction).
  - Multiplying two rotation matrices (and simplifying the result using trigonometric identities), we have:

$$R(s)R(t) = \begin{pmatrix} \cos s & \sin s \\ -\sin s & \cos s \end{pmatrix}\begin{pmatrix} \cos t & \sin t \\ -\sin t & \cos t \end{pmatrix} = \begin{pmatrix} \cos(s+t) & \sin(s+t) \\ -\sin(s+t) & \cos(s+t) \end{pmatrix} = R(s+t).$$

Thus, multiplying to rotation matrices gives another rotation matrix.

The above properties suggest that the rotation matrices form a closed structure under multiplication. This structure is called the *special orthogonal group* in two dimensions, denoted $SO(2)$. More about groups in the next chapter.

- **Determinant as area**. What do you notice about the determinants of M_1, M_2 and M_3?

$$\det M_1 = 1, \qquad \det M_2 = 1, \qquad \det M_3 = t.$$

These determinants are precisely the area of the transformed unit square (rotation and shear do not change the area, but the scaling does). Let's see why the determinant tells us about area.
Consider a general linear transformation T represented by the matrix

$$M = \begin{pmatrix} a & b \\ c & d \end{pmatrix}.$$

It is straightforward to show that the four vertices of the unit square $S = [0, 1] \times [0, 1]$ are mapped onto a parallelogram with vertices at $(0, 0)$, (a, c), $(a + b, c + d)$ and (b, d) This parallelogram is shown in the figure below.

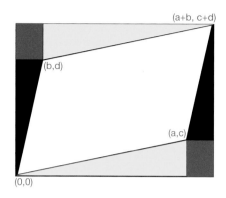

From the figure above, we deduce that:

Area of parallelogram

$$= (a + b)(c + d) - \blacksquare - \square - 2 \times \blacksquare$$
$$= (a + b)(c + d) - bd - ac - 2bc$$
$$= ad - bc.$$

Thus, we see that det M is simply the signed area of the parallelogram ('signed' meaning that the determinant can be negative).

- **Translation.** You may be wondering if translation (*i.e.* shifting the unit square by, say, α unit to the right, and β unit up) can be represented by a matrix. Clearly, translation is not a linear transformation since the origin not preserved, and there is no matrix $\begin{pmatrix} a & b \\ c & d \end{pmatrix}$ such that

$$\forall (x, y) \in \mathbb{R}^2, \quad \begin{pmatrix} x + \alpha \\ y + \beta \end{pmatrix} = \begin{pmatrix} a & b \\ c & d \end{pmatrix} \begin{pmatrix} x \\ y \end{pmatrix}.$$

However, a clever solution is to introduce an extra coordinate and note that

$$\begin{pmatrix} x + \alpha \\ y + \beta \\ 1 \end{pmatrix} = \begin{pmatrix} 1 & 0 & \alpha \\ 0 & 1 & \beta \\ 0 & 0 & 1 \end{pmatrix} \begin{pmatrix} x \\ y \\ 1 \end{pmatrix}. \tag{5.15}$$

The rotation and shear matrices can also be appended with the extra coordinate, allowing them to be multiplied together. This technique, called *homogeneous coordinates*, is important in computer graphics – for example, in OpenGL[4] (see also [127, 207]).

---

[4] http://www.opengl-tutorial.org/beginners-tutorials/tutorial-3-matrices/

transformation.ipynb (for plotting fig. 5.6)	
| | ```python
import numpy as np
import matplotlib.pyplot as plt
from matplotlib.widgets import Slider
%matplotlib
``` |
| Vertices of the unit square $S$ | ```python
O = np.array([0,0])
A = np.array([1,0])
B = np.array([1,1])
C = np.array([0,1])
``` |
| Make space for the slider<br>Set equal aspect ratio (to see a square) | ```python
fig,ax = plt.subplots()
plt.subplots_adjust(bottom=0.15)
plt.axis('square')
``` |
| For plotting a coloured line joining two given points<br>* operator unpacks an array<br>alpha is an optional argument<br>Set arrows to the coordinate-grid size, and remove arrowheads | ```python
def vecplot(start, end, colour, alpha=1):
    vec = end-start
    ax.quiver(*start, *vec, color=colour,
              alpha = alpha, angles='xy',
              scale_units='xy', scale=1,
              headaxislength=0, headwidth=0,
              headlength=0)
    return
``` |
| For plotting $S$ and $T(S)$ given $t$<br>Clear the canvas<br>Plot the original square with alpha= 0.4…each side a different colour | ```python
def transform(t):
    ax.clear()
    vecplot(O,A,'r',0.4)
    vecplot(A,B,'b',0.4)
    vecplot(B,C,'k',0.4)
    vecplot(C,O,'g',0.4)
``` |
| **Specify transformation matrix here** | ```python
    s, c = np.sin(t), np.cos(t)
    M = np.array([[c,-s],[s,c]])
``` |
| The matrix acts on each vertex<br>(The origin is unchanged)<br><br>Plot each side of the transformed figure $T(S)$ | ```python
    At = M@A
    Bt = M@B
    Ct = M@C
    vecplot(O ,At,'r')
    vecplot(At,Bt,'b')
    vecplot(Bt,Ct,'k')
    vecplot(Ct,O ,'g')
``` |
| Specify the plot window | ```python
    ax.axis([-1.5,1.5,-1.5,1.5])
    ax.grid('on')
    return
``` |
| Plot the initial square ($t = 0$)<br>Specify dimensions of the slider<br>Set the range of $t$ and step size | ```python
transform(0)
axt = plt.axes([0.23, 0.05, 0.55, 0.02])
t_slide = Slider(axt, 't', -2*np.pi, 2*np.pi,
        valstep = 0.02, valinit = 0)
``` |
| Update the plot if slider is moved<br>Get a new value of $t$ from the slider<br>Replot | ```python
def update(val):
    t = t_slide.val
    transform(t)
``` |
| | ```python
t_slide.on_changed(update)
plt.show()
``` |
```

5.6 Eigenvalues and eigenvectors

For a given matrix A, suppose that we have a vector $\mathbf{x}$ (with $\mathbf{x} \neq \mathbf{0}$) and a constant $\lambda \in \mathbb{C}$ such that

$$A\mathbf{x} = \lambda\mathbf{x}. \tag{5.16}$$

Then, $\mathbf{x}$ is said to be an *eigenvector* of A with *eigenvalue* λ.

Find the eigenvalues and eigenvectors of the following matrices.

a) $\begin{pmatrix} 1 & 1 \\ 3/2 & 1/2 \end{pmatrix}$ b) $\begin{pmatrix} 1 & 1 \\ 1/2 & 1/2 \end{pmatrix}$ c) $\begin{pmatrix} 1 & 1 \\ -3/2 & 1/2 \end{pmatrix}$.

Solving the eigenvalue problem

Here is the standard way to solve eq. 5.16 by hand. First, we move everything to the LHS, yielding

$$(A - \lambda I)\mathbf{x} = \mathbf{0}, \tag{5.17}$$

where I is the identity matrix of the same size as A. If the matrix $(A - \lambda I)$ were invertible, then we would have $\mathbf{x} = (A - \lambda I)^{-1}\mathbf{0} = \mathbf{0}$. But we require the eigenvector to be nonzero. Therefore, the matrix $(A - \lambda I)$ cannot be invertible, *i.e.*

$$\det(A - \lambda I) = 0. \tag{5.18}$$

Eq. 5.18 is called the *characteristic polynomial* of the matrix A. If A is an $n \times n$ matrix, then $\det(A - \lambda I)$ is a polynomial of order n in λ. Solving for the zeros of this polynomial gives us the eigenvalues.

Let's apply this method to the first matrix $A = \begin{pmatrix} 1 & 1 \\ 3/2 & 1/2 \end{pmatrix}$. We find the characteristic polynomial:

$$\begin{vmatrix} 1 - \lambda & 1 \\ 3/2 & 1/2 - \lambda \end{vmatrix} = 0 \implies 2\lambda^2 - 3\lambda - 2 = 0 \implies \lambda = 2 \text{ or } -\frac{1}{2}.$$

Now let's solve for the eigenvectors. Substituting $\lambda = 2$ into (5.16) and writing the eigenvector $\mathbf{x}$ as $\begin{pmatrix} x \\ y \end{pmatrix}$, we find

$$\begin{pmatrix} 1 & 1 \\ 3/2 & 1/2 \end{pmatrix} \begin{pmatrix} x \\ y \end{pmatrix} = \begin{pmatrix} 2x \\ 2y \end{pmatrix}.$$

The two components of the above equations tell us the same information that $y = x$ (an equation of a line). This means that the eigenvector can be expressed in terms of x as $x \begin{pmatrix} 1 \\ 1 \end{pmatrix}$, where $x \in \mathbb{R}$. It's not surprising that there are infinitely many eigenvectors – after all, we can see in eq. 5.16 that if $\mathbf{x}$ is an eigenvector, so is any multiple of $\mathbf{x}$. As such, it is sufficient to say that:

$$A \text{ has eigenvector } \begin{pmatrix} 1 \\ 1 \end{pmatrix} \text{ with eigenvalue } 2.$$

The same technique tells us that the eigenvector $-\frac{1}{2}$ has solution $y = -\frac{3}{2}x$, *i.e.*

$$A \text{ has another eigenvector } \begin{pmatrix} 2 \\ -3 \end{pmatrix} \text{ with eigenvalue } -\frac{1}{2}.$$

Geometric interpretation

The LHS of the eigenvalue equation $A\mathbf{x} = \lambda\mathbf{x}$ can be thought of as a linear transformation acting on a vector $\mathbf{x}$. For certain special $\mathbf{x}$, the resulting vector $A\mathbf{x}$ happens to be parallel to $\mathbf{x}$ itself, but scaled by a factor λ. Without knowing how to solve the eigenvalue equation, it is possible to search for such a special vector $\mathbf{x}$ numerically.

The code[5] `eigshow.ipynb` creates a little game where the user has to find the directions in which two vectors are parallel. More precisely:

- First, the user picks a candidate for the eigenvector $\mathbf{x}$. Since only the direction of $\mathbf{x}$ matters, it is sufficient to vary $\mathbf{x}$ along the unit circle

$$\mathbf{x} = \begin{pmatrix} \cos\theta \\ \sin\theta \end{pmatrix}, \quad \theta \in [0, 2\pi].$$

  A slider is used to vary the value of the polar angle θ.
- For each $\mathbf{x}$, the transformed vector $A\mathbf{x}$ is also simultaneously plotted.
- The aim is to vary θ until the vectors $\mathbf{x}$ and $A\mathbf{x}$ are parallel. When this happens, we have found the eigenvector $\mathbf{x}$. Furthermore, the length of the vector $A\mathbf{x}$ is the magnitude of the eigenvalue. If $\mathbf{x}$ and $A\mathbf{x}$ point in the same direction, then $\lambda > 0$. If they point in opposite directions, then $\lambda < 0$.

Snapshots of this "*Make the vectors parallel*" game are shown in fig. 5.8, with the slider at three values of θ. The dotted blue circle is the locus of vector $\mathbf{x}$ which we vary using the slider. The dashed red ellipse is the locus of the transformed vector $A\mathbf{x}$.

The two lower panels of fig. 5.8 show the two solutions of this game. The lower left panel shows the first alignment at $\theta = \pi/4 \approx 0.785$. This is consistent with our eigenvector $\begin{pmatrix} 1 & 1 \end{pmatrix}^T$. We also see that the vectors point in the same direction, with the red vector twice the length of the blue vector. This is consistent with the eigenvalue 2 that we found by hand. (You could verify this factor of 2 either by hovering the mouse over the points to see the coordinate readouts, or ask Python to calculate the ratio of the blue and red lines. See exercise 6a.)

Similarly, the lower right panel shows the other alignment at $\theta \approx 2.16$ where the vectors line up in opposite directions. This is consistent with our calculation since the line $y = -\frac{3}{2}x$ makes an angle of $\pi - \tan^{-1}\frac{3}{2} \approx 2.159$ with the positive x-axis. The length of the two vectors can be similarly verified.

In fact, the code also displays the two eigen-directions (in thin green lines). Indeed, we find that the two vectors line up along the green lines.

---

[5] This visualisation is a tribute to the elegant but deprecated MATLAB demo called `eigshow`.

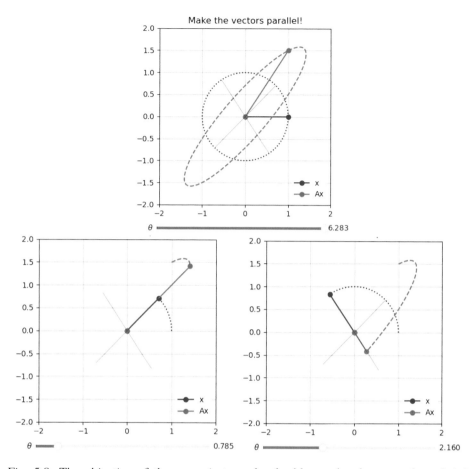

Fig. 5.8: The objective of the game is to make the blue and red vectors ($\mathbf{x}$ and $A\mathbf{x}$) parallel, where $A = \begin{pmatrix} 1 & 1 \\ 3/2 & 1/2 \end{pmatrix}$. The slider changes the polar angle θ for the blue vector $\mathbf{x} = (\cos\theta, \ \sin\theta)^T$. The lower panels show when the parallel configurations are achieved – precisely when the vectors all line up with the eigenvectors (thin green lines). At each of these positions, the length of the red vector is the magnitude of the eigenvalue.

You may be wondering what the equation of the red ellipse is. To work this out, note the result of the transformation

$$\begin{pmatrix} 1 & 1 \\ 3/2 & 1/2 \end{pmatrix} \begin{pmatrix} \cos\theta \\ \sin\theta \end{pmatrix} = \begin{pmatrix} \cos\theta + \sin\theta \\ \frac{3}{2}\cos\theta + \frac{1}{2}\sin\theta \end{pmatrix}.$$

This is a parametric equation of a curve in $\mathbb{R}^2$. If you prefer it in Cartesian form, let $x = \cos\theta + \sin\theta$ and $y = \frac{3}{2}\cos\theta + \frac{1}{2}\sin\theta$. Solving these simultaneous equations for $\cos\theta$ and $\sin\theta$, and using $\cos^2\theta + \sin^2\theta = 1$, we find the Cartesian equation of the ellipse:

$$\frac{5}{2}x^2 - 4xy + 2y^2 = 1.$$

This equation is in *quadric* form which we have previously seen in §3.6.

You will also notice from fig. 5.8 that the maximum and minimum distances of the ellipse from the origin look as though they are along the green eigenvectors, and if that is the case then the maximum and minimum distances would be given by the magnitude of the eigenvalues. This observation is indeed correct. We will explain why this happens when we discuss *change of basis* in the next section.

Zero, imaginary and repeated eigenvalues

If the two eigenvalues of a 2×2 matrix are real and nonzero, the geometric picture is essentially the same as that in fig. 5.8. The transformation takes the blue circle to a red ellipse. The eigenvalues determine the size of the ellipse, and the eigenvectors determine its orientation.

You can then imagine that if one of the eigenvalues is zero (and the other is nonzero), the red ellipse would collapse (or *degenerate*) into a line. This situation is shown in the left panel of fig. 5.9, obtained by replacing the matrix in the code by the second matrix in the question, namely

$$\begin{pmatrix} 1 & 1 \\ 1/2 & 1/2 \end{pmatrix}.$$

A quick calculation shows that the matrix has the following eigenvalues and eigenvectors

$$\lambda_1 = \frac{3}{2}, \ \mathbf{e}_1 = \begin{pmatrix} 2 \\ 1 \end{pmatrix}, \qquad \lambda_2 = 0, \ \mathbf{e}_2 = \begin{pmatrix} -1 \\ 1 \end{pmatrix}.$$

Indeed we see in the figure that the direction of the red line is $\mathbf{e}_1$. The direction $\mathbf{e}_2$ is shown by the green line, although we can't see from this snapshot that the blue and red vectors are actually lined up along $\mathbf{e}_2$. It is worth noting that when one of the eigenvalues is zero, the matrix also has zero determinant and is therefore not invertible.

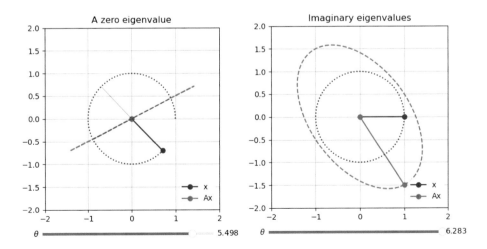

Fig. 5.9: *Left*: For the matrix $\begin{pmatrix} 1 & 1 \\ 1/2 & 1/2 \end{pmatrix}$, the red ellipse previously seen now degenerates into a line due to a zero eigenvalue. *Right*: For the matrix $\begin{pmatrix} 1 & 1 \\ -3/2 & 1/2 \end{pmatrix}$, the vectors never become parallel in this plane because the eigenvalues are not real.

The right panel of fig. 5.9 shows the geometric picture for the final matrix in the question:

$$\begin{pmatrix} 1 & 1 \\ -3/2 & 1/2 \end{pmatrix}.$$

However, this time it is impossible to make the red and blue vectors parallel – moving the slider simply makes the vectors chase each other round the circle. This is because the matrix has complex eigenvalues and eigenvectors, namely

$$\lambda_1 = \frac{3 + \sqrt{23}i}{4}, \; \mathbf{e}_1 = \begin{pmatrix} -1 - \sqrt{23}i \\ 6 \end{pmatrix}, \qquad \lambda_2 = \frac{3 - \sqrt{23}i}{4}, \; \mathbf{e}_2 = \begin{pmatrix} -1 + \sqrt{23}i \\ 6 \end{pmatrix}.$$

We can imagine that the vectors would line up if we add two extra dimensions to the plot to take care of the fact that each component of $\mathbf{x}$ is a complex number.

It's useful to confirm the above calculations with SymPy.

Eigenvalues and eigenvectors with SymPy's `eigenvects`

```
import sympy as sp
A = sp.Matrix([[1,1],[-sp.S(3)/2,sp.S.Half]])
A.eigenvects()
```

The 'Sympify' operator S converts a Python object to a SymPy object. In the above, `S(3)/2` is prevented from being converted to floating point 1.5 (you should try it without Sympifying). Python then gives us the following output on the right (explanation on the left).

SymPy output for `eigenvects`

λ_1	`[(3/4 - sqrt(23)*I/4,`
Its multiplicity	`    1,`
$\mathbf{e}_1$	`    [Matrix([`
	`      [-1/6 + sqrt(23)*I/6],`
	`      [              1]])]),`
λ_2	`  (3/4 + sqrt(23)*I/4,`
Its multiplicity	`    1,`
$\mathbf{e}_2$	`    [Matrix([`
	`      [-1/6 - sqrt(23)*I/6],`
	`      [              1]])])]`

The *multiplicity* of an eigenvalue λ_i means how many times it algebraically counts as a root of the *characteristic polynomial* $\det(A - \lambda I) = 0$. An example of a matrix with repeated real eigenvalues is

$$\begin{pmatrix} 1 & 1/2 \\ 0 & 1 \end{pmatrix}.$$

Solving the characteristic polynomial in the usual way, you will find that the quadratic equation has repeated real root $\lambda_1 = \lambda_2 = 1$ (*i.e.* the eigenvalue has multiplicity 2) with eigenvector $\mathbf{e}_1 = \mathbf{e}_2 = \begin{pmatrix} 1 \\ 0 \end{pmatrix}$. An $n \times n$ matrix with fewer than n linearly independent eigenvectors is called a *defective matrix*. Can you predict how the vector alignment game will look like in this case? (see exercise 6c).

Finally, here is how to do the same eigenvalue/eigenvector calculations with `scipy.linalg`. This is much faster than SymPy and it should be your default method unless symbolic calculations are required. We use this method in `eigshow.ipynb` to plot the eigen-directions in green.

Eigenvalues and eigenvectors with `scipy.linalg.eig`

```
import numpy as np
import scipy.linalg as LA
A = np.array([[1, 1],[-1.5, 0.5]])
lam, w = LA.eig(A)
```

`lam` is now an array of eigenvalues, and `w` is a 2D array whose columns are the normalised eigenvectors of A.

Unlike the SymPy method, `scipy.linalg.eig` can handle large matrices fairly quickly (see exercise 4). Also note that the SymPy method does not normalise the eigenvectors, but SciPy does.

DISCUSSION

• **Dynamical systems revisited**. In chapter 4, we briefly discussed the stability of fixed points in the phase space of a dynamical system. In fact, the stability of fixed points can be determined by eigenvalues of a matrix. In particular, if we *linearise* the dynamical system $\mathbf{x}' = \mathbf{F}(\mathbf{x})$ around a fixed point and consider the linearised system $\mathbf{x}' = A\mathbf{x}$. Then, the signs of the real part of the eigenvalues of A determine the stability of the fixed point. See [75] for more details on linearised stability.

The key results are as follows: if all eigenvalues of A have *negative real parts*, the fixed point is stable (*i.e.* it is a local attractor). If all eigenvalues have positive real parts, then the fixed point is unstable (a local repeller). If the real parts of the eigenvalues have different signs, then it is a saddle point (it is stable in some directions but unstable in other).

For concreteness, let's revisit the Lorenz system in §4.6 where we found that the origin was one of the fixed points. Let's determine whether the origin is stable or unstable. Linearising the system around the origin (*i.e.* ignoring quadratic terms) and using standard parameter values $\sigma = 10, b = 8/3, r = 28$, we obtain the linearised system:

$$\mathbf{x}' = A\mathbf{x}, \quad \text{where } A = \begin{pmatrix} -10 & 10 & 0 \\ 28 & -1 & 0 \\ 0 & 0 & -8/3 \end{pmatrix}.$$

You should verify that the eigenvalues of A are

$$-\frac{8}{3}, -\frac{11 \pm \sqrt{1201}}{2}.$$

There are two negative eigenvalues and one positive eigenvalue. Hence the origin is not a stable fixed point, but rather, a saddle point. This is consistent with the behaviour of trajectories that we saw in the animation: a trajectory flies towards the origin and then away from it, forming the butterfly wings.

• **Cayley-Hamilton Theorem**. The following theorem gives a surprising relationship between a square matrix A and its characteristic polynomial $\det(A - \lambda I) = 0$.

Theorem 5.3 *(Cayley-Hamilton Theorem) Every square matrix satisfies its own characteristic polynomial.*

For instance, we found that the matrix $A = \begin{pmatrix} 1 & 1 \\ 3/2 & 1/2 \end{pmatrix}$ has characteristic polynomial $2\lambda^2 - 3\lambda - 2 = 0$. It is straightforward to verify that

$$2A^2 - 3A - 2I = \begin{pmatrix} 0 & 0 \\ 0 & 0 \end{pmatrix}.$$

For a general proof, see [14]. See exercise 5 for an application.

The theorem is named after the British mathematician *Arthur Cayley* (1821–1895) and the Irish mathematician *Sir William Rowan Hamilton* (1805–1865). Both were prolific mathematicians who made major contributions from a young age: Cayley published at least 28 papers by the age of 25; Hamilton became Professor of Astronomy and Royal Astronomer of Ireland at the age of 22, whilst still an undergraduate.

eigshow.ipynb (for plotting fig. 5.8)	
| | ```
import numpy as np
from scipy import linalg as LA
import matplotlib.pyplot as plt
from matplotlib.widgets import Slider
%matplotlib
``` |
| **Input the matrix here**<br>Find its eigenvectors (w) | ```
A = np.array([[1,1],[1.5,0.5]])
lam, w = LA.eig(A)
``` |
| Leave space for a slider<br>Equal aspect ratio<br><br>Set plotting window | ```
fig, ax = plt.subplots()
plt.subplots_adjust(bottom=0.2)
ax.set_aspect('equal')
ax.set_title('Make the vectors parallel!')
ax.axis([-2,2,-2,2])
ax.grid('on')
``` |
| Plot the blue vector $\mathbf{x} = (1, 0)^T$<br>Plot $A\mathbf{x}$ as a red vector<br>Add legend | ```
Vec1, = ax.plot([0,1], [0,0],'bo-')
Vec2, = ax.plot([0,A[0,0]], [0,A[1,0]],'ro-')
plt.legend(['x', 'Ax'], loc='lower right')
``` |
| Locus of $\mathbf{x}$ (blue dotted circle)<br>Locus of $A\mathbf{x}$ (dashed red ellipse)<br>Plot the directions of the 2 eigenvectors<br>in thin green lines | ```
Curve1,=ax.plot(1, 0,'b:')
Curve2,=ax.plot(A[0,0], A[1,0],'r--')
ax.plot([-w[0,0],w[0,0]],
        [-w[1,0],w[1,0]],'g', lw=0.5)
ax.plot([-w[0,1],w[0,1]],
        [-w[1,1],w[1,1]],'g', lw=0.5)
``` |
| Dimension and location of $\theta$ slider<br>Label slider in LaTeX $\theta$<br>with $\theta \in [0, 2\pi]$ | ```
axt = plt.axes([0.25, 0.1, 0.5, 0.02])
t_slide = Slider(axt, r'$\theta$',
                 0, 2*np.pi, valstep=0.001,
                 valinit=0)
``` |
| Take $\theta$ value from the slider<br>Use 100 values to plot each locus | ```
def update(val):
    t = t_slide.val
    T = np.linspace(0,t,100)
``` |
| Arrays of cos and sin values are<br>used to plot the loci of $\mathbf{x}$ (blue circle)<br>Reshape $\mathbf{x}$ into a $(2 \times 100)$ matrix<br>Calculate $A\mathbf{x}$ (red ellipse)<br>Now we update 4 things:<br>  - The blue vector<br>  - The red vector<br>  - The blue circle<br>  - The red ellipse | ```
    cT = np.cos(T)
    sT = np.sin(T)
    xT=np.array([[cT],[sT]]).reshape(2,len(T))
    AxT= A@xT

    Vec1.set_data([0,cT[-1]], [0,sT[-1]])
    Vec2.set_data([0,AxT[0,-1]],[0,AxT[1,-1]])
    Curve1.set_data(cT, sT)
    Curve2.set_data(AxT[0,:], AxT[1,:])

    fig.canvas.draw_idle()
``` |
| Replot when the slider is changed | ```
t_slide.on_changed(update)
``` |
| | ```
plt.show()
``` |

5.7 Diagonalisation: Fibonacci revisited

> Use matrix diagonalisation to find the expression for the nth term of the Fibonacci sequence defined by
>
> $$F_{n+2} = F_{n+1} + F_n, \qquad F_0 = 0, \ F_1 = 1. \tag{5.19}$$

You may be surprised to know that this problem can be solved elegantly with vectors and matrices despite them not appearing explicitly in the question. There are a few ideas that come together to help us do this.

Idea 1 – A linear recurrence relation like (5.19) *can* be expressed in matrix form.

Idea 2 – The nth term of a sequence can be obtained by finding the nth power of a matrix.

Idea 3 – The nth power of a matrix can easily be computed (by hand) using matrix *diagonalisation* (explained below).

The goal is to obtain F_n in terms of n using these ideas. Let's begin.

Idea 1: Let's find a matrix A such that

$$\begin{pmatrix} F_{n+2} \\ F_{n+1} \end{pmatrix} = A \begin{pmatrix} F_{n+1} \\ F_n \end{pmatrix}. \tag{5.20}$$

In other words, multiplying by the vector by A shifts the sequence forward, revealing the next term. Using the recurrence relation (5.19), we can rewrite the LHS of (5.20) as

$$\begin{pmatrix} F_{n+2} \\ F_{n+1} \end{pmatrix} = \begin{pmatrix} F_{n+1} + F_n \\ F_{n+1} \end{pmatrix} = \begin{pmatrix} 1 & 1 \\ 1 & 0 \end{pmatrix} \begin{pmatrix} F_{n+1} \\ F_n \end{pmatrix}$$

$$\implies A = \begin{pmatrix} 1 & 1 \\ 1 & 0 \end{pmatrix}.$$

Idea 2: Observe in the chain of equalities below that by multiplying $\begin{pmatrix} F_1 \\ F_0 \end{pmatrix}$ repeatedly by A, we will obtain an explicit expression for F_n in terms of n (by simply reading off the lower component).

$$A^n \begin{pmatrix} F_1 \\ F_0 \end{pmatrix} = A^{n-1} \begin{pmatrix} F_2 \\ F_1 \end{pmatrix} = A^{n-2} \begin{pmatrix} F_3 \\ F_2 \end{pmatrix} = \ldots = \begin{pmatrix} F_{n+1} \\ F_n \end{pmatrix}.$$

Idea 3: There is a smarter way to calculate A^n than brute-force matrix multiplication. Here is a very important theorem in introductory linear algebra.

Theorem 5.4 (*Diagonalising a matrix*) *Suppose that A is a $k \times k$ matrix with k linearly independent eigenvectors, $\mathbf{e}_1, \mathbf{e}_2, \ldots \mathbf{e}_k$, with corresponding eigenvalues $\lambda_1, \lambda_2, \ldots \lambda_k$. Let P be a matrix whose ith column is $\mathbf{e}_i$. Then,*

$$A = PDP^{-1}, \quad \text{where } D = \begin{pmatrix} \lambda_1 & 0 & \cdots & 0 \\ 0 & \lambda_2 & \cdots & 0 \\ & & \ddots & \\ 0 & 0 & \cdots & \lambda_k \end{pmatrix}. \tag{5.21}$$

The theorem simply follows from the observation that $AP = \begin{pmatrix} \lambda_1 \mathbf{e}_1 & \ldots & \lambda_n \mathbf{e}_n \end{pmatrix} = PD$. We say that A is *diagonalised* by P.

A very useful application of diagonalisation is in the calculation of integer powers of a matrix. First, note the identity

$$A^2 = (PDP^{-1})(PDP^{-1}) = PD(P^{-1}P)DP^{-1} = PD^2P^{-1}$$

which follows from the associativity of matrix multiplication. Inductively, we have, for all positive integers n,

$$A^n = PD^nP^{-1} = P\begin{pmatrix} \lambda_1^n & 0 & \cdots & 0 \\ 0 & \lambda_2^n & \cdots & 0 \\ & & \ddots & \\ 0 & 0 & \cdots & \lambda_k^n \end{pmatrix}P^{-1}$$

This only involves multiplying 3 rather than n matrices, thanks to the fact that the nth power of a diagonal matrix can be easily calculated by raising each diagonal element to the nth power.

Using our matrix $A = \begin{pmatrix} 1 & 1 \\ 1 & 0 \end{pmatrix}$ and applying the calculation method outlined in the previous section, we obtain the characteristic polynomial

$$\lambda^2 - \lambda - 1 = 0 \tag{5.22}$$

and the following eigenvalues and eigenvectors.

$$\lambda_1 = \frac{1 + \sqrt{5}}{2}, \; \mathbf{e}_1 = \begin{pmatrix} \lambda_1 \\ 1 \end{pmatrix}, \qquad \lambda_2 = \frac{1 - \sqrt{5}}{2}, \; \mathbf{e}_2 = \begin{pmatrix} \lambda_2 \\ 1 \end{pmatrix}.$$

We then use the eigenvectors as columns of P.

$$P = \begin{pmatrix} \lambda_1 & \lambda_2 \\ 1 & 1 \end{pmatrix}, \qquad D = \begin{pmatrix} \lambda_1 & 0 \\ 0 & \lambda_2 \end{pmatrix}, \qquad P^{-1} = \frac{1}{\sqrt{5}}\begin{pmatrix} 1 & -\lambda_2 \\ -1 & \lambda_1 \end{pmatrix}. \tag{5.23}$$

Now we can raise A to the nth power easily.

$$\begin{pmatrix} F_{n+1} \\ F_n \end{pmatrix} = A^n \begin{pmatrix} F_1 \\ F_0 \end{pmatrix} = PD^nP^{-1}\begin{pmatrix} 1 \\ 0 \end{pmatrix} = \frac{1}{\sqrt{5}}\begin{pmatrix} \lambda_1^{n+1} - \lambda_2^{n+1} \\ \lambda_1^n - \lambda_2^n \end{pmatrix}.$$

Finally, note that $\lambda_1 = \phi \approx 1.618$ (the Golden Ratio – see §1.6) and $\lambda_2 = 1 - \phi$. Reading off the lower component of the above equation gives us the expression for F_n in terms of n (this is the so-called *Binet's formula*).

$$F_n = \frac{\phi^n - (1 - \phi)^n}{\sqrt{5}}. \tag{5.24}$$

Exercise 7 challenges you to apply this method on a 4-term recurrence relation.

Visualising diagonalisation

If A represents a linear transformation, then the diagonalisation $A = PDP^{-1}$ can be thought of as breaking A down into a sequence of 3 transformations. Let's use Python to help us visualise these transformations.

But before we get to the code, it would be useful to reexamine the matrix P in eq. 5.23. Recall that if $\mathbf{e}_1$ is an eigenvector of A, then so is any nonzero constant multiple of $\mathbf{e}_1$. In particular, we could fill in the columns of P with *normalised* eigenvectors $\hat{\mathbf{e}}_1$ and $\hat{\mathbf{e}}_2$:

$$\hat{\mathbf{e}}_1 = \frac{1}{\sqrt{\phi+2}} \begin{pmatrix} \phi \\ 1 \end{pmatrix} \qquad \hat{\mathbf{e}}_2 = \frac{1}{\sqrt{3-\phi}} \begin{pmatrix} 1-\phi \\ 1 \end{pmatrix} = \frac{1}{\sqrt{\phi+2}} \begin{pmatrix} -1 \\ \phi \end{pmatrix} \tag{5.25}$$

(where we have used the equation $\phi^2 - \phi - 1 = 0$ to turn higher powers of ϕ into linear functions of ϕ). In this form, it is easy to see that

$$\hat{\mathbf{e}}_1 \cdot \hat{\mathbf{e}}_1 = \hat{\mathbf{e}}_2 \cdot \hat{\mathbf{e}}_2 = 1, \quad \hat{\mathbf{e}}_1 \cdot \hat{\mathbf{e}}_2 = 0.$$

When these properties are satisfied, we say that the set $\{\hat{\mathbf{e}}_1, \hat{\mathbf{e}}_2\}$ is an *orthonormal basis* of $\mathbb{R}^2$.

Now let $P = \begin{pmatrix} \hat{\mathbf{e}}_1 & \hat{\mathbf{e}}_2 \end{pmatrix}$. Let's interpret the transformations represented by P, D and P^{-1}.

- P can be expressed as $\begin{pmatrix} \cos\theta & -\sin\theta \\ \sin\theta & \cos\theta \end{pmatrix}$ where $\theta = \sin^{-1}\left(\frac{1}{\sqrt{\phi+2}}\right)$. In other words, P represents a rotation of $\theta \approx 31.7°$ anticlockwise.
- P^{-1} therefore represents a clockwise rotation of the same angle.
- $D = \begin{pmatrix} \phi & 0 \\ 0 & 1-\phi \end{pmatrix} \approx \begin{pmatrix} 1.618 & 0 \\ 0 & -0.618 \end{pmatrix}$ represents a scaling by a factor of ≈ 1.618 in the x direction (an enlargement) and a scaling by a factor of ≈ -0.618 in the y direction (shrinking and reflecting across the x-axis).

Fig. 5.10 summarises the sequence of transformations PDP^{-1} performed on the shape shown on the top left panel (a unit circle with a line joining the origin to $(1,0)$). These pictures are the output of the code `diagonalise.ipynb` which produces an interactive plot with a slider that transforms the circle stage by stage, starting with P^{-1}. The final transformed shape is the tilted ellipse shown in the top right panel.

From our investigation in the previous section, and the code output in this section, one might make the following conjecture: the semi-major and semi-minor axes of the tilted ellipse are parallel to the eigenvectors $\hat{\mathbf{e}}_1, \hat{\mathbf{e}}_2$, with length ratio $\lambda_1 : \lambda_2$. The picture below summarises this conjecture.

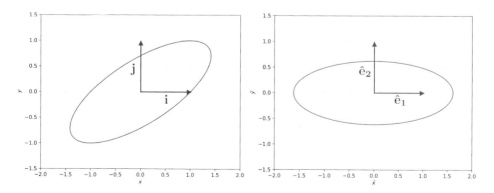

Fig. 5.11: Change of basis from the standard basis (left) to the basis comprising eigenvectors $\hat{\mathbf{e}}_1, \hat{\mathbf{e}}_2$ (right). The tilted ellipse $x^2 - 2xy + 2y^2 = 1$ on the left takes a much simpler form in the eigenbasis (eq. 5.26).

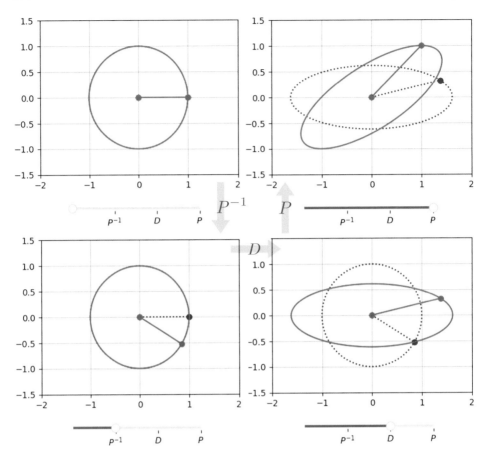

Fig. 5.10: Visualising matrix diagonalisation. Breaking down the linear transformation represented by A into a sequence of 3 transformations PDP^{-1} (done in reverse order), namely: a) P^{-1} (clockwise rotation), b) D (enlargement in the x direction and shrinking+reflection in the y direction), and c) P (anticlockwise rotation). The blue dashed shape in each panel shows the pre-transformed shape in the previous stage.

| DISCUSSION |

- **Change of basis.** Let's examine why the size and orientation of the ellipse are determined by the eigenvalues and eigenvectors of $A = PDP^{-1}$. Let the vector $\mathbf{x}$ be written in two ways: 1) using the standard basis as $\mathbf{x} = x\mathbf{i} + y\mathbf{j}$ and 2) the eigenbasis as $\mathbf{x} = \tilde{x}\hat{\mathbf{e}}_1 + \tilde{y}\hat{\mathbf{e}}_2$. Equating the two expressions, we find

$$\underbrace{\begin{pmatrix} \mathbf{i} & \mathbf{j} \end{pmatrix}}_{I} \begin{pmatrix} x \\ y \end{pmatrix} = \underbrace{\begin{pmatrix} \hat{\mathbf{e}}_1 & \hat{\mathbf{e}}_2 \end{pmatrix}}_{P} \begin{pmatrix} \tilde{x} \\ \tilde{y} \end{pmatrix} \implies \begin{pmatrix} \tilde{x} \\ \tilde{y} \end{pmatrix} = P^{-1}\mathbf{x}.$$

In other words, we can think of P^{-1} as the *change-of-basis matrix* which gives the coordinates of $\mathbf{x}$ in the eigenbasis. In the new coordinates, the ellipse has a simple equation

$$\left(\frac{\tilde{x}}{\lambda_1}\right)^2 + \left(\frac{\tilde{y}}{\lambda_2}\right)^2 = 1. \tag{5.26}$$

This change of basis also explains why the ellipse seen in fig. 5.8 is aligned along the eigenvectors with the size determined by the eigenvalues.

- **The spectral theorem**. In the previous section, we found the eigenvectors for the matrix $\begin{pmatrix} 1 & 1 \\ 3/2 & 1/2 \end{pmatrix}$ (the green lines in fig. 5.8). The eigenvectors are not orthogonal. On the other hand, in this section we found that the eigenvectors for $\begin{pmatrix} 1 & 1 \\ 1 & 0 \end{pmatrix}$ are orthogonal (and therefore, after normalisation, orthonormal). Is there a way to tell whether a matrix has orthogonal eigenvectors? Here is a very important theorem in linear algebra.

  Theorem 5.5 (*Spectral theorem*) *Every real symmetric matrix has real eigenvalues and orthogonal eigenvectors.*

  See [194] for proof. The spectral theorem explains why the matrix $A = \begin{pmatrix} 1 & 1 \\ 1 & 0 \end{pmatrix}$ gave rise to an orthonormal eigenbasis (because it is symmetric).
  The orthonormality of the eigenvectors is equivalent to the expression $P^T P = I$ (where the columns of P are the eigenvectors of A). In other words, $P^{-1} = P^T$, and we could have diagonalised the matrix as $A = PDP^T$.

- **Method of characteristics**. We have shown how the Fibonacci recurrence relation could be solved using matrix diagonalisation. In practice, there is a more compact method called the *method of characteristics* which skips writing down the matrix and condenses the method to its bare essential.
  The method goes like this. To solve $F_{n+2} - F_{n-1} - F_n = 0$ with $(F_0, F_1) = (0, 1)$:

  1. Write down the *characteristic polynomial*

  $$m^2 - m - 1 = 0.$$

  2. Solve it to obtain two roots:

  $$m_\pm = \frac{1 \pm \sqrt{5}}{2}.$$

  3. The solution for F_n is

  $$F_n = C_1(m_+)^n + C_2(m_-)^n,$$

  where C_1 and C_2 are constants to be determined.
  4. Substitute the 'initial condition' $(F_0, F_1) = (0, 1)$ to find:

  $$\left. \begin{array}{r} C_1 + C_2 = 0 \\ m_+ C_1 + m_- C_2 = 1 \end{array} \right\} \implies C_1 = \frac{1}{\sqrt{5}}, \; C_2 = -\frac{1}{\sqrt{5}}.$$

This yields the Binet's formula (5.24) as expected. See [104] for details of why this method produces the same result as the diagonalisation method.

diagonalise.ipynb (for plotting each panel of fig. 5.10)	
| | ```
import numpy as np
from scipy import linalg as LA
import matplotlib.pyplot as plt
from matplotlib.widgets import Slider
%matplotlib
``` |
| **Input the matrix here** | ```
A= np.array([[1,1],[1,0]])
``` |
| The eigenvalues and matrix $P$<br>We also need $P^{-1}\dots$<br>and the diagonal matrix $D$ | ```
lam, P = LA.eig(A)
Pinv = LA.inv(P)
D = np.array([[lam[0],0],[0,lam[1]]])
``` |
| Leave space for a slider<br>Equal aspect ratio<br>Set plotting window | ```
fig, ax = plt.subplots()
plt.subplots_adjust(bottom=0.2)
ax.set_aspect('equal')
ax.axis([-2,2,-1.5,1.5])
ax.grid('on')
``` |
| Polar angle $\in [0, 2\pi]$<br>The unit circle **x** to be transformed<br>Reshape **x** into a $(2 \times 100)$ matrix<br>$P^{-1}\mathbf{x}$<br>$DP^{-1}\mathbf{x}$<br>$PDP^{-1}\mathbf{x}$ | ```
T = np.linspace(0,2*np.pi,100)
X = np.array([[np.cos(T)],[np.sin(T)]])\
    .reshape(2,len(T))
PinvX= Pinv@X
DPinvX= D@PinvX
PDPinvX= P@DPinvX
``` |
| The original shape (circle with line segment) in blue dotted line<br>The transformed shape in red solid line | ```
Vec1,  = ax.plot([0,X[0,0]], [0,X[1,0]],'bo:')
Curve1,= ax.plot(X[0,:], X[1,:],'b:')
Vec2,  = ax.plot([0,X[0,0]], [0,X[1,0]],'ro-')
Curve2,= ax.plot(X[0,:], X[1,:],'r-')
``` |
| Slider dimension and location<br>Divide the diagonalisation into 3 steps<br>with smooth transition in between<br>Hide slider values ($t \in [0, 3]$)<br>Add custom labels to our slider<br>At $t$ values 1, 2, 3<br>set the labels as $P^{-1}$, $D$, $P$ respectively | ```
axt = plt.axes([0.25, 0.1, 0.5, 0.02])
t_slide = Slider(axt, '',0, 3, valstep=0.001,
                    valinit=0)
t_slide.valtext.set_visible(False)
axt.add_artist(axt.xaxis)
slider_ticks = [1,2,3]
slider_label = [r'$P^{-1}$', r'$D$', r'$P$']
axt.set_xticks(slider_ticks, slider_label)
``` |
| Generate data at a given $t$ value<br>Do smooth linear interpolation between:<br>a) Original state (**x**) at $t = 0$<br>b) Transformed state ($P^{-1}\mathbf{x}$) at $t = 1$<br>Interpolation between the next 2 states:<br>a) Original ($P^{-1}\mathbf{x}$) at $t = 1$<br>b) Transformed ($DP^{-1}\mathbf{x}$) at $t = 2$<br>And the final 2 states:<br>a) Original ($DP^{-1}\mathbf{x}$) at $t = 2$<br>b) Transformed ($PDP^{-1}\mathbf{x}$) at $t = 3$ | ```
def Frame(t):
    if t<1:
        Xt=X
        Ft=t*PinvX+(1-t)*Xt
    elif 1<=t and t<2:
        Xt=PinvX
        Ft=(t-1)*DPinvX+(2-t)*Xt
    else:
        Xt=DPinvX
        Ft=(t-2)*PDPinvX+(3-t)*Xt
    return Xt, Ft

# Code continues on the next page
``` |

diagonalise.ipynb (continued)

	``python
	def update(val):
	    t = t_slide.val```

Take t value from the slider	
Data for the original/transformed shapes	```python Xt, Ft = Frame(t)```
Update the original (blue) shape	```python Curve1.set_data(Xt[0,:], Xt[1,:]) Vec1.set_data([0,Xt[0,-1]], [0,Xt[1,-1]])```

Update the transformed (red) shape	```python Curve2.set_data(Ft[0,:], Ft[1,:])
	    Vec2.set_data([0,Ft[0,-1]], [0,Ft[1,-1]])
	    fig.canvas.draw_idle()```

Replot when the slider is changed	```python t_slide.on_changed(update) plt.show()```

5.8 Singular-Value Decomposition

a) Perform singular-value decomposition (SVD) on the matrix

$$A = \begin{pmatrix} 1 & 1 & 0 \\ 0 & 0 & 1 \end{pmatrix},$$

and obtain its rank-1 approximation.

b) Use SVD to perform image reduction using an image of your choice.

In the previous section, we saw how a square matrix A (satisfying certain condition) could be decomposed into $A = PDP^{-1}$ where D is a diagonal matrix containing the eigenvalues of A (theorem 5.4). In this section, we will generalise the theorem to *any* matrix.

Some observations.

(a) Given any matrix A, the matrices AA^T and $A^T A$ are symmetric matrices.
    (This follows from the property $(AB)^T = B^T A^T$. Put $B = A^T$.)
(b) By the spectral theorem (theorem 5.5), the eigenvectors $\mathbf{u_i}$ of AA^T can be chosen to be an orthonormal set. (Same for the eigenvectors $\mathbf{v}_i$ of $A^T A$.)
(c) The matrices AA^T and $A^T A$ share the same eigenvalues.
(d) The eigenvalues of AA^T and $A^T A$ are non-negative.

Here's a quick proof of (c). Suppose AA^T has eigenvalues λ_i, then, multiplying the eigenvalue relation by A^T, we have:

$$AA^T \mathbf{u_i} = \lambda_i \mathbf{u_i} \implies A^T A(A^T \mathbf{u_i}) = \lambda_i (A^T \mathbf{u_i}). \tag{5.27}$$

In other words, $A^T A$ has eigenvalues λ_i corresponding to eigenvectors $\mathbf{v}_i = A^T \mathbf{u_i}$.

For (d), since the eigenvectors $\{\mathbf{u_i}\}$ of AA^T are orthonormal, we have (dropping the subscript for clarity):

$$1 = \mathbf{u} \cdot \mathbf{u} = \mathbf{u}^T \mathbf{u} \implies \lambda = \mathbf{u}^T (\lambda \mathbf{u}) = \mathbf{u}^T (AA^T \mathbf{u}) = (A^T \mathbf{u})^T (A^T \mathbf{u}) = |A^T \mathbf{u}|^2 \geq 0.$$

All these properties suggest that even though A may be an odd-shaped matrix with no discernible special properties or symmetry, we can instead turn to AA^T and $A^T A$, which are square, symmetric matrices, and study their 'eigen-properties'. This would then shed light on the information content of A itself.

The following theorem makes use of the above properties. It is one of the most important theorems in modern applications of linear algebra. For proof, see [103, 194].

Theorem 5.6 (*Singular-value decomposition*) *Let A be an m × n matrix of rank r. Then, we can write*

$$A = U\Sigma V^T,$$

where

- *U is an m × m orthogonal matrix whose columns are the normalised eigenvectors of AA^T (these are called the 'left singular vectors')*
- *V is an n × n orthogonal matrix whose columns are the normalised eigenvectors of $A^T A$ (the 'right singular vectors')*

• Σ *is an* $m \times n$ *diagonal matrix*

$$\Sigma = \left(\begin{array}{c|c} D_{r \times r} & \mathbf{0} \\ \hline \mathbf{0} & \mathbf{0} \end{array}\right).$$

$D = diag\,(\sigma_1, \sigma_2, \ldots \sigma_r)$, *where* σ_i *(the singular values) are the square roots of the nonzero eigenvalues of* AA^T, *arranged in decreasing order of magnitude.*

The rank of A is precisely the number of singular values in Σ. One can see this from the decomposition $A = U\Sigma V^T$ as follows: note that U and V are invertible matrices ($U^{-1} = U^T$). Multiplication by invertible matrices does not change the rank of a matrix. Therefore, we must have rank(A) = rank(Σ).

SVD by hand

.

Let perform SVD on the matrix $A = \begin{pmatrix} 1 & 1 & 0 \\ 0 & 0 & 1 \end{pmatrix}$. Follow these steps.

Step 1: Find the eigenvalues and eigenvectors of AA^T or $A^T A$ (whichever is easier). We find

$$AA^T = \begin{pmatrix} 2 & 0 \\ 0 & 1 \end{pmatrix}, \qquad A^T A = \begin{pmatrix} 1 & 1 & 0 \\ 1 & 1 & 0 \\ 0 & 0 & 1 \end{pmatrix}.$$

Obviously it is easier to work with AA^T. The eigenvalues can be read off along the diagonal and the (normalised) eigenvectors are straightforward to write down. They are:

$$\lambda_1 = 2, \ \mathbf{u_1} = \begin{pmatrix} 1 \\ 0 \end{pmatrix} \qquad \lambda_2 = 1, \ \mathbf{u_2} = \begin{pmatrix} 0 \\ 1 \end{pmatrix}$$

This gives us the first orthogonal matrix

$$U = \begin{pmatrix} 1 & 0 \\ 0 & 1 \end{pmatrix}.$$

We also have the singular values $\sigma_i = \sqrt{\lambda_i}$. Hence,

$$\Sigma = \begin{pmatrix} \sqrt{2} & 0 & 0 \\ 0 & 1 & 0 \end{pmatrix}$$

Make sure to the eigenvalues are ordered in decreasing magnitude, and that the matrix Σ is padded with zeros so that it has the correct dimension.

Step 2: Find the eigenvectors of $A^T A$. We don't necessarily have to do this by solving the characteristic polynomial. Instead, we can use what we know from eq. 5.27 that $\mathbf{v}_i = A^T \mathbf{u_i}$. We get two eigenvectors straightaway (remember to normalise them).

$$A^T \mathbf{u_1} = \begin{pmatrix} 1 & 0 \\ 1 & 0 \\ 0 & 1 \end{pmatrix} \begin{pmatrix} 1 \\ 0 \end{pmatrix} \implies \mathbf{v}_1 = \frac{1}{\sqrt{2}} \begin{pmatrix} 1 \\ 1 \\ 0 \end{pmatrix}.$$

$$A^T \mathbf{u_2} = \begin{pmatrix} 1 & 0 \\ 1 & 0 \\ 0 & 1 \end{pmatrix} \begin{pmatrix} 0 \\ 1 \end{pmatrix} \implies \mathbf{v_2} = \begin{pmatrix} 0 \\ 0 \\ 1 \end{pmatrix}.$$

However, we expect 3 mutually orthogonal eigenvectors $\{\mathbf{v}_1, \mathbf{v}_2, \mathbf{v}_3\}$. In this case, we can simply write down what the third basis vector should be by inspection (perhaps by thinking about the configuration of $\{\mathbf{v}_1, \mathbf{v}_2, \mathbf{v}_3\}$ in $\mathbb{R}^3$). One option is

$$\mathbf{v}_3 = \frac{1}{\sqrt{2}} \begin{pmatrix} 1 \\ -1 \\ 0 \end{pmatrix},$$

(or you could also choose its negative). If it is too difficult to figure out the missing vector(s) by inspection, one could use row-reduction to help identify what other linearly independent vectors should be added to the basis. Then apply the *Gram-Schmidt process* (discussed in §5.10) to obtain an orthonormal basis.

In any case, we have now obtained the other orthogonal matrix V.

$$V = \begin{pmatrix} 1/\sqrt{2} & 0 & 1/\sqrt{2} \\ 1/\sqrt{2} & 0 & -1/\sqrt{2} \\ 0 & 1 & 0 \end{pmatrix}.$$

Step 3: Complete the decomposition $A = U\Sigma V^T$.

$$A = \begin{pmatrix} 1 & 0 \\ 0 & 1 \end{pmatrix} \begin{pmatrix} \sqrt{2} & 0 & 0 \\ 0 & 1 & 0 \end{pmatrix} \begin{pmatrix} 1/\sqrt{2} & 1/\sqrt{2} & 0 \\ 0 & 0 & 1 \\ 1/\sqrt{2} & -1/\sqrt{2} & 0 \end{pmatrix}.$$

Low-rank approximation

Since there are two singular values, A has rank 2. From its SVD, we can also express A in terms of the eigenvectors $\mathbf{u}_i$, $\mathbf{v}_i$ and its 2 singular values as:

$$A = \sigma_1 \mathbf{u_1}(\mathbf{v}_1)^T + \sigma_2 \mathbf{u_2}(\mathbf{v}_2)^T$$

$$= \sqrt{2} \begin{pmatrix} 1 \\ 0 \end{pmatrix} \begin{pmatrix} \frac{1}{\sqrt{2}} & \frac{1}{\sqrt{2}} & 0 \end{pmatrix} + \begin{pmatrix} 0 \\ 1 \end{pmatrix} \begin{pmatrix} 0 & 0 & 1 \end{pmatrix}$$

$$= \begin{pmatrix} 1 & 1 & 0 \\ 0 & 0 & 0 \end{pmatrix} + \begin{pmatrix} 0 & 0 & 0 \\ 0 & 0 & 1 \end{pmatrix}.$$

In the last line, the first piece is deemed to contain more information than the second, since it is associated with a larger singular value. Thus, we say that the rank-1 approximation of A is $\begin{pmatrix} 1 & 1 & 0 \\ 0 & 0 & 0 \end{pmatrix}$ (which seems like an obvious answer).

SVD with Python

In practice, SVD is usually deployed to find row-rank approximations of much larger matrices. We take a look at an example where SVD can be used to compress an image.

A grey-scale image can be represented by an $m \times n$ matrix in which each number represents the brightness of a pixel ranging from 0 (black) to 255 (white). An example is shown in fig. 5.12. The photo of *Epiphylum oxypetalum* (its flowers bloom for only one night) is represented by a 921×1464 matrix. The code `svd.ipynb` applies SVD to this photo and lets the user adjust the number of singular values N used for the rank-N approximation.

Note that the singular values are arranged in decreasing magnitude, and for this photo, the magnitude of σ_i drops off sharply (note the logarithmic scales). The plot suggest that the information contained in the term $\sigma_i \mathbf{u_i}(\mathbf{v}_i)^T$ for $i \gtrsim 100$ is far less crucial than those with $i < 100$. This explains why there is not much visual gain when increasing N from 100 to 921, as we can see from the figure.

One might ask: what is the saving in using the rank-N approximation rather than the full image? Well, the SVD has to store N eigenvectors $\mathbf{u}_i$ (each containing m numbers), N eigenvectors $\mathbf{v}_i$ (each containing n numbers), and of course the N singular values themselves. In total, the amount of data needed is $Nm + Nn + N = N(m + n + 1)$, in contrast with nm (the full data). In the photo compression in fig. 5.12, $(m, n) = (921, 1464)$, so using 100 singular values, we are retaining only around 18% of the full data, yet the reconstructed image is of satisfactory quality.

Finally a few observations about the code `svd.ipynb`.

- The key to the SVD operation is SciPy's `linalg.svd` function:

$$U, \ s, \ VT \ = \ svd(A)$$

  Note that the output s is a one-dimensional array of singular values (and not an $m \times n$ array).
- The compression step (resizing) using N singular values is

$$A = \sum_{i=1}^{N} \sigma_i \mathbf{u_i}(\mathbf{v}_i)^T. \tag{5.28}$$

  This rank-N approximation is done by the function `resize_image` in the code.
- The slider has been transformed so that it is logarithmic and its position matches the logarithmic x-axis of the singular-value plot. We also override the display on the slider so that it shows the value of N and not $\log_{10} N$.
- You can use any jpeg image – just place it in the same folder as the code. We recommend using the Pillow (PIL) package[6] to read in the image as a matrix and convert it to greyscale. This is achieved by the line:

$$A \ = \ Image.open('flowers.jpeg').convert('L')$$

  The code can also be modified to perform SVD on colour images. A walkthrough is given in exercise 9c.

---

[6] https://pillow.readthedocs.io/en/stable/

Fig. 5.12: A photo of *Epiphylum oxypetalum* ('Queen of the night') flowers found in Northern Thailand. The image is reduced using singular-value decomposition. The top left panel shows the size of the singular values arranged in decreasing order (note the logarithmic scales). The slider adjusts N, the number of singular values retained. The reconstructed images are shown for various N.

DISCUSSION

• **A visual summary of SVD**.

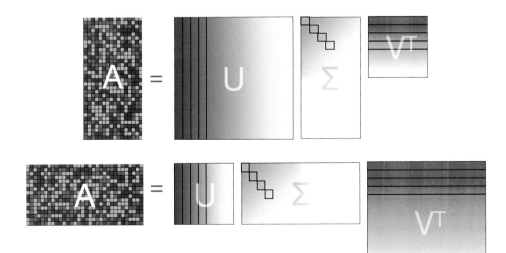

- **Real-world applications of SVD** are in situations where key information needs to be extracted from a large 2-dimensional data. In addition to image/data compression, SVD is also used in machine-learning tasks, such as:

  – Outlier detection and de-noising a dataset;
  – Linear regression (*i.e.* finding the best approximate solution to the system $A\mathbf{x} = \mathbf{b}$);
  – Recommender systems on video/shopping/social-media platforms;
  – Facial recognitions [150].

  More about the role of SVD in these applications in [2, 34, 195].

svd.ipynb (for plotting the top 2 panels of fig. 5.12)	
| | ```python
import numpy as np
import matplotlib.pyplot as plt
from matplotlib.widgets import Slider
from PIL import Image
from scipy.linalg import svd
%matplotlib
``` |
| PIL = Python Imaging Library | |
| Use your own jpeg image (we convert it to greyscale) | ```python
A = Image.open('flowers.jpeg').convert('L')
``` |
| Perform SVD | ```python
U, s, VT = svd(A)
``` |
| $N$ = number of singular values<br>Rank-N approximation (eq. 5.28) | ```python
def resize_image(N):
    return U[:,:N] @ np.diag(s[:N]) @ VT[:N,:]
``` |
| Plot two figures side by side | ```python
fig,(ax1, ax2) = plt.subplots(1,2,
                    figsize=(10,4))
``` |
| Space for the slider<br>Set the range of the index $i$ of $\sigma_i$ | ```python
plt.subplots_adjust(bottom=0.2)
ax1.set_xlim(1,len(s))
ax1.title.set_text('Singular values')
ax1.grid('on')
``` |
| Plot $\sigma_i$ against $i$ on log scales (red) | ```python
ax1.loglog(s,'ro-', markersize=3)
``` |
| Initial choice of $N$<br>Do rank-$N$ approximation<br>Display the resized image | ```python
N = 10
image = resize_image(N)
ax2.imshow(image, cmap='gray')
``` |
| Position the slider below the $x$-axis | ```python
axt = plt.axes([0.125, 0.1, 0.355, 0.05])
N_slide = Slider(axt, 'N',
``` |
| Apply $\log_{10}$ to the slider | ```python
        0, np.log10(len(s)),
        valstep=0.05, valinit=1)
``` |
| The correct value of $N$ on the slider | ```python
N_slide.valtext.set_text(N)
``` |
| Take $N$ from the slider (undo log)<br>Display the correct value of $N$<br>Do rank-$N$ approximation and display | ```python
def update(val):
    N = round(10**N_slide.val)
    N_slide.valtext.set_text(N)
    image = resize_image(N)
    ax2.imshow(image, cmap='gray')
    fig.canvas.draw_idle()
``` |
| Re-display the photo on the right when the slider is changed | ```python
N_slide.on_changed(update)

plt.show()
``` |
```

5.9 The Rank-Nullity Theorem

The *rank-nullity theorem* states that for an $m \times n$ matrix A,

$$\text{rank}(A) + \text{nullity}(A) = n. \qquad (5.29)$$

Verify the theorem for the following matrices.

$$A = \begin{pmatrix} 1 & -1 & -1 \\ 1 & 1 & 1 \\ 0 & 0 & 1 \end{pmatrix} \quad B = \begin{pmatrix} 1 & -1 & -1 \\ 0 & 0 & 1 \\ 0 & 0 & 1 \end{pmatrix} \quad C = \begin{pmatrix} 1 & -1 & -1 \\ 2 & -2 & -2 \\ 3 & -3 & -3 \end{pmatrix} \quad D = (3 \times 3) \text{ zero matrix.}$$

The rank-nullity theorem is one of the most important theorems in university-level linear algebra. In short, it follows from the observation that if (the RREF of) an $m \times n$ matrix has r leading 1s, then the solution of the equation $A\mathbf{x} = \mathbf{0}$ requires $n - r$ free variables. For proof of the theorem, see any of the textbooks recommended in the introduction of this chapter.

An equivalent rephrasing of the rank-nullity theorem is as follows: for an $m \times n$ matrix A,

$$\dim(\text{column space}) + \dim(\text{nullspace}) = n.$$

The theorem can also be expressed in linear-transformation terminology. Recall that the matrix A represents a linear transformation $T : V \to W$. In terms of T, the rank-nullity theorem says:

$$\dim(\text{im } T) + \dim(\text{ker } T) = \dim V.$$

Let's do some calculations before we demonstrate the theorem with Python. Take the matrix A in the question. Performing row reduction, we find

$$A = \begin{pmatrix} 1 & -1 & -1 \\ 1 & 1 & 1 \\ 0 & 0 & 1 \end{pmatrix} \xrightarrow{\text{RREF}} \begin{pmatrix} 1 & 0 & 0 \\ 0 & 1 & 0 \\ 0 & 0 & 1 \end{pmatrix}.$$

Hence, A has rank 3. The column space is the whole of $\mathbb{R}^3$.

To find the nullity, we solve $A\mathbf{x} = \mathbf{0}$ and count the number of free variables. In this case, since A is full-rank, it is invertible, and therefore we have the unique solution $\mathbf{x} = A^{-1}\mathbf{0} = \mathbf{0}$ (*i.e.* no free variables). The nullspace consists of a single vector $\mathbf{0}$, and the nullity is therefore 0.

We conclude that the rank-nullity theorem is satisfied, since 3 (rank) +0 (nullity) = 3 (number of columns).

For matrix B, we can tell by inspection that it has rank 2 (the last row is redundant). To be more precise, the column space is

$$\text{span} \left\{ \begin{pmatrix} 1 \\ 0 \\ 0 \end{pmatrix}, \begin{pmatrix} -1 \\ 1 \\ 1 \end{pmatrix} \right\}.$$

Geometrically, this is the plane $y - z = 0$. You can check this by taking the cross product of the two vectors.

To find the nullity, solving $B\mathbf{x} = \mathbf{0}$, we find the solutions

$$\mathbf{x} = c \begin{pmatrix} 1 \\ 1 \\ 0 \end{pmatrix}, \quad c \in \mathbb{R}.$$

The nullity is therefore 1 (a straight line). This is again consistent with the rank-nullity theorem $(2 + 1 = 3)$.

For matrix C, the rows are simply multiples of one another, and therefore the rank is 1. The column space is simply the line in the direction of the vector $\begin{pmatrix} 1 & 2 & 3 \end{pmatrix}^T$.

For the nullity, solving $C\mathbf{x} = \mathbf{0}$ gives the solutions

$$\mathbf{x} = c_1 \begin{pmatrix} 1 \\ 1 \\ 0 \end{pmatrix} + c_2 \begin{pmatrix} 1 \\ 0 \\ 1 \end{pmatrix}, \quad c_1, c_2 \in \mathbb{R}.$$

The nullity is therefore 2. This agrees with the rank-nullity theorem $(1 + 2 = 3)$.

The nullspace of C is a plane spanned by two vectors. To find the equation of the plane, you *could* find the cross product of the two basis vectors to find the equation

$$x - y - z = 0.$$

Alternatively, the normal $\begin{pmatrix} 1 & -1 & -1 \end{pmatrix}^T$ is also parallel to a row (*any* row) of C, so you could just read of the equation of the 2D nullspace from inspecting the rows of C. This property follows from thinking about the multiplication in the equation $A\mathbf{x} = \mathbf{0}$ and noticing that the rows of A must be perpendicular to the column vector $\mathbf{x}$ in the nullspace. This property is sometimes expressed as the following theorem.

Theorem 5.7 *Let A be an $m \times n$ matrix. The row space and nullspace of A are orthogonal complements in $\mathbb{R}^n$.*

Finally, the case for the zero matrix D is trivial: the column space is the zero vector, whilst the nullspace is the whole of $\mathbb{R}^3$. The rank-nullity theorem is again satisfied $(0 + 3 = 3)$.

Visualising the rank-nullity theorem

For visualisation, it helps to associate each matrix with a linear transformation T. Let $\mathbf{x}$ be the position vectors of points inside and on the unit sphere $x^2 + y^2 + z^2 = 1$. We will take the domain of T to be the sphere. We will then apply the matrix to find the image of the sphere under T. Finally, we will visualise the kernel of T by considering which part of the solid sphere is mapped onto the zero vector by T.

The visualisation of these spaces are shown in fig. 5.14. Each row of figures corresponds to each of the 4 given matrices. The key observation here is that the dimensions of the image and kernel in each row always add up to 3.

It is important to keep in mind that the spheres (and the ellipsoid) in this figure are all meant to represent 3D solids (and not just the shells). For example, note that the kernels are subset of the domain, so the green line in the second row, and the disc in the third row, are all contained in the spherical domain.

The zero vector has also been added to all figures as a black dot. It is clearly part of the domain (the centre of the sphere). Furthermore, since we always have $T(\mathbf{0}) = \mathbf{0}$, the zero vector is also in both $\mathrm{im}(T)$ and $\mathrm{ker}(T)$.

Coding highlights (a quaternion appears)

The code `ranknullity.ipynb` produces each row of fig. 5.14. It is a rather long piece of code since we are producing 3 figures at the same time. Each subplot is interactive, in the sense that you can spin them and zoom in/out.

To visualise the kernel, the code uses `scipy.linalg.null_space` to produce the (normalised) basis of the nullspace. Then, the nullity is obtained by counting how many vectors are in the basis. It is then easy to plot the point, line and sphere if the nullity is 0,1 and 3.

Visualising the kernel when nullity equals 2 is the most complicated part of the code. The function `kernel2D` plots the slice of the sphere corresponding to $\ker(T)$. More precisely, the kernel is a unit disc containing two given unit vectors emanating from the origin.

To do this, we create the unit disc $\{(x, y, z) \in \mathbb{R}^3 : x^2 + y^2 \leq 1, z = 0\}$ and rotate it to match the kernel. The procedure we have chosen is interesting in the sense that it is a rotation technique that is commonly used in computer games, namely, using a *quaternion*. Here is a summary of what the function `kernel2D` does.

- The unit disc in the x-y plane is created (using polar coordinates). Note that it has unit normal $\mathbf{k} = (0, 0, 1)$.
- The normal to the kernel is simply any row of the given matrix (thanks to theorem 5.7). Normalise it to get the vector $\hat{\mathbf{N}}$
- The rotation required is that which maps $\mathbf{k}$ to $\hat{\mathbf{N}}$. The rotation angle is $\theta = \cos^{-1} \mathbf{k} \cdot \hat{\mathbf{N}}$. The axis of rotation is $\mathbf{k} \times \hat{\mathbf{N}}$. Normalise this axis to get the unit vector $\mathbf{I}$.
- The quaternion that performs this rotation can be written in the so-called polar form as

$$\mathbf{q} = \cos \frac{\theta}{2} + \mathbf{I} \sin \frac{\theta}{2}.$$

It does not matter if you are not familiar with quaternions at the moment (we will discuss them in some detail in §6.6). The point here is that to perform a spatial rotation of angle θ around an arbitrary axis $\mathbf{I}$, a process using a quaternion offers the easiest solution. A quaternion has an equivalent 3×3 matrix form, which we also use in the code to multiply to vectors to rotate them in the usual way.

- The rotated unit disc can then be plotted along with the 2 basis vectors of $\ker(T)$.

The code should work for any 3×3 matrix you throw at it.

The rank-nullity theorem also works for non-square matrices, although the code will require a little modification. You will explore the visualisation for non-square matrices in exercise 10.

DISCUSSION

- **The four fundamental subspaces**. In Gilbert Strang's famous book *Introduction to Linear Algebra* [194], you can find a beautiful pictorial summary of the key results on row and column spaces, including the rank-nullity theorem and theorem 5.7. A version of that picture is produced below.

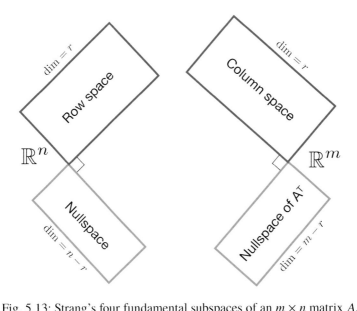

Fig. 5.13: Strang's four fundamental subspaces of an $m \times n$ matrix A.

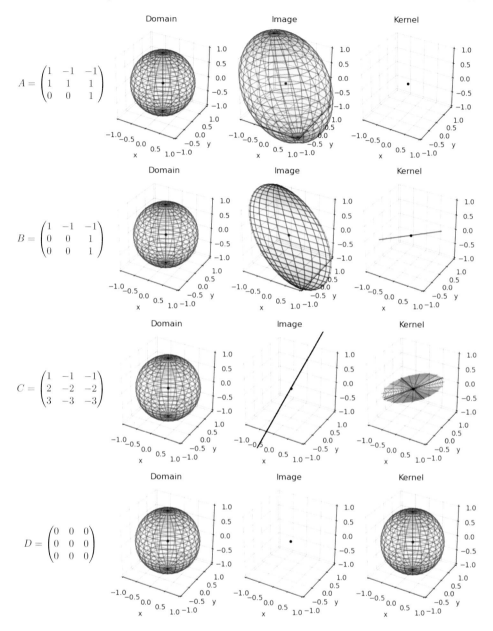

Fig. 5.14: Visualising the rank-nullity theorem: dim(Image) + dim(Kernel) = 3.

ranknullity.ipynb (for plotting each row of fig. 5.14)	
| | ```
import numpy as np
import matplotlib.pyplot as plt
import scipy.linalg as LA
``` |
| Function for 3D rotation | ```
from scipy.spatial.transform import Rotation
%matplotlib
``` |
| **Input the matrix here** | `A= np.array([[1,-1,-1],[1,1,1],[0,0,1]])` |
| Create a grid of $\theta$ and $\phi$ values (spherical coordinates) $(\theta, \phi)$ is a point on the unit sphere | ```
Theta= np.linspace(0, 2*np.pi, 25)
Phi = np.linspace(0, np.pi, 25)
theta, phi = np.meshgrid(Theta, Phi)
``` |
| Cartesian coordinates of points on the surface of the unit sphere | ```
X = np.cos(theta)*np.sin(phi)
Y = np.sin(theta)*np.sin(phi)
Z = np.cos(phi)
``` |
| We will plot 3 figures in a row | `fig = plt.figure()` |
| **First plot** = the unit sphere | ```
### First plot: Domain
ax = fig.add_subplot(131, projection='3d')
``` |
| A function for plot cosmetics Equal aspect ratio | ```
def plot_axes(name):
    ax.set_box_aspect((1,1,1))
    ax.set_xlim(-1,1)
    ax.set_ylim(-1,1)
    ax.set_zlim(-1,1)
    ax.set_xlabel('x')
    ax.set_ylabel('y')
    ax.set_zlabel('z')
``` |
| Display name as plot title Add a black dot at the origin | ```
    ax.set_title(name)
    ax.plot(0,0,0,'k.')
``` |
| A function for plotting a surface Set transparency Add black contour lines | ```
def plot_3D(X, Y, Z):
    ax.plot_surface(X, Y, Z, alpha=0.2,
        edgecolor='k', cmap='rainbow')
``` |
| Prep the first plot titled 'Domain' Plot the sphere | ```
plot_axes('Domain')
plot_3D(X, Y, Z)
``` |
| **Second plot** = the image of the sphere under $A$ | ```
### Second plot: Image
ax = fig.add_subplot(132, projection='3d')
``` |
| A transformation with matrix $M$ Reshape meshgrid data into 3 rows ($x=$ top row, then $y$, then $z$ = bottom row) Multiply the data by $M$ Extract the transformed data and undo the reshaping (revert to meshgrid data format) | ```
def transform(M, x, y, z):
    xyz = np.vstack((np.ravel(x), np.ravel(y),
                     np.ravel(z)))
    Mxyz = M@xyz
    MX = Mxyz[0,:].reshape(x.shape)
    MY = Mxyz[1,:].reshape(y.shape)
    MZ = Mxyz[2,:].reshape(z.shape)
``` |
| Return the transformed data | `    return MX, MY, MZ` |
| Transform the sphere with $A$ Prep the second plot Plot the transformed sphere | ```
AX, AY, AZ = transform(A, X, Y, Z)
plot_axes('Image')
plot_3D(AX, AY, AZ)
``` |
| | `### Code continues on the next page` |

ranknullity.ipynb (continued)	
Third plot shows which part of the (solid) sphere is mapped to **0**	```### Third plot: Kernel
ax = fig.add_subplot(133, projection='3d')```
Let SciPy find the basis of the nullspace How many vectors are in the basis?	```NS = LA.null_space(A)
nullity = NS.shape[1]```
A function for plotting 2D kernel Create a unit disc in the x-y plane using polar coordinates (r, θ) The disc will be rotated and mapped to the kernel Meshgrid data format of the unit disc	```def kernel2D():
    r = np.linspace(0, 1, 20)
    theta = np.linspace(0, 2*np.pi, 20)
    R, Theta = np.meshgrid(r, theta)

    xx = R*np.cos(Theta)
    yy = R*np.sin(Theta)
    zz = 0*xx``` |
| Each row of A is normal to the kernel (see thm. 5.7) If the normal is parallel to **k**, then... the kernel is the untransformed disc | ```   normal = A[0,:]

    if normal[0]==0 and normal[1]==0 and\
        normal[2]!=0:
        RX = xx
        RY = yy
        RZ = zz``` |
| Otherwise, Calculate the unit normal to the kernel Rotaton axis for mapping disc to kernel Normalise the rotation axis Angle θ between **k** and the normal | ```   else:
        N = normal/LA.norm(normal)
        V = np.cross([0,0,1], N)
        axis = V/LA.norm(V)
        th = np.arccos(N[2])
        cos2, sin2 = np.cos(th/2), np.sin(th/2)
        Isin2 = sin2*axis``` |
| The quaternion $\cos(\theta/2) + \mathbf{I}\sin(\theta/2)$ SciPy takes the quaternion to do rotation Transform the unit disc using the matrix form of the quaternion | ```   Quat = np.append(Isin2, cos2)
        rot = Rotation.from_quat(Quat)
        RX, RY, RZ= transform(rot.as_matrix(),
                                xx, yy, zz)``` |
| Add ($\pm$) basis vectors of the nullspace in green Plot the kernel as a rotated disc | ```   ax.plot([-NS[0,0], NS[0,0]],
        [-NS[1,0],NS[1,0]],[-NS[2,0],NS[2,0]],'g-')
    ax.plot([-NS[0,1], NS[0,1]],
        [-NS[1,1],NS[1,1]],[-NS[2,1],NS[2,1]],'g-')
    plot_3D(RX, RY, RZ)``` |
| 1D kernel = a line parallel to the basis vector 2D kernel = a rotated unit disc 3D kernel = the (solid) unit sphere | ```if nullity == 1:
    ax.plot([-NS[0,0], NS[0,0]],
        [-NS[1,0],NS[1,0]],[-NS[2,0],NS[2,0]],'g-')
if nullity == 2:
    kernel2D()
if nullity == 3:
    plot_3D(X, Y, Z)``` |
| Prep the third plot | ```plot_axes('Kernel')``` |
| | ```plt.tight_layout()
plt.show()``` |

5.10 Gram-Schmidt process and orthogonal polynomials

Given the polynomials
$$1, \; x, \; x^2, \; x^3, \; x^4, x^5.$$

Use the Gram-Schmidt process to obtain an orthonormal basis of $\mathcal{P}_5$, namely, the vector space of polynomial degree at most 5 defined on $[-1, 1]$, equipped with the inner product defined by

$$\langle f, g \rangle = \int_{-1}^{1} f(x)g(x)\, dx. \tag{5.30}$$

The *Gram-Schmidt process* is a procedure for obtaining an orthonormal basis $\{e_1, e_2, \ldots e_n\}$ for a vector space V, starting with a given linearly independent basis $\{v_1, v_2, \ldots v_n\}$. The procedure goes as follows (see [14, 103] for proof).

1. Starting with v_1, we normalise it to get e_1.

$$e_1 = \frac{v_1}{|v_1|}.$$

2. Write $v_2 = v_2^{\parallel} + v_2^{\perp}$, where $v_2^{\parallel}$ and $v_2^{\perp}$ are the components of v_2 parallel and perpendicular to e_1 respectively. Note that $v_2^{\parallel}$ is the projection of v_2 along the direction e_1, *i.e.*

$$v_2^{\parallel} = (v_2 \cdot e_1)e_1.$$

3. Obtain the component $v_2^{\perp}$ and e_2 as follows.

$$v_2^{\perp} = v_2 - (v_2 \cdot e_1)e_1 \implies e_2 = \frac{v_2^{\perp}}{|v_2^{\perp}|}.$$

4. Continue this process by subtracting the projections of v_i along the directions of $e_1, e_2 \ldots e_{i-1}$.

$$v_i^{\perp} = v_i - (v_i \cdot e_1)e_1 - (v_i \cdot e_2)e_2 \ldots - (v_i \cdot e_{i-1})e_{i-1} \implies e_i = \frac{v_i^{\perp}}{|v_i^{\perp}|}. \tag{5.31}$$

Continue until e_n is obtained.

The procedure for two vectors is shown in fig. 5.15 below.

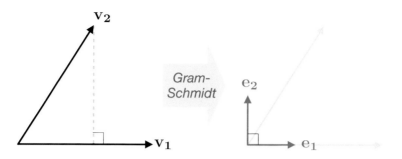

Fig. 5.15: The Gram-Schmidt process for 2 vectors.

The Gram-Schmidt process is named after the Danish mathematician *Jørgen Gram* (1850–1916) and the German mathematician *Erhard Schmidt* (1876–1959) although the process was already used by Laplace in his 1812 publication (Gram's published his results in 1883). See [128] for an interesting overview of the history and applications of the Gram-Schmidt process.

The Gram-Schmidt process can be applied to vectors in an abstract vector space. So far we have only discussed vectors in the vector space $\mathbb{R}^3$, where the dot product is defined in the usual way. In this question, however, we are working in the vector space $\mathcal{P}_5$ in which the 'vectors' are polynomials up to degree 5. The 'dot' or *inner product* between two polynomials f and g is also defined differently as an integral of the product fg over $[-1, 1]$. This inner product naturally gives rise to the 'length' or *norm* of a vector f as

$$|f| = \langle f, f \rangle^{1/2} = \left(\int_{-1}^{1} [f(x)]^2 \, dx \right)^{1/2}.$$

Let's follow the Gram-Schmidt recipe given the polynomials

$$v_0 = 1, \quad v_1 = x, \quad v_2 = x^2, \quad v_3 = x^3, \quad v_4 = x^4, \quad v_5 = x^5.$$

(Note that $\mathcal{P}_n$ has dimension $n + 1$.) We have relabelled the indices so that each subscript matches the degree of the polynomial.

1. Normalising v_0 gives

$$e_0 = \left(\int_{-1}^{1} dx \right)^{-1/2} = \frac{1}{\sqrt{2}}.$$

2. The component of v_1 parallel to e_0 is

$$v_1^{\|} = \langle v_1, e_0 \rangle e_0 = \left(\int_{-1}^{1} \frac{x}{\sqrt{2}} \right) \left(\frac{1}{\sqrt{2}} \right) = 0.$$

3. Subtracting the projection, we find

$$v_1^{\perp} = v_1 - 0 \implies e_1 = x \left(\int_{-1}^{1} x^2 \, dx \right)^{-1/2} = \sqrt{\frac{3}{2}} x.$$

4. Continuing the process, we find the remaining polynomials in the orthonormal basis:

$$e_2 = \frac{1}{2}\sqrt{\frac{5}{2}}\left(3x^2 - 1\right),$$

$$e_3 = \frac{1}{2}\sqrt{\frac{7}{2}}\left(5x^3 - 3x\right),$$

$$e_4 = \frac{1}{8}\sqrt{\frac{9}{2}}\left(35x^4 - 30x^2 + 3\right),$$

$$e_5 = \frac{1}{8}\sqrt{\frac{11}{2}}\left(63x^5 - 70x^3 + 15x\right).$$

Note that the functions e_i are alternately odd and even functions.

Any polynomial of degree at most 5 can be written as a linear combination of these polynomials. For example, take $f(x) = x^3 + 1$, then it is easy to show that

$$x^3 + 1 = \frac{2}{5}\sqrt{\frac{2}{7}}\,e_3 + \frac{3}{5}\sqrt{\frac{2}{3}}\,e_1 + \sqrt{2}\,e_0.$$

Coding

In the code `gramschmidt.ipynb`, we use SymPy to automate the Gram-Schmidt process, print out the orthonormal functions $\{e_i\}$, and plot them on the same set of axes, as shown in fig. 5.16. We note that SymPy's results for e_i agree with our calculations. The code can be easily extended to find the orthonormal basis of $\mathcal{P}_n$ for any n.

The code can also modified to perform the Gram-Schmidt process on other vector spaces with different inner products. See exercise 11.

DISCUSSION

- **Legendre polynomials**. The basis functions of polynomials degree at most n on $[-1, 1]$ (assuming the inner product (5.30)) are called *Legendre polynomials*, denoted $P_i(x)$ ($i = 0, 1, 2, \ldots n$). Like the basis functions e_i, they obey the orthogonal condition

$$\int_{-1}^{1} P_i(x)P_j(x)\,dx = 0 \text{ if } i \neq j.$$

However, instead of the unit-length requirement, the Legendre polynomials obey the *standardisation condition*

$$P_i(1) = 1,$$

for all integers i to up n (note from the graphs in fig. 5.16 that the orthonormal functions e_i do not satisfy this condition). The first 6 Legendre polynomials are

$$P_0 = 1, \quad P_1 = x, \quad P_2 = \frac{1}{2}\left(3x^2 - 1\right), \quad P_3 = \frac{1}{2}\left(5x^3 - 3x\right),$$

$$P_4 = \frac{1}{8}\left(35x^4 - 30x^2 + 3\right), \quad P_5 = \frac{1}{8}\left(63x^5 - 70x^3 + 15x\right).$$

At university, you will meet the Legendre polynomials when studying physical problems associated with spherical symmetry. In particular, they appear in *spherical harmonics*, $Y_\ell^m(\theta, \phi)$, which are the basis functions for expressing a function defined on the unit sphere. See [84] for details of how the Legendre polynomials and spherical harmonics appear in quantum mechanics.

Adrien-Marie Legendre (1752–1833) was a French mathematician who made major contributions to mathematical physics. There appears to be only one surviving likeness of him in the form of an unflattering caricature[7].

- **Orthogonal polynomials**. The Legendre polynomials are one family of many classical *orthogonal polynomials* that occur in mathematics. These polynomials typically satisfy the following properties.

  - They are orthogonal with respect to some inner product.
  - They satisfy a 3-term recurrence relation. For example, the Legendre polynomials satisfy

  $$(n + 1)P_{n+1}(x) = (2n + 1)xP_n - nP_{n-1}.$$

  - They satisfy a 2nd-order differential equation. For example, P_n satisfies the Legendre differential equation

  $$(1 - x^2)P_n''(x) - 2xP_n'(x) + n(n + 1)P_n(x) = 0.$$

  - They satisfy a degree-n differential expression known as *Rodrigues' formula*. For the Legendre polynomial, we have

  $$P_n(x) = \frac{1}{2^n n!} \frac{d^n}{dx^n} \left(x^2 - 1\right)^n.$$

See [41] for an introduction to orthogonal polynomials and their properties. See also Exercise 11c which explores the *Hermite polynomials*.

---

[7] See https://bit.ly/3ypUePq for an intriguing tale of mistaken identity.

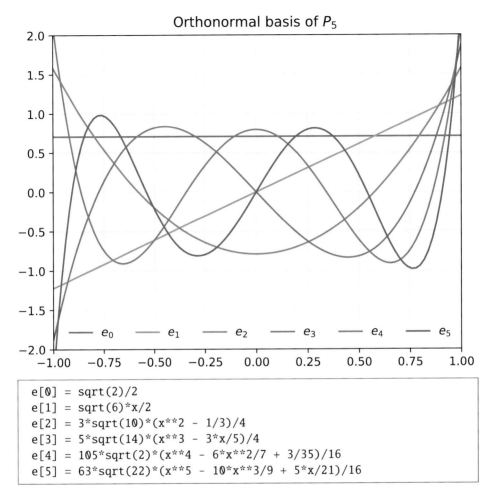

```
e[0] = sqrt(2)/2
e[1] = sqrt(6)*x/2
e[2] = 3*sqrt(10)*(x**2 - 1/3)/4
e[3] = 5*sqrt(14)*(x**3 - 3*x/5)/4
e[4] = 105*sqrt(2)*(x**4 - 6*x**2/7 + 3/35)/16
e[5] = 63*sqrt(22)*(x**5 - 10*x**3/9 + 5*x/21)/16
```

Fig. 5.16: The orthonormal basis for $\mathcal{P}_5$ obtained using the Gram-Schmidt process, along with the printout generated by the code gramschmidt.ipynb.

gramschmidt.ipynb (for plotting fig. 5.16)

```
import numpy as np
import sympy as sp
import matplotlib.pyplot as plt
%matplotlib
```

The linearly independent vectors $\mathbf{v}_i$ given
Initialise $\mathbf{e}_i$ (to be calculated using Gram-Schmidt process)
```
x = sp.Symbol('x')
v = sp.Matrix([[1,x,x**2,x**3,x**4,x**5]])
e = sp.zeros(1,len(v))
```

Define the inner product $\langle f, g \rangle$
```
def Dot(f,g):
    return sp.integrate(f*g, (x,-1,1))
```

Define the norm $|f|$
```
def norm(f):
    return sp.sqrt(Dot(f,f))
```

Begin Gram-Schmidt: obtain $\mathbf{e}_0$
Display result
```
e[0] = v[0]/norm(v[0])
print(f'e[0] = {e[0]}')
```

x values for plotting
$\mathbf{v}_0$ (constant function)
Plot it
```
xarray = np.linspace(-1,1,100)
yarray = e[0]*np.ones_like(xarray)
plt.plot(xarray,yarray)
```

Calculating the remaining $\mathbf{e}_i$
$\mathbf{v}_i^{\perp}$
The main Gram-Schmidt algorithm
Formula 5.31

Display the expression
Turn a SymPy expression into a lambda function
$\mathbf{v}_i$
Plot the result from Gram-Schmidt
```
for i in range(1,len(v)):
    vperp = v[i]
    for j in range(1,i+1):
        vperp -= Dot(v[i], e[j-1])*e[j-1]
    e[i] = vperp/norm(vperp)
    print(f'e[{i}] = {e[i]}')
    f = sp.lambdify(x,e[i])
    yarray = f(xarray)
    plt.plot(xarray,yarray)
```

Insert the legend in LaTeX...

at the bottom of the plot, horizontal style
```
plt.xlim((-1, 1))
plt.ylim((-2, 2))
plt.legend([r'$e_0$',r'$e_1$',r'$e_2$',
            r'$e_3$',r'$e_4$',r'$e_5$'],\
        ncol=len(v),loc='lower right')
plt.title(r'Orthonormal basis of $P_5$')
plt.grid('on')
plt.show()
```

5.11 Exercises

1 (*Visualising linear systems*) Use the code `planes.ipynb` to help you answer these questions.

    a. In the code, add a dot at the intersection point of the 3 planes whenever $\alpha = -1$. Suggestion: To add a single point with coordinates $(1, 2, 3)$ with Plotly, use

```
dot = go.Scatter3d(x=[1],y=[2],z=[3])
```

    b. Explore the number of solutions to the linear system

$$\begin{pmatrix} 2 & 1 & 1 \\ 0 & \alpha & 1 \\ \alpha & -2 & 1 \end{pmatrix} \begin{pmatrix} x \\ y \\ z \end{pmatrix} = \begin{pmatrix} 4 \\ 1 - \alpha \\ 3 + 2\alpha \end{pmatrix}$$

for different values of α. Visualise the results with Plotly and give a mathematical explanation.

Suggestion: First attack the system analytically to work out the critical α values. In the code, two planes should change with the α slider.

    c. We saw an underdetermined system of equations in §5.3 (it has more unknowns than equations). An *overdetermined* system, on the other hand, has more equations than unknown.

In each case, give an example of a system of 4 equations in 3 unknowns such that
- there exists a unique solution,
- there are infinitely many solutions,
- there are no solutions.

Visualise these different scenarios as 4 planes in Plotly.

2 (*Timing linear solvers*) Use the code `solvetimes.ipynb` as a starting point for these questions.

    a. Does the choice of the matrix A matter? Try changing the random matrix to, say, one of SciPy's special matrices such as `toeplitz`, `circulant` or `hankel`, or Numpy's `vander` (Vandermonde matrix). Does the conclusion still hold?

    b. Repeat the time-comparison exercise but this time between SciPy's `solve` and SymPy's `linsolve`. Don't be too ambitious with the matrix size n (unless you plan to leave the code running and go watch a film). Why does it take so long?

    c. Plot the lower panel of fig. 5.5 (runtime for Cramer's rule). This may take a while. You really don't have to go up to $n = 1000$ to see what's going on. Why does it take so long?

3 (*Visualising linear transformations*) Use the code `transformation.ipynb` as a starting point for these questions.

    a. Replace the square by a triangle with vertices at $(1, 2)$, $(-1, -1)$, $(1, -1)$ and rotate it about the origin.

    b. Replace the square by the unit circle $x^2 + y^2 = 1$ and apply a horizontal shear to it.

    c. Replace the transformation in the code by a matrix which represents a vertical shear with shear factor t.

d. Replace the matrix M in the code by each of the following matrices.

i) $\begin{pmatrix} 1 & t \\ t & 1 \end{pmatrix}$, ii) $\begin{pmatrix} 2t & 0 \\ 0 & t \end{pmatrix}$ iii) $\begin{pmatrix} |t|\cos t & -|t|\sin t \\ |t|\sin t & |t|\cos t \end{pmatrix}$

Predict what these transformations represent. Confirm your answers with the code.

e. Using homogeneous coordinates (see eq. 5.15), write down the 3×3 matrix which represents the translation $(x, y) \rightarrow (x + t, y)$. Apply it to the unit square in the code (where t can be adjusted using the slider). (Don't forget to transform the origin.)

4 (*Wigner's semicircle law*)

a. Write a code that generates a *symmetric* matrix M of size 2000×2000 whose entries are random numbers between 0 and 1 (see `solvetimes.ipynb` for a method to initialise such a matrix).

b. Use SciPy to find the eigenvalues of M. What is the largest eigenvalue λ_{max}? Verify that all the eigenvalues are real (this is the spectral theorem).

c. Remove λ_{max} from the array of eigenvalues. Plot a histogram of the eigenvalues in this array.
Suggestion: Use Matplotlib's `hist` function with, say, 50 bins.
You may be surprised to see that the histogram is (roughly) a semicircle. This phenomenon is a consequence of a more general theorem in random matrix theory called *Wigner's semicircle law*.

5 (*Cayley-Hamilton theorem*) Let $A = \begin{pmatrix} -1 & -1 & 1 & 2 \\ 2 & -1 & -1 & 1 \\ 1 & 2 & -1 & -1 \\ -1 & 1 & 2 & -1 \end{pmatrix}$.

Using SymPy, show that its characteristic polynomial is

$$\lambda^4 + 4\lambda^3 + 12\lambda^2 - 4\lambda - 13 = 0.$$

Using the Cayley-Hamilton theorem, express A^{-1} as a cubic polynomial in A, and hence evaluate A^{-1}.
Verify that your answer agrees with the SymPy command `A**-1`.

6 (*Exploring the* `eigshow` *game*) Use `eigshow.ipynb` to help you answer these questions.

a. Add a display on the plot showing the length of the red vector as the slider moves. Suggestion: Use `ax.text`. Consult the code in §6.6.

b. Write down a matrix A that maps the unit circle to the ellipse $x^2 + 16y^2 = 4$. Verify your answer with the code. (There are several answers.)

c. (*Defective matrix*) Show (by hand) that the matrix $\begin{pmatrix} 1 & 1/2 \\ 0 & 1 \end{pmatrix}$ has an eigenvalue with multiplicity 2. Find its eigenvector.
Use this matrix in the code and play the vector-alignment game. Check that what you see is consistent with your calculations.
Find the Cartesian equation of the red ellipse.

7 (*Solving a recurrence relation*) Use matrix diagonalisation to solve for a_n in terms of n, where a_n satisfies the recurrence relation

$$a_{n+3} = -a_{n+2} + 4a_{n+1} + 4a_n, \qquad a_0 = 2, \ a_1 = 3, \ a_2 = 11.$$

Plot the sequence (a_n) up to $n = 20$ by working out the terms iteratively. Verify that your nth-term expression agrees with the values obtained iteratively.

Solve the recurrence relation again using the method of characteristics (see Discussion in §5.7). Feel free to use SymPy to help with fiddly calculations.

8 (*Diagonalising matrices*) Use `diagonalise.ipynb` to help you answer the following question.

Consider the following matrices

$$A = \begin{pmatrix} 1 & -1/2 \\ -1/2 & 1 \end{pmatrix} \qquad B = \begin{pmatrix} 1 & 1 \\ 1 & 1 \end{pmatrix} \qquad C = \begin{pmatrix} 1 & -1/2 \\ 1 & 1 \end{pmatrix} \qquad M = \begin{pmatrix} 1 & 1 \\ 0 & 1 \end{pmatrix}.$$

a. Diagonalise matrices A, B, C by hand. Check your answers with SymPy using the command:

```
import sympy as sp
A = sp.Matrix([[1, -sp.S.Half], [-sp.S.Half,1]])
P, D = A.diagonalize()
```

Explain why matrix M is not diagonalisable.

b. Put each of the matrices into the code, and describe the diagonalisation as a sequence of 3 transformations.

Be as precise as you can. For example, if the transformation is a rotation, calculate the angle of rotation.

c. For matrix C, explain why the visualisation produced by the code is inadequate.

d. For matrix M, explain why the code does show a sequence of (strange) transformations even though the matrix is not diagonalisable.

9 (*Singular-value decomposition*) Use `svd.ipynb` as a starting point for the following questions.

a. The SVD process approximates an $m \times n$ matrix by a rank-N matrix. Above the compressed photo produced by the code, set the title of the photo to show an estimate of the percentage of data retained in the rank-N approximation. Use the formula

$$\frac{N(m + n + 1)}{mn} \times 100\%$$

as discussed in the text.

Does it make sense that this number could be higher than 100% when N is sufficiently large?

b. (*Marchenko-Pastur distribution*) Perform SVD on a 2000×5000 matrix A whose entries are random numbers between 0 and 1 (see `solvetimes.ipynb` for a method to initialise such a matrix).

Let s be the array of singular values. Remove the largest eigenvalue from s (see question 4). Then, plot the distribution of the remaining singular values as a histogram.

Try it with (large) random matrices of different dimensions. Note that the shape of the distribution remains similar. This is called the *Marchenko-Pastur distribution*. This observation is important in filtering noise from data (small singular values are associated with noise).

c. (*Compressing a colour image*) One way to store a colour image is to associate each pixel with a tuple (r, g, b) called the RGB (red-green-blue) values. Each colour value is an 8-bit unsigned integers ('uint8') ranging from 0 to 255.

Let's modify the code so that it performs SVD on a colour image. Here are some pointers and useful commands for working with colour images in PIL.

- Put your colour image (say `tree.jpg`) in the same folder as the code (for smoother sliding, use a smaller photo). Open it with the command

$$A \ = \ \texttt{Image.open('tree.jpg').convert('RGB')}$$

- Obtain the 3 matrices representing each channel using the command `Ar`, `Ag`, `Ab = A.split()`
- Perform SVD on each matrix. Let's call the results `Imr`, `Img`, `Imb`. These are arrays of floating point numbers, as you can verify using the command `Imr.dtype`.
- Convert the results back to an array of unsigned integers using the command `Imr = Imr.astype('uint8')`.
- Convert the arrays to PIL's Image type using `R=Image.fromarray(Imr)`. Similarly, obtain images `G` and `B`
- Finally, merge the 3 compressed images with the command

$$\texttt{Image.merge('RGB', (R, G, B))}$$

For the left-hand plot, plot the singular values of all 3 channels on the same graph.

10 (*Rank-nullity theorem*) Use `ranknullity.ipynb` as a starting point for these questions.

a. Recall matrices B and C in the question, namely:

$$B = \begin{pmatrix} 1 & -1 & -1 \\ 0 & 0 & 1 \\ 0 & 0 & 1 \end{pmatrix} \qquad C = \begin{pmatrix} 1 & -1 & -1 \\ 2 & -2 & -2 \\ 3 & -3 & -3 \end{pmatrix}.$$

Demonstrate theorem 5.7) visually. In other words, for each matrix, plot its row space and its kernel to demonstrate orthogonality.

b. Modify the code to visualise the rank-nullity theorem for 3×2 matrices like

$$\begin{pmatrix} 1 & 1 \\ 2 & -1 \\ 1 & 0 \end{pmatrix}.$$

Produce a set of figures similar to fig. 5.14.

c. Repeat the above exercise for the 2×3 matrix

$$\begin{pmatrix} 1 & 1 & 0 \\ 2 & -1 & 1 \end{pmatrix}.$$

11 (*Gram-Schmidt process*) Use `gramschmidt.ipynb` to help you answer the following questions.

  a. Starting with the vectors

$$\mathbf{v}_1 = \begin{pmatrix} 1\ 0\ 1\ 0 \end{pmatrix}^T, \qquad \mathbf{v}_2 = \begin{pmatrix} 1\ 2\ 0\ -1 \end{pmatrix}^T,$$

add two more vectors $\mathbf{v}_3$ and $\mathbf{v}_4$ such that the set $\{\mathbf{v}_1, \mathbf{v}_2, \mathbf{v}_3, \mathbf{v}_4\}$ is linearly independent.

Then, use the Gram-Schmidt process to construct an orthonormal basis $\{\mathbf{e}_1, \mathbf{e}_2, \mathbf{e}_3, \mathbf{e}_4\}$ of $\mathbb{R}^4$.

Suggestion: You can check your answer with SymPy's own Gram-Schmidt function. Look up `sympy.matrices.GramSchmidt`.

Hence, express the vector $\begin{pmatrix} 1\ 2\ 3\ 4 \end{pmatrix}^T$ as a linear combination of the orthonormal basis $\{\mathbf{e}_i\}$.

  b. Write a Python code to automate and visualise the Gram-Schmidt process for vectors in $\mathbb{R}^3$. In particular, given a set of 3 vectors $\{\mathbf{v}_1, \mathbf{v}_2, \mathbf{v}_3\}$, your code should:

- Test whether the set is linearly independent. If not, throw an error with a `raise` command.
- Perform the Gram-Schmidt process and produce the orthonormal basis $\{\mathbf{e}_1, \mathbf{e}_2, \mathbf{e}_3\}$. Print the results.
- Plot $\mathbf{e}_1, \mathbf{e}_2, \mathbf{e}_3$ and $\mathbf{v}_1, \mathbf{v}_2, \mathbf{v}_3$ on the same set of axes to visualise the basis pre- and post-Gram-Schmidt.

Suggestion: Use SymPy to work with exact expressions up to the print stage.

  c. (*Hermite polynomials*) Given the polynomials

$$1,\ x,\ x^2,\ x^3,\ x^4, x^5.$$

Use the Gram-Schmidt process to obtain an orthonormal basis of $\mathcal{P}_5$, namely, the vector space of polynomial degree at most 5 defined on $\mathbb{R}$, equipped with the inner product defined by

$$\langle f, g \rangle = \int_{-\infty}^{\infty} f(x)g(x)\, e^{-x^2}\, \mathrm{d}x.$$

Use SymPy to display the orthonormal polynomials and plot them on the same set of axes (in the style of fig. 5.16).

Note: If the polynomials are not subject to normalisation, but instead, to the condition that the coefficient of the highest power of x^n is 2^n, then the polynomials obtained are called *Hermite polynomials* which have applications in physics and probability. See [157] for a review.

Suggestion: Here is how you can evaluate $\int_{-\infty}^{\infty} e^{-x^2}\, \mathrm{d}x$ in SymPy.

```
import sympy as sp
x = sp.Symbol('x')
sp.integrate(sp.exp(-x**2), (x, -sp.oo, sp.oo))
```

Abstract Algebra and Number Theory

Fig. 6.1: A monument to *Muhammad ibn Musa al-Khwarizmi* (c.781–850), the Persian scholar many consider to be the father of algebra. As his work spread in 12th century Europe, his latinised name gave rise to the word *algorithm* (Image source: [213]).

6.1 *Basic concepts in abstract algebra*

The word *algebra* takes on a very different meaning from how it is used in school. Algebra means much more than the rules of shuffling around symbols and expanding or factorising expressions in x and y. At university, algebra is a pure mathematics subject concerned with abstract structures like *groups*, *rings* and *fields*, as well as their properties and mappings between them. Abstract algebra allows integers, matrices, real and complex numbers to be viewed with a bird's eye view as particular types of abstract structures (*e.g.* $\mathbb{R}$ is a type of field). Algebraists often describe their subject as elegant and beautiful. A good algebraic proof is as succinct and refined as a haiku.

Visualising abstraction may sound a little contrary – certainly a typical abstract algebra textbook contains far fewer figures than a book on, say, vector calculus (with the exception of [38]). But some visualisation is possible (and useful), and in this chapter we will attempt to visualise some aspects of abstract algebra with the help of Python. We will focus on group theory and produce visualisations that help us understand of some of the most important groups you will meet in group theory.

Here are some good introductory books on abstract algebra: [12, 19, 68, 70, 149]. The student guide [5] is particularly readable. Ref. [81] is interesting overview of the history of abstract algebra.

Groups

Groups are one of many types of structures in abstract algebra. But what makes them special is their boundless mathematical richness despite being defined by only 4 properties below. Groups occur abundantly not only in mathematics but also in other areas of science, thanks to their close association with *symmetry*. For instance, chemists use groups to describe the structure of molecules and crystals [111]. In physics, groups are essential in the studies of fundamental forces and elementary particles [106].

A *group* is a set G together with a binary operation $*$ satisfying the following requirements (called the group axioms).

- (*Closure*) For all $a, b \in G$, $a * b \in G$.
- (*Associativity*) For all $a, b, c \in G$, $(a * b) * c = a * (b * c)$.
- (*Existence of identity*) $\exists e \in G$ such that for all $a \in G$, $a * e = e * a = a$. We call e the *identity* of G .
- (*Existence of inverse*) $\forall a \in G$, $\exists a^{-1} \in G$ such that $a * a^{-1} = a^{-1} * a = e$. We call a^{-1} the *inverse* of a.

We say that "$(G, *)$ is a group", or "G is a group under $*$", or simply "G is a group" if it is clear from the context what the binary operation is.

Example: $G = \{-1, 1\}$ is a group under multiplication. Closure can be quickly verified. Associativity follows from that of multiplication of real numbers. The identity is 1, and each element is the inverse of itself.

Often the symbol for the binary operation is not shown, *i.e.* ab means $a * b$, and a^2 means $a * a$. Be careful, this notation can cause confusion when the operation is addition.

Here are some important group-theoretical terms that we will need in this chapter.

- **Finite group** A group with finitely many elements.
- **Order of a group** The number of elements in the group. The order of G is denoted $|G|$.
- **Order of an element** Let $g \in G$. The order of g is the smallest number $n \in \mathbb{N}$ such that $g^n = e$.

Example: $G = \{-1, 1\}$ under multiplication is a finite group with 2 elements, so $|G| = 2$. The order of the identity element is (always) 1. The order of -1 is 2 since $(-1)^2 = 1$.

In §6.3-6.6, we will explore important groups including the cyclic group C_n, the dihedral group D_n, the symmetric group S_n, the alternating group A_n, and the quaternion group Q.

Group theory in Python

Compared to other specialised algebra packages, SymPy has a limited group-theory capability[1] although we will only need a limited number of group-theory functions. In particular, we will only use SymPy to express group elements in terms of *permutations*. We will introduce permutations properly and learn how to work with them in SymPy in §6.5.

You are encouraged to explore the following free specialist computer packages that are more tailored to algebraic calculations.

- Magma
  `http://magma.maths.usyd.edu.au/magma/`
- GAP (Groups, Algorithms, Programming).
  `https://www.gap-system.org`
- Sage
  `https://www.sagemath.org`

6.2 *Basic concepts in number theory*

Number theory is, in the broadest term, the study of integers. On one hand, it can be viewed as a subset of abstract algebra (for example, $\mathbb{Z}$ under addition can be viewed as a group). However, techniques in number theory also involve aspects of other mathematical fields such as combinatorics, analysis and calculus, thus making number theory a rich tapestry that is often considered a standalone subject.

Within number theory there are several subdisciplines, two of which will be touched on in this chapter. First, *elementary number theory* introduces students to basic concepts (*e.g.* gcd, divisibility, modulo arithmetic) and tools of number theory (*e.g.* Euclidean algorithm, Chinese Remainder Theorem, Quadratic Reciprocity Law).

There are many excellent books on elementary number theory, including [35, 57, 105].

Analytic number theory, on the other hand, uses techniques in analysis (real, complex and numerical) to understand integers, particularly prime numbers. Analytic number theory is usually introduced at senior undergraduate level, but it is useful for beginning undergraduates to see how concepts such as integration and power series can help us understand prime numbers. Most excitingly, in this chapter we will discuss the *Prime Number Theorem* and the *Riemann zeta function*, and only basic knowledge of high-school calculus and complex numbers will be assumed.

For those who want a deeper dive, [11, 167] are particularly good books on analytic number theory.

Here are some basic number-theory terms that we will need in this chapter.

- **Divisor** A divisor d (also called a factor) of an integer N is a positive integer such that N/d is an integer.
  In this case, we say that "N is divisible by d" or "d divides N" and write $d \mid N$. If d does not divide N, we write $d \nmid N$.
  Example: $2 \mid 4$ but $2 \nmid 5$
- **Greatest common divisor** The greatest common divisor of two integers a, b, denoted $\gcd(a, b)$, is the greatest integer d such that $d \mid a$ and $d \mid b$. This can be extended to more than 2 integers.

---

[1] `https://docs.sympy.org/latest/modules/combinatorics/named_groups.html`

Example: $\gcd(2,4) = \gcd(2,4,6) = 2$.

- **Prime number** A prime number is a positive integer which is divisible only by 1 and itself. The smallest prime number is 2.
  A positive number greater than 2 that is not prime is called a **composite number**.
- **Coprime** Two integers are coprime (or relatively prime) if they share no common divisors except 1. Equivalently, a and b are coprime iff $\gcd(a, b) = 1$.
  A set of 3 integers $\{a, b, c\}$ is said to be *(setwise) coprime* if $\gcd(a, b, c) = 1$.
  A set of 3 integers $\{a, b, c\}$ is said to be *pairwise coprime* if $\gcd(a, b) = \gcd(b, c) = \gcd(a, b) = 1$.
  Setwise and pairwise coprimality can be extended to more integers.
  Example: 2, 3 are coprime. 4, 5, 6 are coprime, but not pairwise coprime since $\gcd(4, 6) = 2$.

We will also need a fundamental result in number theory.

Theorem 6.1 *(Fundamental Theorem of Arithmetic) Every integer $N \geq 2$ can be written as a product of prime powers:*

$$N = p_1^{k_1} p_2^{k_2} \cdots p_m^{k_m}$$

where $k_i \in \mathbb{N}$. This prime factorisation is unique up to the ordering of the primes p_i.

Example: $223344 = 2^4 \cdot 3^3 \cdot 11 \cdot 47$.

Modular arithmetic

Modular arithmetic reduces the task of studying all integers into a simpler task of studying a finite set of integers. Modular arithmetic is actually something that's already familiar to you: the 12-hour clock operates under "modulo 12", *i.e.* 13 o'clock = 1 o'clock. We write $13 = 1 \pmod{12}$. Remainders play a key role in modular arithmetic. Much of elementary number theory is about studying integers in the modular world and hence revealing deep, beautiful results about integers, some of which are explored in §6.7-6.9.

Let $a, b \in \mathbb{Z}$ and $n \in \mathbb{N}$. We say that "a is congruent to b modulo n" iff $(a - b)$ is divisible by n. In symbols, we write

$$a = b \pmod{n}.$$

(Some books use the symbol $\equiv$ instead of $=$.) The above is called a *congruence* rather than an equation.

The following theorem shows that many usual rules of arithmetic still hold in modulo n.

Theorem 6.2 *(Rules of modular arithmetic) Let $n \in \mathbb{N}$ and $a, b, c, d \in \mathbb{Z}$. The following properties hold.*

- *(Adding and multiplying to both sides) $a = b \pmod{n} \implies a + c = b + c \pmod{n}$ and $ac = bc \pmod{n}$.*
- *(Adding and multiplying congruences) $a = b \pmod{n}$ and $c = d \pmod{n} \implies a + c = b + d \pmod{n}$ and $ac = bd \pmod{n}$.*
- *(Exponentiating both sides) $a = b \pmod{n} \implies a^k = b^k \pmod{n}$ for all $k \in \mathbb{N}$.*
- *(Cancellation) $ka = kb \pmod{n}$ and $\gcd(k, n) = 1 \implies a = b \pmod{n}$.*

Another way to interpret the cancellation property is to use the following concept:

- Given an integer $k \in \mathbb{N}$, its *multiplicative inverse* (modulo n) is an integer k^{-1} such that $k^{-1}k = 1$.

Multiplying the congruence $ka = kb \pmod{n}$ by k^{-1} on both sides gives $a = b$ $\pmod{n}$.

The multiplicative inverse of k exists if and only if $\gcd(k, n) = 1$ (we explain why in §6.7).

Examples:

- $3 = 0 \pmod{3} \implies 4 = 1 \pmod{3}$ (by adding 1 to both sides), and $6 = 0 \pmod{3}$ (by multiplying both sides by 2).
- $4 = 1 \pmod{3}$ and $2 = -1 \pmod{3} \implies 6 = 0 \pmod{3}$ (by adding two congruences) and $8 = -1 \pmod{3}$ (by multiplying two congruences).
- $2 = -1 \pmod{3} \implies 4 = 1 \pmod{3}$ (squaring both sides)
- $8 = 2 \pmod{3}$ and $\gcd(2, 3) = 1 \implies 4 = 1 \pmod{3}$ (cancelling a factor of 2). On the other hand, $6 = 3 \pmod{3}$ but one cannot cancel a factor of 3 from both sides since $\gcd(3, 3) \neq 1$.
- Instead of using the cancellation property, note that $2 \cdot 2 = 1 \pmod{3}$, so the multiplicative inverse $2^{-1} = 2$. Multiplying the congruence $8 = 2 \pmod{3}$ by 2^{-1} on both sides gives $4 = 1 \pmod{3}$.

Generating prime numbers

In our exploration of number theory, it will be essential to generate a list of prime numbers up to a given integer N. An ancient method for doing this is the *Sieve of Eratosthenes* which we now discuss.

Sieve of Eratosthenes *The following algorithm produces a list of prime numbers up to a given integer N*

1. *Write down a list of all integers from 2 up to and including N*
2. *Retain the smallest number on the list (2). Then cross off all multiples of that number.*
3. *Retain the next smallest number on the list (3). Then cross off all multiples of that number.*
4. *Repeat while the smallest numbers are the list are all $\leq \sqrt{N}$*
5. *The numbers on the list that are not crossed out are all the primes up to and including N*

The method is named after *Eratosthenes* (~276–194BC), a Greek all-round scholar who made important contributions to mathematics, astronomy, geography, drama and literature.

Let's see the Sieve in action. Suppose we want a list of prime numbers up to 50.

Start with the list of integers from 2 up to 50. First we retain 2 and cross off all its multiples.

$$
\begin{array}{cccccccccc}
2 & 3 & 4 & 5 & 6 & 7 & 8 & 9 & 10 \\
11 & 12 & 13 & 14 & 15 & 16 & 17 & 18 & 19 & 20 \\
21 & 22 & 23 & 24 & 25 & 26 & 27 & 28 & 29 & 30 \\
31 & 32 & 33 & 34 & 35 & 36 & 37 & 38 & 39 & 40 \\
41 & 42 & 43 & 44 & 45 & 46 & 47 & 48 & 49 & 50
\end{array}
$$

Next we retain 3 and cross off all its multiples.

$$2 \quad 3 \quad 4 \quad 5 \quad 6 \quad 7 \quad 8 \quad 9 \quad 10$$
$$11 \quad 12 \quad 13 \quad 14 \quad 15 \quad 16 \quad 17 \quad 18 \quad 19 \quad 20$$
$$21 \quad 22 \quad 23 \quad 24 \quad 25 \quad 26 \quad 27 \quad 28 \quad 29 \quad 30$$
$$31 \quad 32 \quad 33 \quad 34 \quad 35 \quad 36 \quad 37 \quad 38 \quad 39 \quad 40$$
$$41 \quad 42 \quad 43 \quad 44 \quad 45 \quad 46 \quad 47 \quad 48 \quad 49 \quad 50$$

Next we retain 5 and cross off all its multiples.

$$2 \quad 3 \quad 4 \quad 5 \quad 6 \quad 7 \quad 8 \quad 9 \quad 10$$
$$11 \quad 12 \quad 13 \quad 14 \quad 15 \quad 16 \quad 17 \quad 18 \quad 19 \quad 20$$
$$21 \quad 22 \quad 23 \quad 24 \quad 25 \quad 26 \quad 27 \quad 28 \quad 29 \quad 30$$
$$31 \quad 32 \quad 33 \quad 34 \quad 35 \quad 36 \quad 37 \quad 38 \quad 39 \quad 40$$
$$41 \quad 42 \quad 43 \quad 44 \quad 45 \quad 46 \quad 47 \quad 48 \quad 49 \quad 50$$

Next we retain 7 and cross off all its multiples. This is the last step since $\sqrt{50} \approx 7.07$. Everything that's not crossed out is prime (we circle these numbers), and the sieving is complete.

$$②\quad ③\quad 4\quad ⑤\quad 6\quad ⑦\quad 8\quad 9\quad 10$$
$$⑪\quad 12\quad ⑬\quad 14\quad 15\quad 16\quad ⑰\quad 18\quad ⑲\quad 20$$
$$21\quad 22\quad ㉓\quad 24\quad 25\quad 26\quad 27\quad 28\quad ㉙\quad 30$$
$$㉛\quad 32\quad 33\quad 34\quad 35\quad 36\quad ㊲\quad 38\quad 39\quad 40$$
$$㊶\quad 42\quad ㊸\quad 44\quad 45\quad 46\quad ㊼\quad 48\quad 49\quad 50$$

The reason we can stop sieving once we have reached $\sqrt{N}$ is as follows. If there were a composite number $K > \sqrt{N}$ left on the list at this point, then writing $K = ab$, we note that both a and b must be greater than $\sqrt{N}$ (otherwise K would have been sieved out). But this makes $ab > \sqrt{N} \cdot \sqrt{N} = N$, a contradiction since the largest number on our list is N.

Number theory in Python

Modulo arithmetic	
100 (mod 11) Output = 1	`100 % 11`
2^5 (mod 11) Output =10	`pow(2, 5, 11)`
2^{-1} (mod 11) (multiplicative inverse) Output = 6	`pow(2,-1, 11)`
Generate primes with SymPy A list of all primes $p < 19$ Output = [2, 3, 5, 7, 11, 13, 17]	`from sympy import sieve, prime` `list(sieve.primerange(19))`
A list of primes $6 \le p < 23$ Output = [7, 11, 13, 17, 19]	`list(sieve.primerange(6, 23))`
A list of the first 9 prime numbers Output = [2, 3, 5, 7, 11, 13, 17, 19, 23]	`list(sieve.primerange(prime(9) + 1))`

SymPy comes with many more functions that are useful for number theory[2]. More advanced functions will be introduced in the text.

Magma, GAP and Sage also have extensive number-theory libraries that are more comprehensive than SymPy, and you are encouraged to explore them.

6.3 Groups I – Cyclic group

Produce the Cayley tables for each of the following groups.
a) $(\mathbb{Z}_{10}, +)$ (integers under addition modulo 10)
b) $(\mathbb{Z}_{11}^*, \times)$ (integers coprime to 11 under multiplication modulo 11)
Are these groups i) Abelian? ii) cyclic? iii) isomorphic?

The group $(\mathbb{Z}_{10}, +)$ consists of integers $\{0, 1, 2, 3, \ldots, 9\}$ under addition modulo 10, which simply gives the last digit of the result. It is straightforward to check that all the group axioms are satisfied, with 0 being the identity element. We can tabulate the result of the binary operation $a + b$ as a table called *Cayley table* (after Arthur Cayley whom we met in Chapter 5). The following short code generates the required table in Python.

Cayley table for addition mod 10

```
n=10
for row in range(0, n):
    print(*(f"{(row + col)%n :3}" for col in range(0, n)))
```

The key syntax is the % operator which applies modulo n to the result. More precisely, a%b gives the remainder when a is divided by b. It also works for non-integers.

The number 3 adjusts the width of each column, and the operator * unpacks the numbers generated in each row for printing. The output of this code is shown below on the left.

0	1	2	3	4	5	6	7	8	9
1	2	3	4	5	6	7	8	9	0
2	3	4	5	6	7	8	9	0	1
3	4	5	6	7	8	9	0	1	2
4	5	6	7	8	9	0	1	2	3
5	6	7	8	9	0	1	2	3	4
6	7	8	9	0	1	2	3	4	5
7	8	9	0	1	2	3	4	5	6
8	9	0	1	2	3	4	5	6	7
9	0	1	2	3	4	5	6	7	8

+	1	2	3	4	5	6	7	8	9
1	1+1	1+2	1+3	1+4	1+5	1+6	1+7	1+8	1+9
2	2+1	2+2	2+3	2+4	2+5	2+6	2+7	2+8	2+9
3	3+1	3+2	3+3	3+4	3+5	3+6	3+7	3+8	3+9
4	4+1	4+2	4+3	4+4	4+5	4+6	4+7	4+8	4+9
5	5+1	5+2	5+3	5+4	5+5	5+6	5+7	5+8	5+9
6	6+1	6+2	6+3	6+4	6+5	6+6	6+7	6+8	6+9
7	7+1	7+2	7+3	7+4	7+5	7+6	7+7	7+8	7+9
8	8+1	8+2	8+3	8+4	8+5	8+6	8+7	8+8	8+9
9	9+1	9+2	9+3	9+4	9+5	9+6	9+7	9+8	9+9

The interpretation of the table is shown on the right. Let a_{ij} be the entry in the ith row and jth column. The Cayley table shows the result $a_{i1} + a_{1j} = a_{ij}$. The ordering is important. Here we have suppressed the trivial result $0 + a = a$ (which is just a group axiom).

A more sophisticated version of the same table is shown in fig. 6.2 (bottom right), in which numbers are colour-coded. The code cayley.ipynb generates this figure.

---

[2] https://docs.sympy.org/latest/modules/ntheory.html

The same code can be used to generate the Cayley table for the group $(\mathbb{Z}_{11}^*, \times)$. Here the star indicates that we are only interested in elements of $\mathbb{Z}_{11}$ which are coprime to 11. If p is prime, then

$$\mathbb{Z}_p^* = \{1, 2, 3, \ldots, p-1\}$$

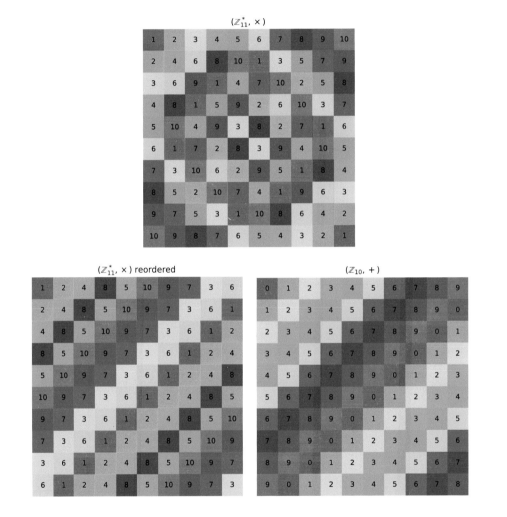

Fig. 6.2: The Cayley table for $(\mathbb{Z}_{11}^*, \times)$ (top) and the reordered table (bottom left). The diagonal structure shows that the group is cyclic and that it is isomorphic to the group $(\mathbb{Z}_{10}, +)$ (bottom right). Raising 2 to the power of all the elements in the lower-right table (mod 11) gives the lower-left table.

Abelian groups

A group $(G, \cdot)$ is said to be *Abelian* if, for all $a, b \in G$, $a \cdot b = b \cdot a$. This means that the Cayley table for an Abelian group would be symmetric along the diagonal going top-left to bottom-right (*i.e.* the same symmetry for a symmetric matrix). Of course, we know that integer addition and multiplication are both commutative, hence $(\mathbb{Z}_{10}, +)$ and $(\mathbb{Z}_{11}^*, \times)$ are both Abelian. But we can also see this from the symmetry of all the tables in fig. 6.2.

The word 'Abelian' comes from the surname of *Niels Henrik Abel* (1802–1829). Abel was a Norwegian mathematician who, despite succumbing to tuberculosis aged only 27, made profound contributions to analysis and algebra, particularly in proving that a general quintic polynomial equation cannot be solved algebraically.

Cyclic groups

A group $(G, \cdot)$ is said to be *cyclic* if there exists an element $g \in G$ such that each element of G has the form g^n for some $n \in \mathbb{Z}$. g is said to be a generator of the group G, and we write $G = \langle g \rangle$.

The group $(\mathbb{Z}_{10}, +)$ is cyclic because every integer $n \in \mathbb{Z}_{10}$ can be expressed as $1 + 1 + \cdots + 1$ (*n* times). Thus, 1 generates the group.

In fact, for all $N \in \mathbb{N}$, $(\mathbb{Z}_N, +)$ is a cyclic group with 1 as a generator. The Cayley table of the group $(\mathbb{Z}_N, +)$ will have the same diagonal stripe structure as $(\mathbb{Z}_{10}, +)$ as shown in fig. 6.2.

The table for the group $(\mathbb{Z}_{11}^*, \times)$ (top figure) does not appear to have the same diagonal structure, but if we reorder the elements of $\mathbb{Z}_{11}^*$ as $(1, 2, 4, 8, 5, 10, 9, 7, 3, 6)$, the diagonal stripes appear (lower left table in the figure). Could this group be cyclic?

Indeed it is. The reordered elements are simply the sequence

$$(2^n \pmod{11})$$

for $n = 0, 1, 2, \ldots, 9$. You should verify that this is the case. This means that all the elements in $\mathbb{Z}_{11}^*$ can be expressed as 2^n. Multiplying two elements gives $2^n 2^m = 2^{n+m}$. The addition of the exponents explains why the same diagonal structure as $\mathbb{Z}_{10}$ appears. In fact, if we raise 2 to the power of each element in the table for $\mathbb{Z}_{10}$, we obtain the reordered table for $\mathbb{Z}_{11}^*$. Therefore, the latter is a cyclic group generated by 2.

Group isomorphism

Let $G = (\mathbb{Z}_{10}, +)$ and $H = (\mathbb{Z}_{11}^*, \times)$. We saw that they have the same Cayley table when the elements of G are 'relabelled' by the function $\phi : G \to H$ such that $\phi(g) = 2^g \pmod{11}$. Let's tabulate this relabelling function explicitly.

$g \in G$	0	1	2	3	4	5	6	7	8	9
$\phi(g) \in H$	1	2	4	8	5	10	9	7	3	6

The relabelling function ϕ has the following special properties.

- **ϕ is a homomorphism**
  let $(G, \cdot)$ and $(H, *)$ be two groups. A function $\phi : G \to H$ is said to be a *homomorphism* if

  $$\phi(g_1 \cdot g_2) = \phi(g_1) * \phi(g_2).$$

  In our case, let $g_1, g_2 \in (\mathbb{Z}_{10}, +)$. We observe that

  $$\phi(g_1 + g_2) = 2^{g_1+g_2} \quad (\text{mod } 11)$$
  $$= 2^{g_1} \times 2^{g_2} \quad (\text{mod } 11)$$
  $$= \phi(g_1) \times \phi(g_2).$$

  Hence the relabelling function ϕ is a homomorphism.

- **ϕ is injective**
  A function $\phi : G \to H$ is said to be a *injective* (or *one-to-one*) if

  $$\phi(g_1) = \phi(g_2) \implies g_1 = g_2.$$

  Equivalently, the contrapositive of the above says that if $g_1 \neq g_2$ then $\phi(g_1) \neq \phi(g_2)$. In other words, if ϕ is injective, then distinct elements in G are mapped to distinct elements in H.
  It is straightforward to check from the table of correspondence that this property holds for the relabelling function ϕ.

- **ϕ is surjective**
  A function $\phi : G \to H$ is said to be a *surjective* (or *onto*) if, for all $h \in H$, $\exists g \in G$ such that $h = \phi(g)$.
  In other words, every element in H is the image of some element in G. Again it is straightforward to check from the correspondence table that this holds.

A function which is both injective and surjective is said to be *bijective*.

Two groups $(G, \cdot)$ and $(H, *)$ are said to be *isomorphic* if there exists a bijective homomorphism $\phi : G \to H$. If this holds, we write $(G, \cdot) \cong (H, *)$, or simply $G \cong H$.

We have shown that $(\mathbb{Z}_{10}, +)$ is isomorphic to $(\mathbb{Z}_{11}^*, \times)$. As the word suggests, two isomorphic groups have the same structure. They have the same number of elements, and their Cayley tables are identical when relabelled by the homomorphism ϕ.

Here is a more general result which you may have already conjectured yourself.

Conjecture: If p is prime, then $(\mathbb{Z}_p^*, \times) \cong (\mathbb{Z}_{p-1}, +)$.

We have proved the isomorphism for $p = 11$. A more dramatic visualisation of this isomorphism is shown in figure 6.3 for $p = 79$. You will be asked to produce these plots in exercise 3. Think about how you can use Python to find a generator for the cyclic group $\mathbb{Z}_{79}^*$.

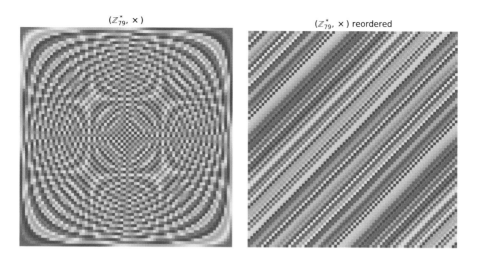

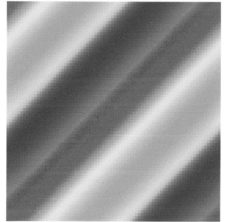

Fig. 6.3: *Top left*: The Cayley table for $(\mathbb{Z}_{79}^*, \times)$. *Top right*: the reordered table showing the cyclic structure. *Bottom*: The Cayley table for $(\mathbb{Z}_{78}, +)$.

DISCUSSION

- **More on cyclic groups**. Here are some interesting results on cyclic groups. Proofs can be found in most introductory texts on group theory.

  – All cyclic groups are Abelian.
  – Any two cyclic groups of order n (*i.e.* containing n elements) are isomorphic. We use the symbol C_n to denote a generic cyclic group of order n.
  – Let $n \in \mathbb{N}$. The group $(\mathbb{Z}_n^*, \times)$ is cyclic if and only if $n = 1, 2, 4, p^k$ or $2p^k$ where $p > 2$ is prime and $k \in \mathbb{N}$.
  – Let $n \in \mathbb{N}$. The cyclic group $(\mathbb{Z}_n, +)$ has $\phi(n)$ generators, where ϕ is *Euler's totient function*, which counts the number of positive integers coprime to n.

  Exercise 2 explores how Python can be used to find generators of a cyclic group.

- **Cyclic groups as rotations and roots of unity**. A cyclic group of order n generated by a can be written as

$$C_n = \{e, a, a^2, a^3, \ldots, a^{n-1}\},$$

where $a^n = e$ (the identity element). One can think of a as a rotation in 2D by angle $2\pi/n$ (denoted $R_{2\pi/n}$) so that a^n corresponds to a full 2π rotation, which of course is the identity transformation. We can write $C_n \cong \langle R_{2\pi/n} \rangle$. The binary operation is the composition of rotations.

We can also use the roots of unity (*i.e.* complex solutions to $z^n = 1$) to represent the same cyclic group. We write $C_n \cong \langle e^{2\pi i/n} \rangle$. The binary operation here is multiplication of complex numbers.

cayley.ipynb (for plotting fig. 6.2)	
	`import numpy as np` `import matplotlib.pyplot as plt`

Integers from 0 to 9
Modulo 10
Display numbers in the table if True
	`N = np.arange(0,10)`
	`Nmod = 10`
	`labels = True`

$f(i, j) = i + j \pmod{Nmod}$	`f = lambda i,j : (i+j) % Nmod`
Generate a square matrix (a_{ij}) with function $f(i, j)$	`array=np.fromfunction(lambda i,j:f(N[i],N[j]),` `                      (len(N),len(N)), dtype=int)`
Display the matrix (amplitude determines colour). Specify range of colormap	`plt.imshow(array, cmap='hsv',` `           vmin = N[0], vmax = len(N))`

Display the matrix elements
x, y = column, row numbers
Centre-justified
	`if labels:`
	`    for ind, X in np.ndenumerate(array):`
	`        plt.text(s = str(X),`
	`        x = ind[1], y = ind[0],`
	`        va='center', ha='center')`

Use LaTeX in title	`plt.title(r'($\mathbb{Z}_{10}, +)$')`
	`plt.axis('off')`
	`plt.show()`

6.4 Groups II – Dihedral group

> Consider a regular hexagon. Let r be the anticlockwise rotation about its centre by angle $\pi/3$. Let s be a reflection about a fixed axis of symmetry. Show that r and s form a group (call it D_6).
> What are the subgroups of D_6?

Let's number the vertices of the hexagon 1 to 6 and colour the sectors of the hexagon as shown below.

Starting with the top-left hexagon, the figure shows a sequence of transformations comprising rotation r and reflection s about the horizontal line through its centre, which we will take to be the fixed axis of symmetry. The elements of D_6, called the *dihedral group of degree 6*, are transformations that are compositions of r and s, plus the identity e (which does nothing). The binary operation of the group is composition of transformations.

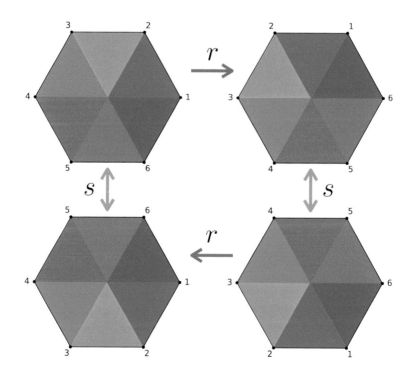

The transformations clearly satisfy

$$r^6 = s^2 = e. \tag{6.1}$$

The sequence in the figure also shows that

$$srsr = e \implies sr = (sr)^{-1} = r^{-1}s^{-1} = r^5 s. \tag{6.2}$$

Therefore, using (6.1) and (6.2), every combination of r and s can be expressed as the form $r^i s^j$ (or, equally, $s^j r^i$), where $i = 0, 1, 2, 3, 4, 5$ and $j = 0, 1$. This means that D_6 has 12 elements (*i.e.* it has order 12). Be careful, some authors denote this group by D_{12}.

The Cayley table for D_6 is shown in table 6.1. Remember the ordering: to calculate $a \times b$, look up a from the leftmost column, and b from the very top row. (We will discuss how to generate this Cayley table in Python in the next section.)

	e	r	r^2	r^3	r^4	r^5	s	rs	r^2s	r^3s	r^4s	r^5s
e	e	r	r^2	r^3	r^4	r^5	s	rs	r^2s	r^3s	r^4s	r^5s
r	r	r^2	r^3	r^4	r^5	e	rs	r^2s	r^3s	r^4s	r^5s	s
r^2	r^2	r^3	r^4	r^5	e	r	r^2s	r^3s	r^4s	r^5s	s	rs
r^3	r^3	r^4	r^5	e	r	r^2	r^3s	r^4s	r^5s	s	rs	r^2s
r^4	r^4	r^5	e	r	r^2	r^3	r^4s	r^5s	s	rs	r^2s	r^3s
r^5	r^5	e	r	r^2	r^3	r^4	r^5s	s	rs	r^2s	r^3s	r^4s
s	s	r^5s	r^4s	r^3s	r^2s	rs	e	r^5	r^4	r^3	r^2	r
rs	rs	s	r^5s	r^4s	r^3s	r^2s	r	e	r^5	r^4	r^3	r^2
r^2s	r^2s	rs	s	r^5s	r^4s	r^3s	r^2	r	e	r^5	r^4	r^3
r^3s	r^3s	r^2s	rs	s	r^5s	r^4s	r^3	r^2	r	e	r^5	r^4
r^4s	r^4s	r^3s	r^2s	rs	s	r^5s	r^4	r^3	r^2	r	e	r^5
r^5s	r^5s	r^4s	r^3s	r^2s	rs	s	r^5	r^4	r^3	r^2	r	e

Table 6.1: Cayley table for the group D_6.

As you can see, when a group gets larger, it can be daunting to understand its structure from the Cayley table. There is a more concise and visually illuminating way to represent this group.

Cayley graph

Fig. 6.4 shows the *Cayley graph* of the group D_6. Essentially the graph is a pair of large concentric hexagonal circuits connected cobweb style. The vertices show the possible configurations of the regular hexagon. The edges are transformations r and s. Each blue arrow represents the rotation r, whilst the (double-headed) red arrow represents the reflection s.

The Cayley graph gives us a more geometric way to understand the group. For example, we observe:

- The identity $r^6 = e$ can be seen from the closed loop around the inner (or outer) hexagon.
- The identity $s^2 = e$ can be seen from traversing forwards and backwards on any double-headed red arrow.
- The identity $srsr = e$ can be seen from any 4-sided loop formed between the outer and inner hexagons.
- Going against the blue arrow means applying r^{-1} (or, equivalently, r^5).

As such, every entry in the Cayley table corresponds to a path on the Cayley graph.

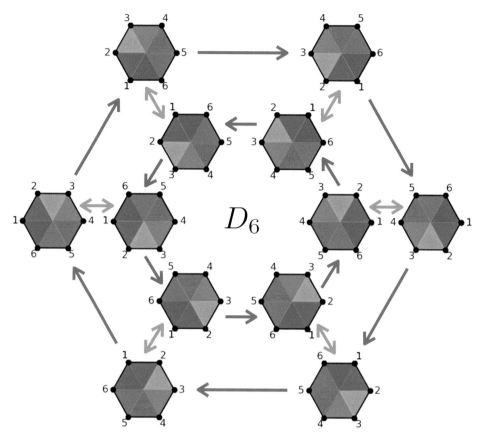

Fig. 6.4: The Cayley graph for the dihedral group D_6. Each blue arrow is an application of r (anti-clockwise rotation by angle $\pi/3$). Each double-headed red arrow is an application of s (reflection across the x-axis). Each hexagon can be produced using the code `dihedral.ipynb`.

Coding techniques

The main components of an elaborate diagram such as fig. 6.4 are the hexagons that have undergone particular transformations. Here we discuss the coding techniques used in producing these hexagons.

If we place the centre of a hexagon at the origin of the x-y plane, then the transformations in D_6 can be represented by the following 2×2 matrices (see §5.5).

$$r = \begin{pmatrix} \cos\frac{\pi}{3} & -\sin\frac{\pi}{3} \\ \sin\frac{\pi}{3} & \cos\frac{\pi}{3} \end{pmatrix}, \qquad s = \begin{pmatrix} 1 & 0 \\ 0 & -1 \end{pmatrix}, \qquad e = I.$$

Once the elements of D_6 are represented as matrices, we can easily code them and multiply them to vertices of a regular hexagon. This is what the code `dihedral.ipynb` does to visualise the effect of a given transformation on the hexagon.

Some highlights from the code.

- Although we only need 2D arrays, we work with 3D arrays since the z coordinate can be used to store the label of each vertex. This allows us to keep track of the orientation of the hexagon as it gets rotated and flipped around.
- We use the Matplotlib function `fill` to create colourful triangular segments.
- You only need to change the line annotated "** **Specify the transformation here**" to see the effect of $r^i s^j$ on the hexagon. In the code, `rot` = r and `ref` = s.
- The same code can be used to study the symmetry of any n-gon simply by changing the value of n. (See exercise 4.)

dihedral.ipynb (for plotting each hexagon in fig. 6.4)	

	`import numpy as np`
	`import matplotlib.pyplot as plt`
	`%matplotlib`

Specify n (number of sides of n-gon)
Rotation angle θ
	`n = 6`
	`theta = 2*np.pi/n`
	`C, S = np.cos(theta), np.sin(theta)`

2D rotation matrix. Use z cooord. to store label	`rot = np.array([[C,-S,0],[S, C,0],[0,0,1]])`

Reflection matrix (about the x-axis)
For changing the z-coord (label)
Specify one vertex at $(1, 0)$. Label it '1'
	`ref = np.array([[1, 0,0],[0,-1,0],[0,0,1]])`
	`k = np.array([0,0,1])`
	`V = np.array([1,0,1])`
	`U = V`

Find the remaining vertices
Next vertex = rotate previous vertex & adjust label. Then, append the coords+label of each new vertex found
	`for i in range(n-1):`
	`    U = rot@U + k`
	`    V = np.vstack((V, U))`

Complete the circuit
*** Specify the transformation here
(For example, this applies r^3 to the n-gon)
Separate out the x, y, z coords of the transformed n-gon
(remember that z is just the label)
	`V = np.vstack((V,np.array([1,0,1]))).T`
	`V = np.linalg.matrix_power(rot,3)@V`
	
	`poly_x = V[0,:]`
	`poly_y = V[1,:]`
	`poly_z = V[2,:]`

Outline the n-gon with black line	`plt.plot(poly_x, poly_y, 'ko-')`

Add labels to the vertices
Place the label slightly away from the point (s = label)
	`for i, X in enumerate(poly_x):`
	`    plt.text(x = 1.1*X, y = 1.1*poly_y[i],`
	`             s = int(poly_z[i]), fontsize=14,`
	`             ha = 'center', va = 'center')`

Let's divide n-gon into triangles
x and y coords of each triangle
 	
Use `fill` function for colouring
	`for i in range(n):`
	`    Tx = np.array([0, poly_x[i], poly_x[i+1]])`
	`    Ty = np.array([0, poly_y[i], poly_y[i+1]])`
	`    plt.fill(Tx, Ty)`

This prevents shape distortion	`plt.axis('square')`
	`plt.axis('off')`
	`plt.show()`

Subgroups

You may have noticed from the Cayley graph that the inner hexagonal circuit (comprising pure rotations) represents a group. More precisely, it is a cyclic group of order 6, *i.e.* $C_6 = \langle r \rangle$. Equally, you might have noticed the same thing in the Cayley table (the top-left quadrant contains all terms without s).

A subset H of a group G is said to be a *subgroup* if H itself is a group (under the binary operation of G).

Every group G has the trivial subgroup $\{e\}$ and G itself as subgroups. We say that H is a proper subgroup of G if H is a subgroup of G and $H \neq G$. In symbols, we write $H < G$.

We have seen that $C_6 < D_6$. The subgroups of D_6 are shown in fig. 6.5. For each subgroup, we list the elements of the subgroup and show the corresponding subset of the Cayley graph.

Some observations:

- There are other possible choices for the elements of the dihedral subgroups D_d. For example, D_1 can be formed by any one of the following 6 possibilities

$$D_1 = \{e, r^i s\}, \text{ where } i = 0, 1, 2, 3, 4, 5.$$

  You can verify that the element $r^i s$ has order 2 (or check the diagonal of the Cayley table). Equally we could use $s r^i$ instead of $r^i s$.

  Try to write down possible choices for the elements of D_2 and D_3 in terms of r and s, and sketch their Cayley graphs.
- The subgroups of D_6 are all of the form C_d or D_d, where d is a divisor of 6.
- The subgroups of D_6 have orders 1, 2, 3, 4, 6 and 12 (the group itself). These are all divisors of 12 (= the order of D_6). Is this a coincidence?

DISCUSSION

- **Lagrange's theorem**. The following important result explains why all the subgroups of D_6 have orders that are divisors of 12.

  Theorem 6.3 *(Lagrange's theorem) Let H be a subgroup of a finite group G. Then* $|H|$ *divides* $|G|$.

  The converse does not hold in general: if d is a divisor of $|G|$, there may not be a subgroup of order d.
- **Subgroups of D_n**. Here is another general result which you may have conjectured.

  Lemma: All the subgroups of D_n are C_d and D_d where $d|n$.

  See [19] for proof.
- **Klein four-group**. The dihedral group $D_2 = \{e, r, s, rs\}$ (where $r^2 = s^2 = (rs)^2$) is also called the *Klein four-group*, after the German mathematician *Felix Klein* (1849–1925) (who also devised the famous *Klein bottle*). The Klein four-group is sometimes given the symbol V or K_4. It has a special place in group theory for being the smallest non-cyclic group.

Subgroup	Elements	Cayley graph
C_1 (trivial)	$\{e\}$	
C_2 $\cong$ D_1	$\{e, r^3\}$ $\{e, s\}$	
C_3	$\{e, r^2, r^4\}$	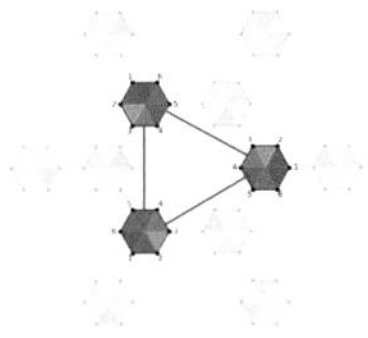
D_2	$\{e, r^3, s, r^3 s\}$	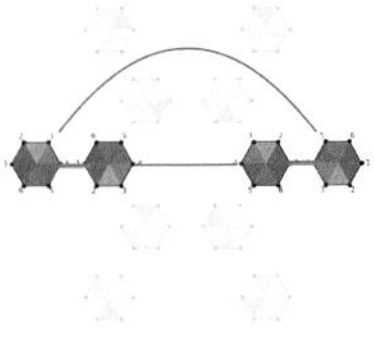
D_3	$\{e, r^2, r^4, s, r^2 s, r^4 s\}$	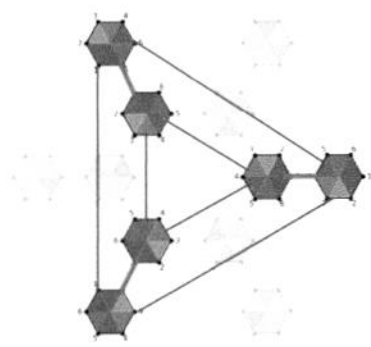
C_6	$\{e, r, r^2, r^3, r^4, r^5\}$	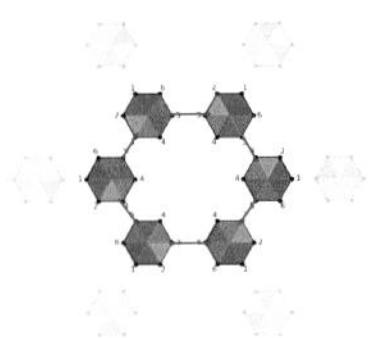

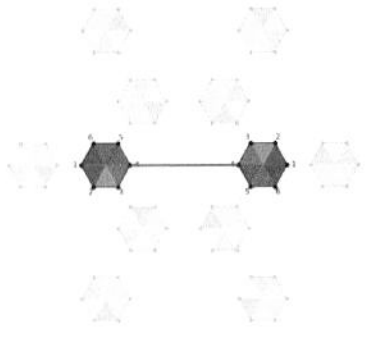

Fig. 6.5: Proper subgroups of D_6. There may be more than one way to represent the elements of each subgroup.

6.5 Groups III – Symmetric and alternating groups

> List the elements of S_4, the symmetric group of 4 objects.
> Which of the following are normal subgroups of S_4?
> a) S_3,   b) The alternating group A_4,   c) D_3,   d) D_2.
> Describe the quotient group S_4/A_4.

The elements of the symmetric and alternating groups are *permutations* which we will introduce here along with the main results. Details can be found in any of the recommended texts on algebra.

Suppose we have 4 objects labelled 0 to 3 lined up in that order. A permutation is a function which rearranges of these object. For example, here are two permutations f and g defined in terms of their rearranging action.

$$f(0\ 1\ 2\ 3) = (2\ 0\ 3\ 1), \qquad g(0\ 1\ 2\ 3) = (1\ 2\ 0\ 3).$$

Another way to notate a permutation is to use a 2-layer notation:

$$f = \begin{pmatrix} 0\ 1\ 2\ 3 \\ 1\ 3\ 0\ 2 \end{pmatrix}, \qquad g = \begin{pmatrix} 0\ 1\ 2\ 3 \\ 2\ 0\ 1\ 3 \end{pmatrix}.$$

The top row lists the original positions whilst the bottom row lists the new position of each element directly above.

Yet another notation which is more widely used (and one which we will use) is the following 'cycle' notation.

$$f = \begin{pmatrix} 0\ 1\ 3\ 2 \end{pmatrix}, \qquad g = \begin{pmatrix} 0\ 2\ 1 \end{pmatrix}\begin{pmatrix} 3 \end{pmatrix}.$$

Each pair of bracket is called a cycle where the object originally in position n goes to position in the next entry. For example, $\begin{pmatrix} 0\ 1\ 3\ 2 \end{pmatrix}$ means, the object originally in position 0 goes to position 1, the object originally in position 1 goes to position 3, the object originally in position 3 goes to position 2, and the object originally in position 2 goes to the object in position 0.

In the cycle notation, $\begin{pmatrix} 0\ 1\ 3\ 2 \end{pmatrix}$ is the same permutation as $\begin{pmatrix} 2\ 0\ 1\ 3 \end{pmatrix}$, $\begin{pmatrix} 3\ 2\ 0\ 1 \end{pmatrix}$ and $\begin{pmatrix} 1\ 3\ 2\ 0 \end{pmatrix}$.

The cycle notation for g can also be written as $\begin{pmatrix} 0\ 2\ 1 \end{pmatrix}$, omitting (3) since 3 is left unchanged by the permutation. However, we will see that when working in SymPy, sometimes it is necessary to include singletons (1-cycles) like (3).

The symmetric group

We can compose two permutations, for example,

$$fg = \begin{pmatrix} 0\ 1\ 3\ 2 \end{pmatrix}\begin{pmatrix} 0\ 2\ 1 \end{pmatrix} = \begin{pmatrix} 2\ 3 \end{pmatrix}, \qquad gf = \begin{pmatrix} 0\ 2\ 1 \end{pmatrix}\begin{pmatrix} 0\ 1\ 3\ 2 \end{pmatrix} = \begin{pmatrix} 1\ 3 \end{pmatrix}. \qquad (6.3)$$

where we perform the operation on the rightmost cycle first and move leftwards. As you can see, this operation is non-commutative.

Each permutation has an inverse which undoes the permutation. For example,

$$f^{-1} = \begin{pmatrix} 0 & 2 & 3 & 1 \end{pmatrix}, \qquad g^{-1} = \begin{pmatrix} 0 & 1 & 2 \end{pmatrix},$$

(just write the cycles backward). You should check that

$$f f^{-1} = f^{-1} f = g g^{-1} = g^{-1} g = e,$$

where e is the identity permutation (*i.e.* doing nothing). To follow SymPy's convention, we can also express the identity permutation on the set $\{0, 1, 2, 3, \ldots, n\}$ as (n). In our case, $f f^{-1} = g g^{-1} = (3)$.

All permutations on n objects form a group called the *symmetric group*, S_n. This group has $n!$ elements.

Odd/even permutations and the alternating group

Every cycle can be written as a product of cycles of length 2 (these are called *transpositions*). Here is one method to decompose f and g into transpositions. Study the following pattern:

$$f = \begin{pmatrix} 0 & 1 & 3 & 2 \end{pmatrix} = \begin{pmatrix} 0 & 1 \end{pmatrix} \begin{pmatrix} 1 & 3 \end{pmatrix} \begin{pmatrix} 3 & 2 \end{pmatrix} \qquad g = \begin{pmatrix} 0 & 2 & 1 \end{pmatrix} = \begin{pmatrix} 0 & 2 \end{pmatrix} \begin{pmatrix} 2 & 1 \end{pmatrix}.$$

Another method uses the following pattern:

$$\begin{pmatrix} 0 & 1 & 3 & 2 \end{pmatrix} = \begin{pmatrix} 0 & 2 \end{pmatrix} \begin{pmatrix} 0 & 3 \end{pmatrix} \begin{pmatrix} 0 & 1 \end{pmatrix} \qquad \begin{pmatrix} 0 & 2 & 1 \end{pmatrix} = \begin{pmatrix} 0 & 1 \end{pmatrix} \begin{pmatrix} 0 & 2 \end{pmatrix}.$$

Both of these methods decompose an n-cycle into $(n-1)$ transpositions. There are other ways to do this.

An *odd permutation* is one which can be expressed as a product of an odd number of transpositions.

An *even permutation* can be expressed as a product of an even number of transpositions. The state of being odd or even is called the *parity* of a permutation.

The cycle notation makes it easy to identify if a permutation is odd or even – a cycle of odd length is even, and a cycle of even length is odd. Note that the identity permutation is an even permutation.

All even permutations of S_n form the *alternating group*, A_n, with $n!/2$ elements. The alternating group A_n is clearly a subgroup of S_n.

Permutations in SymPy

Detailed documentation on working with permutations in SymPy can be found on SymPy's website[3]. We highlight some important syntax here. See the accompanying code snippets in the box *Working with permutations in SymPy*.

- The composite permutation fg means g is evaluated before f. However, **in SymPy, the order is reversed!** fg should be coded as `g*f`.
- In SymPy, $(0\ 2\ 1)\ (3)$ is different from $(0\ 2\ 1)$. SymPy interprets the latter as a permutation of only 3 objects $\{0, 1, 2\}$. You can check this using the command `g.size` where g is $(0\ 2\ 1)$ or $(3)\ (0\ 2\ 1)$.
  The documentation recommends that "*it is better to start the cycle with the singleton*".
- SymPy comes with ready-made lists of elements of S_n, A_n, C_n (cyclic group) and D_n (dihedral group). Elements are represented as permutations (see Discussion).

For example, Table 6.2 shows the 4! = 24 elements of S_4 generated using SymPy.

The code `permutation.ipynb` generates the Cayley tables for S_4 as shown in fig. 6.6. With minor modifications, the table for A_4 can be obtained whilst retaining the same indices as those in the table for S_4 (exercise 5(a)i).

Some observations from the Cayley tables:

- A_4 is a subgroup of S_4. It is precisely half the size of S_4.
- S_3 (the permutation group of 3 objects) appears in the top left (6 × 6) block of the S_4 table. This arises since we can leave the element 0 alone and perform permutations on $\{1, 2, 3\}$. Similarly, S_2 takes up the top 2 × 2 block, since we can leave both 0 and 1 alone and permute $\{2, 3\}$. In short, $S_2 < S_3 < S_4$.
- There appear to be 3 other distinct 6 × 6 sub-blocks along the left edge of the S_4 table. Each sub-block is closed (*i.e.* they don't contain elements external to the block).
- Similarly, there appear to be four distinct 3 × 3 sub-blocks along the left ledge of the A_4 table.

index	element	parity	index	element	parity
0	e	E	12	(0 2 1)	E
1	(2 3)	O	13	(0 2 3 1)	O
2	(1 2)	O	14	(0 2)	O
3	(1 2 3)	E	15	(0 2 3)	E
4	(1 3 2)	E	16	(0 2)(1 3)	E
5	(1 3)	O	17	(0 2 1 3)	O
6	(0 1)	O	18	(0 3 2 1)	O
7	(0 1)(2 3)	E	19	(0 3 1)	E
8	(0 1 2)	E	20	(0 3 2)	E
9	(0 1 2 3)	O	21	(0 3)	O
10	(0 1 3 2)	O	22	(0 3 1 2)	O
11	(0 1 3)	E	23	(0 3)(1 2)	E

Table 6.2: Elements of S_4 and their indices (0-23) in the Python list. The even-parity elements form the group A_4.

[3] https://docs.sympy.org/latest/modules/combinatorics/permutations.html

S_4

0	1	2	3	4	5	6	7	8	9	10	11	12	13	14	15	16	17	18	19	20	21	22	23
1	0	4	5	2	3	7	6	10	11	8	9	18	19	20	21	22	23	12	13	14	15	16	17
2	3	0	1	5	4	12	13	14	15	16	17	6	7	8	9	10	11	19	18	22	23	20	21
3	2	5	4	0	1	13	12	16	17	14	15	19	18	22	23	20	21	6	7	8	9	10	11
4	5	1	0	3	2	18	19	20	21	22	23	7	6	10	11	8	9	13	12	16	17	14	15
5	4	3	2	1	0	19	18	22	23	20	21	13	12	16	17	14	15	7	6	10	11	8	9
6	7	8	9	10	11	0	1	2	3	4	5	14	15	12	13	17	16	20	21	18	19	23	22
7	6	10	11	8	9	1	0	4	5	2	3	20	21	18	19	23	22	14	15	12	13	17	16
8	9	6	7	11	10	14	15	12	13	17	16	0	1	2	3	4	5	21	20	23	22	18	19
9	8	11	10	6	7	15	14	17	16	12	13	21	20	23	22	18	19	0	1	2	3	4	5
10	11	7	6	9	8	20	21	18	19	23	22	1	0	4	5	2	3	15	14	17	16	12	13
11	10	9	8	7	6	21	20	23	22	18	19	15	14	17	16	12	13	1	0	4	5	2	3
12	13	14	15	16	17	2	3	0	1	5	4	8	9	6	7	11	10	22	23	19	18	21	20
13	12	16	17	14	15	3	2	5	4	0	1	22	23	19	18	21	20	8	9	6	7	11	10
14	15	12	13	17	16	8	9	6	7	11	10	2	3	0	1	5	4	23	22	21	20	19	18
15	14	17	16	12	13	9	8	11	10	6	7	23	22	21	20	19	18	2	3	0	1	5	4
16	17	13	12	15	14	22	23	19	18	21	20	3	2	5	4	0	1	9	8	11	10	6	7
17	16	15	14	13	12	23	22	21	20	19	18	9	8	11	10	6	7	3	2	5	4	0	1
18	19	20	21	22	23	4	5	1	0	3	2	10	11	7	6	9	8	16	17	13	12	15	14
19	18	22	23	20	21	5	4	3	2	1	0	16	17	13	12	15	14	10	11	7	6	9	8
20	21	18	19	23	22	10	11	7	6	9	8	4	5	1	0	3	2	17	16	15	14	13	12
21	20	23	22	18	19	11	10	9	8	7	6	17	16	15	14	13	12	4	5	1	0	3	2
22	23	19	18	21	20	16	17	13	12	15	14	5	4	3	2	1	0	11	10	9	8	7	6
23	22	21	20	19	18	17	16	15	14	13	12	11	10	9	8	7	6	5	4	3	2	1	0

A_4

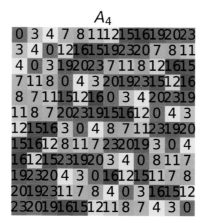

Fig. 6.6: Cayley tables for S_4 and A_4.

Working with permutations in SymPy

Defining permutation in SymPy	```from sympy.combinatorics \``` ``` import Permutation as P```

$f = (0\ 1\ 3\ 2)$
$g = (0\ 2\ 1)(3)$
	```f = P(0,1,3,2)```
	```g = P(3)(0,2,1)```
	```# Careful: P(0,2,1)(3) gives an error!```

Composition In SymPy, f*g is evaluated as "f then g"!!! Output: f*g = (1 3), g*f = (2 3) (The opposite of mathematical notation (6.3))	```f*g``` ```g*f```
Inverse Output: g**-1 = (0 1 2)(3)	```g**-1```
Order Output: 4 (try f**4)	```f.order()```
Parity Output: 1 for f 0 for g	```f.parity()``` ```g.parity()```
Decomposition into transpositions Output: (0 2)(0 3)(0 1)	```f.transpositions()```
Are two permutations equal? Output: True	```P(0,1)(0,2) == P(0,2)(2,1)```

Symmetric and alternating groups
Generate these groups in SymPy
(Also available: dihedral and cyclic)
Make output pretty
	```from sympy.combinatorics.generators \```
	``` import symmetric, alternating```
	```from sympy import init_printing as IP```
	```IP(perm_cyclic=True, pretty_print=True)```

All elements of S_4 as a list All elements of A_4 as a list	```list(symmetric(4))``` ```list(alternating(4))```

permutation.ipynb (for plotting fig. 6.6)

For generating S_4	```import numpy as np```
	```import matplotlib.pyplot as plt```
	```from sympy.combinatorics.generators import \```
	``` symmetric```
	```%matplotlib```

Put elements of S_4 into a list Index each element (Table 6.2) (Choose False for large tables)	```G = list(symmetric(4))``` ```labels = True```
Compose the permutations and find which element of S_4 is the result	```f = lambda i,j : G.index(G[j]*G[i])```
Create a 2D square array using the function f(i,j)	```array = np.fromfunction(np.vectorize(f),``` ``` (len(G),len(G)), dtype=int)```
Display the matrix (index determines colour). Specify range of colormap	```plt.imshow(array, cmap='hsv',``` ``` vmin = 0, vmax = len(G))```

Display the index (not just the colour)
x, y = column, row numbers
Centre-justify labels
	```if labels:```
	``` for ind, X in np.ndenumerate(array):```
	``` plt.text(s = str(X),```
	``` x = ind[1], y = ind[0],```
	``` va='center', ha='center', fontsize=9)```

Use LaTeX in title	```plt.title(r'S_4', fontsize=12)```
	```plt.axis('off')```
	```plt.show()```

The Cayley table suggests that there are substructures within the group structure of S_4. Ideas that will help us make sense of these substructures are *cosets, normal subgroups* and *quotient groups*. Let's discuss these ideas one by one.

Coset

Let H be a subgroup of G and take a fixed element $g \in G$. A *left coset* of H in G is the set

$$gH := \{gh : h \in H\}.$$

The notation gH means that we left-multiply all elements of H by a fixed element g and collect the results. Note that if $g \in H$, then $gH = H$ due to the closure of H.

Similarly, a *right coset* of H in G is the set

$$Hg := \{hg : h \in H\}.$$

Let's examine the possible left cosets of S_3 in S_4. The result depends on the fixed element $g \in S_4$.

- Case **A** – if $g \in S_3$, then $gS_3 = S_3 = \{\text{elements with indices } 0 - 5 \text{ in Table 6.2}\}$.
- Case **B** – if $g = (0, 1)$, then $gS_3 = \{\text{elements with indices } 6 - 11.\}$
- Case **C** – if $g = (0, 2, 1)$, then $gS_3 = \{\text{elements with indices } 12 - 17.\}$
- Case **D** – if $g = (0, 3, 2, 1)$, then $gS_3 = \{\text{elements with indices } 18 - 23.\}$

You can check that for all $g \in S_4$, any left coset is one of the above types. We can use SymPy to help us quickly calculate the indices of the elements in the cosets of S_3.

	`from sympy.combinatorics.generators \`
	`    import symmetric`
	`from sympy.combinatorics \`
	`    import Permutation as P`
S_4	`G  = list(symmetric(4))`
S_3	`H =  G[0:6]`
Left coset $(0\ 1)S_3$.	`coset = [h*P(3)(0,1) for h in H]`
Note SymPy's reverse ordering	
Produce a set of indices (ignore repeats)	`set([G.index(x) for x in coset])`

The above calculations show that the 24 elements of S_4 appear to organise themselves into 4 substructures (left cosets) - each containing 6 elements. This explains the 4 sub-blocks on the left edge of the Cayley table for S_4: They are simply the left cosets of S_4.

We say that *cosets partition a group, i.e.* every element of a group belongs to exactly one of the cosets.

The top panel of fig. 6.7 shows the result of partitioning S_4 by the left cosets of the subgroup S_3. There are 4 cosets with 6 elements each. The same procedure can be done with the right cosets of S_3. In the figure we also show the partitioning using other subgroups such as A_4, D_3 and D_2. (Exercise 5b asks you to obtain these partitioning with SymPy.)

Similarly, in the Cayley table for A_4, the top-left 3×3 sub-block is the subgroup C_3, and the 3 sub-blocks below are its left cosets.

Partitioning S_4 by cosets of various subgroups

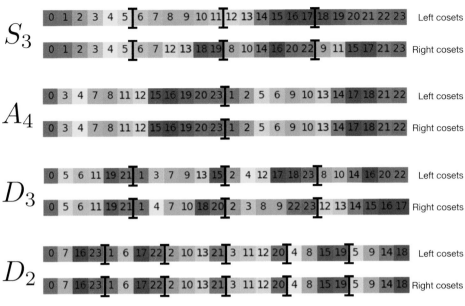

Fig. 6.7: Partitioning S_4 by left and right cosets of its subgroups S_3, A_4, D_3 and D_2.

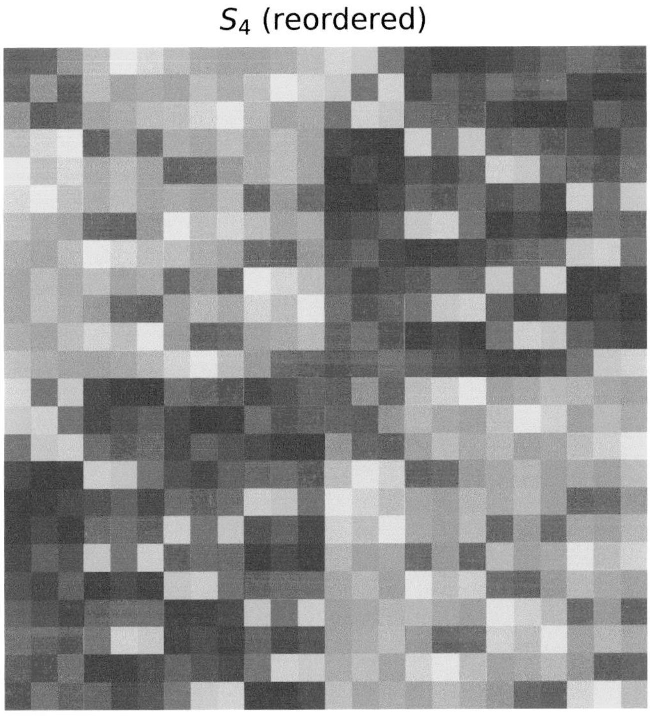

S_4 (reordered)

Fig. 6.8: The Cayley table for S_4, reordered so that A_4 appears on the top-left block. The 4-block structure illustrates the isomorphism $S_4/A_4 \cong \mathbb{Z}_2$.

Normal subgroup

A subgroup H is said to be a *normal* subgroup of G (written $H \triangleleft G$) if, for all $g \in G$, the left and right cosets are equal, *i.e.*

$$gH = Hg.$$

This condition can also be expressed as $gHg^{-1} = H$.

From fig. 6.7, we see that A_4 and D_2 are normal subgroups of S_4, whilst S_3 and D_3 are not normal subgroups S_4

Quotient group

The cosets of a subgroup $H < G$ partition the group into substructures. If H is also a normal subgroup, then those substructures themselves behave like group elements. This allows a large group to be simplified into a simpler, smaller group.

Let $H \triangleleft G$. The set of cosets $\{gH : g \in G\}$ form a group called the *quotient group*, denoted $(G/H, *)$, where the 'coset multiplication' $*$ is defined as

$$(g_1 H) * (g_2 H) = (g_1 g_2)H.$$

For example, take the normal subgroup $A_4 \triangleleft S_4$. We established that there are two cosets

$$(\text{even})A_4 \qquad \text{and} \qquad (\text{odd})A_4,$$

where (even) and (odd) are any even or odd permutations in S_4. The quotient group S_4/A_4 has 2 elements, namely, the two cosets. The results of the coset multiplications are:

$*$	$(\text{even})A_4$	$(\text{odd})A_4$
$(\text{even})A_4$	$(\text{even})A_4$	$(\text{odd})A_4$
$(\text{odd})A_4$	$(\text{odd})A_4$	$(\text{even})A_4$

$+ \pmod 2$	0	1
0	0	1
1	1	0

Table 6.3: The Cayley tables for $(S_4/A_4, *)$ (left) and $(\mathbb{Z}_2, +)$ (right).

We see that the Cayley table for the quotient group looks like that for $(\mathbb{Z}_2, +)$ (addition modulo 2). We conclude that the two groups are isomorphic.

$$S_4/A_4 \cong \mathbb{Z}_2.$$

Let's visualise this isomorphism in SymPy. In fig. 6.8, we produce the Cayley table for S_4 but the elements have been rearranged so that the first 12 are the even permutations followed by the 12 odd ones. This produces a 4-block structure that (when you zoom out) looks like the Cayley table for $\mathbb{Z}_2$. This is one way we can think about what quotient groups are: zooming out of a Cayley table for a group, we see another Cayley table for its quotient group.

In exercise 5b, you will perform a similar reordering to study the quotient group S_4/D_2.

DISCUSSION

- **Cayley's theorem**. The cyclic group C_n and dihedral group D_n can all be represented as permutation groups. Just draw the corresponding polygon with numbered vertices to see that a transformation maps one number to another like a permutation.

  For example, the dihedral group D_3 (symmetry group for an equilateral triangle) can be thought of as, S_3, the permutations of 3 symbols. Note that both have 6 elements. In fact, we have the isomorphism $D_3 \cong S_3$ (take another look at fig. 6.7 to see if you can spot the isomorphism). The following result goes even further.

Theorem 6.4 *(Cayley's theorem) Every finite group of order n is isomorphic to a subgroup of S_n.*

In other words, we can use permutations to describe every finite group!

6.6 Quaternions

> Let P be the point $(2, 0, 3)$ and $\mathbf{v} = \mathbf{i} + \mathbf{j} + \mathbf{k}$. The point P is rotated about the axis $\mathbf{v}$ by 6 radians. Find the coordinates of the rotated point P'.

It is generally uncommon for quaternions to be studied in a typical first-year mathematics curriculum, but had history of the late 1800s panned out differently, students would probably be studying quaternions instead of vectors. In this section, we will give a summary of the mathematics of quaternions and use them to perform rotation in $\mathbb{R}^3$.

We use of the letter i for the imaginary unit, and also $\mathbf{i}$ for the unit vector $(1, 0, 0)$. These two concepts, as we will see, are combined in the world of quaternions.

We can think of quaternions as a generalisation of the imaginary number i to 3 dimensions. Let $\mathbf{i}, \mathbf{j}, \mathbf{k}$ be mathematical objects and define their multiplication via the formula

$$\mathbf{i}^2 = \mathbf{j}^2 = \mathbf{k}^2 = \mathbf{ijk} = -1. \tag{6.4}$$

These objects also satisfy non-commutative multiplication rules that are reminiscent of the vector cross product.

$$\mathbf{ij} = -\mathbf{ji} = \mathbf{k}, \qquad \mathbf{jk} = -\mathbf{kj} = \mathbf{i}, \qquad \mathbf{ki} = -\mathbf{ik} = \mathbf{j}. \tag{6.5}$$

The story of the conception of quaternions is now legend. Sir William Rowan Hamilton had a flash of inspiration whilst out walking in Dublin in October 1843 and carved their algebraic formulation (eq. 6.4) on Broom Bridge. However, the French mathematician *Olinde Rodrigues* (1795–1851) was the first to discover the formulation of quaternion rotation a few years earlier (recall that we met Rodrigues and his famous 'formula' in §5.10).

The *quaternion group Q* consists of 8 elements $\{\pm 1, \pm\mathbf{i}, \pm\mathbf{j}, \pm\mathbf{k}\}$ with the following multiplication rule.

	1	-1	$\mathbf{i}$	$-\mathbf{i}$	$\mathbf{j}$	$-\mathbf{j}$	$\mathbf{k}$	$-\mathbf{k}$
1	1	-1	$\mathbf{i}$	$-\mathbf{i}$	$\mathbf{j}$	$-\mathbf{j}$	$\mathbf{k}$	$-\mathbf{k}$
-1	-1	1	$-\mathbf{i}$	$\mathbf{i}$	$-\mathbf{j}$	$\mathbf{j}$	$-\mathbf{k}$	$\mathbf{k}$
$\mathbf{i}$	$\mathbf{i}$	$-\mathbf{i}$	-1	1	$\mathbf{k}$	$-\mathbf{k}$	$-\mathbf{j}$	$\mathbf{j}$
$-\mathbf{i}$	$-\mathbf{i}$	$\mathbf{i}$	1	-1	$-\mathbf{k}$	$\mathbf{k}$	$\mathbf{j}$	$-\mathbf{j}$
$\mathbf{j}$	$\mathbf{j}$	$-\mathbf{j}$	$-\mathbf{k}$	$\mathbf{k}$	-1	1	$\mathbf{i}$	$-\mathbf{i}$
$-\mathbf{j}$	$-\mathbf{j}$	$\mathbf{j}$	$\mathbf{k}$	$-\mathbf{k}$	1	-1	$-\mathbf{i}$	$\mathbf{i}$
$\mathbf{k}$	$\mathbf{k}$	$-\mathbf{k}$	$\mathbf{j}$	$-\mathbf{j}$	$-\mathbf{i}$	$\mathbf{i}$	-1	1
$-\mathbf{k}$	$-\mathbf{k}$	$\mathbf{k}$	$-\mathbf{j}$	$\mathbf{j}$	$\mathbf{i}$	$-\mathbf{i}$	1	-1

Table 6.4: The Cayley table for the quaternion group Q.

In general, a quaternion $\mathbf{q}$ can be expressed as a linear combination of $1, \mathbf{i}, \mathbf{j}$ and $\mathbf{k}$.

$$\mathbf{q} = s + x\mathbf{i} + y\mathbf{j} + z\mathbf{k}, \qquad s, x, y, z \in \mathbb{R}.$$

We call s the *real part* of $\mathbf{q}$, and $\mathbf{v} := x\mathbf{i} + y\mathbf{j} + z\mathbf{k}$ the *vector* part. If $s = 0$, we call $\mathbf{q}$ a *pure quaternion*.

Let $\mathbf{q_1} = s_1 + x_1\mathbf{i} + y_1\mathbf{j} + z_1\mathbf{k}$ and $\mathbf{q_2} = s_2 + x_2\mathbf{i} + y_2\mathbf{j} + z_2\mathbf{k}$. We can add two quaternions componentwise.

$$\mathbf{q_1} + \mathbf{q_2} = (s_1 + s_2) + (x_1 + x_2)\mathbf{i} + (y_1 + y_2)\mathbf{j} + (z_1 + z_2)\mathbf{k}.$$

We can also multiply two quaternions using the multiplication rules (6.4)-(6.5). The result is a little messy.

$$
\begin{aligned}
\mathbf{q_1}\mathbf{q_2} &= (s_1 + x_1\mathbf{i} + y_1\mathbf{j} + z_1\mathbf{k})(s_2 + x_2\mathbf{i} + y_2\mathbf{j} + z_2\mathbf{k}) \\
&= (s_1 s_2 - x_1 x_2 - y_1 y_2 - z_1 z_2) + (s_1 x_2 + s_2 x_1 + y_1 z_2 - y_2 z_1)\mathbf{i} + \cdots \\
&\quad (s_1 y_2 + s_2 y_1 - x_1 z_2 + x_2 z_1)\mathbf{j} + (s_1 z_2 + s_2 z_1 + x_1 y_2 - x_2 y_1)\mathbf{k}.
\end{aligned}
\tag{6.6}
$$

Thankfully there is a less painful way to express the above result using the cross product. Write $\mathbf{q_1} = s_1 + \mathbf{v_1}$ and $\mathbf{q_2} = s_2 + \mathbf{v_2}$ (where $\mathbf{v_1}$ and $\mathbf{v_2}$ are the vector parts). The multiplication rule can be written more compactly as

$$\mathbf{q_1}\mathbf{q_2} = (s_1 s_2 - \mathbf{v_1} \cdot \mathbf{v_2}) + (s_1 \mathbf{v_2} + s_2 \mathbf{v_1} + \mathbf{v_1} \times \mathbf{v_2}). \tag{6.7}$$

Conjugate, modulus and inverse

- The *conjugate* of a quaternion $\mathbf{q} = s + \mathbf{v}$ is defined as $\bar{\mathbf{q}} = s - \mathbf{v}$.
- The *modulus* of $\mathbf{q} = s + \mathbf{v}$ is defined as $|\mathbf{q}| = (\mathbf{q}\bar{\mathbf{q}})^{1/2} = \left(s^2 + |\mathbf{v}|^2\right)^{1/2}$. (Verify this using eq. 6.7.)
- The *inverse* of a quaternion $\mathbf{q}$ is defined as $\mathbf{q}^{-1} = \dfrac{\bar{\mathbf{q}}}{|\mathbf{q}|^2}$.

  Every nonzero quaternion has an inverse which satisfies $\mathbf{q}\mathbf{q}^{-1} = 1$.
- If $|\mathbf{q}| = 1$, we say that $\mathbf{q}$ is a *unit* quaternion. Its inverse is its conjugate: $\mathbf{q}^{-1} = \bar{\mathbf{q}}$.

Polar form

We can associate a quaternion with an angle through the following theorem.

Theorem 6.5 (*Polar form of a quaternion*) *Any unit quaternion $\mathbf{q}$ can be expressed in the form*

$$\mathbf{q} = \cos(\theta) + \sin(\theta)\,\mathbf{I}, \tag{6.8}$$

for some angle θ and some unit vector $\mathbf{I}$. This is called the polar form of $\mathbf{q}$.

If $\mathbf{q}$ is not a unit quaternion, we can apply the above theorem to $\mathbf{q}/|\mathbf{q}|$. Thus the polar form of $\mathbf{q}$ is simply $\mathbf{q} = |\mathbf{q}|\,(\cos(\theta) + \sin(\theta)\,\mathbf{I})$.

Rotation in $\mathbb{R}^3$ using quaternions

The following important theorem states that a rotation in $\mathbb{R}^3$ about an axis through the origin can be represented by a unit quaternion.

Theorem 6.6 *(Rotation in $\mathbb{R}^3$) Let $\mathbf{q} = \cos(\theta/2) + \sin(\theta/2)\mathbf{I}$ be a (unit) quaternion and $\mathbf{p} = 0 + \mathbf{v}$ be a pure quaternion. When the point P with position vector $\mathbf{v}$ is rotated by angle θ about the axis $\mathbf{I}$ (a unit vector through the origin), the image is given by the vector part of*

$$\mathbf{p}' = \mathbf{q}\mathbf{p}\mathbf{q}^{-1}. \tag{6.9}$$

This is the key theorem behind the usefulness of quaternions in computer graphics. Be careful: the angle in the quaternion $\mathbf{q}$ is *half* the rotation angle.

You may be wondering why we bother with quaternions when surely the same rotation can be performed by a 3×3 matrix. Indeed, when a point with position vector $\mathbf{v}$ is rotated through an angle θ about the axis (α, β, γ) (a unit vector), the result is $R_\theta \mathbf{v}$ where

$$R_\theta = \begin{pmatrix} \alpha^2(1-c)+c & \alpha\beta(1-c)+\gamma s & \alpha\gamma(1-c)-\beta s \\ \alpha\beta(1-c)-\gamma s & \beta^2(1-c)+c & \beta\gamma(1-c)+\alpha s \\ \alpha\gamma(1-c)+\beta s & \beta\gamma(1-c)-\alpha s & \gamma^2(1-c)+c \end{pmatrix}. \tag{6.10}$$

Here $c \equiv \cos\theta$ and $s \equiv \sin\theta$. However, you can see that this requires working with 9 entries in the matrix. Quaternions require fewer operations and is therefore useful in computer applications where a large number of elements may be undergoing rotations at the same time. Quaternions are also used when objects need to be able to rotate very freely (*e.g.* in a flight simulator or in determining the orientation of mobile phones) as quaternions, unlike matrices, do not suffer from *gimbal lock*[4], a loss of degree of freedom which could lead to erratic behaviour.

Python implementation

The code `quaternion.ipynb` rotates a given point P with position vector $\mathbf{v}$ by an angle θ around a given axis $\mathbf{I}$ (a unit vector). The code produces an interactive display with a slider for the angle θ, as shown in fig. 6.9. In the code, we follow this recipe.

1. Write $\mathbf{q} = \cos(\theta/2) + \sin(\theta/2)\mathbf{I}$.
2. Find its inverse $\mathbf{q}^{-1} = \bar{\mathbf{q}}$
3. Write $\mathbf{p} = 0 + \mathbf{v}$ and calculate the 'sandwich' $\mathbf{p}' = \mathbf{q}\mathbf{p}\mathbf{q}^{-1}$. (Formula 6.7 reduces this to two cross products.)
4. The image is the vector part of $\mathbf{p}'$

For instance, when the point $(2, 0, 3)$ is rotated about the axis $(1, 1, 1)$ by $\theta = 6$ radians, we find the image point (to 3 SF) at

$$(1.503, 0.228, 3.270). \tag{6.11}$$

We can check our results with SciPy's own quaternionic rotation (we previously used this method in the code `ranknullity.ipynb` in §5.9). The following code snippet uses variables in the code `quaternion.ipynb` and so should only be run after.

---

[4] See *GuerrillaCG*'s video `https://www.youtube.com/watch?v=zc8b2Jo7mno` for a clear explanation.

3D Rotation with SciPy (use this after running `quaternion.ipynb`)

```
from scipy.spatial.transform import Rotation
t = 6
c, s = np.cos(t/2), np.sin(t/2)
Is = s*Axnorm
Q = np.append(Is, c)
rot = Rotation.from_quat(Q)
rot.as_matrix()@v
```

You should find that the output of the above snippet agrees with our answer (6.11).

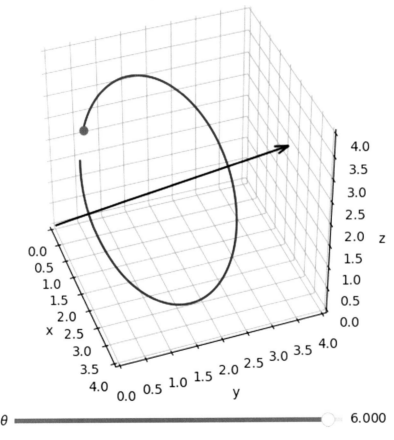

Rotated point at (1.503, 0.228, 3.270)

Fig. 6.9: Rotation in $\mathbb{R}^3$ implemented using the code `quaternion.ipynb`. The red dot is rotated about the black vector by angle θ (adjusted using the slider).

For further reading on quaternions, see [50] (and [7]) for the history of (and mathematical introduction to) quaternions, including proofs of the theorems discussed in this section. Ref. [29] gives an overview of the vector-versus-quaternion debate, whilst [206] gives a tour of the modern applications of quaternions to computer graphics.

DISCUSSION

- **Pauli matrices**. The unit quaternions $1, \mathbf{i}, \mathbf{j}, \mathbf{k}$ have an interesting representation in terms of the following 2×2 matrices

$$1 \mapsto I, \quad \mathbf{i} \mapsto -i\sigma_1, \quad \mathbf{j} \mapsto -i\sigma_2, \quad \mathbf{k} \mapsto -i\sigma_3,$$

$$\text{where} \quad \sigma_1 = \begin{pmatrix} 0 & 1 \\ 1 & 0 \end{pmatrix}, \quad \sigma_2 = \begin{pmatrix} 0 & -i \\ i & 0 \end{pmatrix}, \quad \sigma_3 = \begin{pmatrix} 1 & 0 \\ 0 & -1 \end{pmatrix}$$

The matrices $\sigma_1, \sigma_2, \sigma_3$ are called *Pauli matrices*, after the Austrian physicist *Wolfgang Pauli* (1900–1958) who used them study the spin of particles in quantum mechanics. We have not shown why quaternion multiplication is associative. This could be checked with SymPy, say (exercise 6b), but a nice consequence of the Pauli-matrix representation is that associativity follows automatically from that of matrix multiplication.

- **Rings and fields**. Apart from groups, you will also come across other important algebraic structures, such as rings and fields.

  A *ring* is an extension of the idea of a group to include addition and multiplication (but not division).

  More precisely, a ring is a a nonempty set R with two binary operations $+$ and $\times$, satisfying the following ring axioms. (We will write $a \times b$ as ab.)

  (1) R is an Abelian group under $+$.
  (2) $\times$ is associative.
  (3) $\exists 1 \in R$ such that, $\forall a \in R, a1 = 1a = a$.
  (4) $\times$ is distributive over $+$, *i.e.* $\forall a, b, c \in R$, $a(b+c) = ab+ac$ and $(a+b)c = ac+bc$.

  A *division ring* satisfies an additional axiom.

  (5) $\forall a \neq 0, \exists a^{-1} \in R$ such that $aa^{-1} = a^{-1}a = 1$.

  The set of all quaternions with the operations $+$ and $\times$ form a division ring and is given the symbol $\mathbb{H}$.

  $$\mathbb{H} = \{s + x\mathbf{i} + y\mathbf{j} + z\mathbf{k} : s, x, y, z \in \mathbb{R}\}.$$

  Finally, a *field* is a division ring in which $\times$ is commutative.

  At university, much of your time in the first (or second) year will be spent studying why the real numbers $\mathbb{R}$ is said to be a *complete ordered field*. See any textbooks on real analysis for details.

quaternion.ipynb (for plotting fig. 6.9)	
| | ```python
import numpy as np
import matplotlib.pyplot as plt
from matplotlib.widgets import Slider
%matplotlib
``` |
| **Input 1**: Coordinates of point to be rotated<br>**Input 2**: Axis of rotation<br><br>Normalised axis<br>Quaternion $\mathbf{p} = 0 + \mathbf{v}$ | ```python
v   = np.array((2,0,3))
Ax  = np.array((1,1,1))
L= np.sqrt(Ax@Ax)
Axnorm = Ax/L
P = np.append(0,v)
``` |
| Function for multiplying quaternions<br>scalar parts<br>vector parts<br><br><br>Formula 6.7 | ```python
def qmult(q1, q2):
    s1, s2 = q1[0], q2[0]
    v1 = q1[1:]
    v2 = q2[1:]
    S = s1*s2 - v1@v2
    V = s1*v2 + s2*v1 + np.cross(v1,v2)
    return np.append(S,V)
``` |
| Quaternion rotation (Thm. 6.6)<br><br><br>$\mathbf{q} = \cos(\theta/2) + \sin(\theta/2)\mathbf{I}$<br>$\mathbf{q}^{-1} = \bar{\mathbf{q}}$<br>Formula 6.9 – only the vector part is needed | ```python
def qrotate(t):
    c, s = np.cos(t/2), np.sin(t/2)
    sI = s*Axnorm
    Q  = np.append(c, sI)
    Qinv = np.append(c, -sI)
    return qmult(Q, qmult(P, Qinv))[1:]
``` |
| Resolution of the $\theta$ slider<br>All $\theta$ values to perform rotation<br>Empty lists to be filled with coords. of rotated pts<br><br>Perform the rotation for all $\theta \in [0, 2\pi]$<br>Collect results | ```python
tstep = 0.001
T = np.arange(0,2*np.pi, tstep)
X, Y, Z = ([] for i in range(3))

for i, t in enumerate(T):
    x,y,z = qrotate(t)
    X.append(x)
    Y.append(y)
    Z.append(z)
``` |
| <br><br><br><br>Set plotting window<br>Equal aspect ratio | ```python
fig = plt.figure()
ax = fig.add_subplot(111, projection='3d',
    xlabel = 'x', ylabel = 'y', zlabel = 'z',
    xlim = (0,4), ylim = (0,4), zlim= (0,4))
ax.set_box_aspect((1,1,1))
``` |
| Plot axis of rotation (black arrow)<br>(reduce arrow head size)<br>Plot rotation arc in blue<br>Plot rotated point in red<br>Display coords. of red point<br>to 3 SF. | ```python
vec  =ax.quiver(0,0,0, *Ax, color = 'k',
        length=2*L, arrow_length_ratio=0.05)
Cir, =ax.plot(X[0],Y[0],Z[2],'b',markersize=3)
Pnt, =ax.plot(*v, 'ro', markersize=6)
Coord=ax.text(0,1,5,'Initial point at '\
        f'({X[0]:.3f}, {Y[0]:.3f}, {Z[0]:.3f})')
``` |
| Position and size of $\theta$ slider | ```python
axt = plt.axes([0.26, 0.02, 0.5, 0.02])
t_slide = Slider(axt, r'$\theta$', 0, 2*np.pi,
                valstep=tstep, valinit=0)
``` |
| | ```python
# Code continues on the next page
``` |
```

quaternion.ipynb (continued)	
| | ```
def update(val):
    t = t_slide.val
    i = int(t/tstep)
``` |
| Get θ from slider | |
| Recalculate coords of red point | ```
    x,y,z = qrotate(t)
``` |
| Update blue arc | `    Cir.set_data_3d(X[0:i], Y[0:i], Z[0:i])` |
| Update red dot | `    Pnt.set_data_3d(X[i:i+1],Y[i:i+1],Z[i:i+1])` |
| Update coordinate display | ```
    Coord.set_text('Rotated point at '\
            f'({x:.3f}, {y:.3f}, {z:.3f})')
    fig.canvas.draw_idle()
``` |
| Update plot when slider is moved | ```
t_slide.on_changed(update)
plt.show()
``` |

6.7 Elementary number theory I: Multiplicative inverse modulo *n*

> Solve for *x* if
> $$13x = 1 \quad (\text{mod } n),$$
> where $n = 88, 130, 168, 263, 303$.

Let's deal with the case $n = 88$ by hand. First, we apply the *Euclidean algorithm* to find $\gcd(13, 88)$. The algorithm goes as follows

$$88 = 13(6) + 10 \tag{6.12}$$
$$13 = 10(1) + 3 \tag{6.13}$$
$$10 = 3(3) + 1 \tag{6.14}$$

The key component in the algorithm is the remainder (in bold) in each line of division. The algorithm stops because the next step would produce no remainder. The algorithm shows that

$$\gcd(13, 88) = 1.$$

Now we reverse the steps with a sequence of back-substitutions, starting with the last line.

$$1 = 10 - 3(3)$$
$$= 10(1) - [13 - 10(1)](3) \quad (\text{using } 6.13)$$
$$= 13(-3) + 10(4)$$
$$= 13(-3) + [88 - 13(6)](4) \quad (\text{using } 6.12)$$
$$= 13(-27) + 88(4).$$

From the last line, it follows that

$$13(-27) = 1 \quad (\text{mod } 88) \implies 13^{-1} = -27 = 61 \quad (\text{mod } 88).$$

You can see that if $\gcd(13, n) \neq 1$, then the back-substitution process cannot start with $1 = \dots$. In fact, we have:

Theorem 6.7 *Let $n \in \mathbb{N}$ and $a \in \mathbb{Z}$. Then a has a multiplicative inverse modulo n iff $\gcd(a, n) = 1$.*

The theorem implies that $13x = 1 \pmod{130}$ has no solution.

Thankfully, $\gcd(13, 168) = \gcd(13, 263) = \gcd(13, 303) = 1$, and the Euclidean algorithm yields:

$$13^{-1} = 13 \quad (\text{mod } 168), \qquad 13^{-1} = 81 \quad (\text{mod } 263), \qquad 13^{-1} = 70 \quad (\text{mod } 303).$$

Finding the multiplicative inverse of *a* mod *n* in Python only requires a single command: `pow(a, -1, n)`. Yet there is an interesting visualisation that we can do. Let's arrange the nonzero integers mod *n* $(1, 2, \dots, n - 1)$ in a circle and join up pairs of multiplicative inverses.

The code `inversewheel.ipynb` produces such a circle[5] (fig. 6.10). The key to this plot is to evenly distribute the integers mod n on the unit circle, but only displaying those that are coprime to n. The point labelled k has the Cartesian coordinate

$$(\cos\theta, \sin\theta), \quad \text{where } \theta = 2k\pi/n.$$

When n is large, as in the case for 263 and 303, we omit the labels as they become unhelpfully tiny.

The patterns in the circles are mesmerizing, and they exhibit symmetry.

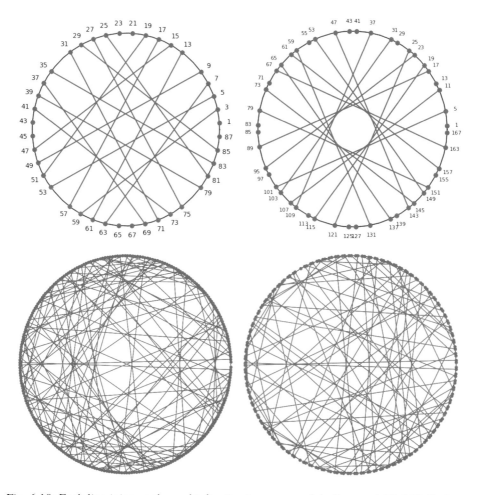

Fig. 6.10: Each line joins up the multiplicative inverses modulo (top row) 88, 168, (bottom row) 263, 303. Dots without connecting lines are inverses of themselves. The symmetry in the patterns is discussed in the text.

---

[5] The visualisation was inspired by Peter Karpov's fascinating blog `http://inversed.ru/Blog_1.htm`

Two observations in the patterns seen in fig. 6.10 can be made:

- All 4 circles have a horizontal line of symmetry
- The patterns for $n = 88$ and 168 have an additional symmetry across the vertical line.

The symmetry about the horizontal line occurs for all n. This follows from the simple observation that if $ab = 1 \pmod n$, then

$$(n - a)(n - b) = n^2 - na - nb + ab$$
$$= 1 \pmod n.$$

If, in particular, $n = 4k$, then there is another symmetry about the vertical line. This follows from the observation that

$$(2k - a)(2k - b) = 4k^2 - 2k(a + b) + ab$$
$$= 1 \pmod{4k},$$

where we have used the fact that a and b are necessarily odd (being coprimes to an even number). Therefore, $a + b$ is even, and so the term $2k(a + b)$ is divisible by $4k$.

These symmetries mean that the code can be slightly improved by omitting unnecessarily calculations. Exercise 7c asks you to implement such an improvement.

DISCUSSION

- The **The Euclidean algorithm** is a fundamental tool in number theory and is worth mastering. The algorithm is described in *Elements* ($\sim$300 BC), the legendary series of 13 books written by the Greek mathematician *Euclid* (c.325–265 BC). However, the origin of the algorithm probably predates Euclid.

- **Bézout's identity** is a useful result related to the Euclidean algorithm. It states that if $\gcd(a, b) = d$, then there exist integers x, y such that

$$ax + by = d.$$

We have seen that $\gcd(13, 88) = 1$ and found with back substitution that $13(-27) + 88(4) = 1$.

The French mathematician *Étienne Bézout* (1730–1783) showed that the result holds not just for integers, but also in certain types of rings. Bézout's identity is the key to proving the next result.

inversewheel.ipynb (for plotting fig. 6.10)	

	```import numpy as np```
	```import matplotlib.pyplot as plt```
	```%matplotlib```

Modulo what integer? Numbers around the wheel? (set False if **n** large)	```n = 88``` ```label = True```
Plot unit circle in blue	```t = np.linspace(0,2*np.pi)``` ```plt.plot(np.cos(t), np.sin(t), 'b')```
List of integers coprime to n List of their multiplicative inverses	```A= [a for a in range(1,n+1) if np.gcd(a,n)==1]``` ```B= [pow(a,-1,n) for a in A]```

Angular positions for a and a^{-1}
(x, y) coordinates of a on the wheel
(x, y) coordinates of a^{-1}
Join up a and a^{-1} with a red line
 	
Label a (slight offset)
Text size and alignment
	```for i, a in enumerate(A):```
	``` tA, tB = 2*np.pi*a/n, 2*np.pi*B[i]/n```
	``` xA, yA = np.cos(tA), np.sin(tA)```
	``` xB, yB = np.cos(tB), np.sin(tB)```
	``` plt.plot([xA, xB], [yA, yB], 'ro-')```
	``` if label:```
	``` plt.text(x= 1.1*xA, y= 1.1*yA, s = a,```
	``` fontsize=10, ha='center', va='center')```

Equal aspect ratio to see a circle	```plt.axis('square')```
	```plt.axis('off')```
	```plt.show()```

6.8 Elementary number theory II: Chinese Remainder Theorem

Find x if

$$x = 3 \quad (\text{mod } m),$$
$$x = 5 \quad (\text{mod } n),$$

where (m, n) equals: i) $(7, 8)$ ii) $(6, 8)$ iii) $(4, 8)$.

Let's first solve the case $(m, n) = (7, 8)$ by hand.

$$x = 3 \quad (\text{mod } 7) \implies x = 3 + 7n \ (n \in \mathbb{Z}). \tag{6.15}$$

We then substitute this into the other equation, giving

$$3 + 7n = 5 \quad (\text{mod } 8) \implies -n = 2 \quad (\text{mod } 8) \implies n = -2 + 8m \ (m \in \mathbb{Z}).$$

(We have used $7 = -1 \ (\text{mod } 8)$.) Putting this back into (6.15), we find the solutions

$$x = -11 + 56m \ (m \in \mathbb{Z}).$$

There are infinitely many integer solutions, but only one in the range $0 \le x < 56$, namely

$$x = 45.$$

Let's generalise the problem slightly to the following. Find x if

$$x = a \quad (\text{mod } 7),$$
$$x = b \quad (\text{mod } 8), \tag{6.16}$$

where $a = 0, 1, 2, \ldots, 6$ and $b = 0, 1, 2, \ldots, 7$. Let's use Python to help us find the solutions to all 56 systems.

To solve the system of congruences (6.16) with SymPy, we use the following syntax

```
from sympy.ntheory.modular import solve_congruence
solve_congruence((a, 7), (b, 8))
```

The code `crt.ipynb` produces the grid in fig. 6.11 (top table) in which the solution to the system (6.16) is located in row $(a + 1)$ and column $(b + 1)$. Note the solution $x = 45$ for the original problem in which $(a, b) = (3, 5)$ is at position (row, column)= $(4, 6)$. The coding technique is similar to that for producing Cayley tables in §6.3.

You will have noticed that all integers from 0 up to (but excluding) $56 = 7 \times 8$ appear somewhere on the table. This means that the system (6.16) can be solved for all choices of (a, b).

However, when a similar table is produced for $(m, n) = (6, 8)$ and $(4, 8)$ (lower tables in fig. 6.11), we see that for some values of (a, b), there are no solutions. For instance, the system

$$x = 3 \quad (\text{mod } 6),$$
$$x = 5 \quad (\text{mod } 8), \tag{6.17}$$

has solution $x = 21 \pmod{48}$, but the system

$$x = 3 \pmod 7,$$
$$x = 5 \pmod 8, \tag{6.18}$$

has no solution (the entry in the table shows a black square).

It is also interesting to note from fig. 6.11 that

$$\gcd(7, 8) = 1 \implies 56 = 7 \times 8 \text{ solvable cases,}$$
$$\gcd(6, 8) = 2 \implies 24 = 6 \times 8/2 \text{ solvable cases,}$$
$$\gcd(4, 8) = 4 \implies 8 = 4 \times 8/4 \text{ solvable cases.}$$

Given these observations, perhaps we might conjecture the following:

Conjecture 1 Let $\gcd(a, b) = 1$. Then the system of congruences.

$$x = a \pmod m,$$
$$x = b \pmod n, \tag{6.19}$$

has a unique solution modulo mn.

Conjecture 2 If $\gcd(m, n) = 1$, there is a one-to-one correspondence between

- the pairs of integers (a, b) where $a = 0, 1, 2, \ldots, m - 1$ and $b = 0, 1, 2, \ldots, n - 1$,

and

- the set of integers $\{0, 1, 2, \ldots, mn - 1\}$.

Conjecture 3 Consider the system (6.19). If $\gcd(m, n) = d$, then there are $\frac{mn}{d}$ choices of the pairs of integers (a, b) for which the system is solvable, where $0 \leq a < m$ and $0 \leq b < n$.

DISCUSSION

- **Chinese Remainder Theorem**. Conjecture 1 is a special case of the following theorem

Theorem 6.8 *(Chinese Remainder Theorem) Let $n_1, n_2, \ldots, n_i$ be pairwise coprime integers. The system of congruences*

$$x = a_1 \pmod{n_1},$$
$$x = a_2 \pmod{n_2},$$
$$\vdots$$
$$x = a_k \pmod{n_k},$$

has a unique solution modulo $n_1 n_2 \cdots n_k$.

The theorem indeed originates from China: a system of congruences appeared in the ancient text *Sun Zi Suan Jing* (Master Sun's Mathematical Manual - ca. 3rd century BC). The identity of *Sun Zi* ('Master Sun') still remains unclear, but perhaps explicitly labelling a theorem as "Chinese" is now antiquated exoticism (after all there aren't theorems that are labelled English or German). In any case, the name stuck.

x=(#row-1) (mod 7) and x = (#col-1) (mod 8)

0	49	42	35	28	21	14	7
8	1	50	43	36	29	22	15
16	9	2	51	44	37	30	23
24	17	10	3	52	45	38	31
32	25	18	11	4	53	46	39
40	33	26	19	12	5	54	47
48	41	34	27	20	13	6	55

x=(#row-1) (mod 6) and x = (#col-1) (mod 8)

0		18		12		6	
	1		19		13		7
8		2		20		14	
	9		3		21		15
16		10		4		22	
	17		11		5		23

x=(#row-1) (mod 4) and x = (#col-1) (mod 8)

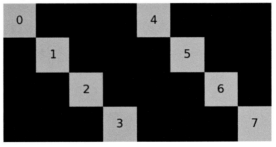

Fig. 6.11: Solutions of the simultaneous equations $x = a$ (mod m) and $x = b$ (mod n) where (a, b) are determined by the row and column numbers in each table. *Top*: $(m, n) = (7, 8)$. *Middle*: $(m, n) = (6, 8)$. *Bottom*: $(m, n) = (4, 8)$. A black square means there is no solution.

• **The Pigeonhole Principle**. Conjecture 2 is a consequence of the following.

Theorem 6.9 (*The Pigeonhole Principle*) *Let $n \in \mathbb{N}$. If $n + 1$ pigeons roost in n holes, then there must be a hole containing more than one pigeon.*

This simple, almost trivial, observation is of utmost importance in mathematics. An overview of its wide-ranging applications can be found, for instance, in [82].

For the system of congruences at hand, the pigeons are the mn integers $\{0, 1, 2, \ldots, mn - 1\}$ (these are the possible values of x). The pigeonholes are the $m \times n$ grid like the top table in fig. 6.11. If $\gcd(m, n) = 1$, the Chinese Remainder Theorem guarantees that no two pigeons share the same hole. The Pigeonhole Principle then implies Conjecture 2.

• **When is a system of congruences solvable?** Conjecture 3 is a consequence of the following result, which is a generalisation of the Chinese Remainder Theorem to the case when $n_1, n_2, \ldots, n_k$ are not necessarily coprime.

Theorem 6.10 (*Solvability criterion*) *The system of congruences*

$$x = a_1 \quad (\text{mod } n_1),$$
$$x = a_2 \quad (\text{mod } n_2),$$
$$\vdots$$
$$x = a_k \quad (\text{mod } n_k),$$

is solvable iff $a_i - a_j$ is divisible by $\gcd(n_i, n_j)$ $(i \neq j)$.

For example, the system

$$x = a \quad (\text{mod } 6),$$
$$x = b \quad (\text{mod } 8),$$

is solvable if $a - b$ is divisible by 2. In terms of the table in fig. 6.11, $a = \#\text{row}-1$ and $b = \#\text{col}-1$. Thus, the system is solvable iff #row and #col have the same parity. This explains the chessboard pattern in the middle table.

Try to explain the pattern in the bottom table in fig. 6.11 using the solvability criterion, which in fact predicts a diagonal pattern in the table for all (m, n). For proof of the criterion, see [105].

crt.ipynb (for plotting fig. 6.11)	
| | ```
import numpy as np
import matplotlib.pyplot as plt
from sympy.ntheory.modular import\
    solve_congruence
%matplotlib
``` |
| Use SymPy's number theory module | |
| We want to solve<br>$x = i$ (mod m0)<br>$x = j$ (mod m1)<br>for all $0 \leq i <$ m0, $0 \leq j <$ m1 | ```
m0 = 7
m1 = 8
``` |
| A big negative number yields a black square if unsolvable | ```
neg = -100
``` |
| For each entry in the table<br>Solve the system<br>If unsolvable<br><br>If solvable<br>return solution $0 \leq x <$ m0 $\cdot$ m1 | ```
def f(i,j):
    u = solve_congruence((i, m0), (j, m1))
    if u==None:
        return neg
    else:
        return int(u[0])
``` |
| Populate the array using f<br>Array size = m0×m1 | ```
array=np.fromfunction(np.vectorize(f),
                      (m0,m1), dtype=int)
``` |
| Display the array<br>neg will show up as black | ```
plt.imshow(array, cmap='magma',
           vmin = neg, vmax = m0*m1)
``` |
| Labelling<br>(if solvable)<br>Display the solution<br><br>Text alignment and size | ```
for ind, x in np.ndenumerate(array):
    if x!=neg:
        plt.text(s = str(x),
        x = ind[1], y = ind[0],
        va='center', ha='center', fontsize=12)
``` |
| | ```
plt.title(f'x=(#row-1) (mod {m0}) and '
          f'x = (#col-1) (mod {m1})',
          fontsize=12)
plt.axis('off')
plt.show()
``` |

6.9 Elementary number theory III: Quadratic residue and the reciprocity law

Let $p \in \mathbb{N}$ and $a \neq 0$ (mod p). a is said to be a *quadratic residue* of p if $a = b^2$ (mod p) for some b, otherwise it is said to be a quadratic non-residue.
The *Legendre symbol* $\left(\frac{a}{p}\right)$ indicates whether a is a quadratic residue or non-residue. It is defined as follows.

$$\left(\frac{a}{p}\right) = \begin{cases} 0 & \text{if } a = 0 \pmod{p}, \\ 1 & \text{if } a \text{ is a quadratic residue of } p, \\ -1 & \text{if } a \text{ is a quadratic non-residue of } p. \end{cases}$$

Find all prime numbers p such that   a) $\left(\frac{p}{3}\right) = 1$,   b) $\left(\frac{3}{p}\right) = 1$.

The first question asks what prime number p is a square mod 3. Since the only possible squares in mod 3 are

$$1^2 = 2^2 = 1 \pmod{3},$$

we conclude that all primes such that $p = 1$ (mod 3) are quadratic residues of 3, and those such that $p = 2$ (mod 3) are quadratic non-residues.

Fig. 6.12 (upper table) shows the values of the Legendre symbol $\left(\frac{p}{q}\right)$ for a range of odd primes p and q. This table can be generated using the code `legendre.ipynb`. The key calculation of the Legendre symbol is performed using SymPy. For example, to calculate $\left(\frac{83}{3}\right)$:

```
from sympy.ntheory import legendre_symbol
legendre_symbol(83, 3)
```

We can verify our findings on $\left(\frac{p}{3}\right)$ against the leftmost column of the table. Indeed, we see that

$$\left(\frac{p}{3}\right) = \begin{cases} 1 & \text{if } p = 1 \pmod{3}, \\ -1 & \text{if } p = 2 \pmod{3}. \end{cases} \tag{6.20}$$

The question becomes much more difficult if we switch the roles of 3 and p, since in mod p, there are potentially many possible squares. For example, when $p = 17$, all possible squares are

$1^2 = 1$ (mod 17) $2^2 = 4$ (mod 17) $3^2 = 9$ (mod 17) $4^2 = 16$ (mod 17)
$5^2 = 8$ (mod 17) $6^2 = 2$ (mod 17) $7^2 = 15$ (mod 17) $8^2 = 13$ (mod 17)
$9^2 = 13$ (mod 17) $10^2 = 15$ (mod 17) $11^2 = 2$ (mod 17) $12^2 = 8$ (mod 17)
$13^2 = 16$ (mod 17) $14^2 = 9$ (mod 17) $15^2 = 4$ (mod 17) $16^2 = 1$ (mod 17)

This gives us a list of 16 quadratic residues (of which 8 are unique), namely

$$1, 4, 9, 16, 8, 2, 15, 13, 13, 15, 2, 8, 16, 9, 4, 1. \tag{6.21}$$

Since 3 is not on this list, $\left(\frac{3}{17}\right) = -1$.

It is interesting to note (as Euler did in 1751) that only half the integers in the set $\{1, 2, 3, \ldots, 16\}$ are quadratic residues, and half are non-residues. The list of residues also appear to be arranged symmetrically about the midpoint of the list. (We will revisit this in the Discussion section.)

If only the calculation of $\left(\frac{3}{17}\right)$ could be as easy as $\left(\frac{17}{3}\right)$. Well, Gauss proved that it is!

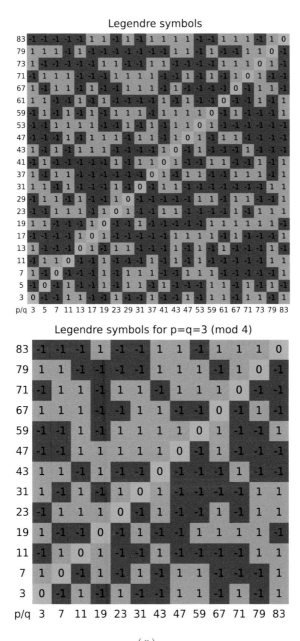

Fig. 6.12: Values of the Legendre symbol $\left(\frac{p}{q}\right)$ (p is tabulated against q), for all odd primes $p, q \leq 83$ (upper table). The lower table displays the Legendre symbols for primes of the form $4k + 3$. The lower table is antisymmetric about the green diagonal, in agreement with the quadratic reciprocity law.

Quadratic reciprocity law

The upper table in fig. 6.12 looks *almost* symmetric about the green diagonal (going bottom left to top right). If we isolate those entries such that $\left(\frac{p}{q}\right) \neq \left(\frac{q}{p}\right)$, we get the lower table. These are the entries that spoil the symmetry.

But we notice that these entries themselves form a totally antisymmetric table about the green diagonal (meaning that $\left(\frac{p}{q}\right) = -\left(\frac{q}{p}\right)$ in this table). Furthermore, these always occur when $p = q = 3 \pmod 4$.

What we have observed is a consequence of the following.

Quadratic reciprocity law *If p and q are distinct odd primes, then*

$$\left(\frac{q}{p}\right) = \begin{cases} -\left(\frac{p}{q}\right) & \text{if } p = q = 3 \pmod 4, \\ \left(\frac{p}{q}\right) & \text{otherwise.} \end{cases}$$

The quadratic reciprocity law is one of the most fundamental results in number theory. It may surprise you that it is also probably the 'most proved' result in mathematics. In 1801, Gauss published two proofs and later gave 6 further proofs. A tally of the number of proofs is kept on F. Lemmermeyer's website[6]. At the time of writing, there are well over 300 proofs listed, although many are variations of one another. The historical development and generalisations of the quadratic reciprocity law can be found in [126].

Understanding any proof of the reciprocity law will require some work. References [4, 57, 105] give detailed discussions of a particularly accessible proof based on counting lattice points in a rectangle. This proof is probably one that will be most accessible to undergraduates. The proof relies on *Gauss's Lemma* which is discussed in exercise 11.

Now let's apply the reciprocity law to our problem.

$$\left(\frac{3}{p}\right) = \begin{cases} \left(\frac{p}{3}\right) & \text{if } p = 1 \pmod 4 \\ -\left(\frac{p}{3}\right) & \text{if } p = 3 \pmod 4 \end{cases} \quad \text{(using the quadratic reciprocity law)}$$

$$= \begin{cases} 1 & \text{if } p = 1 \pmod 3 \text{ and } p = 1 \pmod 4 \\ -1 & \text{if } p = 2 \pmod 3 \text{ and } p = 1 \pmod 4 \\ -1 & \text{if } p = 1 \pmod 3 \text{ and } p = 3 \pmod 4 \\ 1 & \text{if } p = 2 \pmod 3 \text{ and } p = 3 \pmod 4 \end{cases} \quad \text{(using eq. 6.20)}$$

$$= \begin{cases} 1 & \text{if } p = 1 \pmod{12} \\ -1 & \text{if } p = 5 \pmod{12} \\ -1 & \text{if } p = 7 \pmod{12} \\ 1 & \text{if } p = 11 \pmod{12} \end{cases} \quad \text{(using the Chinese Remainder Theorem)}$$

$$= \begin{cases} 1 & \text{if } p = 1 \text{ or } p = 11 \pmod{12} \\ -1 & \text{if } p = 5 \text{ or } p = 7 \pmod{12}. \end{cases}$$

You should check this result against the last row of the upper table in fig. 6.12.

---

[6] https://www.mathi.uni-heidelberg.de/~flemmermeyer/qrg_proofs.html

DISCUSSION

- Let's prove the following observation which we made earlier.
  Lemma: Let p be an odd prime. Then half of the numbers in the list

$$1, 2, \ldots, p - 1$$

are quadratic residues of p.

Proof: Observe that for all $a \in \mathbb{Z}$, we have the identity

$$(p - a)^2 = a^2 \quad (\text{mod } p). \tag{6.22}$$

Letting $a = 1, 2, \ldots, (p-1)/2$ gives us $(p-1)/2$ values of a^2. It remains to show that these values of a^2 are all distinct modulo p.
Take a_1 and a_2 from the set $S = \{1, 2, \ldots, (p-1)/2\}$. Suppose that $a_1^2 = a_2^2 \ (\text{mod } p)$ but $a_1 \neq a_2 \ (\text{mod } p)$. Then $(a_1 - a_2)(a_1 + a_2) = 0 \ (\text{mod } p)$, from which we conclude that $a_1 + a_2 = 0 \ (\text{mod } p)$. In other words, $a_1 + a_2$ is a multiple of p. But this is impossible because $0 < a_i < p/2$, and therefore $0 < a_1 + a_2 < p$.
Therefore, when all the elements of S are squared, the results are $(p-1)/2$ different integers mod p. Hence, we have $(p-1)/2$ quadratic residues.
This explains why we obtained 8 quadratic residues of 17 (below eq. 6.20. In fact, the identity 6.22 also explains why the quadratic residues are arranged symmetrically about the midpoint of the list (6.21).

- **Euler's criterion**. The lemma above only tells us that there are $(p-1)/2$ quadratic residues of p, but it does not tell us *which* integers are quadratic residues. The following result gives us more clues.

Theorem 6.11 *(Euler's criterion) Let p be an odd prime and $a \neq 0 \ (\text{mod } p)$. Then,*

$$\left(\frac{a}{p}\right) = a^{\frac{p-1}{2}} \quad (\text{mod } p).$$

So a is a quadratic residue if $a^{\frac{p-1}{2}} = 1 \ (\text{mod } p)$. Try verifying this result using any entry in the tables in fig. 6.12.

legendre.ipynb (for plotting fig. 6.12)	
| | ```
import numpy as np
import matplotlib.pyplot as plt
``` |
| For generating primes | ```
from sympy import sieve
``` |
| For calculating Legendre symbol | ```
from sympy.ntheory import legendre_symbol
%matplotlib
``` |
The user to specify this number	`Pmax = 85`
L = list of all primes from 3 up to Pmax	`L = list(sieve.primerange(3,Pmax))`
Calculate the Legendre symbol in each entry	`f = lambda i,j: legendre_symbol(L[i], L[j])`
| Generate an array of symbols $(p/q)$ | ```
array = np.fromfunction(np.vectorize(f),
                (len(L),len(L)), dtype=int)
``` |
| These 3 lines attach an extra row and column of L onto the array, acting as the labels on the $x$ and $y$ axes. The number 2 acts as the origin and will be replaced later. | ```
array = np.vstack((L, array))
u = [2]+ L
array = np.column_stack((u,array))
``` |
| Display the array as a colourful table. The terrain colormap is the key to success. | ```
plt.imshow(array, cmap='terrain',
        vmin = -1, vmax = 2, origin='lower')
``` |
| Labelling each entry in the table The origin | ```
for ind, x in np.ndenumerate(array):
    if x==2:
        plt.text(s='p/q', x = 0, y = 0,
        va='center', ha='center',fontsize=7)
``` |
| Elsewhere in the table | ```
    else:
        plt.text(s = str(x),
        x = ind[1], y = ind[0],
        va='center', ha='center',fontsize=7)
``` |
| | ```
plt.title('  Legendre symbols', fontsize=10)
plt.axis('off')
plt.show()
``` |
```

6.10 Analytic Number Theory I: The Prime Number Theorem

Define the *prime-counting function*, $\pi(x)$, as the number of prime numbers less than or equal to x.
For large x, $\pi(x)$ has been shown to take the following asymptotic forms.

$$\text{a)} \quad \pi(x) \sim \frac{x}{\ln x},$$

$$\text{b)} \quad \pi(x) \sim \mathrm{Li}(x) := \int_2^x \frac{dt}{\ln t},$$

(where $f(x) \sim g(x)$ means f and g are asymptotically equal, *i.e.* $\lim_{x \to \infty} f(x)/g(x) = 1$).
Investigate the accuracy of these expressions.

These results are collectively known as the *Prime Number Theorem* (PNT). The PNT was proved independently in 1896 by *Jacques Hadamard* (1865–1963) and *Charles Jean de la Vallée Poussin* (1866–1962), although the result was already known to Gauss around a century earlier.

The prime-counting function $\pi(x)$ is one of many important number-theoretic functions in analytic number theory. Here is a plot of $\pi(x)$ for x up to 100. We can see, for instance, that there are 25 prime numbers less than 100.

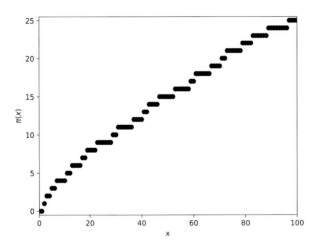

Fig. 6.13: The prime-counting function $\pi(x)$ for $x \in [0, 100]$

In SymPy, we can calculate, say, $\pi(100)$, using the syntax below.

```
from sympy import primepi
primepi(100)
```

The PNT predicts the behaviour of $\pi(x)$ when x is large, and a linear plot like fig. 6.13 is not going to be very useful. In fig. 6.14 (top panel) we plot $\pi(x)$ on logarithmic scales up to $x = 10^8$, along with the PNT estimates (a) and (b). They all seem to coincide pretty well on this domain. Increasing x further, you will start to notice the exponential increase in the time taken for SymPy to compute $\pi(x)$. The latest computational progress for $\pi(x)$ can be found on the OEIS website[7].

Let's study the plot of the ratios:

$$\frac{\pi(x)}{x/\ln x} \quad \text{and} \quad \frac{\pi(x)}{\text{Li}(x)}.$$

(middle panel of fig. 6.14). Both ratios appear to approach 1 as $x \to \infty$ (which is what the PNT predicts), but $\pi(x)/\text{Li}(x)$ seems to converge to 1 more rapidly.

From this middle panel, we can also deduce that

$$\pi(x) > \frac{x}{\ln x}, \quad x \in (10, 10^8). \tag{6.23}$$

In fact this inequality holds for all $x > 10$. See [63] for the proof of this and other related estimates (including the PNT). We can also see that

$$\pi(x) < \text{Li}(x), \quad x \in (10, 10^8).$$

However this does not always hold. Intensive computations have shown that $\pi(x)$ can exceed $\text{Li}(x)$ when x is *very* large (of order 10^{316} [22, 39]). Interestingly, it has been shown[8] that $\text{Li}(x) - \pi(x)$ switches sign infinitely many times!

Finally let's take a look at the differences for the two estimates:

$$\pi(x) - \frac{x}{\ln x} \quad \text{and} \quad \text{Li}(x) - \pi(x).$$

(bottom panel of fig. 6.14). A common misconception is to say that if $f(x)/g(x) \to 1$ then $f(x) - g(x) \to 0$. You can see that this isn't the case for our functions. (Can you think of a simpler counterexample?) You can see that the differences grow, but as described above, for infinitely many values of very large x, the graph $\text{Li}(x) - \pi(x)$ will dip below the x-axis (and then rises above it again).

In summary, the PNT gives a fascinating and perhaps unexpected connection between the distribution of prime numbers and the logarithm. A particularly useful takeaway from the PNT is that the size of the nth prime number, when n is large, is roughly

$$p_n \approx n \ln n.$$

(Yet another form of the PNT is discussed in exercise 12.)

---

[7] https://oeis.org/A006880

[8] In J. E. Littlewood's 1914 paper *Sur la distribution des nombres premiers* on page 1868 of this online facsimile of the original journal: https://archive.org/details/comptesrendusheb158acad

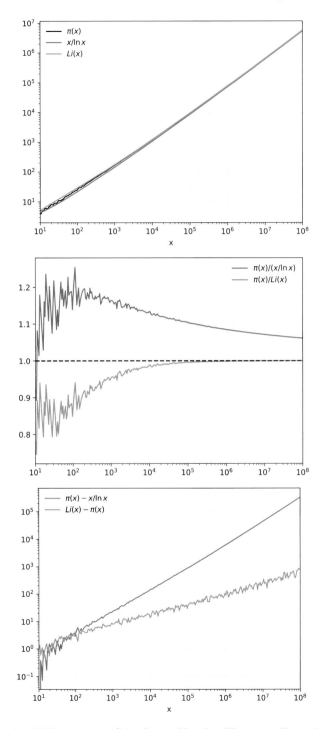

Fig. 6.14: Graphical illustrations of the Prime Number Theorem. *Top*: $\pi(x)$, $x/\ln x$ and $\mathrm{Li}(x) = \int_2^x \mathrm{d}t/t$ plotted logarithmically against x. *Middle*: The ratios $\pi(x)/(x/\ln x)$ and $\pi(x)/\mathrm{Li}(x)$ both approach 1 for large x. *Bottom*: The differences $\pi(x) - x/\ln x$ and $\mathrm{Li}(x) - \pi(x)$ grow at different rates.

- **The infinitude of primes**. The inequality $\pi(x) > x/\ln x$ (which holds for $x > 10$) implies that there are infinitely many primes (since $\lim_{x \to \infty} \frac{x}{\ln x} = \infty$). The fact that there are infinitely many prime numbers was already proven by Euclid in $c.300$ BC. A varied collection of interesting proofs of the infinitude of primes can be found in the first chapter of [4] – a highly recommended read.

- **Computers and prime numbers**. Computers have become an integral part of the understanding of prime numbers. For instance, the largest primes continue to be discovered by computers[9], and questions like "when does $\pi(x)$ exceed $\mathrm{Li}(x)$?" certainly require intensive computing and the use of clever algorithms. See [49] for an excellent introduction to this field of study known as *computational number theory*.

  Big questions in number theory continue to be explored with computers, including perhaps the biggest of them all, the *Riemann Hypothesis*, which we will touch on shortly.

---

pnt.ipynb (for plotting the middle panel of fig. 6.14)

	```import numpy as np```
	```import matplotlib.pyplot as plt```
For integration	```from scipy.integrate import quad```
For calculating $\pi(x)$	```from sympy import primepi```
	```%matplotlib```
Specify domain to count primes	```Nmin, Nmax = 1e1, 1e8```
300 equally spaced points in log space	```Nlist = np.logspace(np.log10(Nmin),``` ``` np.log10(Nmax), 300)```
List of prime counts $\pi(x)$	```P = [primepi(n) for n in Nlist]```

$\mathrm{Li}(x) = \int_2^x dt/\ln t$	```def li(x):```
	 ``` X, err = quad(lambda t: 1/np.log(t), 2, x)```
	 ``` return X```

Make `li(x)` accept array input	```Li = np.vectorize(li)```
PNT (a) $\pi(x) \sim x/\ln x$ PNT (b) $\pi(x) \sim \mathrm{Li}(x)$	```PNT1 = Nlist/(np.log(Nlist))``` ```PNT2 = Li(Nlist)```
Ratios of $\pi(x)$ and its asymptotic forms	```err1 = P/PNT1``` ```err2 = P/PNT2```

Plot the ratios (log x scale)...
 	
 and the asymptote $y = 1$ (dashed)
	```plt.semilogx(Nlist, err1,```
	 ``` Nlist, err2,```
	 ``` Nlist, np.ones_like(Nlist),'k--')```
	 ```plt.xlim(Nmin,Nmax)```
	 ```plt.grid('on')```
	 ```plt.xlabel('x')```
	 ```plt.legend([r'$\pi(x)/(x/\ln x)$',```
	 ``` r'$\pi(x)/Li(x)$'])```
	 ```plt.show()```

---

[9] You too can join the search at `https://www.mersenne.org`

6.11 Analytic Number Theory II: The Riemann zeta function

Plot the Riemann zeta function $\zeta(s)$ for $s \in \mathbb{R}$.

Bernhard Riemann (1826–1866) was a German mathematician who did groundbreaking work in analysis, differential geometry and number theory. A student of Gauss, Riemann also laid the mathematical foundation for Einstein's theory of general relativity. He died aged only 39.

One of Riemann's most celebrated mathematical results is the extension of the *zeta function*

$$\zeta(s) = \sum_{n=1}^{\infty} \frac{1}{n^s}, \quad \mathrm{Re}(s) > 1, \tag{6.24}$$

(previously studied by Euler) to the complex numbers. In this section, we will focus on the case when s is real. We will consider complex s in the next section.

Region $s > 1$

The series (6.24) converges if $s > 1$ (it is a *p*-series discussed in §1.4). The graph of $\zeta(s)$ on this domain is plotted in fig. 6.15 (top panel). Panel A zooms in on the domain $[2, 6]$, and panel B shows a logarithmic plot near $s = 1$.

To plot these graphs, we use SciPy to calculate $\zeta(s)$. SymPy can also give exact expressions for $\zeta(s)$ (if known).

Evaluating the zeta function with SciPy and SymPy	
Using SciPy Output: [1.0173430619844492, 1.2020569031595942]	`from scipy.special import zeta` `[zeta(6), zeta(3)]`
Using SymPy Output: $\left[\dfrac{\pi^6}{945}, \zeta(3)\right]$	`from sympy import zeta` `[zeta(6), zeta(3)]`

Let's write down some observations on the graph in panel (A). We will discuss and justify some of these observations as we go along.

A1 For $s > 1$, ζ is a decreasing function. You might even conjecture from graph (A) that

$$\lim_{s \to \infty} \zeta(s) = 1.$$

This makes sense since, for large s, the series (6.24) looks like 1+(small contributions).

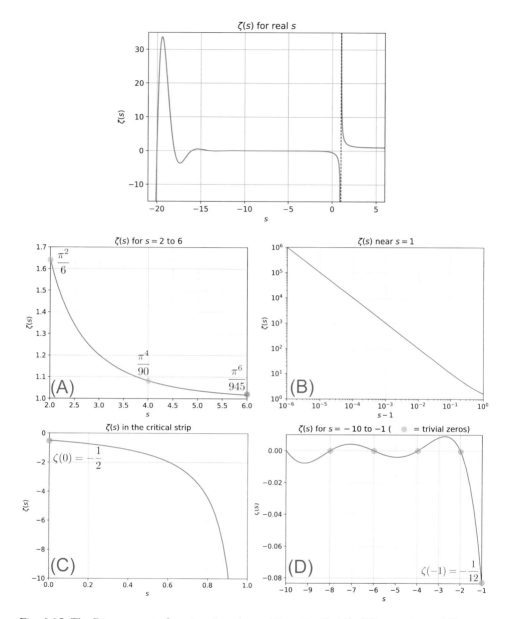

Fig. 6.15: The Riemann zeta function ζ on the real line. Panels (A)–(D) zoom in on different parts of the top panel. Key observations from each of these graphs are discussed in the text.

A2 We proved (using Fourier series and Parseval's Theorem in chapter 2) that $\zeta(2) = \pi^2/6$ (§2.9) and $\zeta(4) = \pi^2/90$ (exercise 14). A similar technique shows that $\zeta(6) = \pi^6/945$. We can continue using this technique to obtain ζ at any positive even integer. In fact, Euler showed that

$$\zeta(2n) = (-1)^{n+1} \frac{(2\pi)^{2n}}{2(2n)!} B_{2n}, \tag{6.25}$$

where B_{2n} are the *Bernoulli numbers* that can be calculated from the series expansion

$$\frac{x}{e^x - 1} = \sum_{k=0}^{\infty} \frac{B_k}{k!} x^k.$$

We can let SymPy do the expansion and manually pick out the Bernoulli numbers.

Power series expansion with SymPy

```
import sympy as sp
x = sp.symbols('x')
f = x/(sp.exp(x)-1)
sp.series(f, x, n=10)
```

Output: $1 - \dfrac{x}{2} + \dfrac{x^2}{12} - \dfrac{x^4}{720} + \dfrac{x^6}{30240} - \dfrac{x^8}{1209600} + O(x^{10})$

Alternatively, SymPy can give us the Bernoulli numbers directly.

```
[sp.bernoulli(n) for n in range(10)]
```

Output: [1, -1/2, 1/6, 0, -1/30, 0, 1/42, 0, -1/30, 0]

Note that $B_{\text{odd} \geq 3} = 0$.

A3 $\zeta(3) \approx 1.202057$

Unlike $\zeta(2n)$ (which are all rational multiples of π^{2n}), it remains unknown whether ζ at a given odd positive integer is rational or irrational, with the exception of $\zeta(3)$. The French mathematician *Roger Apéry* (1916–1994) proved that $\zeta(3)$ is irrational.

Region near $s = 1$

B1 In the graph of ζ (top panel of fig. 6.15), we see a vertical asymptote at $s = 1$, which looks suspiciously like that of $f(s) = \frac{1}{s-1}$ at $s = 1$.

If the asymptote of ζ at $s = 1$ really behaves like that of $\frac{1}{s-1}$, then, multiplying ζ by $(s - 1)$ would remove the singularity, *i.e.* the graph of $y = (s - 1)\zeta(s)$ would have no asymptotes. Indeed, the graph in fig. 6.16 shows that this is the case.

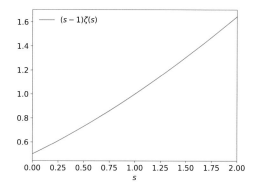

Fig. 6.16: The graph of the function $y = (s - 1)\zeta(s)$, showing that the singularity at $s = 1$ has been removed.

Of course, Python did not calculate $y = (s - 1)\zeta(s)$ at $s = 1$ exactly (this would produce a NaN) but evaluating $y(s)$ in the neighbourhood of 1 suggests that $\lim_{s=1} y(s) = 1$. Morever, the graph of $y = (s - 1)\zeta(s)$ looks smooth (differentiable) everywhere on $\mathbb{R}$. This suggests that y could be expressed a power series around $s = 1$.

$$y(s) := (s - 1)\zeta(s) = a_0 + a_1(s - 1) + a_2(s - 1)^2 + a_3(s - 1)^3 + \cdots$$

Since $\lim_{s \to 1} y(s) = 1$, we have $a_0 = 1$, and we can write the zeta function near $s = 1$ as a power series

$$\zeta(s) = \frac{1}{s - 1} + a_1 + a_2(s - 1) + a_3(s - 1)^2 + \cdots$$

In the language of complex analysis, we say that the zeta function has a *simple pole* at $s = 1$. Near $s = 1$ (where s may even be a complex number), one can prove (in a complex analysis course) that the zeta function can be expressed as

$$\zeta(s) = \frac{1}{s - 1} + \sum_{n=0}^{\infty} \frac{\gamma_n}{n!}(1 - s)^n, \qquad (6.26)$$

where the coefficients γ_n are called the *Stieltjes constants* (with $\gamma_0 = \gamma$, the Euler-Mascheroni constant). Series such as (6.26) are called *Laurent series* (a generalisation of the Taylor series to complex numbers, with terms of negative degree).

B2 The graph in fig. 6.15 (B) shows the behaviour of ζ just to the right of the asymptote (plotted on log scales). A straight line is seen over the region where $s - 1$ is small. The line passes through all the grid points so neatly that we can deduce the relation

$$\zeta(10^{-k} + 1) \approx 10^k, \quad k \in \mathbb{N}$$

We can now understand that this is simply a consequence of the Laurent expansion (6.26) with $s = 10^{-k} + 1$. The first term on the RHS dominates as we go nearer to $s = 1$. Think about what you would see on the other side of the asymptote, *i.e.* if $\zeta(s)$ is plotted against $1 - s$ on log scales (exercise 13b).

Region $0 \le s < 1$

The region $0 \le s < 1$ is an important one for the zeta function: it is a subset of the *critical strip* $\{s + it, s \in [0, 1), t \in \mathbb{R}\}$ in the complex plane. The critical strip plays a key role in the Riemann Hypothesis discussed in the next section. For now, let's explore ζ on the real segment $s \in [0, 1)$.

Here are some observations on the graph 6.15 (C):

C1 The function ζ is decreasing on the interval $[0, 1)$.

But wait - if the power series (6.24) is only valid for $s > 1$, how does one calculate $\zeta(s)$ on the domain $[0, 1)$? The idea is to derive a different expression altogether that is valid on this domain.

First, consider the identity $1 + (-1)^n = 2$ if n is even (and 0 otherwise). Dividing every term by n^s and summing over n, we obtain

$$\sum_{n=1}^{\infty} \frac{1}{n^s} + \sum_{n=1}^{\infty} \frac{(-1)^n}{n^s} = \sum_{n=1}^{\infty} \frac{2}{(2n)^s}$$

$$\Longrightarrow \zeta(s) + \sum_{n=1}^{\infty} \frac{(-1)^n}{n^s} = 2^{1-s}\zeta(s)$$

$$\Longrightarrow \zeta(s) = \frac{1}{1 - 2^{1-s}} \sum_{n=1}^{\infty} \frac{(-1)^{n+1}}{n^s}, \quad s > 0. \tag{6.27}$$

The alternating series on the RHS (called the *Dirichlet eta function*) converges for $s > 0$ (this can be verified using the *alternating-series test* of convergence).

In complex analysis, this extension of the domain of a function is called *analytic continuation*.

A technical note: Although in theory the series (6.27) holds for $s > 0$, the convergence is impractically slow, particularly near $s = 0$. In practice, other forms of the analytic continuation are normally used (exercise 15).

C2 $\zeta(0) = -\frac{1}{2}$. Be careful!

$$\textbf{Do not write:} \quad 1 + 1 + 1 + \cdots = -\frac{1}{2},$$

because the power series expression (6.24) is not valid here (and neither is (6.27)). We will see why $\zeta(0) = -\frac{1}{2}$ next.

Region $s < 0$

Analytic continuation into the region $s < 0$ shows that

$$\zeta(s) = 2^s \pi^{s-1} \sin\frac{\pi s}{2}\Gamma(1 - s)\zeta(1 - s), \quad s < 0, \tag{6.28}$$

This was proved by Riemann but conjectured by Euler a century earlier (see [15] for Euler's line of argument). Equation 6.28 is called the *functional equation* for the zeta function.

In (6.28), Γ is the *gamma function* defined by

$$\Gamma(s) = \int_0^\infty t^{s-1}e^{-t}\,dt, \quad s > 0,$$ (6.29)

and analytically continued to $s < 0$ by its own functional equation

$$\Gamma(1-s)\Gamma(s) = \frac{\pi}{\sin \pi s}, \quad s \neq \mathbb{Z}.$$ (6.30)

The gamma function satisfies $\Gamma(n) = (n-1)!$ for integer $n \in \mathbb{N}$. As such, the gamma function is said to extend the factorial from integers to real numbers.

Here is how to evaluate the gamma function in SciPy and SymPy. More about the gamma function in exercise 14.

Evaluating the gamma function with SciPy and SymPy	
Using SciPy Output: [6.0, 0.8862269254527579]	`from scipy.special import gamma` `[gamma(4), gamma(3/2)]`
Using SymPy Output: $\left[6, \dfrac{\sqrt{\pi}}{2}\right]$	`from sympy import gamma, S` `[gamma(4), gamma(S(3)/2)]`

Although the functional equation for ζ (6.28) is not valid at $s = 0$, we can investigate the limit as $s \to 0$. The RHS has the term $\zeta(1-s)$ which, near $s = 0$ can be replaced by the Laurent series (6.26):

$$\zeta(1-s) = -\frac{1}{s} + O(1).$$

Using the functional equation, we have

$$\lim_{s\to 0}\zeta(s) = \frac{1}{\pi}\lim_{s\to 0}\sin\left(\frac{\pi s}{2}\right)\Gamma(1-s)\left(-\frac{1}{s} + O(1)\right)$$

$$= -\frac{1}{2}\lim_{s\to 0}\frac{\sin(\pi s/2)}{\pi s/2}$$

$$= -\frac{1}{2}$$

where we have used $\Gamma(1) = 1$ and the identity $\lim_{x\to 0}\sin x/x = 1$. This proves observation **C2**.

Finally, let's make some observation on the graph 6.15 (D) which shows the zeta function on the domain $[-10, -1]$

D1 $\zeta(-1) = -\frac{1}{12}$.

This can be obtained by substituting $s = -1$ into the functional equation, giving

$$\zeta(-1) = -\frac{1}{2\pi^2}\Gamma(2)\zeta(2) = -\frac{1}{12}.$$

You may have seen the provocative equation below somewhere online. But be careful!

$$\textbf{Do not write:} \quad 1 + 2 + 3 + \cdots = -\frac{1}{12}.$$

This kind of 'equation' is nonsensical since the zeta function does not take the power-series form (6.24) on this domain. Any 'derivations' you may have seen are usually based on cavalier handling of divergent series.

D2 $\zeta(-2n) = 0$ for all $n \in \mathbb{N}$.

This is immediate upon substituting $s = -2n$ into the functional equation. These are called the *trivial zeros* of the zeta function. There are other (non-trivial) zeros in the critical strip, which we will explore in the next section.

D3 As $s < 0$ increases in magnitude, ζ oscillates with increasing amplitude. It diverges as $s \to -\infty$.

This too can be explained using the functional equation. Using the result $\lim_{s \to \infty} \zeta(s) = 1$, it follows that ζ takes the following asymptotic form as $s \to -\infty$ (ignoring constants)

$$\zeta(s) \sim \frac{\Gamma(1-s)}{(2\pi)^{-s}} \sin \frac{\pi s}{2}.$$

$\Gamma(1-s)$ grows faster than $(2\pi)^{-s}$ (the former grows like the factorial) and thus the coefficient of the sine term grows without bound as $s \to -\infty$. This results in an amplified oscillation as $s \to -\infty$.

DISCUSSION

- **Complex analysis**. Although we did not deal with complex numbers in this section, we did touch on a few topics that are specific to complex analysis, namely, poles, removable singularities, Laurent series, and analytic continuation. A course in complex analysis (usually in the 2nd or 3rd year) will reveal the fascinating world of complex functions. For those who can't wait, there are plenty of good textbooks on introductory complex analysis. Here are a few recommended reads: [3, 16, 172, 191].

- **Prime numbers and the zeta function** are connected via *Euler's product*

$$\zeta(s) = \prod_{p \text{ prime}} (1 - p^{-s})^{-1}, \quad s > 1. \tag{6.31}$$

Here's a heuristic argument for why this holds. Each term in the product on the RHS is the infinite sum of a geometric series

$$\frac{1}{1 - p^{-s}} = 1 + \frac{1}{p^s} + \frac{1}{p^{2s}} + \frac{1}{p^{3s}} + \cdots$$

Taking the product of the RHS over all primes $p = 2, 3, 5, \ldots$, we have

$$\left(1 + \frac{1}{2^s} + \frac{1}{2^{2s}} + \cdots\right)\left(1 + \frac{1}{3^s} + \frac{1}{3^{2s}} + \cdots\right)\left(1 + \frac{1}{5^s} + \frac{1}{5^{2s}} + \cdots\right).$$

If we were to expand out all the brackets in these multiplications, we would get

$$1 + \frac{1}{2^s} + \frac{1}{3^s} + \frac{1}{2^{2s}} + \frac{1}{5^s} + \frac{1}{2^s 3^s} + \cdots = \sum_{n=1}^{\infty} \frac{1}{n^s},$$

where we have used the Fundamental Theorem of Arithmetic.

As a bonus, from the fact that $\zeta(s) \to \infty$ as $s \to 1$, we deduce that there are infinitely many primes.

A very readable and accessible overview of Euler's work on the zeta function can be found in [15].

In the next section, we continue to explore the zeta function and uncover more connections to the prime numbers.

6.12 Analytic Number Theory III: The Riemann Hypothesis

Locate the first 10 non-trivial zeroes of the Riemann zeta function.

This section requires basic knowledge of complex numbers.

Let's consider the zeta function as a map ζ from a two-dimensional domain (a subset of the complex plane $\mathbb{C}$) to the two-dimensional codomain $\mathbb{C}$.

The zeta function $\zeta(s)$ for complex s takes on different analytic continuations on different domains (just like the version for real s discussed in the previous section).

- The power-series form

$$\zeta(s) = \sum_{n=1}^{\infty} \frac{1}{n^s}, \quad \text{Re}(s) > 1. \tag{6.32}$$

- The alternating-series (Dirichlet-eta) form

$$\zeta(s) = \frac{1}{1 - 2^{1-s}} \sum_{n=1}^{\infty} \frac{(-1)^{n+1}}{n^s}, \quad \text{Re}(s) > 0. \tag{6.33}$$

- The analytic continuation into $\text{Re}(s) < 0$ via the functional equation

$$\zeta(s) = 2^s \pi^{s-1} \sin \frac{\pi s}{2} \Gamma(1 - s)\zeta(1 - s), \quad \text{Re}(s) < 0, \tag{6.34}$$

In terms of Python, SciPy's `zeta` (at the time of writing) cannot handle a complex argument. However, SymPy's `zeta` can. Here's how to calculate $\zeta(\pm i)$.

```
Evaluating the zeta function with SymPy
```
```
from sympy import zeta
print(f'zeta(i) ={complex(zeta(1j))}')
print(f'zeta(-i)={complex(zeta(-1j))}')
```

This gives the following output.

```
zeta(i) =(0.0033002236853241027-0.4181554491413217j)
zeta(-i)=(0.0033002236853241027+0.4181554491413217j)
```

You might then suspect that

$$\zeta(\bar{s}) = \overline{\zeta(s)} \tag{6.35}$$

i.e. the complex conjugate can 'come out' of the argument. Indeed, this is true, although the proof requires some advanced complex analysis[10]. The upshot is that we only need to study how ζ maps the upper half of the complex plane $\text{Im}(s) \geq 0$. The mapping of the domain $\text{Im}(s) < 0$ is then completely determined thanks to eq. 6.35.

An interesting question is, where are the zeros of the map ζ? In other words, we want to find all $s \in \mathbb{C}$ such that $\zeta(s) = 0$. We have already found the trivial zeros in the previous section at $s = -2n$, $n \in \mathbb{N}$. Are there any more zeros?

One way to roughly locate where the zeros are is to plot the amplitude $|\zeta(s)|$ for s on the complex plane. One could do this as a heatmap as shown in fig. 6.17 (exercise 16 discusses

---

[10] see *Schwarz Reflection Principle* in [16] for instance

how to produce this figure). We see a dark region near the trivial zero at $z = -2$. In addition, we see some suspicious dark spots where the amplitude is close to zero. In this plotting window ($0 \leq \mathrm{Im}(s) < 50$), we see 10 potential nontrivial zeros. Zooming into the figure reveals that these dark spots all occur where $\mathrm{Re}(s)$ is roughly $1/2$.

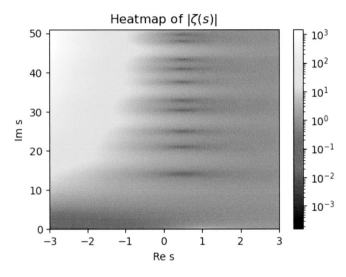

Fig. 6.17: The heatmap of $|\zeta(s)|$. The 10 dark horizontal stripes correspond to 10 nontrivial zeros of $\zeta(s)$ in the region $0 < \mathrm{Im}(s) < 50$. There is also a dark area around the trivial zero at $z = -2$.

All nontrivial zeros are in the critical strips $0 \leq \mathrm{Re}(s) \leq 1$ (this follows from Euler's product formula 6.31 and the functional equation 6.34). Riemann went further and made the following conjecture.

> **Riemann Hypothesis**: *All nontrivial zeroes of $\zeta(s)$ lie along the line $Re(s) = 1/2$.*

The Riemann Hypothesis is one of the most intriguing unsolved problem in mathematics, featuring both on David Hilbert's famous list of 23 problems[11] and the Clay Institute's list of Millennium Prize Problems[12]. We discuss some of its far-reaching implications later.

Now that we know roughly where the nontrivial zeros are, let's locate them at a higher resolution. Suppose we accept the Riemann Hypothesis, we can ask, what is the image of the line $\mathrm{Re}(s) = 1/2$ under the zeta function map? As before, we only need to consider part of the line where $\mathrm{Im}(s) \geq 0$, since we see from eq. 6.35 that if s is a zero of ζ, then so is the conjugate $\bar{s}$.

Instead of simply plotting the image of the line $\mathrm{Re}(s) = 1/2$ (which, as you will see, turns out to be a complicated swirl of circles), let's create an animation showing a real-time mapping as we travel up the line $\frac{1}{2} + it$ where t increases from 0 to 50.

The code `zetaanim.ipynb` produces this animation, and three frames from the animation is shown in fig. 6.18. Each pair of figures show the domain (left) and the image (right) under

---

[11] https://tinyurl.com/yckch9e6 (simonsfoundation.org)
[12] https://www.claymath.org

ζ. The animation shows that for $\text{Im}(s) \in [0, 50]$, $\zeta(s)$ passes through zero 10 times (count as you watch!), corresponding to the 10 dark spots in the heatmap (fig. 6.17).

The animation gives us 10 nontrivial zeros to a handful of decimal places (depending on the resolution in your animation). You can check that they are all consistent with known results shown in table 6.5[13].

Zero number	Imaginary part (4 dec. pl.)
1	14.1347
2	21.0220
3	25.0109
4	30.4249
5	32.9351
6	37.5862
7	40.9187
8	43.3271
9	48.0052
10	49.7738

Table 6.5: The first 10 nontrivial zeros of $\zeta(s)$, where $s = \frac{1}{2} + it$. The second column gives the values of t to 4 decimal places.

| DISCUSSION |

- **Riemann Hypothesis and prime numbers**. One of the key implications of the Riemann Hypothesis is that the Prime Number Theorem (§6.10) can be stated more precisely as

$$|\pi(x) - \text{Li}(x)| = O(\sqrt{x} \ln x).$$

This was proven by the Swedish mathematician *Helge von Koch* (1870–1924) (of the Koch snowflake fame). Even more precisely, the PNT under the assumption of the Riemann Hypothesis can also be stated as

$$|\pi(x) - \text{Li}(x)| < \sqrt{x} \ln x, \quad x \geq 2.01,$$

(see [49] for details). Another implication of the Riemann Hypothesis is that the gap between two consecutive prime numbers, p_n and p_{n+1} satisfies [48]

$$p_{n+1} - p_n = O(\sqrt{p_n} \ln p_n).$$

In short, the Riemann Hypothesis gives us a good understanding on the distribution of prime numbers and gaps between prime numbers.

References [47, 142] are excellent accessible reviews of the Riemann Hypothesis and its connections to prime numbers[14].

- **Computer verifications of RH**. To date, billions of zeros of $\zeta(s)$ have all been verified to lie on the line $\text{Re}(s) = \frac{1}{2}$ (e.g. [166]). The numerical techniques used to compute the zeros (such as the widely used *Odlyzko-Schönhage* and the *Riemann-Siegel* algorithms) are interesting by themselves. Reviews of computational techniques in connection to the Riemann Hypothesis can be found in [30, 95].

---

[13] Billions of nontrivial zeros are given here https://www.lmfdb.org/zeros/zeta/

[14] See also this video from Quanta magazine https://www.youtube.com/watch?v=zlm1aajH6gY

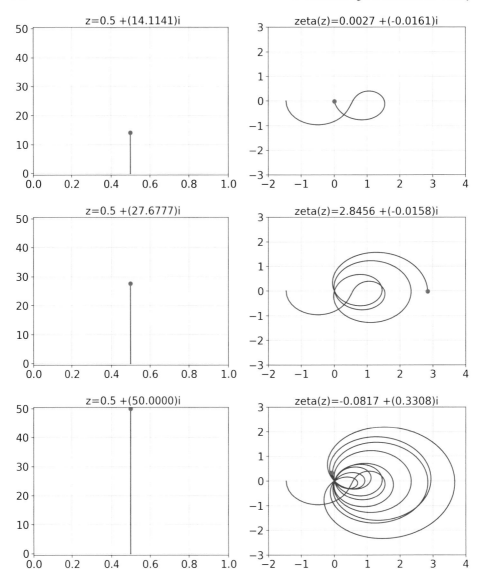

Fig. 6.18: Three frames from the animation produced by `zetaanim.ipynb`. Each pair of panels show the domain (left) and the image (right) in the complex plane under the Riemann-zeta function $z \mapsto \zeta(z)$. The domain is the line $\frac{1}{2} + it$ where t increases from 0 to 50. The image curve passes through zero 10 times in this animation.

zetaanim.ipynb (for producing the animation shown in fig. 6.18)

| For calculating $\zeta(z)$ with SymPy | ```
import numpy as np
import matplotlib.pyplot as plt
from matplotlib.animation import FuncAnimation
from sympy import zeta
%matplotlib
``` |
| Domain of $\zeta$ = line segment in $\mathbb{C}$ from $z = \frac{1}{2}$ to $\frac{1}{2} + 50i$ <br> List of the real parts of domain <br> and imaginary parts of domain <br> List of the real parts of image under $\zeta$ <br> and imaginary parts of image | ```
Tmax = 50
T  = np.linspace(0, Tmax, 1000)
Z  = [0.5 + t*1j for t in T]
Dom_r = np.real(Z)
Dom_i = np.imag(Z)
Img_r = []
Img_i = []
``` |
| Apply the complex map $\zeta$ <br> Store results in 2 lists | ```
for z in Z:
    fz = complex(zeta(z))
    Img_r.append(np.real(fz))
    Img_i.append(np.imag(fz))
``` |
| We'll plot 2 figures side-by-side <br> Increase font size a little <br> Templates for left plot (domain): <br> 1) Trace (line segment) in blue <br> 2) Current point $z$ (red dot) <br> 3) Current $z$ value (a complex number) <br> 4) Positioning of the text display | ```
fig,(ax1,ax2)=plt.subplots(1,2,figsize=(12,4))
plt.rcParams.update({'font.size': 16})

curve1, = ax1.plot([],[], 'b', markersize=3)
dot1, = ax1.plot([],[],'ro', markersize=6)
text1 = 'z=%.1f +(%.4f)i'
tit1 = ax1.text(0.27, 1.03*Tmax,'')
ax1.set_ylim(-0.5, 0.5+Tmax)
ax1.set_xlim(0, 1)
ax1.grid('on')
``` |
| Similar templates for the right plot (the image) <br><br> May need to adjust these if Tmax or the domain is changed | ```
curve2, = ax2.plot([],[], 'b', markersize=3)
dot2, = ax2.plot([],[],'ro', markersize=6)
text2 = 'zeta(z)=%.4f +(%.4f)i'
tit2 = ax2.text(-1.2, 3.1,'')
ax2.set_ylim(-3, 3)
ax2.set_xlim(-2, 4)
ax2.grid('on')
``` |
| In the $i$th frame, plot these: <br> Trace of domain in $\mathbb{C}$ <br> Trace of image in $\mathbb{C}$ <br> $z$ as a dot <br> $\zeta(z)$ as a dot <br> Display $z$ value <br> Display $\zeta(z)$ value | ```
def animate_frame(i, Dom_r,Dom_i,Img_r,Img_i):
    curve1.set_data(Dom_r[:i], Dom_i[:i])
    curve2.set_data(Img_r[:i], Img_i[:i])
    dot1.set_data(Dom_r[i], Dom_i[i])
    dot2.set_data(Img_r[i], Img_i[i])
    tit1.set_text(text1 % (Dom_r[i],Dom_i[i]))
    tit2.set_text(text2 % (Img_r[i],Img_i[i]))
    return curve1,curve2, dot1,dot2, tit1,tit2
``` |
| Animate the domain and its image <br><br> How many frames? <br> How many microseconds between frame? | ```
ani = FuncAnimation(fig, animate_frame,
        fargs=(Dom_r, Dom_i, Img_r, Img_i),
        frames = len(T),
        interval = 20)

plt.show()
``` |
```

6.13 Exercises

1 (*Sieve of Eratosthenes*) How long does it take for SymPy to generate a list of prime numbers from 2 to N? Plot (logarithmically) the time taken (in seconds) against N up to 10^8. You should find that the time taken scales exponentially.
Suggestion: SymPy's sieve stores previous results. To reset the sieve, use `sieve._reset()`. To time the code, see `solvetimes.ipynb` in §5.4.

2 (*Primitive roots*)

   a. Given an integer $n \geq 2$, generate a list of integers coprime to n. Try doing this using list comprehension.

   b. Let $n \geq 2$ be an integer. Let a be coprime to n.
Write a code that determines whether a is a generator of the cyclic group $(\mathbb{Z}_n^*, \times)$.
Suggestion: Generate the set $\langle a \rangle$ as a Python `set` object. Then test whether the set equals G.

   c. Write a code to find all generators of the cyclic group $(\mathbb{Z}_n^*, \times)$.
These generators are called the *primitive roots* modulo n.
For example, you should find that there are 8 primitive roots modulo 17.

3 (*Cayley table*) Use `cayley.ipynb` as a starting point.

   a. Produce the three plots in fig. 6.3.
Suggestions: To reorder the elements, first show that 3 is a generator for the cyclic group $(\mathbb{Z}_{79}^*, \times)$. Perhaps do the previous question first.

   b. Produce the Cayley table for $(\mathbb{Z}_{24}^*, \times)$. There should be 8 elements. Deduce that this group is not cyclic.

4 (*Cayley graph*) Consider the dihedral groups D_4 and D_7.
For each dihedral group, use `dihedral.ipynb` to produce the coloured polygons and assemble them to form a Cayley graph in the style of fig. 6.4.
List all their proper subgroups and identify the corresponding subset of the Cayley graphs that represents each subgroup.
Suggestion: Unless you have to produce a whole lot of Cayley graphs, it's probably too time-consuming to write a code that assembles all the polygons into a Cayley graph. Instead, assemble the polygons and insert arrows between them manually using, say, Keynote or Powerpoint.

5 (*Permutations groups*) Use `permutation.ipynb` as a starting point for these questions.

   a. i. Produce the Cayley table for A_4 as shown in fig. 6.6.
Reorder the elements of A_4 so that the elements of the Klein four-group

$$\{(e),\ (0\ 1)(2\ 3),\ (0\ 2)(1\ 3),\ (0\ 3)(1\ 2)\}$$

appear in the top-left corner of the Cayley table.

     ii. Express each element of D_4 as a permutation (do this by hand first) and check with SymPy.
Use Python to create the Cayley table in the style of fig. 6.6. Make sure to order the elements so that you obtain precisely Table 6.1. Examine the table and hence deduce that

$$D_4/C_4 \cong \mathbb{Z}_2.$$

b. Use SymPy to verify the partitioning of S_4 using the Klein four-group D_2 (as shown in fig. 6.7).
By reordering the elements of S_4, produce the Cayley table of S_4 like fig. 6.8 (but with the table for D_2 in the top left corner). Hence, deduce the isomorphism

$$S_4/D_2 \cong S_3.$$

Suggestion: The Python command `sorted` turns a set into a list of increasing integers.

6 (*Quaternions*)

a. Modify `quat.ipynb` so that the code rotates a line segment joining two points about a given axis.
b. Write a code to verify that for the set of quaternions $\{\pm 1, \pm i, \pm j, \pm k\}$, multiplication is associative.
Suggestion: Use SymPy to initialise quaternions – see `https://docs.sympy.org/latest/modules/algebras.html`.
c. (Do question 5b first) From table 6.4, spot the normal subgroups of Q. Hence, produce a Cayley table that visualises the isomorphism

$$Q/\mathbb{Z}_2 \cong D_2.$$

7 (*Patterns from multiplicative inverses*) Use `inversewheel.ipynb` as a starting point in these questions.

a. Investigate the pattern of the circle modulo n (fig. 6.10) for other values of n.
Why do some values of n result in busier patterns than others? (e.g. 203 vs. 204).
b. Modify the code so that it reports the number of lines in each circle. For example, with $n = 203$, the code should report that there are 82 lines.
When n is prime, make a conjecture on the number of lines. (Note: this observation leads to *Wilson's Theorem* – see any recommended textbooks on elementary number theory).
c. In the code, each line joining a and a^{-1} is drawn twice (once starting from a and once starting from a^{-1}).
Modify the code to avoid this redundancy. You should use the symmetries that we found in §6.7.

8 (*Chinese Remainder Theorem*) Use `crt.ipynb` to help you with this problem.
Consider the following problem from the ancient Chinese text *Sunzi Suanjing* (ca. 3rd century BC).
"*There are an unknown number of things. If we count by threes, the remainder is 2. If we count by fives, the remainder is 3. If we count by sevens, the remainder is 2. Find the number of things.*"

a. Solve the problem by hand.
b. Solve the problem using SymPy (this takes one line of code).
c. Generalise the problem so that the remainders are a b and c when counted by threes, fives and sevens. Is the system always solvable for all integers (a, b, c)?
d. Demonstrate the connection between the solution and the Pigeonhole Principle by displaying all possible solutions as slices of a 3D array (say, show 3 slices of 5×7 arrays).

9 (*Legendre symbol and quadratic reciprocity*)

    a. Use SymPy to verify that the 1000th prime is 7919.

       Find $\left(\frac{5}{7919}\right)$ by hand and verify your answer with SymPy.

       Suggestion: Use reciprocity law and the property that if $a = b \pmod{p}$ then $\left(\frac{a}{p}\right) = \left(\frac{b}{p}\right)$. Alternatively, find a general formula for $\left(\frac{5}{p}\right)$ by following the steps for $\left(\frac{3}{p}\right)$ as discussed in §6.9.

    b. Use `legendre.ipynb` as a starting point.

       i. Render the squares for which $\left(\frac{p}{q}\right) \neq \left(\frac{q}{p}\right)$ the same colour and remove the labels. Explain the symmetry of the resulting table.

      ii. Produce the lower table in fig 6.12 (*i.e.* the antisymmetric table).

10 (*Chebyshev bias*) Plot the ratio

$$\frac{\text{Total number of primes } p < N \text{ such that } \left(\frac{3}{p}\right) = 1}{\text{Total number of primes} < N}$$

against N in the range 10 to 10^6.

You should find that the ratio approaches a number just below 50% but, surprisingly, does not go above it.

You have just verified the phenomenon known as *Chebyshev bias*. See [80] for detail. Does the bias occur for, say, $\left(\frac{5}{p}\right)$? What about $\left(\frac{4}{p}\right)$? Experiment and make a conjecture.

11 (*Gauss's Lemma*) Many proofs of the reciprocity law rely on the following lemma (published by Gauss in his third proof of the reciprocity law). The lemma helps us determine whether a number is a quadratic residue.

Gauss's Lemma *Let p be an odd prime and $a \neq 0 \pmod{p}$. Consider the list of numbers*

$$a, 2a, 3a, \ldots, \frac{p-1}{2}a.$$

Reduce these numbers modulo p. If there are n numbers that are greater than $(p-1)/2$, then

$$\left(\frac{a}{p}\right) = (-1)^n.$$

Write a code that verifies Gauss's Lemma and produces the following output given, say, $a = 83, p = 19$.

```
List = [7 14 2 9 16 4 11 18 6]
There are 4 numbers in the list > 9.
Thus (83/19) = (-1)^4 = 1.
```

(For proof of Gauss's Lemma, see any of the recommended texts on elementary number theory.)

12 (*More on the Prime Number Theorem*)

    a. By modifying `pnt.ipynb`, plot the top and bottom panels in fig. 6.14.

    b. The PNT is equivalent to the statement that

$$\pi(x) \sim \frac{x}{\ln x - a}, \tag{6.36}$$

for some constant a.

Let's investigate the accuracy of the above approximation using different values of a (for large x).

   i. In 1808, Legendre found empirically that $a = 1.08366$. Do you agree? Try plotting the difference between the two sides of eq. 6.36 for x up to 10^8.

   ii. In this range of x, find the optimal value of the constant a using SciPy's `curve_fit` function[15].

   iii. Show that the optimal value of a decreases as the range of x is extended to 10^9 and then 10^{10}.

   (In fact, Chebyshev proved that the constant approches 1 as $x \to \infty$.)

13 (*The zeta function for real s*) Enjoy these easy mini exercises!

   a. Using eq. 6.25, find an expansion for $\zeta(10)$ in terms of π. Check your answer with SymPy.

   b. Plot the graph of $|\zeta(s)|$ against $1 - s$ on log scales (similar to fig. 6.15 but on the other side of the singularity at $s = 1$).

   On the same plot, insert the Laurent series (6.26), keeping up to the γ_0 term. You should find excellent agreement between the two curves.

   c. Plot the absolute difference between $\zeta(4)$ and the Euler's product

$$\prod_{p \text{ prime}} (1 - p^{-4})^{-1}$$

as a function of the number of primes in the product (say, up to 30 primes).

How many primes are needed in the product to achieve 6 dec. pl. accuracy? (Recall that n-dec. pl. accuracy is achieved when the absolute difference is $< 0.5 \times 10^{-p}$).

14 (*The gamma function*)

   a. Use the functional equation for Γ (eq. 6.30) to find the exact expressions for $\Gamma(1/2)$, $\Gamma(3/2)$ and $\Gamma(-1/2)$.

   Check that your results agree with the SymPy's symbolic output.

   b. Plot $y = \Gamma(x)$ on the domain $[-5, 5]$. Show that it agrees with $(x - 1)!$ when $x \in \mathbb{N}$.

   c. When x is a large, positive number, the gamma function takes the following asymptotic form

$$\Gamma(x) \approx \sqrt{2\pi} x^{x-\frac{1}{2}} e^{-x}. \tag{6.37}$$

This approximation is called *Stirling's approximation* after the Scottish mathematician, *James Stirling* (1692–1770). This formula was actually discovered by de Moivre, but the fact that numerical factor equals $\sqrt{2\pi}$ was due to Stirling.

   i. Plot the $\Gamma(x)$ and Stirling's approximation on the same set of axes for x between 1 and 4. You should find excellent agreement between the two curves.

   ii. Plot the fractional error in using the Stirling's approximation to approximate $\Gamma(x)$. Do this for x up to 100 and use `semilogy`. You should find that the absolute fractional error decreases very slowly.

   Calculate the fractional error in using Stirling's approximation to calculate $100!$ (Answer: -8.25×10^{-4}).

---

[15] https://docs.scipy.org/doc/scipy/reference/generated/scipy.optimize.curve_fit.html

15 (*Computing the zeta function efficiently*)

a. Plot the analytic continuation of the zeta function $\zeta(s)$ for real $s \in (0, 0.5)$ using eq. 6.27 (involving the Dirichlet eta function). Use 10^5 terms.
Compare the graph of the analytic continuation with SciPy's zeta function. You should see that the eta-function expression is noticeably off from $\zeta(s)$ near $s = 0$. Try increasing the number of terms.

b. There are several alternative formulae for the analytic continuation of ζ that are useful for computational purposes. See [30] for an overview.
One of these formulae is the following:

$$\zeta(s) = \frac{1}{\Gamma(s/2)} \left(\frac{\pi^{s/2}}{s(s-1)} + S_1 + S_2 \right), \tag{6.38}$$

$$S_1 := \sum_{n=1}^{\infty} n^{-s} \Gamma\left(s/2, \pi n^2\right),$$

$$S_2 := \pi^{s-\frac{1}{2}} \sum_{n=1}^{\infty} n^{s-1} \Gamma\left((1-s)/2, \pi n^2\right),$$

where $\Gamma(a, z)$ is the *incomplete gamma function*

$$\Gamma(a, z) = \int_z^{\infty} t^{a-1} e^{-t} \, dt.$$

The above integral expression is valid for $\mathrm{Re}(s) > 0$, but can be extended to other values of s via analytic continuation (though you will not need the extended expression explicitly in this question).
Use the analytic continuation 6.38 to evaluate $\zeta(0.5)$ using 10^5 terms in the sums. You should find that it gives essentially the same result as SciPy's $\zeta(0.5)$.
On log-log scales, plot the absolute errors

$$|\text{Eq. 6.27} - \zeta(0.5)| \quad \text{and} \quad |\text{Eq. 6.38} - \zeta(0.5)|$$

against the number of terms used (from 1 to 10^5). You should see that the formula (6.38) converges *much* more quickly than (6.27). in fact, only a handful of terms are needed to achieve machine-epsilon-level accuracy.
Suggestion: Do everything in SciPy. SciPy has a built-in function `gammaincc`[16], but be careful - it has a different normalisation from the incomplete gamma function defined above.

---

[16] https://docs.scipy.org/doc/scipy/reference/generated/scipy.special.gammaincc.html

16 (*Locating nontrivial zeros of ζ*) Reproduce fig. 6.17 (or a similar plot which shows the locations of the nontrivial zeros of ζ in the complex plane).
Suggestion: Use `meshgrid` to populate a grid of (x, y) coordinates, then define $z = |\zeta(x + iy)|$ (use SymPy's `zeta`). Plot the heatmap with `pcolormesh`. Use logarithmic colour mapping for the z values. Here's one way to do this.

```
fig, ax = plt.subplots()
ax.pcolormesh(x, y, z, norm=colors.LogNorm(vmin=z.min(), vmax=z.max()),
              shading='gouraud',cmap='hot')
```

17 (*Visualising ζ as a complex map*) Use `zetaanim.ipynb` to help answer the following questions.

  a. Modify the code to show the images of the lines $\mathrm{Re}(s) = 0.5$ and $\mathrm{Re}(s) = 0.8$ simultaneously. You should see that the trajectories evolve in sync, coming very close together at times.

  b. Plot the image of the circle $|z| = R$ under the ζ mapping. Experiment with various values of R. Take care to avoid the singularity at $z = 1$.
     You should find a range of interesting images, ranging from circles to cardioid-like loops.

  c. Plot the image of the square with vertices at $z = \pm R + \pm iR$ under the ζ mapping (where $R \neq 1$). Experiment with various values of R.
     You should see that a right angle get mapped to a right angle.
     Note: Complex functions that preserve angles are called *conformal maps*.

  d. Consider the *Möbius transformation* defined as

$$f(z) = \frac{az + b}{cz + d}, \quad a, b, c, d \in \mathbb{C}.$$

     We wish to find the image of the unit disc $|z| \leq 1$ under the Möbius transformation for some choices of parameters a, b, c, d.
     Plot all the images of the circles $|z| = R$ where R takes several values between 0 and 1, using the following values of (a, b, c, d).
     • $(1, -i/2, i/2, 1)$
     • $(0, 1, 1, 0)$
     • $(1, i, 1, -i)$
     (Answers: the unit disc; complement of the unit disc; a half plane.)
     You should notice that all circles are mapped onto either a circle or a line. See [154] and this video[17] for more on the Möbius transformations and its applications.

---

[17] https://www.youtube.com/watch?v=0z1fIsUNhO4

Probability

Fig. 7.1: Portrait of *Blaise Pascal* (1623–1662), a French mathematician, physicist and philosopher who laid the mathematical foundation for probability. Image from [210]

The study of probability arose out of the desire to understand (and perhaps gain an upper hand in) games of chance and gambling. Today, probability deeply pervades our everyday life, and has forged connections to surprising areas of mathematics. For instance, the probabilistic number theory and probabilisitic group theory are active research fields with applications in modern cryptography. The mathematical foundation of probability relies on real analysis, particularly measure theory. In other areas of science, probability is essential in the understanding of quantum mechanics, molecular dynamics, genetic mutation, and econometrics.

At the heart of probability is the concept of *randomness* which is difficult to replicate with pen and paper. It is rather disheartening that many students, perhaps in their introduction to probability in school, only ever see phrases like *"a ball is drawn at random from an urn"*, *"a pair of fair dice is thrown 20 times"* or *"5 students out of 100 are randomly chosen to form a committee"*. Such phrases normally appear in exercises, and some calculations are then performed, but the *randomness* is never seen.

S. Chongchitnan, *Exploring University Mathematics with Python,*

Thankfully, Python allows us to invoke and visualise randomness instantly. Millions of simulations involving games of coins, dice, cards and randomly selecting objects can be performed with little effort. The understanding of probability through simulation and visualisation is the spirit of this chapter.

We will use Python's randomisation capability to perform a large number (typically millions) of simulations and render visualisations of the results. This approach gives us insights into probability that are difficult to gain with the old pen-and-paper approach. Along the way, we will encounter some truly fascinating and often counterintuitive results in probability.

There are of course many excellent textbooks on probability that are suitable for beginning undergraduates. These include [8, 52, 179], and the classic series by Feller [62]. Good books on probability with focus on numerical simulations include [109, 156, 205] and Nahin's books [151, 152] which are particularly accessible and entertaining. For a problem-based approach, [77, 86] are highly recommended.

7.1 *Basic concepts in combinatorics*

Combinatorics is, in simple terms, the study of counting different ways that something can happen. Two useful concepts in combinatorics are *permutation* and *combination*.

A *permutation* is a rearrangement of objects where order matters. For example, the number of ways a word can be formed given n letters (we assume that a word does not have to make sense). There are, for instance, 6 permutations of 3 letters A, B, C, where letter repetition is not allowed.

$$ABC, ACB, BCA, BAC, CAB, CBA.$$

A *combination*, on the other hand, is a way to group together objects where order does not matter. For example, there are 6 ways to form a two-object subset out of the set of letters $\{A, B, C, D\}$.

$$\{A, B\}, \{A, C\}, \{A, D\}, \{B, C\}, \{B, D\}, \{C, D\}$$

Here are some key results in basic combinatorics.

Proposition 7.1 (*Key results in basic combinatorics*)

1. *Given n different letters, there are n! unique words that can be formed using those letters (where letter repetition is not allowed).*
2. *Given n letters, of which k_1 are alike, k_2 are alike... k_r are alike, then there are*

$$\frac{n!}{k_1! k_2! \dots k_r!}$$

   unique words that can be formed using those letters without repetition.
3. *Given n different letters, the number of unique r-letter words (where $0 \le r \le n$) that can be formed without repetition is*

$$^{n}P_r := \frac{n!}{(n-r)!}.$$

Here P stands for permutation.

4. *Given n different objects, the number of different subsets with r objects (where $0 \leq r \leq n$) is ^nC_r, also written $\binom{n}{r}$, defined as*

$$\binom{n}{r} := \frac{n!}{r!(n-r)!}. \tag{7.1}$$

Here C stands for combination. The symbol $\binom{n}{r}$ is called the binomial coefficient.

To calculate $\binom{n}{r}$ with Python, we can use SciPy's built-in command `binom(n,r)`. This will be important when we come to discuss Pascal's triangle in §7.4.

7.2 Basic concepts in probability

Axioms of probability

Consider an experiment in which the set of all possible outcomes is denoted Ω (called the *sample space*). The *probability* that the outcome $A \subseteq \Omega$ occurs, denoted $\Pr(A)$, satisfies the following 3 axioms.

1. $0 \leq \Pr(A) \leq 1$.
2. $\Pr(\Omega) = 1$.
3. Let $A, B, C, \ldots$ (countably many, possibly infinite) be mutually exclusive outcomes, meaning that no two outcomes can occur simultaneously. Then,

$$\Pr(A \cup B \cup C \cup \ldots) = \Pr(A) + \Pr(B) + \Pr(C) + \cdots$$

These are called the *Kolmogorov axioms* after the Russian mathematician *Andrey Nikolaevich Kolmogorov* (1903-1987). These axioms give rise to the following elementary results.

Proposition 7.2 *Let Ω be a sample space and let $A, B \subseteq \Omega$.*

1. *The probability that A does not occur, denoted $\Pr(A^c)$, equals $1 - \Pr(A)$.*
2. *If $A \subset B$, then $\Pr(A) \leq \Pr(B)$.*
3. *$\Pr(A \cup B) = \Pr(A) + \Pr(B) - \Pr(A \cap B)$.*

Conditional probability

Let $\Pr(B) > 0$. The *conditional probability* that A occurs given that B has occurred is denoted $\Pr(A|B)$. It can be expressed as

$$\Pr(A|B) = \frac{\Pr(A \cap B)}{\Pr(B)}. \tag{7.2}$$

If $\Pr(A|B) = \Pr(A)$, or in other words, if

$$\Pr(A \cap B) = \Pr(A)\Pr(B), \tag{7.3}$$

then A and B are said to be *independent* events. For example, a fair die showing 6 when thrown in London, and a fair die showing 6 when thrown in Singapore, are independent events.

Here is another useful result in connection with conditional probability.

Proposition 7.3 (*Law of total probability*)

$$Pr(A) = Pr(A \cap B) + Pr(A \cap B^c)$$
$$= Pr(A|B) Pr(B) + Pr(A|B^c) Pr(B^c).$$

This follows from eq. 7.2 and the axioms of probability. We will calculate conditional probabilities in our investigation into the famous Monty Hall problem (§7.7).

Random variables

A *random variable* X is a function defined on the sample space Ω. For example, let Ω be all possible results of 3 successive throws of a fair die. Let X be the sum of the 3 throws. Then X is a random variable such that $3 \leq X \leq 18$. Each possible value of X is associated with a probability value.

A random variable may be discrete or continuous.

Discrete random variables

A *discrete random variable* X takes on a countable number of possible values $x_1, x_2, x_3, \ldots$ (possibly infinitely many). Examples include outcomes of throwing coins and dice, card games, and general counts of people or things satisfying some conditions.

For a discrete random variable X, the probabilities $Pr(X = x_i)$ can be represented graphically using a histogram such as those in figs. 7.6, 7.9, 7.12 and 7.13.

The function

$$f(x_i) = Pr(X = x_i)$$

is called a *probability mass function* (pmf) or a *probability distribution function* (or simply, a distribution).

It follows that a probability distribution function satisfies the normalisation condition

$$\sum_{i=1}^{\infty} f(x_i) = 1.$$

In this chapter, we will encounter a number of discrete probability distributions, including the Bernoulli, binomial and Poisson distributions. To visualise these distributions, we will use Matplotlib's `hist` function[1] to produce histograms. Another method is to use the *Seaborn* library – this will be explained in chapter 8.

---

[1] https://matplotlib.org/stable/api/_as_gen/matplotlib.pyplot.hist.html

Continuous random variables

A *continuous random variable* X can take on an uncountable number of possible values. Examples include outcomes associated with physical observables (time, length, mass, temperature...) and general quantities that can take any values on some interval $[a, b] \subseteq \mathbb{R}$.

Consider the interval $[a, b] \subseteq \mathbb{R}$. The probability that $X \in [a, b]$ can be expressed as

$$\Pr(a \leq X \leq b) = \int_a^b f(x)\,dx,$$

where f is called the *probability density function* (pdf).

A pdf can be represented as a graph $y = f(x)$. An example is shown in fig 7.12 in which the red curves show a continuous probability distribution known as the *normal distribution*.

It is important to note that a single value of $f(x)$ does not correspond to a probability. We can only deduce probability values from the integral of the pdf. In other words, the *area* under the graph of the pdf $y = f(x)$ between $x \in [a, b]$ tells us about the probability $\Pr(a \leq X \leq b)$.

Analogous to the discrete case, it follows that the pdf of a continuous random variable satisfies the normalisation condition

$$\int_{-\infty}^{\infty} f(x)\,dx = 1.$$

In this chapter, we will encounter a number of continuous probability distributions, including the uniform, normal and exponential distributions.

Cumulative distribution function

The *cumulative distribution function* (cdf) of a random variable X with probability distribution function f is defined by

$$F(x) := \Pr(X \leq x)$$

$$= \begin{cases} \sum_{x_i \leq x} f(x_i), & \text{if } X \text{ is discrete,} \\ \int_{-\infty}^{x} f(t)\,dt, & \text{if } X \text{ is continuous.} \end{cases} \tag{7.4}$$

The graph of a cdf is an increasing function from $F(x) = 0$ to 1. It satisfies the limits

$$\lim_{x \to -\infty} F(x) = 0, \qquad \lim_{x \to \infty} F(x) = 1.$$

An example of a cdf can be seen in fig. 7.9 (lower panel).

It is conventional to use a lowercase letter to denote a probability mass or density function, and a capital letter to denote the cumulative distribution function.

Expectation (mean)

The *expectation value* of a random variable X with probability distribution function f is defined by

$$E[X] := \begin{cases} \sum_{i=1}^{\infty} x_i f(x_i), & \text{if } X \text{ is discrete,} \\ \int_{-\infty}^{\infty} x f(x) \, dx, & \text{if } X \text{ is continuous.} \end{cases} \tag{7.5}$$

(assuming that the sum or integral converges). The expectation value $E[X]$ is sometimes written $\langle X \rangle$. The expectation value of X is also called the *mean* of X, denoted μ.

The expectation value of a function g of a random variable X can itself by considered a random variable $g(X)$. Therefore, we have:

$$E[g(X)] = \begin{cases} \sum_{i=1}^{\infty} g(x_i) f(x_i), & \text{if } X \text{ is discrete,} \\ \int_{-\infty}^{\infty} g(x) f(x) \, dx, & \text{if } X \text{ is continuous.} \end{cases} \tag{7.6}$$

Let a and b be constant. It is straightforward to show that E is a linear operator, meaning that for two random variables X and Y,

$$E[aX + bY] = aE[Y] + bE[Y]. \tag{7.7}$$

Variance

The *variance* of a random variable X, denoted $\text{Var}(X)$, is defined by

$$\text{Var}(X) := E\left[(X - \mu)^2\right] \tag{7.8}$$

where $\mu = E[X]$. Intuitively, the variance measures the spread of the values of X around the mean.

A little simplification gives us a useful alternative formula for the variance.

$$\text{Var}(X) = E[X^2] - \mu^2. \tag{7.9}$$

The variance of X is denoted σ^2. Its square root, $\sigma = \sqrt{\text{Var}(X)}$ is called the *standard deviation*. It is used to quantify the error or uncertainty in some repeated experimental measurements or simulations.

Here is a useful property which will come in handy in our investigation into Monte Carlo simulations (§7.10). Let a and b be constant. Then,

$$\text{Var}(aX + b) = a^2 \text{Var}(X). \tag{7.10}$$

This follows immediately from the definition of the variance. This shows that Var is not a linear operator. In fact, for two random variables X and Y, we have

$$\text{Var}(aX + bY) = a^2 \text{Var}(X) + b^2 \text{Var}(Y) + 2ab \, \text{Cov}(X, Y), \tag{7.11}$$

$$\text{where} \quad \text{Cov}(X, Y) = E[XY] - E[X]E[Y].$$

More about the *covariance* $\text{Cov}(X, Y)$ in §8.7.

Example: The uniform distribution

An important probability distribution is the *uniform distribution*. Let's discuss the case where X is a continuous random variable with pdf

$$f(x) = \begin{cases} \frac{1}{b-a} & \text{if } a \leq x \leq b, \\ 0 & \text{otherwise,} \end{cases}$$

for some constants a, b such that $a < b$. We say that X is distributed uniformly between a and b, and write

$$X \sim U(a, b).$$

The tilde can be interpreted as the phrase "*follows the distribution*".

It is straightforward to check that

$$\int_{-\infty}^{\infty} f(x)\, dx = 1, \qquad E[X] = \frac{a+b}{2}, \qquad \text{Var}(X) = \frac{1}{12}(b-a)^2,$$

and that the cumulative distribution is given by

$$F(x) = \begin{cases} 0 & \text{if } x < a, \\ \frac{x-a}{b-a} & \text{if } a \leq x \leq b, \\ 1 & \text{if } x > b. \end{cases}$$

The uniform distribution is especially important for computer simulations. In particular, we will usually work with Python's default random number generator which gives a small random number with the distribution $U(0, 1)$.

7.3 *Basics of random numbers in Python*

Here is a quick guide on how to generate random numbers in Python using either the NumPy or Python method. Take care of the different endpoints and types of output in each method.

	NumPy method	**Python method**
	`rng = np.random.default_rng()`	`import random`
Pick a random number in [0,1)	`rng.random()`	`random.random()`

Pick N random numbers in [0,1)	`rng.random(N)`
	
	`# An array`
		`[random.random() for i in range(N)]`
		
		`# A list`

A matrix of random numbers	`rng.random((M,N))`
	
	`# An MxN array`	

Pick a random number from a uniform distribution $U(a, b)$	`# Result in [a,b)`
	`a + (b-a)*rng.random()`
	
	`# OR`
	
	`rng.uniform(a,b)`
		`# Result in [a,b]`
		`random.uniform(a,b)`

Pick a random integer between m and n	`# Result in [m,n)`
	
	`rng.integers(m,n)`
		`# Result in [m,n)`
		
		`random.randrange(m,n)`
		
		`# OR, for result in [m,n]`
		
		`random.randint(m,n)`

7.4 Pascal's triangle

Study the structure of Pascal's triangle modulo 2.

Pascal's triangle is shown on the left panel of fig. 7.2. Starting at the top of the triangle (call it row 0), the entries in row n are $\binom{n}{0}, \binom{n}{1}, \binom{n}{2}, \ldots, \binom{n}{n}$. The rows are stacked to form a triangular pattern.

It follows from the binomial theorem that row n contains the coefficients in the expansion $(1 + x)^n$, since

$$(1 + x)^n = \binom{n}{0} + \binom{n}{1}x + \binom{n}{2}x^2 + \cdots + \binom{n}{n}x^n. \tag{7.12}$$

Although the triangle bears the name of Pascal who studied it in 1654, it is likely to have been known to the Indian poet-mathematician *Pingala* ($\sim$ 300–200 BC), and later mathematicians in Persia and China.

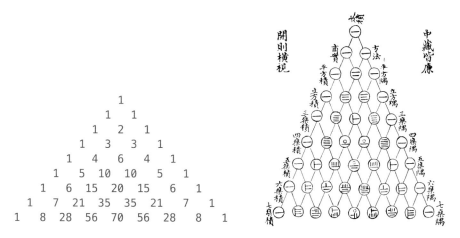

Fig. 7.2: *Left*: Pascal's triangle generated using Python. *Right*: A page from a 14th century Chinese manuscript attributing the triangle to the mathematician *Jia Xian* (ca. 1010–1070). Image from [211].

Here are some well-known properties of Pascal's triangle.

1. **Recursion** Two adjacent entries in the same row add up to the entries between them in the row below.

$$\binom{n}{k} + \binom{n}{k + 1} = \binom{n + 1}{k + 1}. \tag{7.13}$$

This is easily proved by algebraically simplifying the LHS.
2. **Symmetry** The triangle is symmetric about the vertical line drawn down the centre of the triangle.

$$\binom{n}{k} = \binom{n}{n-k}. \tag{7.14}$$

This follows from the definition (7.1).

3. **The sum of the nth row** is 2^n. This follows from setting $x = 1$ in eq. 7.12.
   A fun observation: setting $x = 10$ in eq. 7.12., we deduce that the first 5 rows of Pascal's triangle are simply the digits of $11^0, 11^1, 11^2, 11^3, 11^4$.

4. **Triangular numbers** The 3rd diagonal reads

$$1, 3, 6, 10, 15, 21, 28, \ldots$$

These are the *triangular numbers*, $t_n = 1 + 2 + 3 + \cdots + n = n(n+1)/2$.

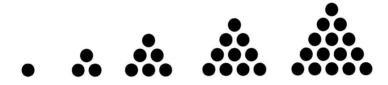

The next diagonal along contains the *tetrahedral numbers*, $T_n = t_1 + t_2 + \cdots + t_n = n(n+1)(n+2)/6$ and so on.

5. **Fibonacci numbers** Left justify the triangle. Then, the sum of each diagonal produces a Fibonacci number (see exercise 1c).

There are many other exotic patterns within the triangle[2].

Computing $\binom{n}{k}$ with Python

SciPy has a built-in command `binom`. For example, to calculate $\binom{9}{4}$, we can use the following syntax.

```
from scipy.special import binom
binom(9,4)
```

This also works with non-integers (SciPy replaces $n!$ by $\Gamma(n-1)$ - see eq. 6.29).

Nevertheless, it is instructive to try to write the code that can compute $\binom{n}{k}$ accurately ourselves. The (home-made) function `binom(n,k)` (at the top of `pascal.ipynb`) shows how $\binom{n}{k}$ can be computed recursively. The idea is to calculate each bracketed term on the RHS of the expression below

$$\binom{n}{k} = \left(\frac{n}{1}\right)\left(\frac{n-1}{2}\right)\left(\frac{n-2}{3}\right)\cdots\left(\frac{n-(k-1)}{k}\right)$$

and multiply them recursively going from left to right. The 'floor division' operator `//` ensures that the result of the division is an integer (otherwise we would see lots of unnecessary .0 when generating, say, Pascal's triangle in fig. 7.2).

---

[2] See https://tinyurl.com/y396zt8k (cut-the-knot.org)

Sierpiński's triangle

The code `pascal.ipynb` produces a large Pascal's triangle in modulo 2 (*i.e.* we retain only the parity of the entries). The result is shown in the top panel of fig. 7.4. The white squares are 0's and the black squares are 1's. The code prints these symbols as Unicode characters. An introductory guide to Unicode in Python can be found here[3].

An intriguing triangular pattern emerges...

On the right of the same figure, we show the *Sierpiński triangle*, a fractal in which the white triangles are iteratively generated in the black spaces using the following pictorial rule (5 iterations are shown).

Fig. 7.3: Constructing the Sierpiński triangle. *Wacław Sierpiński* (1882–1969) was a prolific Polish mathematician who made important contributions to set theory and number theory. Several fractals bear his name.

Of course the binary Pascal's triangle is not a true fractal in the sense that one cannot infinitely zoom in to see a self-similar structure. Nonetheless, zooming *out* of the triangle does reveal an infinite self-similar structure. You may find it surprising that a self-similar structure emerges out of Pascal's triangle. The proof is a little fiddly but is accessible to undergraduates - see [18, 218] for example.

Without going into the detail of the proof, one can get a feel of why a self-repeating pattern emerges. From the top panel of fig. 7.4, we see that now and then a line appears with mostly 0's (white squares) with 1's (black squares) at either end, *i.e.*

$$1\ 0\ 0\ 0\ \ldots\ 0\ 0\ 0\ 1$$

Doing a little counting in the figure will convince you that this type of row appears in lines $\binom{n}{k}$, where $n = 2^1, 2^2, 2^3, 2^4, \ldots$. This observation is equivalent to the following statement.

Lemma: If $n = 2^m$ (where $m \in \mathbb{N}$), then $(1 + x)^n = 1 + x^n \pmod 2$.

This result is straightforward to prove by induction. This means that as we go down Pascal's triangle, we will keep seeing a line corresponding to one side of an upside-down triangle whenever $n = 2^m$.

The appearance of the such a line then completely determines the lines below (due to the recursive property of binomial coefficients). Below two squares of the same colour, we must have a white square, since $1 + 1 = 0 + 0 = 0 \pmod 2$. Below two squares of opposite colours, we have a black square. These rules then generate the repeating pattern, leading to self-similarity on large scale.

In summary, the structure of Pascal's triangle modulo 2 resembles that of Sierpiński's triangle with one key difference: Pascal's triangle shows self-similarity when zoomed out, whilst Sierpiński's triangle shows self-similarity when zoomed in.

pascal.ipynb (for plotting the top panel of fig. 7.4)	
Generate $n + 1$ lines of Pascal's triangle Modulo what integer?	`n = 30` `mod = 2`

Binomial coefficients computed recursively
 	
// means floor division
	`def binom(n,k):`
	`    a = 1`
	`    for i in range(k):`
	`        a = a*(n-i)//(i+1)`
	`    return a`

Convert a number to a symbol
 	
\u25A1 is an empty square (0)
 	
\u25A0 is a black square (1)
	`def symb(u):`
	`    if u==0:`
	`        return "\u25A1"`
	`    else:`
	`        return "\u25A0"`

Printing $n + 1$ lines
 	
Prepend empty spaces to centre-justify
 	
Compute binomial cofficients mod n
Convert result to a symbol
Print the line
	`for i in range(n+1):`
	`    for j in range(0, n-i+1):`
	`        print(' ', end='')`
	`    for j in range(i+1):`
	`        u = binom(i,j) % mod`
	`        s = symb(u)`
	`        print(' ', s, sep='', end='')`
	`    print()`

Fig. 7.4: *Top*: Pascal's triangle in mod 2, showing 64 rows, generated using `pascal.ipynb`. Filled (black) squares are ones and empty (white) squares are zeroes. *Bottom*: The Sierpiński Triangle (after 5 iterations).

DISCUSSION

- **Cellular automata**. The recursive property of Pascal's triangle allows one line to determine the next. This recursive property is typical of *cellular automata*. In the simplest one-dimensional version, a set of rules determines how one line of squares (which can either be black or white) evolves to the next line below. For example, the so-called *Rule 90* dictates the following evolution.

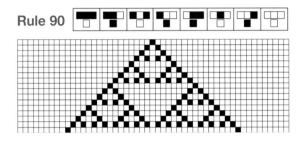

Fig. 7.5: *Rule 90* in Wolfram's classification of elementary cellular automata. Starting with a single black cell initially (top row of the grid), the binary Pascal's (Sierpiński) triangle appears. Figure adapted from `http://atlas.wolfram.com/01/01/90/01_01_1_90.html`.

Starting with a line containing a single black square, Rule 90 generates the binary Pascal's (Sierpiński) triangle. We can see why this holds: assign value 1 to a black and 0 to a white square. Given 3 adjacent squares with values p, q, r, Rule 90 simply says that the value of the square below q is $(p + r) \pmod 2$.

Cellular automata were extensively studied by Stephen Wolfram in the 1980's [217]. A well-known game using cellular automata is the *Game of Life*, invented by John Conway[4].

- **Turtle**. A popular Python package for drawing lines and shapes is *Turtle*, which was used to produce the Sierpiński triangle in fig. 7.4. See `https://docs.python.org/3/library/turtle.html` for an introduction to Turtle.

---

[4] `https://conwaylife.com`

7.5 Coin tossing

> I toss a fair coin 10 times. In each throw, the coin shows either heads (H) or tails (T).
> a) What is the probability that I see at least 5 heads?
> b) What is the probability that I see the sequence HH?
> c) What is the probability that I see the sequence TH?

Parts (b) and (c) seem like they should have the same answer. After all, H and T occur with equal probability. However, we will see that this intuition turns out to be wrong...

Simulating coin tosses

There are several ways to simulate coin tosses in Python. We give 3 ways below.

Three methods to simulate 10 coin tosses and count the number of heads	

Method 1: Python's random module	<pre># ------------------------------
	import random
	heads= 0
	for trial in range(10):
	    roll = random.randint(0, 1)
	    if roll == 1: heads += 1
	heads</pre>

Pick between 0 (heads) or 1 (tails) Add 1 to count if heads appears	

Method 2: NumPy's RNG	<pre># ------------------------------
	import numpy as np
	rng = np.random.default_rng()
	rng.binomial(10, 0.5)</pre>

Sample from the binomial distribution with $n = 10$, $p = 0.5$	

Method 3: SciPy's stats module	<pre># ------------------------------
	from scipy.stats import bernoulli
	import numpy as np
	dist = bernoulli(0.5)
	rolls = dist.rvs(10)
	np.count_nonzero(rolls==1)</pre>

Bernoulli trials with Pr(success) = 0.5 Draw 10 'random variates' with replacement Count how many successes occurred	

Bernoulli trials are independent events each of which is either a success with probability p, or a failure with probability $1 - p$. For example, obtaining heads in a toss of a fair coin is a Bernoulli trial with $p = 0.5$.

The Bernoulli trial is named after Jacob Bernoulli (whose Lemniscate we studied in §3.4).

The probability distribution of the two possible outcomes of a Bernoulli trial is called the *Bernoulli distribution*. Let 1 and 0 correspond to success and failure respectively. We can then express the pmf of the Bernoulli distribution as

$$f(k) = \begin{cases} p & \text{if } k = 1, \\ 1 - p & \text{if } k = 0. \end{cases} \tag{7.15}$$

Binomial distribution

Part (a) asks for the probability that we have at least 5 successes (heads) in 10 Bernoulli trials.

Let's first consider the probability of obtaining *exactly* 5 heads which do not necessarily appear successively. We could have, for example, any of the following sequences

$$HHHHHTTTTT$$
$$THTHTHTHTH$$
$$TTHHTHHHTT$$

How many such sequences are there? Well, it is simply a matter of choosing 5 positions (out of 10 possibilities) for which the heads can occur. The answer is therefore $\binom{10}{5} = 252$ possibilities.

In each of these possibilities, heads occur 5 times with probability $\frac{1}{2^5}$ and tails occur 5 times also with probability $\frac{1}{2^5}$. Therefore, the probability that exactly 5 heads occur in 10 throws is

$$\Pr(5 \text{ heads}) = \binom{10}{5} \cdot \frac{1}{2^5} \cdot \frac{1}{2^5} = \frac{63}{256} \approx 0.2461 \quad (4 \text{ dec. pl.}).$$

We can generalise the above observation as follows: the probability that k successes are observed in n Bernoulli trials, where each success occurs with probability p, is given by

$$\Pr(k \text{ successes observed}) = \binom{n}{k} p^k (1-p)^{n-k}. \tag{7.16}$$

The RHS is the pmf of the *binomial distribution*, denoted $B(n, p)$, with pmf

$$f(k) = \binom{n}{k} p^k (1-p)^{n-k}.$$

Note that when $n = 1$, the binomial distribution reduces to the Bernoulli distribution (7.15).

In our scenario of tossing a fair coin 10 times, the probability of obtaining exactly 5 heads follows the binomial distribution with $n = 10$, $k = 5$ and $p = 0.5$. We can then repeat the calculation for the probability of obtaining exactly 6, 7, 8, 9, 10 heads.

Therefore, to answer part (a), the probability of obtaining at least 5 heads is given by

$$\Pr(\geq 5 \text{ heads}) = \left(\binom{10}{5} + \binom{10}{6} + \binom{10}{7} + \binom{10}{8} + \binom{10}{9} + \binom{10}{10} \right) \frac{1}{2^{10}}$$
$$= \frac{319}{512}$$
$$\approx 0.6230 \quad (4 \text{ dec. pl.}).$$

(Alternatively, we could calculate $1 - \Pr(< 5 \text{ heads})$.)

Let's see if this result agrees with simulation. The code `coin1.ipynb` generates 10^5 simulations of 10 throws of the coin (using **Method 1**), and counts the number of heads in each 10-throw simulation. The histogram showing the count is shown in fig. 7.6. The height of the bars is normalised so that the blue area is 1. In other words, the histogram can be interpreted as a probability mass function.

The more throws we perform, the closer the distribution gets to the binomial distribution

$$\Pr(k \text{ heads}) = \binom{10}{k} \frac{1}{2^{10}}. \tag{7.17}$$

In this particular simulation with 10^5 throws, we found the following output.

```
Pr(5 heads)= 0.24625
Pr(at least 5 heads)= 0.62296
```

These values are in good agreement with our calculations.

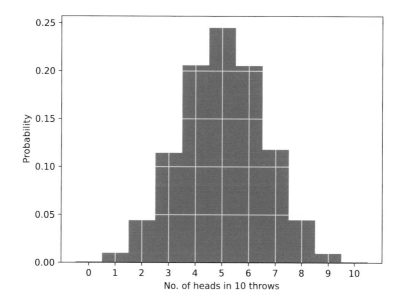

Fig. 7.6: The probability mass function for the total number of heads seen in 10 throws. We simulated 10^5 ten-throw experiments using the code $\texttt{coin1.ipynb}$. The result agrees with the binomial distribution (7.17).

coin1.ipynb (for plotting fig. 7.6 and calculating Pr(at least 5 heads))	
	``` import numpy as np import matplotlib.pyplot as plt import random %matplotlib ```
Number of 10-throw experiments to simulate This will be filled with heads counts	``` exprmt = int(1e5) throws = 10 Htally = [] ```

Throw the coin lots of times
 	
Use **Method 1** to simulate throws
 	
Record the number of heads
	``
	for i in range(exprmt):
	    heads= 0
	    for trial in range(throws):
	        roll = random.randint(0, 1)
	        if roll == 0: heads += 1
	    Htally.append(heads)
	``

Plot 11 bars (centred on integers) Plot histogram (prob=bar height) density=True normalises the distribution	``` bins = np.arange(throws+2)-0.5 prob, b1, b2 = plt.hist(Htally, bins= bins, density = True) plt.xticks(range(11)) plt.xlabel('No. of heads in 10 throws') plt.ylabel('Probability') plt.grid('on') plt.show() ```
Obtain probabilities from histogram	``` print(f'Pr(5 heads)= {prob[5]}') print(f'Pr(at least 5 heads)= {sum(prob[5:])}') ```

A surprise (and Fibonacci appears)

Now let's address parts (b) and (c), namely the probability of seeing HH or TH in 10 throws. First, let's examine this problem analytically.

Let $f(n)$ be the number of ways of tossing a coin n times such that HH does *not* appear. In other words, $f(n)$ is the number of sequences of length n, comprising H and T, such that HH does not appear.

In a combinatorial problem like this, a tree diagram can help us see that $f(n)$ can be computed recursively. Note that when T appears, the result of the next throw is always part of the admissible sequence. But when H appears, only T is the next admissible outcome.

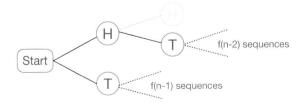

Fig. 7.7: Tossing a coin n times in such a way that HH does not appear, represented as a tree diagram.

Since there are altogether $f(n)$ sequences, and the paths in the diagram are mutually exclusive, we deduce that

$$f(n) = f(n-1) + f(n-2),$$

which is of course the Fibonacci recursion. The first 2 terms are $f(1) = 2$ (namely H,T) and $f(2) = 3$ (TT, TH, HT). This means that $f(n) = F_{n+2}$, the $(n+2)$th Fibonacci number (recall that $F_1 = F_2 = 1$). In particular, note that $f(10) = F_{12} = 144$ (see §1.6 for computation of F_n).

We can now calculate the probability of seeing the sequence HH in 10 throws as follows.

$$\Pr(HH) = 1 - \Pr(\text{no HH})$$
$$= 1 - \frac{144}{2^{10}}$$
$$= \frac{55}{64}$$
$$= 0.859375.$$

Now let's repeat our analysis for the sequence TH. Let $g(n)$ be the probability that TH does *not* appear in n tosses of a coin. Following the same method, we deduce that

$$g(n) = 1 + g(n-1),$$

with the initial terms $g(1) = 2$ and $g(2) = 3$. It follows that $g(n) = n + 1$, and in particular $g(10) = 11$. Therefore,

$$\Pr(TH) = 1 - \Pr(\text{no TH})$$
$$= \frac{1013}{1024}$$
$$= 0.9892578125.$$

The tree-diagram analysis helps us make sense of the counter-intuitiveness of the answer: it is more likely to observe TH rather than HH because of the simple fact that there are many more sequences without HH compared to those without TH, *i.e.* $f(n) > g(n)$.

Let's turn to the simulation. The code `coin2.ipynb` confirms the results of our recursion analysis. In one particular run of 10^5 ten-throw simulations, we found

```
Pr(HH seen)= 0.85923
Pr(TH seen)= 0.98916
```

In the code, we use **Method 3** to produce 10-throw sequences, and convert them to binary strings. In each string, we search for the appearance (if any) of HH (11) or TH (01).

A bonus from the simulation is that we can plot the distribution of the number of throws needed for either HH or TH to first appear, and compare them quantitatively. Fig. 7.8 shows these (unnormalised) distributions. For example, in the 10-throw sequence

$$\text{TTTHTHTHHH}$$

we say that HH appears in position 8, and that TH appears in position 3. Let's assign position 0 to the case when HH or TH does not appear in the sequence.

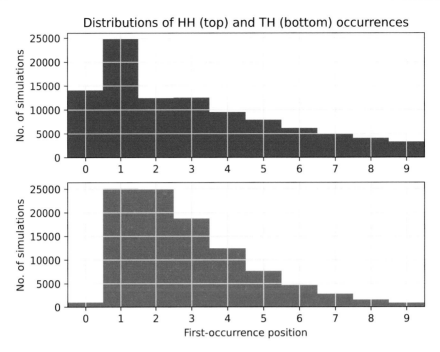

Fig. 7.8: Distributions of the first occurrences (in a 10-throw sequence) of HH (top) and TH (bottom) in 10^5 simulations. Position 0 indicates that the sequence does not appear in the simulation.

Interestingly we see from the simulations that both HH and TH are equally likely to occur in position 1 (the probability is $\frac{1}{4}$ since there are only 4 possible combinations for the first two tosses). However, we are far more likely to observe TH than HH in position 2. At the opposite end, in position 9 it is far more likely to observe HH rather than TH. Try to work out these probabilities by hand.

We conclude that although we have random events that are equally likely to occur, the *sequences* of such events are not necessarily equally likely to occur. Exercise 2 further explores this somewhat counterintuitive idea.

coin2.ipynb for plotting fig. 7.8 and calculating the probability of seeing HH or TH	

For performing Bernoulli trials	```python
	import numpy as np
	import matplotlib.pyplot as plt
	from scipy.stats import bernoulli
	%matplotlib
	``

Use **Method 3** to simulate throws	```python
	dist = bernoulli(0.5)
	``

Number of experiments
How many throws per experiment?
Running tallies of HH and TH occurrences
Lists for storing occurence positions
	``python
	exprmt = int(1e5)
	throws = 10
	HHfound, THfound = 0, 0
	PHH = []
	PTH = []
	``

Perform all throws at once!
Convert result to a binary string
The position where HH or TH first occurs
 	
Store the positions
(Start counting from 1)
seq.find = −1 ⟹ sequence not found
If found, add to tally
	``python
	for i in range(exprmt):
	    rolls = dist.rvs(throws)
	    seq = ''.join(str(r) for r in rolls)
	    posHH = seq.find('11')
	    posTH = seq.find('01')
	    PHH.append(posHH + 1)
	    PTH.append(posTH + 1)
	    if posHH !=-1:
	        HHfound += 1
	    if posTH !=-1:
	        THfound += 1
	``

Pr(HH seen)
Pr(TH seen)
	``python
	HHprob = HHfound/exprmt
	THprob = THfound/exprmt
	
	print(f'Pr(HH seen)= {HHprob}')
	print(f'Pr(TH seen)= {THprob}')
	``

Plot 2 stacked figures
Adjust the bins to centre-justify the bars
 	
Plot the two histograms
HHtally and THtally are the frequencies
 	
Common settings for both histograms
	``python
	fig,(ax1, ax2) = plt.subplots(2,1)
	bins = np.arange(0, throws+1) - 0.5
	ax1.set_title('Distributions of HH (top) and'
	              ' TH (bottom) occurrences')
	HHtally, b1, b2 = ax1.hist(PHH, bins= bins,
	                           color = 'b')
	THtally, c1, c2 = ax2.hist(PTH, bins= bins,
	                           color = 'r')
	for X in [ax1,ax2]:
	    X.set_xticks(range(0,throws))
	    X.set_xlim(-0.5, throws-0.5)
	    X.set_ylabel('No. of simulations')
	    X.grid('on')
	ax2.set_xlabel('First-occurrence position')
	fig.show()
	``

DISCUSSION

- **Penney's game**. We could modify the scenario a little and turn the coin tosses into a 2-player game. Player A picks a sequence of length 3 (HHT, say) and Player B picks a different sequence. The player whose sequence appears first wins. It can be shown that Player B can always win. Exercise 2 investigates a winning strategy.

  This game is called called *Penney's game* (or Penney ante) after Walter Penney who described it in the journal for recreational mathematics in 1969. See [71, 155] for a more thorough analysis.

- **Random number generators** (RNG). At the heart of probabilistic simulations lies an RNG that allows random numbers to be produced. However, current binary-based computers are deterministic, and so we can never really produce truly random numbers, but *pseudo-random* numbers at best. Pseudo-RNG are also periodic - although the period can be extremely long.

  Understanding RNGs and measuring the randomness of pseudo-random numbers are important topics in computing, numerical analysis and cryptography. See [102] for a deep dive into the world of random numbers. The website `https://www.random.org` provides *true* random numbers by using atmospheric noise.

- **Mean and variance**. It is worth stating the mean and variance of the Bernoulli and binomial distributions with probability of success p. Let $q = 1 - p$.

	μ	σ^2
Bernoulli(p)	p	pq
Binomial(n, p)	np	npq

These results are consistent with the fact that when $n = 1$, the binomial distribution reduces to the Bernoulli distribution.

7.6 The Birthday Problem

> Find the smallest number of people in a group such that the probability of two
> people sharing the same birthday is at least 0.5.
> For simplicity, assume that a year comprises 365 days and that births are equally
> likely to occur on any day of the year.

This is a classic problem in probability in which, for most people, the answer doesn't sit
well with intuition.

Let's first see how this problem could be solved analytically. As is often the case in
probability, it is sometimes easier to calculate the probability of the complementary situation,
i.e. the probability that a group of n people have all distinct birthdays.

For two people, A and B,

$$\text{Pr(2 distinct birthdays)} = \frac{364}{365},$$

since, having fixed A's birthday, there are 364 possibilities for B's birthday. Continuing this
way with n people,

$$\text{Pr}(n \text{ distinct birthdays}) = \frac{364}{365}\frac{363}{365}\cdots\frac{365-(n-1)}{365} \tag{7.18}$$

$$\implies \text{Pr(at least 2 people share a birthday)} = 1 - \frac{364}{365}\frac{363}{365}\cdots\frac{365-(n-1)}{365}. \tag{7.19}$$

It remains to solve for n when the above probability exceeds 0.5. This can be done
numerically with Python (exercise 3). Alternatively, using the linear approximation $e^{-x} \approx$
$1 - x$ for small x, we find:

$$\text{Pr(at least 2 people share a birthday)}$$

$$= 1 - \left(\frac{365-1}{365}\right)\left(\frac{365-2}{365}\right)\cdots\left(\frac{365-(n-1)}{365}\right)$$

$$= 1 - \left(1 - \frac{1}{365}\right)\left(1 - \frac{2}{365}\right)\cdots\left(1 - \frac{n-1}{365}\right)$$

$$\approx 1 - \exp\left(-\frac{1}{365}(1 + 2 + \cdots + (n-1))\right)$$

$$= 1 - \exp\left(-\frac{n(n-1)}{730}\right). \tag{7.20}$$

If the above expression exceeds 0.5, then it follows that

$$n^2 - n - 730\ln 2 > 0 \implies n > \frac{1 + \sqrt{1 + 2920\ln 2}}{2} \approx 22.999943\ldots.$$

Thus, assuming that the linear approximation is sufficiently accurate, we conclude that in a
group of 23 people, the probability of at least 2 people sharing a birthday exceeds 0.5.

One could generalise the problem and replace the probability 0.5 by other values as
shown in the table 7.1.

Pr(2 people share a birthday)	Min group size
0.33	18
0.5	23
0.9	41
0.99	57

Table 7.1: Probability that two people share the same birthday and the minimum group size.

Simulation

Let's use Python to simulate a scenario in which people enter a room one-by-one whilst each new entrant compares their birthday with the rest of the group. Sometimes there will be a lucky simulation in which only a few people enter the room before a match occurs. Other times, we might need, say, 50 people in the room before a match occurs. Python can help us play out this scenario thousands of times. We can then construct a histogram of the number of people in the room in each successful simulation. This is precisely what the code `birthday.ipynb` does.

Fig. 7.9 shows the histogram obtained by performing 10^5 simulations. We normalise the histogram by dividing the frequency by 10^5, giving us a probability distribution. It is interesting to note that there are indeed rare simulations in which it only takes 2 people in the room to share a birthday. At the other extreme, there are also rare simulations in which close to 100 people are needed before a birthday is shared.

The lower plot is the cumulative probability distribution obtained by summing the height of the successive bars. This gives the probability of a shared birthday as the group size increases. The code reports the group size at which the probability exceeds 0.5. Indeed the answer is 23.

Rerunning the simulations will produce slightly different distributions, but the answer to the birthday problem remains 23 (unless the number of simulations is too small). It is reassuring to see that although we have not used any pre-calculated probabilistic formulae in the code, yet, with a large number of ensembles, the simulation result agrees with our calculation.

Finally, we note that the peak of the probability distribution (the mode) is around 20. This can be interpreted as saying that, as people enter a room one by one, the 20th person is most likely to be the first one to share a birthday with someone already in the room. See exercise 3 for a way to obtain the mode by hand.

DISCUSSION

- **Only 23?** The answer '23 people' is probably far lower than most people would have initially guessed. The reason is that many people misinterpret the question as the number of people needed for someone to have the same birthday as them. In this case, if you enter a room with 22 other people, there are 22 birthday comparisons to be made. The probability can be calculated as

$$\text{Pr(Same birthday as yours)} = 1 - \left(\frac{364}{365}\right)^{22} = 0.059 \ \ (3 \text{ dec. pl.})$$

As expected, this is rather small. Solving a quick inequality shows that we need 253 people to compare birthdays with before the probability of finding someone with the same birthday as yours exceeds 0.5.

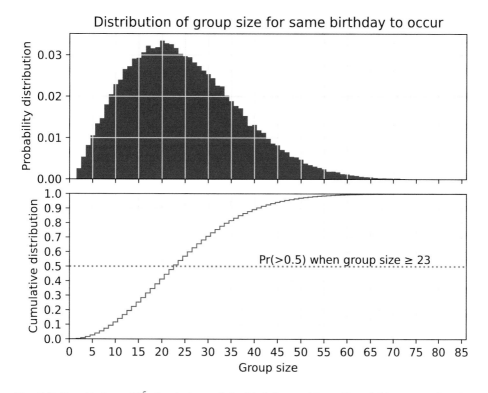

Fig. 7.9: Results from 10^5 simulations of the birthday problem. *Top*: A histogram showing the distribution of the group sizes in which two people share the same birthday. The height of the bars is normalised so that the area under this curve is 1. *Bottom*: The cumulative distribution, showing that the probability of two people sharing the same birthday in a group of 23 exceeds 0.5.

However, this isn't what the birthday problem is about. In a group of 23 people, there are many more birthday comparisons, namely $\binom{23}{2} = 253$, any of which could be a potential match. The pitfall is that 'same birthday' does not mean 'same birthday as yours'.

- **Generalisations**. There are many interesting generalisations of the birthday problem. For instance, what is the average number of people in a group such that their birthdays take up all 365 days of the year?

  This is called the *coupon collector's problem* and can be rephrased as follows: suppose there are some mystery boxes, each containing a random type of coupon. On average, how many boxes must be opened to find all n types of coupons?

  The answer turns out to be nH_n, where $H_n = \sum_{k=1}^{n} \frac{1}{k}$ is the nth harmonic number (see also §1.5). For $n = 365$, we need, on average, a group of 2364 people to fill up a whole year's calendar of birthdays. See for example [179] for calculation details.

  This and other generalisations of the birthday problem will be investigated in exercise 3.

birthday.ipynb (for plotting fig. 7.9)	

```
import numpy as np
import matplotlib.pyplot as plt
%matplotlib
```

Number of simulations to perform
Initialise RNG
Min. group size from each simulation

```
sims = int(1e5)
rng = np.random.default_rng()
grpsize = []
```

Simulations
Start with an empty room

Add one person (random birthday)
Add their birthday to the list
All unique birthdays on the list so far
Repeat until a shared birthday appears
Note how many people are in the room

```
for i in range(sims):
    BD = []
    sameBD = 0
    while sameBD == 0:
        newBD = rng.integers(1, 366)
        BD.append(newBD)
        uniq = set(BD)
        sameBD = len(BD)-len(uniq)
    grpsize.append(len(BD))

Max = max(grpsize)+1
```

Let's plot two stacked histograms

Plot the group-size distribution
Normalise bar height to get probability

```
fig,(ax1, ax2) = plt.subplots(2,1)
bins = np.arange(0, Max)-0.5
prob, a1, a2 = ax1.hist(grpsize, bins,
              density=True, color='purple')
ax1.set_ylabel('Probability distribution')
ax1.set_title('Distribution of group size for'
              ' same birthday to occur')
```

Remove *x* labels of the top histogram

```
plt.subplots_adjust(hspace=0.1)
plt.setp(ax1.get_xticklabels(), visible=False)
```

Plot the cumulative distribution

Show unfilled bars (so we see 'steps')

```
cprob, b1, b2 = ax2.hist(grpsize, bins,
               density=True, cumulative=True,
               histtype='step', color = 'b')
ax2.set_ylabel('Cumulative distribution')
ax2.set_yticks(np.arange(0, 1.01, 0.1))
ax2.set_ylim([0,1])
ax2.set_xlabel('Group size')
```

Add a red horizontal dashed line
Read the position where prob. > 0.5
Add the reading to the figure

```
ax2.axhline(y=0.5, color='r', linestyle=':')
ans = np.searchsorted(cprob, 0.5)
ax2.text(41, 0.52, 'Pr(>0.5) when group size'
              r'$\geq$' f'{ans}')
```

Common settings for both plots

```
for X in [ax1, ax2]:
    X.set_xticks(range(0, Max, 5))
    X.set_xlim(0, Max-2)
    X.grid('on')

plt.show()
```

7.7 The Monty Hall problem

> Monty Hall hosts a TV gameshow in which a contestant wins whatever is behind a
> door that is chosen amongst 3 closed doors. A car is behind one door and a goat is
> behind each of the other 2 doors.
>
> Once the contestant has chosen a door, Monty does not open it but instead opens a
> different door to reveal a goat. Monty then gives the contestant the choice of staying
> with the original choice or switching to the remaining unopened door.
>
> Should the contestant stay or switch to maximise the chance of winning the car?

This famous problem was posed in 1975 [183] by the American statistician Steve Selvin,
who also described its notoriously counterintuitive solution. Monty Hall was a real TV
personality who hosted a gameshow called 'Let's Make a Deal' with elements of the scenario
(but not exactly) as described by Selvin.

Most people would intuitively think that the chance of winning the car after the goat is
revealed is 50%, and so switching would not make a difference.

But imagine scaling the problem up to, say, 100 doors and Monty revealing 98 goats.
You would be quite certain that the car is behind the remaining unopened door rather than
the one you picked initially (unless you were incredibly lucky). So by switching we are
making use of extra information provided by Monty, giving us an advantage of winning the
car. We wish to quantify this advantage.

There are some implicit rules (*i.e.* the assumptions) that determine which door Monty
can open immediately after the contestant has chosen a door.

- Monty knows which door hides the car, but does not open it.
- Monty does not open the door initially chosen by the contestant.
- Monty always offers a switch.
- If more than one 'goat' doors could be opened, Monty chooses one randomly.

Before we perform the simulations, let's see how the problem could be tackled with
probability theory.

Bayes' Theorem

We wish to calculate the probability that the car is behind the chosen door *given* that a goat
is revealed behind another door. It seems sensible to consider conditional probabilities.

First, note that from the definition of conditional probability (7.2), we have

$$\Pr(A|B) = \frac{\Pr(A \cap B)}{\Pr(B)}, \qquad \Pr(B|A) = \frac{\Pr(B \cap A)}{\Pr(A)}.$$

Equating $\Pr(A \cap B) = \Pr(B \cap A)$ we have

Theorem 7.1 (*Bayes' Theorem*)

$$\Pr(A|B) = \frac{\Pr(B|A)\,\Pr(A)}{\Pr(B)}.$$

Revd. Thomas Bayes (1702–1761) was an English clergyman who laid the foundation to Bayesian statistics, a subject which lies at the heart of modern data analysis and machine learning. We will discuss Bayesian statistics in §8.9.

Now suppose that the contestant has chosen a door (call it door 1) and Monty reveals a goat behind another door (call it door 2). Let C_n be the event that the car is behind the door n. Let G_n be the event that Monty reveals a goat behind door n.

Using Bayes' Theorem, we have the following expression for the probability of winning the car if the contestant were to switch to door 3.

$$
\begin{aligned}
\Pr(C_3|G_2) &= \frac{\Pr(G_2|C_3)\Pr(C_3)}{\Pr(G_2)} \\
&= \frac{\Pr(G_2|C_3)\Pr(C_3)}{\Pr(G_2|C_1)\Pr(C_1) + \Pr(G_2|C_2)\Pr(C_2) + \Pr(G_2|C_3)\Pr(C_3)} \\
&= \frac{1 \times \frac{1}{3}}{\frac{1}{2} \times \frac{1}{3} + 0 \times \frac{1}{3} + 1 \times \frac{1}{3}} = \frac{2}{3}.
\end{aligned}
$$

The second line uses the law of total probability (Prop. 7.3). The third line uses the rules and assumptions of the game. For example, $\Pr(G_2|C_3) = 1$ because the contestant picked door 1 whilst door 3 has the car, so Monty is forced to reveal the goat behind door 2. However, $\Pr(G_2|C_1) = \frac{1}{2}$ because, with the car behind door 1, Monty could reveal the goat behind either door 2 or door 3.

The probability of winning the car without switching is the number of cars divided by the number of doors, which equals $\frac{1}{3}$. It is instructive to confirm this with Bayes' Theorem as follows.

$$
\begin{aligned}
\Pr(C_1|G_2) &= \frac{\Pr(G_2|C_1)\Pr(C_1)}{\Pr(G_2)} \\
&= \frac{\Pr(G_2|C_1)\Pr(C_1)}{\Pr(G_2|C_1)\Pr(C_1) + \Pr(G_2|C_2)\Pr(C_2) + \Pr(G_2|C_3)\Pr(C_3)} \\
&= \frac{\frac{1}{2} \times \frac{1}{3}}{\frac{1}{2} \times \frac{1}{3} + 0 \times \frac{1}{3} + 1 \times \frac{1}{3}} = \frac{1}{3}.
\end{aligned}
$$

Thus, switching gives us an advantage by *doubling* the chance of winning the car.

n-door generalisation

Suppose there are n doors and Monty reveals g goats. A car is behind one of the doors.

A similar calculation shows that the probability of winning a car upon switching to an unopened door is

$$
\frac{n-1}{n(n-g-1)}. \tag{7.21}
$$

Since the no-switch probability of winning is $\frac{1}{n}$, we see that switching improves the winning probability by a factor of $\frac{n-1}{n-g-1} > 1$.

For example, if a single goat is revealed ($g = 1$), the winning probability is $\frac{n-1}{n(n-2)}$. For large n, this probability is comparable with the no-switch probability of $\frac{1}{n}$. This makes intuitive sense: with so many doors, switching makes little difference.

On the other hand, if all but one goat are revealed ($g = n - 2$), the winning probability simplifies to $1 - \frac{1}{n}$. For large n, this probability approaches 1, agreeing with our 100-door discussion earlier. Switching essentially guarantees winning.

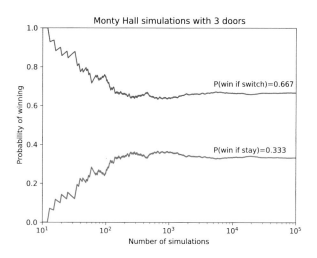

Fig. 7.10: Result from 10^5 simulations of the Monty Hall problem with 3 doors. The long-term behaviour agrees with our calculations using Bayes' Theorem (2/3 and 1/3).

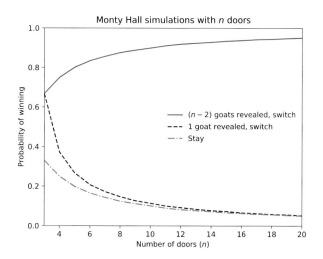

Fig. 7.11: Result from 10^5 simulations of the Monty Hall problem with n doors (up to $n = 20$). The curves are the probability of winning the car if – a) a switch is made when all but one goats have been revealed by Monty (blue solid line), b) a switch is made when 1 goat is revealed (dashed black line), c) no switch is made (dash-dot red line).

Simulation

The code `montyhall.ipynb` simulates a large number of games and keeps track of the probabilities of winning with and without switching doors. The result is shown in fig. 7.10 which shows that the switch and no-switch probabilities get close to $\frac{2}{3}$ and $\frac{1}{3}$ respectively (after about 1000 games). If you run this code yourself, you will obtain a slightly different graph, but the long-term behaviour is the same.

It is worth noting the key syntax in the code: the function `random.choice` is used to randomly pick an element from a list (*i.e.* pick a door).

The code is easily modified to simulate a scenario with multiple doors. Fig. 7.11 shows the long-term behaviour when the number of doors (n) is varied. The solid blue line shows the winning probability if all but one goat are revealed, showing that the probability approaches 1 as expected. The dashed black line shows the case when only one goat is revealed, showing that the winning probability approaches $1/n$, *i.e.* it approaches the no-switch scenario shown as the red (dash-dot) line.

In exercise 4 you will perform simulations to investigate other generalisations of the Monty Hall problem. For even more outrageous variations, see [176].

> ### DISCUSSION
>
> - **Bayes' Theorem in the real world**. Bayes' Theorem can help us make sense of everyday probabilities in the face of evidence. This is particularly important when making medical decisions based on empirical evidence.
>
>   A classic example is in understanding the reliability of, say, a Covid test kit. Suppose that at a certain time, 1% of the population are actually infected with Covid. If you are tested positive for Covid using a home-test kit that says it is "99% accurate", what is the probability that you actually have Covid?
>
>   Most people would instinctively say that you *almost certainly* have Covid. But let's see what Bayes' Theorem says.
>
>   Let $\Pr(\text{Covid})$ be the probability that you are actually infected with Covid and $\Pr(\text{No Covid}) = 1 - \Pr(\text{Covid})$.
>
>   Let $\Pr(+)$ be the probability that the test kit shows a positive result (suggesting that you probably have Covid) and similarly let $\Pr(-) = 1 - \Pr(+)$ (we ignore inconclusive results). Since the test is 99% accurate, we have
>
> $$\Pr(+|\text{Covid}) = 0.99 \qquad \text{and} \qquad \Pr(-|\text{No Covid}) = 0.99$$
>
> $$\implies \Pr(\text{Covid}|+) = \frac{\Pr(+|\text{Covid})\,\Pr(\text{Covid})}{\Pr(+)} \qquad \text{(Bayes' Theorem)}$$
>
> $$= \frac{\Pr(+|\text{Covid})\,\Pr(\text{Covid})}{\Pr(+|\text{Covid})\,\Pr(\text{Covid}) + \Pr(+|\text{No Covid})\,\Pr(\text{No Covid})}$$
>
> $$= \frac{0.99 \times 0.01}{0.99 \times 0.01 + 0.01 \times 0.99} = 0.5.$$

The answer seems surprisingly low, but this is the same kind of mathematical counter-intuition that arises in the Monty Hall problem. For another surprise: guess the answer for $\Pr(\text{No Covid}|-)$ and then calculate it with Bayes' Theorem.

montyhall.ipynb (for plotting fig. 7.10)	

	```import matplotlib.pyplot as plt```
	 ```from random import choice```
	 ```%matplotlib```

Specify the number of simulations and the number of doors	```sims = 1e5``` ```Ndoors = 3```
	```listsims = range(int(sims))``` ```round1 = range(Ndoors)```

Scoreboard for each strategy
 Keep track of how winning probabilities
 evolve with the number of simulations
	```staywin, switchwin = 0, 0```
	 ```Pstay = []```
	 ```Pswitch = []```

Let's Make a Deal!
 The winning door which hides a car
 Our initial pick
	```for i in listsims:```
	 ``` car = choice(round1)```
	 ``` pick1 = choice(round1)```

Doors that Monty can pick
 	
 Goat door opened by Monty
	``` monty = [n for n in round1```
	 ``` if n!= car and n!= pick1]```
	 ``` goat = choice(monty)```

Doors available for switching to
 	
 Final door that we switch to
	``` round2 = [n for n in round1```
	 ``` if n!= pick1 and n!=goat]```
	 ``` pick2 = choice(round2)```

Add win (if any) to a scoreboard	``` if pick1==car: staywin += 1``` ``` if pick2==car: switchwin +=1```

Probability of winning for each strategy	``` p1 = staywin/(i+1)```
	 ``` p2 = switchwin/(i+1)```
	 ``` Pstay.append(p1)```
	 ``` Pswitch.append(p2)```

Plot evolving probabilities	```plt.semilogx(listsims, Pstay, 'r',``` ``` listsims, Pswitch, 'b')```

Display probability limits (3 DP)	```plt.text(5e3, 0.36, f'Pr(win if stay)={p1:.3}')```
	 ```plt.text(5e3, 0.7,f'Pr(win if switch)={p2:.3}')```
	 ```plt.xlim([10, sims])```
	 ```plt.ylim([0,1])```
	 ```plt.xlabel('Number of simulations')```
	 ```plt.ylabel('Probability of winning')```
	 ```plt.title('Monty Hall sims. with 3 doors')```
	 ```plt.grid('on')```
	 ```plt.show()```

7.8 The Normal distribution

The figure below shows a vertical board with rows of pegs through which tiny metal balls fall through and are collected at the bottom in vertical columns.
Show that when the number of pegs and balls are large, the height of the columns follows the normal distribution.

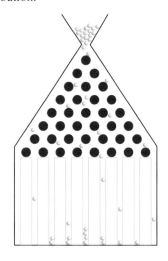

This device was invented by the English polymath *Francis Galton* (1822–1911). His unusually wide range of academic contributions include biology, anthropology, mathematics and statistics amongst other fields. Sadly, his legacy is marred by his strong advocacy of eugenics (having himself coined the term).

Galton's device can be modelled in a simplified way as follows. When a ball hits a peg, it either goes to left or right by a fixed distance Δx with equal probability, then falls vertically to hit the next peg. We assume that each ball encounters N pegs before falling into one of the collecting columns at the bottom. All bounces are disregarded.

At each encounter with a peg, the probability of going left or right is a Bernoulli trial (*i.e.* a coin flip) with $p = \frac{1}{2}$. Let's take take the outcome 0 to be 'left' and '1' to be right. In §7.5, we saw that the sum of N Bernoulli trials follows the binomial distribution (7.16). The probability of a ball making k moves to the right on its journey down is therefore

$$\Pr(k \text{ right moves}) = \binom{N}{k}\left(\frac{1}{2}\right)^k \left(\frac{1}{2}\right)^{N-k} = \frac{N!}{(N-k)!k!}\left(\frac{1}{2}\right)^N.$$

Let $x = 0$ correspond to the centre of the board. If a ball moves k times to the right, then it must move $N - k$ times to the left (assuming no balls get stuck). Thus the final distance it lands a the bottom of the board is $x = k\Delta x - (N - k)\Delta x = (2k - N)\Delta x$. We wish to know the distribution of x as we release more balls from the top.

Simulation

The code *normal.ipynb* simulates the board with N rows of pegs and N_{balls} balls. The resulting distributions of the final landing distance x with $N_{balls} = 20000$ and $\Delta x = 1$ are shown in fig. 7.12, assuming $N = 10, 50, 100$ and 500.

The (normalised) histograms follow the binomial distribution, which is of course a discrete distribution. But as N increases the distribution is well approximated by the bell-like continuous curve shown in red.

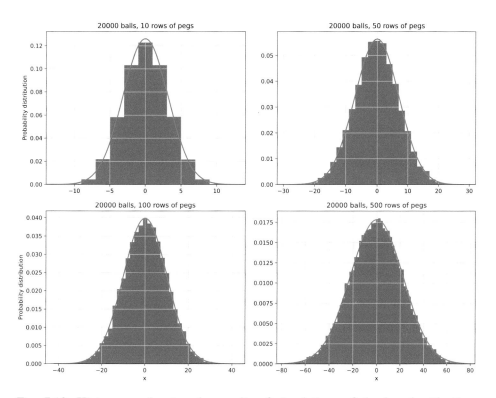

Fig. 7.12: Histograms showing the results of simulations of the board with $N = 10, 50, 100, 500$ rows of pegs, using 20000 balls. The resulting distribution of the final landing distance x is the binomial distribution. The red curve in each panel shows the approximation by the normal distribution $\mathcal{N}(0, N)$.

Approximation by normal distribution

The red curves in fig. 7.12 show pdf of the *normal distribution*, denoted $\mathcal{N}(\mu, \sigma^2)$. The normal distribution, also known as the *Gaussian distribution*, is a continuous probability distribution with pdf

$$f(x) = \frac{1}{\sqrt{2\pi\sigma^2}} e^{\frac{-(x-\mu)^2}{2\sigma^2}}, \tag{7.22}$$

where μ is the mean and σ^2 is the variance of the distribution. The distribution is a bell-shaped curve which is symmetrical about $x = \mu$ (where it peaks). The pdf approaches 0 as $x \to \pm\infty$.

The fact that the normal distribution is a good approximation to the binomial distribution for large N is due to the following theorem (see [62] for proof).

Theorem 7.2 (De Moivre-Laplace) *When N is large and k is around Np, we have the approximation*

$$\binom{N}{k} p^k q^{N-k} \sim \frac{1}{\sqrt{2\pi\sigma^2}} e^{\frac{-(k-\mu)^2}{2\sigma^2}},$$
$$\text{where} \quad \mu = Np \quad \text{and} \quad \sigma^2 = Npq.$$

In other words, for a large number of trials N, the binomial distribution $B(N, p)$ approaches the normal distribution $\mathcal{N}(\mu, \sigma^2)$ with the same mean and variance as those of $B(N, p)$.

A special case of this theorem was discovered by *Abraham de Moivre* (1667–1754), a French mathematician who fled religious persecution in France into exile in England. He was a close friend of Newton and is remembered for his contributions to probability and the development of complex numbers. De Moivre's initial results on the large N limit were later generalised by *Pierre-Simon, Marquis de Laplace* (1749–1827), a prolific French scientist who made profound contributions to probability, mechanics and astronomy. Exercise 5 investigates the accuracy of the de Moivre-Laplace approximation as N and k vary.

The theorem tells us about the asymptotic distribution of k, the number of moves to the right. However, we are interested in the distribution of the landing distance x, so a little conversion is needed. Earlier, we found that

$$x = (2k - N)\Delta x. \tag{7.23}$$

Assuming $p = q = \frac{1}{2}$, according to Theorem 7.2, the expected number of right moves $\mu := E[k] = \frac{N}{2}$. Taking the mean of eq. 7.23, we have $\mu_x = (2\mu - N)\Delta x = 0$. This is not surprising since overall we expect as many right moves as left moves, resulting in the mean $\mu_x = 0$.

Next, to find the variance σ_x^2, Theorem 7.2 gives $\sigma^2 = N/4$, so $E[k^2] = (E[k])^2 + \sigma^2 = \frac{N^2+N}{4}$. Using eq. 7.23, we find

$$\sigma_x^2 = E[x^2] - (E[x])^2 = (4E[k^2] - 4NE[k] + N^2)(\Delta x)^2 = N(\Delta x)^2.$$

In conclusion, the red curves in fig. 7.12 are the pdfs of the normal distribution $\mathcal{N}(0, N(\Delta x)^2)$, which provides a good approximation to the binomial distribution when N is large.

Let's do a quick accuracy test of this approximation. For example, with $\Delta x = 1$ and $N = 50$ pegs, the probability that a ball lands at the centre of the board (*i.e.* requiring $k = 25$ right moves) is

$$\Pr(k = 25) = \binom{50}{25}\left(\frac{1}{2}\right)^{50} \approx 0.1123.$$

On the other hand, using Theorem 7.2 with $\mu = 25$ and $\sigma^2 = 25/2$, we find

$$\Pr(k = 25) \approx \frac{1}{5\sqrt{\pi}} \approx 0.1128,$$

which seems reasonably accurate. This approximation says that the height of the central histogram bar is roughly the height of the peak of the normal distribution.

Alternatively, we could approximate the area of the central histogram bar (centred at $x = 0$) as the area under the pdf of the normal distribution $\mathcal{N}(0, 50)$ from $x = -1$ to $x = 1$. This gives

$$\Pr(-1 < x < 1) = \frac{1}{\sqrt{100\pi}} \int_{-1}^{1} e^{\frac{-x^2}{100}} \, \mathrm{d}x \approx 0.1125,$$

which is a better approximation. This is an example of *continuity correction*, which broadly refers to measures that can be used to improve the accuracy when using a continuous distribution to approximate a discrete distribution.

> **DISCUSSION**

- **The Central Limit Theorem**. Theorem 7.2 states that the sum of N independent Bernoulli random variables converges to the normal distribution. In fact, in §8.3, we will show that the sum of N independent random variables from *any* probability distribution also converges to the normal distribution. This is the *Central Limit Theorem*, one of the most fundamental results in probability and statistics. The theorem also holds even if the independent random variables are drawn from different distributions (satisfying certain conditions).

- **The normal distribution in the real world**. Many observables in the real world are the net results of various factors drawn from different probability distributions. Thanks to the Central Limit Theorem, the normal distribution appears to underlie many real-world phenomena. For example:

  – Weight distribution of babies at birth (see [215]);
  – Average height of adults [177];
  – Global temperature anomaly [90].

normal.ipynb (for plotting fig. 7.12)	
| | ```python
import numpy as np
import matplotlib.pyplot as plt
from scipy.stats import bernoulli
%matplotlib
``` |
| Number of balls<br>Number of rows of pegs | ```python
Nballs = 25000
Nrows = 500
``` |
| Each ball-peg encounter is a Bernoulli trial<br>Ball moves left or right by distance dx<br>For storing final position of each ball | ```python
Ber = bernoulli(0.5)
dx = 1
X = []
``` |
| Left/right (0/1) moves for each ball<br>Number of moves to the right<br>Final position of each ball (eq. 7.23)<br>Store it | ```python
for i in range(Nballs):
    D = Ber.rvs(Nrows)
    rights = sum(D)
    x = (2*rights - Nrows)*dx
    X.append(x)
``` |
| Note extreme distances | ```python
Max = max(X) + 4*dx
Min = min(X) - 4*dx
``` |
| $\sigma^2$<br>Normal-distribution fit $\mathcal{N}(0,\ N(\Delta x)^2)$ | ```python
xN = np.linspace(Min, Max,100)
sig2 = Nrows*dx**2
yN = np.exp(-0.5*xN**2/sig2)/\
     np.sqrt(2*np.pi*sig2)
``` |
| Centre bars at multiples of dx<br>Plot histogram of column height | ```python
fig, ax = plt.subplots(1,1)
bins = np.arange(Min, Max, 2*dx)-dx
ax.hist(X, bins, density=True, color='teal')
ax.set_xlabel('x')
ax.set_ylabel('Probability distribution')
ax.set_title(
    f'{Nballs} balls, {Nrows} rows of pegs')
ax.set_xlim([-round(Max),round(Max)])
``` |
| Overlay normal-distribution fit | ```python
ax.plot(xN, yN, 'r', linewidth=2)
plt.grid('on')
plt.show()
``` |

7.9 The Poisson distribution

Place N dots randomly within a grid of 10×10 squares. Count how many dots fall in each square.
Find the probability distribution of the number of the dots in a square.

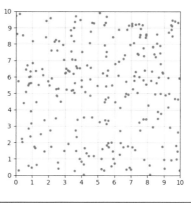

We are interested in the probability of the occurrence of an event per some spatial unit. We assume that the occurrences are random and independent of one another.

Let's start by simulating the problem.

Simulation

In the code `poisson.ipynb`, dots are randomly distributed in a 10×10 square grid by choosing `Ndots` pairs of real numbers (x, y). Taking the integer parts of (x, y) tells us the coordinates of the bottom left corner of the square that each dot falls into. Each simulation then produces a grid of dot counts, and therefore a histogram showing the distribution of dot counts per square. We repeat the simulation 500 times.

Fig. 7.13 shows the histograms of the normalised dot counts per square for `Ndots` = $100, 250, 500, 1000$. The normalisation is obtained by dividing the actual dot count by the total number of dots, and thus the height of the bars can be interpreted as the probability of finding `Ndots` dots in a square.

The results show a consistently skewed (asymmetric) distribution. More specifically, the distribution is *positively skewed*, meaning that it leans towards the left with a long tail. Number counts are more concentrated towards lower values.

The Poisson distribution

Consider events that are random, independent and mutually exclusive (*i.e.* two events cannot occur simultaneously). The probability that k such events occur over a fixed time or space interval (such as those that we see in fig. 7.13) is given by the distribution

$$\Pr(k \text{ events occur}) = \frac{\lambda^k}{k!} e^{-\lambda},\tag{7.24}$$

for some parameter $\lambda > 0$. We call this distribution the *Poisson distribution*, after the French mathematician *Siméon-Denis Poisson* (1781–1840). See [89] for an interesting account of the history of the Poisson distribution. Poisson is remembered today for his contributions to probability, complex analysis and various areas of mathematical physics (recall that we met Poisson's equation in §3.9).

The Poisson distribution is discrete and is determined by a single parameter λ. Calculating the mean μ and variance σ^2, we find that

$$\mu = \sigma^2 = \lambda.$$

This means that λ is simply the mean number of events occuring over a time or space interval.

In our simulations, the mean number of dots per square is $\mu = \texttt{Ndots}/100$. Equating this to λ and plotting the Poisson distribution (black lines) on top of the histograms in fig. 7.13, we observe a very good fit to the data.

Note that we join up the points for ease of visual comparison with the histograms – keeping in mind that the Poisson distribution is of course discrete.

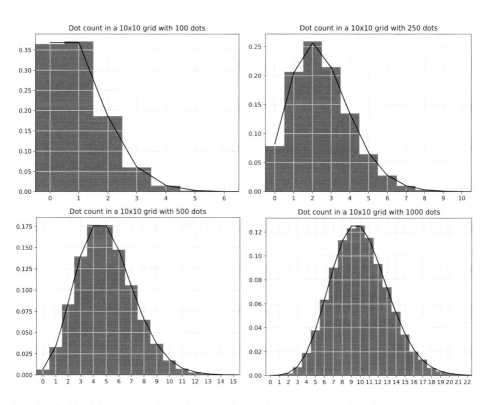

Fig. 7.13: The histograms show the probability that N dots are found in a square within a 10×10 grid, assuming that the dots are distributed randomly in the grid. The plots show the probability for $N = 100, 250, 500$ and 1000. Each figure was obtained from 500 simulations using the code `poisson.ipynb`. The black line shows the Poisson distribution with $\lambda = N/100$. Discrete points are joined up for visual clarity.

DISCUSSION

- **The Poisson distribution in the real world**. The Poisson distribution is useful for modelling the probability of rare events occurring over a time or space interval. Real-world applications include:

  - The number of bombs (per square km) dropped in London during WWII (see [44]);
  - The number of heart attacks per week in a certain area [188];
  - The number of distant quasars observed per square degree of the sky [198].

- **Relationship between Poisson and binomial distributions**. The binomial distribution for rare events is well approximated by the Poisson distribution. More precisely, if $n \to \infty$ and $p \to 0$ whilst $np = \lambda$ (constant), then for $k = 0, 1, 2, \ldots$

$$\binom{n}{k} p^k (1-p)^{n-k} \to \frac{\lambda^k}{k!} e^{-\lambda}. \tag{7.25}$$

In practice, this approximation is useful when p is small, n large, and np is moderate.

- **Relationship between Poisson and normal distributions**. We observe in fig. 7.13 that as the number of dots increases (e.g. when `Ndots = 1000`), the asymmetry of the distribution is less pronounced, and the distribution resembles the normal distribution. This is due to the following limit: when λ is large, we have

$$\frac{\lambda^k}{k!} e^{-\lambda} \approx \frac{1}{\sqrt{2\pi\lambda}} e^{-\frac{(x-\lambda)^2}{2\lambda}}, \tag{7.26}$$

where the RHS is the pdf of the normal distribution $\mathcal{N}(\lambda, \lambda)$. Combining this observation with (7.25), we deduce that the normal distribution is a good approximation to the binomial approximation for large N, which is of course the de Moivre-Laplace theorem (Theorem 7.2).

In exercise 6, you will explore the accuracy of approximations (7.25) and (7.26).

- **The exponential distribution**. If the number of rare events occurring over a time or space interval follows the Poisson distribution, then the interval between two successive occurrences follows the *exponential distribution*

$$\Pr(\text{interval length } = x) = \begin{cases} \frac{1}{\beta} e^{-x/\beta} & x > 0 \\ 0 & x \le 0, \end{cases} \tag{7.27}$$

with parameter $\beta > 0$. In exercise 7, you will explore the probability of events occurring over a time interval and show that the inter-arrival times follow the exponential distribution.

poisson.ipynb (for plotting fig. 7.13)

|  | ```python
import numpy as np
import matplotlib.pyplot as plt
from math import factorial
%matplotlib
``` |
Number of simulations	`sims = 500`
Number of dots in each simulation	`Ndots = 100`
There are grid×grid squares	`grid = 10`
Initialise random number generator	`rng = np.random.default_rng()`
List of dot counts in the squares	`Tally= []`
In each simulation	`for sim in range(sims):`
Initialise grid of number counts	`    Count = np.zeros((grid,grid))`
Scatter the dots randomly, storing their	`    dotx = rng.uniform(0, grid, Ndots)`
(x, y) coords	`    doty = rng.uniform(0, grid, Ndots)`
At each scattered dot...	`    for x,y in zip(dotx,doty):`
Get the bottom left coord. of the square	`        n, m = int(x), int(y)`
Add to the number count in the grid	`        Count[m,n] += 1`
Flatten the grid to a long 1D array	`    C = Count.flatten()`
Store each simulation result	`    Tally = np.concatenate((Tally,C))`
Highest number count	`Max = max(Tally)`
	`xbins = np.arange(Max+2)`
Centre bars at integer values	`bins = xbins-0.5`
Plot histogram of normalised counts	`prob = plt.hist(Tally, bins=bins,`
	`                density=True, color='deeppink')`
Mean number of dots per square	`mean = Ndots/(grid**2)`
Poisson distribution fit	`Poisson =[mean**x*np.exp(-mean)/\`
	`                factorial(int(x)) for x in xbins]`
	`plt.title(f'Dot count in a {grid}x{grid} grid'`
	`            f' with {Ndots} dots')`
Plot Poisson distribution (joining	`plt.plot(xbins, Poisson, 'k')`
discrete points with black line)	`plt.xticks(xbins)`
	`plt.xlim([-0.5, Max+0.5])`
	`plt.grid('on')`
	`plt.show()`
```

7.10 Monte Carlo integration

Evaluate the following integrals.

a) $\displaystyle\int_0^1 \sqrt{1-x^2}\,dx,$ b) $\displaystyle\int_0^1 \frac{\tan^{-1}x}{x}\,dx,$ c) $\displaystyle\int_0^\infty e^{-x}\ln x\,dx,$

d) $\displaystyle\int_0^1 \int_0^1 \int_0^1 \left\{\frac{x}{y}\right\}\left\{\frac{y}{z}\right\}\left\{\frac{z}{x}\right\}\,dx\,dy\,dz,$ where $\{x\}$ denotes the fractional part of x.

Monte Carlo methods (named after the famous casino) use random numbers and their distributions to solve problems that would otherwise be too difficult or time-consuming to tackle. Here we will investigate how random numbers can help us perform numerical integration.

$\boxed{(a)}$ The integral $\int_0^1 \sqrt{1-x^2}\,dx$ represents the area of a quarter circle of radius 1, so we know the exact answer is $\pi/4 = 0.785398\ldots$. One way to evaluate this area is to pick many random points in the unit square $[0, 1]^2$ and work out the fraction of points within the circular sector. In other words, we throw darts randomly at the unit square and compute the ratio

$$\frac{\text{Darts landing below } y = \sqrt{1-x^2}}{\text{Total number of darts}}.$$

Fig. 7.14 show the result when we throw $N = 10^3$, 10^4 and 10^5 darts at the unit square. In exercise 8, you will quantify the error in this approximation and show that it shrinks like $N^{-1/2}$.

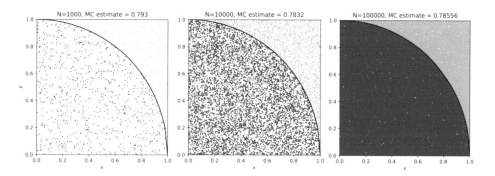

Fig. 7.14: Monte Carlo estimates of $\int_0^1 \sqrt{1-x^2}\,dx$ using the darts method. The area under the curve is approximately the number of darts landing inside the area (shown in blue) divided by the total number of darts.

$\boxed{(b)}$ The darts method requires inequality tests to be performed with every dart used. Here is a more efficient Monte Carlo method that does not require explicit inequality tests.

Consider the integral

$$\int_a^b f(x)\,dx.$$

Suppose that on (a, b), $f(x)$ has mean value $\langle f(x)\rangle$. The area represented by the integral can be expressed as the area of the rectangle with base $(b-a)$ and height $\langle f(x)\rangle$.

$$\int_a^b f(x)\,dx = (b-a)\langle f(x)\rangle.$$

We can find a numerical estimate of the mean $\langle f(x)\rangle$ by sampling over N values of x_i, where N is large:

$$\langle f(x)\rangle \approx \frac{1}{N}\sum_{i=1}^{N} f(x_i).$$

Therefore, we have obtained the fundamental equation for 1D *Monte-Carlo integration*.

$$\int_a^b f(x)\,dx = (b-a)\langle f(x)\rangle \approx \frac{b-a}{N}\sum_{i=1}^{N} f(x_i). \tag{7.28}$$

Let's apply this method to the integral $\int_0^1 (\tan^{-1} x)/x\,dx$, drawing $N = 10^7$ values of x_i from the uniform distribution $U(0, 1)$. Here's the code.

```
import numpy as np
rng = np.random.default_rng()

f = lambda x: np.arctan(x)/x
xi = rng.random(int(1e7))
np.mean(f(xi))
```

Output: 0.9159685092587398

Of course, every run will give a slightly different answer. But how accurate is this answer? Recall the geometric series

$$\frac{1}{1+x^2} = 1 - x^2 + x^4 - x^6 + \cdots$$

valid for $|x| < 1$. Integrating both sides from 0 to x gives

$$\tan^{-1} x = x - \frac{x^3}{3} + \frac{x^5}{5} - \frac{x^7}{7} + \cdots$$

Dividing by x and integrating again from 0 to 1 gives

$$\int_0^1 \frac{\tan^{-1} x}{x}\,dx = \int_0^1 \left(1 - \frac{x^2}{3} + \frac{x^4}{5} - \frac{x^6}{7}\right) + \cdots$$

$$= 1 - \frac{1}{3^2} + \frac{1}{5^2} - \frac{1}{7^2} + \cdots$$

$$= G,$$

where the *Catalan's constant* $G = 0.91596559\ldots$. *Eugène Charles Catalan* (1814–1894) was a Belgian mathematician whose name we now associate, in addition to the constant

G, with *Catalan's numbers* in combinatorics, and *Catalan's conjecture* (now proven) in number theory.

Knowing the exact value of the integral allows us to plot the magnitude of the fractional error in the Monte Carlo integration as a function of N (the number of random points sampled). In fig. 7.15, each blue dot is a Monte Carlo estimate of the integral.

We have also plotted the best-fit line (in red) through the log data points (`logx`, `logy`). In Python, we can do this by finding the best-fit polynomial using the following syntax.

```
Finding the best-fit polynomial with NumPy
from numpy.polynomial import Polynomial as P
poly = P.fit(logx, logy, 1).convert()
```

Here `poly` is a polynomial object of order 1 (a line), calculated using the least-square method (see mathematical details in §8.6). The polynomial can be called as a function, and also has an attribute `poly.coef` – an array whose ith element is the coefficient of x^i.

The code reports that the gradient of the line is approximately -0.5. This tells us that the error shrinks like $N^{-1/2}$.

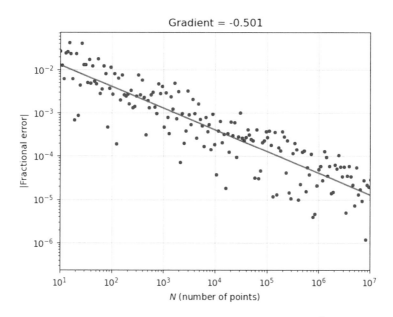

Fig. 7.15: The magnitude of the fractional error in the estimate of $\int_0^1 (\tan^{-1} x)/x \, dx$ using Monte Carlo integration (eq. 7.28) plotted as a function of the number of sampled points N. Note the log scales. The gradient of the best-fit line tells us that the error shrinks like $N^{-1/2}$.

The magnitude of the fractional error is a long way from machine epsilon, unlike the numerical integration methods discussed in chapter 2. For instance, recall that with N points, Simpson's Rule shrinks like N^{-4} which is a lot faster than $N^{-1/2}$.

Whilst those numerical integration methods converge faster, they do require the number of sampled points to scale with the dimension of the problem. For example, to evaluate a triple integral with quad, a 3D grid of N^3 points are required to maintain the high accuracy. However, the accuracy of Monte Carlo integration using N points is always $\approx N^{-1/2}$

regardless of dimension. In mathematical speak, we say that Monte Carlo integration does not suffer from the *curse of dimensionality*.

(c) Monte Carlo integration can also be adapted to evaluate improper integrals. Let's look at two methods to evaluate $I := \int_0^\infty e^{-x} \ln x \, dx$.

Method 1 - a new variable: We can transform the integral so that the domain $[0, \infty)$ is mapped to, say, $[0, 1]$. One method that we discussed in §2.8 was to split the interval $[0, \infty) = [0, 1] \cup [1, \infty)$ and use the variable $u = 1/x$ to map $[1, \infty)$ to $[0, 1]$. The result of this transformation is given in eq. 2.17, which, in this case, reads

$$I = \int_0^1 \left(e^{-x} - \frac{e^{-1/x}}{x^2} \right) \ln x \, dx. \tag{7.29}$$

This can then be evaluated with the same Monte Carlo code as in part (b).

Method 2 - a new probability distribution: Recall that if x_i is drawn from a probability distribution $p(x)$, then, for a given function f, the mean value of $f(x)$ is

$$\langle f(x) \rangle = \int f(x) p(x) \, dx,$$

Comparing this with the integral at hand, we see that $\int_0^\infty e^{-x} \ln x \, dx$ can be considered the mean value of $\ln x$, where x is drawn from $p(x) = e^{-x}$ with $x \in (0, \infty)$. Fortunately, we have seen precisely such a probability distribution: it is the exponential distribution (7.27) with $\beta = 1$.

Conveniently, NumPy has such a ready-to-use exponential distribution from which we can easily sample random numbers. In the code below, the integral is evaluated using 10^7 numbers drawn from the exponential distribution with $\beta = 1$.

```
import numpy as np
rng = np.random.default_rng()
xi = rng.exponential(1, int(1e7))
np.mean(np.log(xi))
```

Output: -0.577228988705456

An eagle-eyed reader might recognise this as being awfully close to $-\gamma$, where $\gamma = 0.57721566\ldots$, the Euler-Mascheroni constant which we met in our discussion of the harmonic series (§1.5). Indeed the exact answer turns out to be $-\gamma$. The proof, using clever integration trickery, can be found in [153] (eq. 5.4.3).

The accuracy for both methods I and II is $O(N^{-1/2})$, which you will verify in exercise 9. In the Discussion, we will justify where $N^{-1/2}$ comes from.

(d) In practice, Monte Carlo integration is normally used to evaluate multiple integrals where traditional quadratures suffer from the curse of dimensionality. The same fundamental equation works for any dimensions:

$$\int_\Omega f(\mathbf{x}) \, d\mathbf{x} \approx \frac{\text{Vol}(\Omega)}{N} \sum_{i=1}^{N} f(\mathbf{x}_i) + O(N^{-1/2}), \tag{7.30}$$

even if f is a discontinuous function such as the integrand in this question.

Here is the code which evaluates $\int_0^1 \int_0^1 \int_0^1 \{x/y\}\{y/z\}\{z/x\}\,dx\,dy\,dz$, using 10^7 sampled points. Note how the modulo operator % is used to find the fractional part of a number.

```
import numpy as np
rng = np.random.default_rng()

f = lambda x,y,z: (x/y %1)*(y/z %1)*(z/x %1)
N = int(1e7)
xi = rng.random(N)
yi = rng.random(N)
zi = rng.random(N)
np.mean(f(xi, yi, zi))
```

Output: 0.09585272990649539

Quite shockingly, in [208] (problem 49) the exact answer was obtained (for an even more general problem) as $1 - \frac{3}{4}\zeta(2) + \frac{1}{6}\zeta(3)\zeta(2) \approx 0.09585017\ldots$, where ζ is the Riemann zeta function. The working only requires undergraduate mathematics and is a beautiful feat of integration.

Finally, it seems daunting to think about how the function $f(x,y,z) = \{x/y\}\{y/z\}\{z/x\}$ *looks* like, but perhaps we can get a sense of its discontinuities by plotting $z = \{x/y\}$ (see fig. 7.20 in exercise 9).

DISCUSSION

- **Why $O(N^{-1/2})$?** We now discuss why the error of Monte Carlo integration shrinks like $N^{-1/2}$.

  Consider the identically distributed random variables X_i ($i = 1, 2, \ldots, N$), where $X_i = f(x_i)$ (*i.e.* the given function evaluated at a random point x_i). From eq. 7.28, observe that the integral $\int_a^b f(x)\,dx$ is a good approximation of the mean of the random variable

$$Y := \frac{b-a}{N}\sum_{i=1}^{N} X_i,$$

in the large N limit.

The error of Monte Carlo integration is determined by (the square root of) the variance of Y. Using basic properties of the variance (including eq. 7.10), we find

$$\mathrm{Var}(Y) = \frac{(b-a)^2}{N^2}\sum_{i=1}^{N}\mathrm{Var}(X_i) = \frac{(b-a)^2}{N^2}\cdot N\sigma^2 = \frac{(b-a)^2\sigma^2}{N},$$

where σ^2 is the variance of each X_i. Taking the square root of the above equation gives the magnitude of the error which scales like $N^{-1/2}$ (and also quantifies the spread of the blue points in fig. 7.15). The proof for higher dimensions is identical - just replace $(b-a)$ by the volume of the integration domain.

- **Monte Carlo integration in graphics rendering**. Monte Carlo integration is an important tool for realistic rendering in animations and games. In particular, it is used in finding approximate solutions to the *rendering equation* [107], a fundamental equation in computer graphics. The rendering equation is an integral equation which expresses the radiance of each pixel by tracing the paths of light rays that reach that pixel. For details of the role of Monte Carlo integration in the rendering techniques of raytracing and pathtracing, see [46, 164].

montecarlo.ipynb (for plotting fig. 7.15)	
For plotting the line of best fit	```import numpy as np
import matplotlib.pyplot as plt
from numpy.polynomial import Polynomial as P
from sympy import S
%matplotlib```
Function f to be integrated	```f = lambda x: np.arctan(x)/x```
List of number of points (integers) to sample (up to 10^7)	```Nlist = np.round(np.logspace(1,7,200))```
Initialise random number generator	```rng = np.random.default_rng()```
For storing the fractional errors	```Err = []```
Catalan's constant (from SymPy)	```G = S.Catalan```
Sample N random points $x_i \in [0, 1)$ $\langle f(x_i) \rangle$ Store |fractional error|	```for N in Nlist:
xi = rng.random(int(N))
est = np.mean(f(xi))
Err.append(float(abs(est/G -1)))```
Best-fit polynomial degree 1 (log data) y coord of best-fit line	```logx = np.log10(Nlist)
logy = np.log10(Err)
poly = P.fit(logx, logy, 1).convert()
yfit = 10**(poly(logx))```
Plot fractional error as blue dots Overlay best fit line in red	```plt.loglog(Nlist, Err, 'b.',
Nlist, yfit, 'r')
plt.xlim(10, 1e7)
plt.xlabel(r'N (number of points)')
plt.ylabel('|Fractional error|')```
Display the gradient (coefficient of x) to 3 dec. pl.	```plt.title(f'Gradient = {poly.coef[1]:.3}')
plt.grid('on')
plt.show()```

7.11 Buffon's needle

> A needle of length ℓ is dropped onto a large sheet of paper with parallel horizontal
> lines separated by distance d. Find the probability that the needle intersects a line.

Georges-Louis Leclerc, Comte de Buffon (1707–1788), a French scientist and natural historian, posed this problem in 1777. Let's first investigate this scenario with Monte Carlo simulation, then discuss the solution using a probabilistic calculation.

Simulation

We begin by fixing the needle length $\ell = 0.05$ and the inter-line distance $d = 0.1$. Clearly the two parameters can be simultaneously scaled up and down without affecting the probability (*i.e.* only the ratio ℓ/d matters). Let's drop our needle onto a 'paper' which is a unit square $[0, 1] \times [0, 1]$. Using a large number of needles in the simulation, the probability of intersection can be estimated by

$$\Pr(\text{intersection}) \approx \frac{\text{Number of needles that intersect a line}}{\text{Total number of needles}}. \qquad (7.31)$$

A visualisation of such a simulation is shown in the top panel of fig. 7.16 in which 100 needles were used. Note that there are 11 potential lines that each needle can intersect (a needle is shown in red if it intersects a line).

To 'throw' a needle, one might start by randomly placing the 'head' of the needle at a point $(x_{\text{head}}, y_{\text{head}})$, where x_{head} and y_{head} are independent random numbers, each one being drawn from $U(0, 1)$. We then sample the angle θ that the needle makes (with respect to the horizontal) from $U(0, 2\pi)$. The 'tail' of the needle is then at

$$(x_{\text{tail}}, y_{\text{tail}}) = (x + \ell \cos \theta, y + \ell \sin \theta).$$

One way to detect a needle-line intersection is to check whether y_{head} and y_{tail} satisfy the condition

$$\lfloor y_{\text{head}}/d \rfloor \neq \lfloor y_{\text{tail}}/d \rfloor. \qquad (7.32)$$

where $\lfloor \cdot \rfloor$ denotes the floor function (see Discussion of §1.3). This condition says that there is an intersection iff the two ends of the needle lie above different horizontal lines.

In the code `buffon.ipynb`, we throw 10^6 needles onto the grid one at a time whilst keeping track of how the fraction (7.31) evolves. The result (with $\ell/d = 0.5$) is shown in the lower panel of fig. 7.16. The fraction fluctuates rather wildly before converging to approximately 0.318 after an impractically large number of needles (which should discourage teachers everywhere from attempting this activity in class).

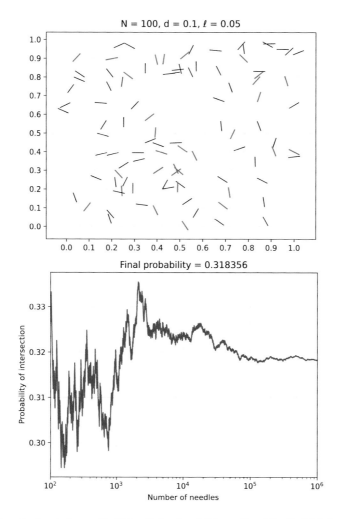

Fig. 7.16: *Top*: A simulation with $\ell/d = 0.5$ using 100 needles, 32 of which intersect a line (shown in red). *Bottom*: The evolution of the fraction of needles that intersect lines (eq. 7.31) as the number of needles increases to 10^6.

Analysis using joint probability distribution

Let's try to make analytical progress on the problem using the idea of joint probability distribution.

Let X and Y be continuous random variables drawn from a single probability distribution called the *joint probability distribution* $f(X, Y)$ (also called a *bivariate* distribution). The probability that $x_1 < X < x_2$ and $y_1 < Y < y_2$ is obtained by integrating the joint distribution as in the one dimensional case, but in this case a double integral is needed.

$$\Pr(x_1 < X < x_2 \text{ and } y_1 < Y < y_2) = \int_{y_1}^{y_2} \int_{x_1}^{x_2} f(x, y) \, \mathrm{d}x \, \mathrm{d}y.$$

We assume that f has been normalised so that $\iint_{\mathbb{R}^2} f(x, y) \, \mathrm{d}x \, \mathrm{d}y = 1$.

Two random variables X and Y are independent if and only if there exist univariate probability distributions g and h such that

$$f(X, Y) = g(X)h(Y). \tag{7.33}$$

See [179] for proof. Let's identify two independent random variables in the Buffon's needle problem.

First, let X be the size of the acute angle that a needle makes with respect to the horizontal (where we will not need to distinguish between the head and tail of a needle). Then,

$$X \sim g(X) := U(0, \pi/2).$$

Let Y be the nearest distance of the midpoint of a needle measured from the nearest horizontal line. Then,

$$Y \sim h(Y) := U(0, d/2).$$

It is reasonable that X and Y are independent. Therefore, the joint probability distribution becomes

$$f(X, Y) = g(X)h(Y) = \begin{cases} \frac{4}{\pi d} & \text{if } 0 \le X \le \pi/2 \text{ and } 0 \le Y \le d/2, \\ 0 & \text{otherwise.} \end{cases}$$

The diagram on the left shows an example of a needle making an angle X with a horizontal grid line (in blue), with its midpoint distance Y from the nearest line. We deduce that a needle intersects a horizontal line if and only if

$$Y < \min\left(\frac{\ell}{2}\sin X, \frac{d}{2}\right).$$

The behaviour of the function $y = \min\left(\frac{\ell}{2}\sin x, \frac{d}{2}\right)$ depends on whether the needle is short ($\ell < d$) or long ($\ell > d$). The graphs of this function are shown in fig. 7.17 as ℓ crosses the boundary between short and long. The area under the graph is the integration domain in the integral $\iint f(x, y)\, dx\, dy$.

Fig. 7.17: The graph of $y = \min\left(\frac{\ell}{2}\sin x, \frac{d}{2}\right)$ for the 3 cases: $\ell < d$, $\ell = d$, and $\ell > d$ (left to right). The area under the graph is the integration domain of $\iint f(x, y)\, dx\, dy$.

If the needle is short ($\ell \leq d$), then we automatically have $Y \leq \frac{d}{2}$. In this case, the probability of intersection is simple to work out.

$$\text{Pr(intersection if } \ell \leq d) = \int_0^{\pi/2} \int_0^{\frac{\ell}{2}\sin x} f(x, y) \, dy \, dx$$

$$= \frac{2\ell}{\pi d} \int_0^{\pi/2} \sin x \, dx$$

$$= \frac{2\ell}{\pi d}.$$

On the other hand, if the needle is long, we have the situation in the right-most panel in fig. 7.17. Breaking up the integration domain into two regions, we have:

$$\text{Pr(intersection if } \ell > d) = \int_0^{\sin^{-1}(d/\ell)} \int_0^{\frac{\ell}{2}\sin x} f(x, y) \, dy \, dx + \int_{\sin^{-1}(d/\ell)}^{\pi/2} \int_0^{d/2} f(x, y) \, dy \, dx$$

$$= \frac{2\ell}{\pi d}\left(1 - \sqrt{1 - \left(\frac{d}{\ell}\right)^2}\right) + 1 - \frac{2}{\pi} \sin^{-1}\frac{d}{\ell}$$

$$= \frac{2}{\pi}\left(\frac{\ell}{d} - \sqrt{\left(\frac{\ell}{d}\right)^2 - 1} + \cos^{-1}\frac{d}{\ell}\right),$$

where we have simplified the answer using simple trigonometric identities. Finally, we can combine our results for short and long needles and write the intersection probability in terms of the ratio variable $r := \ell/d$.

$$\text{Pr(intersection if } \ell/d = r) = \begin{cases} \frac{2}{\pi} r, & \text{if } r \leq 1, \\ \frac{2}{\pi}\left(r - \sqrt{r^2 - 1} + \cos^{-1}(1/r)\right), & \text{if } r > 1. \end{cases} \quad (7.34)$$

This is then the solution to the Buffon's needle problem. The graph of this function is shown in fig. 7.18.

Note that when $r = 0.5$, we obtain the probability $1/\pi \approx 0.31831$, in agreement with the simulation result in fig. 7.16. Additionally, in the large r limit (when the needles are extremely long compared to the line spacing), intersections are almost certain since $\lim_{r \to \infty} \text{Pr}(r) = 1$.

DISCUSSION

- **Geometric probability** is concerned with measuring probability using length, area or volume, when there are infinitely many possible outcomes. The Buffon's needle problem was one of the earliest examples of a problem in geometric probability. Another well-known problem in geometric probability is Bertrand's chord problem (exercise 11), infamous for paradoxically having infinitely many solutions.

- **Buffon's 'noodles'.** In 1860, the mathematician J-É. Barbier gave an elegant solution to the Buffon's needle problem without the use of integrals. His proof, which is discussed in [4,77], also established the remarkable fact that the average number of line crossings is $2\ell/(\pi d)$ *regardless of the shape of the needle* (hence 'noodle'). We will verify a special case of this result in exercise 10.

- **Covariance matrix**. If random variables X and Y are sampled from the joint probability distribution $f(X, Y)$, we can quantify the extent to which X and Y are correlated using the *covariance* given by

$$\text{Cov}(X, Y) = E[XY] - E[X]E[Y].$$

Note that $\text{Cov}(X, X) = \text{Var}(X)$. The covariance is often expressed as a matrix $\Sigma_{ij} = \text{Cov}(X_i, X_j)$ which is square and symmetric. The covariance matrix plays a key role in data analysis as it quantifies the inter-relationships between pairs of variables. More about this in §8.7.

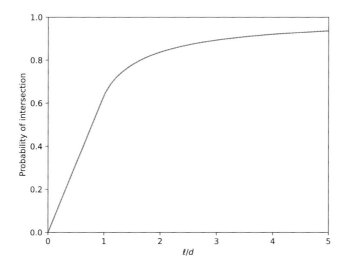

Fig. 7.18: The solution to the Buffon's needle problem. There are two curves on this graph showing the intersection probability as a function of $r = \ell/d$. The red curve shows the simulation result obtained from using 10^6 needles per r value to estimate the intersection probability. The dotted blue curve is the exact answer (7.34), practically indistinguishable from the simulation result.

buffon.ipynb (for plotting the lower panel of fig. 7.16)	
| | ```python
import numpy as np
import matplotlib.pyplot as plt
%matplotlib
``` |
| How many needles to drop in total?<br>Specify $d$ and $\ell$<br>Initialise random number generator | ```python
Nmax = int(1e6)
d, ell = 0.1, 0.05
rng = np.random.default_rng()
``` |
| Running tally of intersection count | ```python
Nlist = range(1, Nmax+1)
Prob =[]
count = 0
``` |
| Drop 1 needle at a time onto the grid<br>$y$ coordinates of needle 'head'<br>Orientation of the needle<br>$y$ coordinates of needle 'tail' | ```python
for N in Nlist:
    yhead = 10*d*rng.random()
    theta = 2*np.pi*rng.random()
    ytail = yhead + ell*np.sin(theta)
``` |
| Check intersection condition (7.32)<br>(use floor division) and add to count | ```python
    if yhead//d != ytail//d:
        count += 1
``` |
| Keep track of evolving fraction of in-tersecting needles | ```python
    prob = count/N
    Prob.append(prob)
``` |
| Plot results from 100 needles onwards<br><br><br><br>Display final estimate to 5 dec. pl. | ```python
plt.semilogx(Nlist[100:], Prob[100:], 'b')
plt.xlabel('Number of needles')
plt.ylabel('Probability of intersection')
plt.xlim(1e2, Nmax)
plt.title(f'Final probability = {prob:5}')
plt.grid('on')
plt.show()
``` |
```

7.12 Exercises

1 Use `pascal.ipynb` as a starting point for these exercises.

  a. In `pascal.ipynb`, a student suggests changing the line

$$a = a*(n-i)//(i+1) \qquad to \qquad a *= (n-i)//(i+1)$$

  (which seems a harmless thing to do). Why does this give the wrong answer?

  b. Produce Pascal's triangle as shown on the left panel of fig. 7.2.
  Suggestion: To print a centre-justified string s in a space with width of 4 characters,
  say, use the syntax `print('{:^4}'.format(s), end='')`
  Then, modify the code so that the function `binom` is not used. Instead, use the
  recursion (7.13). This improves the efficiency and reduces redundant calculations.

  c. (*Fibonacci sequence in Pascal's triangle*) Produce a left-justified version of Pascal's
  triangle. Then, generate a print-out the sums of elements along the 45° diagonals.
  They should read:

```
1
1
1, 1
1, 2
1, 3, 1
1, 4, 3
```

  and so on. Calculate the sum of each row and observe that the Fibonacci sequence
  is obtained.
  For an elementary proof, see [85].

  d. (*Pascal mod 4*) Produce the plot of Pascal's triangle mod 4 with 4 different symbols
  (or colours). Make a conjecture about the emerging pattern.
  Suggestion. To change the colour of a Unicode symbol, look up "*ANSI colour
  codes*".
  The answer can be found in [79].

2 (*Expected waiting time*) Use `coin2.ipynb` as a starting point for the following exercises.

  a. Increase the number of throws in each sequence from 10 to 100. What is changed
  in the histograms? What stays the same?
  In the following problems, keep the number of throws in each sequence as 100.

  b. Verify (using `np.mean`) that the expected (mean) position for HH is 5 (disregard
  sequences in which HH does not appear). Similarly, show that the expected position
  for TH is 3.
  The *expected waiting time* is the mean number of throws that will produce the
  desired sequence in its entirety. Thus, the expected waiting time for HH and TH
  are 6 and 4 respectively.
  Show that the probability that in TH occurs before HH is around 0.75. In other
  words, TH is more likely to occur before HH, rather than the other way round.

  c. Find the expected waiting time for the sequence HTHH to appear. Repeat for THTH.
  Show that the probability that in THTH occurs before HTHH is around 0.64.
  You should find that the waiting time for THTH is longer, yet THTH is more likely
  to occur before HTHH!

d. Amongst the 8 sequences of length 3 (*e.g.* TTT, THT, . . .), conjecture which one has the longest waiting time. Which has the shortest waiting time?
Verify your conjecture with Python. This analysis should give you an edge when playing Penney's game. See [155] for details.

3 (*The Birthday Problem - further explorations*) Use `birthday.ipynb` to help you answer these questions.

a. In our investigation, we have the exact answer (eq. 7.19), the linear approximation (eq. 7.20) and the simulation result (fig. 7.9). Let's see how they compare to one another.

   i. Plot the exact probability that at least two people in a group of n people share the same birthday (Eq. 7.19) as a function of n. Read off where the probability exceeds 0.5. Insert this curve into the cumulative plot in fig. 7.9. Are they in good agreement?

   ii. Plot the fractional error in the linear approximation (eq. 7.20) for $n \leq 100$. How accurate (in terms of decimal places) is the linear approximation for $n = 23$? Show that the maximum error in the approximation occurs at $n = 26$. Is the approximation still useable at that point?

   iii. By differentiating eq. 7.20, obtain an approximate expression for the probability distribution shown on top of fig. 7.9. Plot this curve on top of the histogram. Use the expression to show that the peak of the distribution (the mode) occurs at $n = \sqrt{365} + \frac{1}{2} \approx 20$.

b. (*Generalising the Birthday Problem*) Ref. [77] provides a good analysis of these problems.
Suggestion: These exercises will require nontrivial modifications to the code `birthday.ipynb`. Alternatively, you may like to take a fresh approach to the problem rather than building on `birthday.ipynb`. For example, you can improve the efficiency by starting with a guess, say, n from a small interval $[\alpha, \beta]$. For each n, you could quickly simulate realisations of their birthdays and calculate the required probability.

   i. (*More than 2 people sharing a birthday*) Let $P(n, k)$ be the probability that at least k people in a group of n people share the same birthday. For a given k, we wish to find the smallest n such that $P(n, k) > 0.5$.
   The original birthday problem corresponds to the case $k = 2$.
   Modify `birthday.ipynb` to show that when $k = 3$, the answer is $n = 88$.
   Tabulate k and n for $k \leq 5$.
   [Useful tool: `np.histogram`]

   ii. (*Multiple shared birthdays*) Let $P(n, k)$ be the probability that amongst n people there are at least k unique shared birthdays. For a given k, we wish to find the smallest n such that $P(n, k) > 0.5$.
   The original birthday problem corresponds to the case $k = 1$.
   Modify `birthday.ipynb` to show that when $k = 2$, the answer is $n = 36$.
   Tabulate k and n for $k \leq 5$.

   iii. (*Almost birthday*) Let $P(n, k)$ be the probability that in a group of n people, at least 2 were born within k adjacent days from each other. For a given k, we wish to find the smallest n such that $P(n, k) > 0.5$.
   The original birthday problem corresponds to the case $k = 0$.

Modify `birthday.ipynb` to show that when $k = 1$, the answer is $n = 14$. Tabulate k and n for $k \leq 5$.

[Useful tools: `np.sort`, `np.diff` and the mod operator `%`.]

c. (*Coupon collector's problem*) Suppose there are some mystery boxes, each containing a random type of coupon. If there are n types of coupons, how many boxes must be opened, on average, to find all n types of coupons?

Use Python to generate simulations and show that when $n = 10$, 30 boxes must be opened on average.

Plot the average number of boxes needed against $n = 10, 20, 30, \ldots 100$. On the same plot, overlay nH_n, where H_n is the nth harmonic number. You should find good agreement between the two curves. See [179] for calculation details.

4 (*Monty Hall problem - generalisations*) Let's explore generalisations of the Monty Hall problem. See [176] for theoretical discussions. The code `montyhall.ipynb` may be useful as a starting point for these problems.

a. (*Multiple doors*) Simulate the Monty Hall problem with n doors concealing 1 car, and suppose that g goats are revealed by Monty after the initial pick (where $0 \leq g \leq n - 2$). Reproduce fig. 7.11. Check that the simulation results agree with the exact solution (eq. 7.21).

Suggestion: Use `random.sample()` to select multiple random elements from a list.

b. (*Multiple cars*) Simulate the Monty Hall problem with n doors concealing c cars, and suppose that g goats are revealed by Monty after the initial pick. In this case, one can show that the probability of winning upon switching becomes

$$\frac{c(n-1)}{n(n-g-1)}$$

which is c times the one-car probability (eq. 7.21). Verify that this holds for a few values of c.

c. (*Multiple switches*) Suppose there are 4 doors, behind which there are 3 goats and 1 car.

Suppose that Monty will reveal 2 goats, and we are entitled to 2 switches. In other words, we have the sequence of events

Pick a door → Monty reveals a goat → Switch/stay → Monty reveals another goat → Switch/stay → Car is revealed

Simulate this scenario. Show that the winning probability if we switch twice is 0.625. Show that there is a better strategy.

5 Use `normal.ipynb` as a starting point for these questions.

a. Instead of summing N Bernoulli trials, use `scipy.stats.binom` to obtain the histograms in fig. 7.12.

b. Study the effect of a bias to the left. For example, suppose at each ball-peg encounter, the ball moves to the left with probability 0.6 (perhaps due to a magnet being placed on the left of the board). Plot the resulting distribution. Does the normal approximation still work?

c. This exercise explores the accuracy when approximating the binomial distribution using the normal distribution (Theorem 7.2).

Plot the absolute value of the fractional error as a heat map in the k-N plane assuming $p = 0.5$. You should obtain something like this.

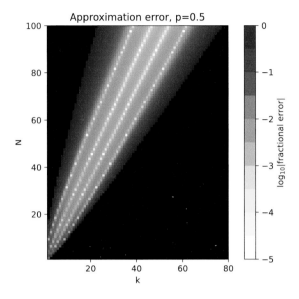

Fig. 7.19: The error in approximating the binomial distribution using the normal distribution with $p = 0.5$. The darkest regions are either invalid ($k > N$) or too large (fractional error >1). The brighter the pixel, the greater the accuracy.

Suggestion: One could use imshow to display $\log_{10}$ of the absolute error as elements of a large matrix. The more negative the entry, the better the approximation. If the error is too large (>1) or if $k > N$, set the element to be 1.

Interestingly, there are 4 (bright yellow) lines around $k \approx Np$ on which the approximation is exceptionally good. Which region do you consider the approximation to be reasonably accurate? (This is a little subjective.) Describe the region using inequalities.

Repeat the analysis for $p = 0.1$ and $p = 0.9$.

d. (Challenging) Study the effect of the distribution if a peg is removed from the Galton board.

6 (*Accuracy of Poisson approximations*)

a. Let's explore accuracy using the Poisson distribution to approximate the binomial distribution (Eq. 7.25).

Do this by plotting the absolute value of the fractional error as a heat map in the p-N plane (for a fixed k) similar to fig. 7.19 in the previous exercise. Now increase k. Incorporate a slider for k going from 0 to N if you feel adventurous.

There is a folk theorem that says that the approximation 7.25 is reasonable if $n > 20$ and $p < 0.05$. Comment on this statement in light of your result.

b. Quantify the accuracy of using the Poisson distribution to approximate the normal distribution (Eq. 7.26). Do this by first fixing k and then plotting the graph of the *absolute error* as a function of λ.

Another folk theorem says that the approximation 7.26 is reasonable for $\lambda > 20$. Comment on this statement in light of your result.

7 (*Characterising radioactivity with the Poisson distribution*|) Use `poisson.ipynb` as a starting point for this question.

a. A lump of radioactive element emits α particles at a constant average rate of λ particles per minute (assume that this element has a very long half-life). In the lab, it was found that a total of N particles were emitted over a period of 100 minutes. In how many 1-minute intervals can we expect to detect k emitted particles? (where $k = 0, 1, 2, \ldots$)

This problem can be regarded as a one-dimensional version of the dot-scattering problem discussed in §7.9. Let's simulate this situation.

Start by randomly choosing N real numbers on the (time) domain [0, 100]. These are the times at which the particles are detected.

By modifying `poisson.ipynb`, produce a histogram of the particle counts in each unit-length (1-minute) interval. Do this for $N = 50, 100, 200$ etc. Show that the distribution of the counts is well-described by the Poisson distribution with $\lambda = N/100$.

b. With $N = 50$, plot a histogram of the inter-arrival times of the particles (*i.e.* time between two successive detections) in units of minutes. Show that the distribution roughly follows the exponential distribution (7.27) with β = mean inter-arrival time.

8 (*Monte Carlo integration - Dart method*)

a. Write a code which produces each of the scatter plots shown in fig. 7.14.

b. Plot the fractional error as a function of the number of points used. Show graphically that the error scales like $N^{-1/2}$.

c. Plot the region in the square $S = \{(x, y) \in [-2, 2] \times [-2, 2]\}$ that satisfies the inequality

$$(x - 2y)^2 \geq (x - y + 1)^2 \cos(x + y).$$

Estimate this area.

9 (*More Monte Carlo integrations*) Use `montecarlo.ipynb` as a starting point for these questions.

a. Show graphically that the accuracy in the evaluation of the integrals (c) and (d) is $O(N^{-1/2})$ (*i.e.* Monte Carlo integration does not suffer from the curse of dimensionality). Your graph should all look like fig. 7.15.

b. Evaluate the following integrals using Monte Carlo integration.

i) $\displaystyle\int_0^\infty \frac{\cos x}{1 + x^2}\, dx$ ii) $\displaystyle\int_{-\infty}^\infty \cos x e^{-x^2/2}\, dx$

iii) $\displaystyle\int_0^1 \int_0^1 \left\{\frac{x}{y}\right\} dx\, dy$ iv) $\displaystyle\int_0^\pi \int_0^\pi \int_0^\pi \frac{dx\, dy\, dz}{3 - \cos x - \cos y - \cos z}.$

Answers: i) $\frac{\pi}{2e}$ (from [208] problem 51).

ii) $\sqrt{\frac{2\pi}{e}}$ (from [153] eq. 3.1.5). Try doing this by sampling x from a normal distribution.

iv) look up *Watson's triple integrals* [209].

iii) $\frac{3}{4} - \frac{\gamma}{2}$. An interesting exercise is to plot a heatmap of the function $z = \left\{\frac{x}{y}\right\}$ on the square $[0, 1]^2$. One way to do this is to modify the code `mandelbrot.ipynb` (§4.8). You should obtain something like fig. 7.20, which may inspire a way to solve the problem by hand.

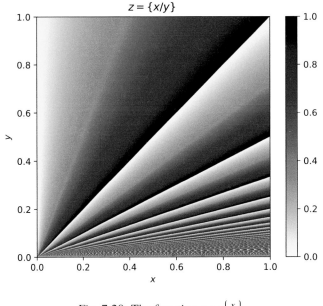

Fig. 7.20: The function $z = \left\{\frac{x}{y}\right\}$.

10 (*More fun with Buffon's needle*) Use `buffon.ipynb` to help you with these questions.

a. Plot the top panel of fig. 7.16. Suggestion: To show only the horizontal gridlines, use `plt.grid(axis='y')`.

b. Fix $r = 0.5$ say. Plot the (absolute) fractional error between the intersection probability obtained from simulation and the exact answer (7.34). Obtain a graph of the error against N and conjecture how the error term shrinks with N.

c. Try experimenting with a scenario where some needles are long and some are short. Do this by drawing ℓ from, say, a uniform distribution $U(0, 2d)$. Find the intersection probability numerically.

d. How long should the needles be to guarantee that the probability of intersection is at least 0.9? You can solve this graphically or use a root-finding method discussed in §1.9.

e. What if we add vertical grid lines as well? (so that the paper is a grid of $d \times d$ squares). Plot the probability as a function of $r = \ell/d$. In the short needle case ($\ell < d$), you should obtain the answer $\ell(4d - \ell)/\pi d^2$. Look up *Buffon-Laplace problem*.

     f. In the origin Buffon's needle problem, verify that the average number of intersections is $2\ell/\pi d$ regardless of whether the needle is long or short.

     g. Instead of needles, throw *circles* of perimeter ℓ onto the grid. Show that the average number of intersections remains the same.

       Plot a graph to show how the probability of intersection depends on ℓ.

11 (*Bertrand paradox*) An equilateral triangle is inscribed in a circle. Suppose a chord is chosen *at random* (where a chord is a straight line joining two points on the circumference). Find the probability that the chord is longer than one side of the triangle.

The French mathematician *Joseph Bertrand* (1822–1900) posed this problem in 1889 and offered 3 apparently correct solutions which are all different. This has come to be known as *Bertrand paradox*. Let's simulate them in Python.

Suppose that the circle has radius 1, so that the inscribed equilateral triangle has side $\sqrt{3}$. Place the origin O at the centre of the circle.

     a. Pick two random points, $P = (\cos\theta_1, \sin\theta_1)$ and $Q = (\cos\theta_2, \sin\theta_2)$ on the circumference of the circle, where θ_1, θ_2 are chosen randomly from $[0, 2\pi)$ (use a uniform distribution). Calculate the distance PQ.

       Show that for a large number of chords, the solution is $1/3$.

     b. Draw a radius at a random angle $\theta \in [0, 2\pi)$. Pick a random point P along the radius and construct a chord passing through P perpendicular to the radius. Note that $|OP| \in [0, 1]$. Measure the length of the chord.

       Show that for a large number of chords, the solution is $1/2$.

     c. Pick a point $(r\cos\theta, r\sin\theta)$ randomly in the circle, where $r \in [0, 1]$ and $\theta \in [0, 2\pi)$. Let this be the midpoint of the chord. Measure the length of the chord.

       Show that for a large number of chords, the solution is $1/4$.

All 3 solutions can be obtained analytically (see, for example, [77]). There are in fact infinitely many solutions, depending on how the chord is chosen *randomly*. See [31,99] for further discussions.

Bonus question: in each method, plot all the chords within the circle. Do the chords appear cover the circle uniformly?

# CHAPTER EIGHT

Statistics

Fig. 8.1: Many will know *Florence Nightingale* (1820–1910) as a pioneer of modern nursing, but she also made major contributions to statistics, particularly in data visualisation. She was the first female fellow of the Royal Statistical Society. Image from [212]

Statistics refers to the science of collection, analysis and interpretation of data. We will use probability theory to help us describe data mathematically, and make inferences about the population given the data.

Introductory statistics is sometimes taught in school (and in many online tutorials) as an algorithmic subject requiring students to follow step-by-step recipes to obtain, say, a z-score, a p-value, or a confidence interval (we will discuss all these concepts shortly). Along the way, students learn how to quickly navigate unwieldy statistical tables, and use statistics functions on fancy calculators.

In this chapter, we will see how Python frees us from the archaic approach described above. In particular,

- Python allows data to be visualised (as histograms, scatter plots, contour plots *etc.*), allowing easy interpretation.
- Statistical inferences follow naturally through visualisation without having to rely on magical formulae and recipes.

S. Chongchitnan, *Exploring University Mathematics with Python*,

- Calculations involving complicated probability distributions can be done easily, doing away with the need for statistical tables.

The material in this chapter comprises some highlights from a traditional pre-university and first-year university course in statistics, including the Central Limit Theorem, hypothesis testing and linear regression. However, our discussions will focus on visualisation and numerical experimentation.

Towards the end of the chapter, we will also discuss two more advanced topics, namely, Bayesian statistics and machine learning. These topics are not part of the standard canon in a typical university mathematics degree (unless you follow a statistics-focussed course), but they have such prominence in today's data-driven world that it would be unthinkable not to include them here. It is only a matter of time before they become a core part of every university's degree in mathematical sciences.

We do not assume that you have already studied some statistics. While we try to be as self-contained as possible, there will be occasions where you may want to consult other references that will explain statistical concepts more clearly and thoroughly.

There are many excellent introductory textbooks on statistics offering modern expositions guided by real-world applications. Amongst these, we recommend [160, 179, 216]. For statistics textbooks led by Python, see [92, 110, 205]. Other references will be recommended along the way.

8.1 *Basic concepts in statistics*

We will need the basic concepts of probability covered in §7.2, including random variables, probability density functions (pdf), cumulative distribution functions (cdf), mean and variance. You may find it helpful to study chapter 7 before proceeding.

Probability distributions

You will need to be familiar with the most common probability distributions such as the uniform, normal and binomial distributions. We will also come across the following distributions (in order of appearance in this chapter).

- Cauchy
- Arcsine
- Bates
- Triangular
- Student's t
- Chi squared (χ^2)
- Chi (χ)
- Gamma
- Beta

For an encyclopedic study of statistical distributions, see [66].

In terms of Python, we will use `scipy.stats` to draw samples (or *random variates*) from different distributions. The samples are stored as arrays. Here are some examples:

Sampling from probability distributions	
	```from scipy.stats import uniform, norm, binom```
i) Sample 5 numbers $x \in [0, 1]$ from the uniform distribution $U(0, 1)$	```X = uniform.rvs(size=5)```
ii) Sample 4 numbers $y \in \mathbb{R}$ from the (standard) normal distribution $N(0, 1)$	```Y = norm.rvs(size=4)```
iii) Sample 3 integers $(0 \leq k \leq 9)$ from the binomial distribution $B(9, k)$ with $p = 0.5$	```K = binom.rvs(9, 0.5, size=3)```
iv) Same as (iii) but the output is reproducible (pick your favourite non-negative integer for ```random_state```) Output = [3, 6, 6]	```K = binom.rvs(9, 0.5, size=3, random_state=74)```

Standard probability distributions

We studied the normal distribution $N(\mu, \sigma^2)$ in §7.8. In the case that $\mu = 0$ and $\sigma = 1$, the pdf ϕ and cdf Φ of $N(0, 1)$ are given by

$$\phi(x) = \frac{1}{\sqrt{2\pi}} e^{-x^2/2}, \tag{8.1}$$

$$\Phi(x) = \frac{1}{\sqrt{2\pi}} \int_{-\infty}^{x} e^{-t^2/2} \, dt. \tag{8.2}$$

These are called the pdf and cdf of the *standard normal distribution*, which plays a very important role in statistics.

If $(\mu, \sigma^2) \neq (0, 1)$, we can *standardise* the distribution by the substitution $z = \frac{x-\mu}{\sigma}$ (called the *z-score*). More precisely, the random variable X follows $N(\mu, \sigma^2)$ if and only if the standardised variable

$$Z = \frac{X - \mu}{\sigma} \tag{8.3}$$

follows $N(0, 1)$.

More generally, we can standardise any pdf $f(x)$ by thinking about the transformation (8.3) geometrically as a scaling (enlargement) of the graph $y = f(x)$ by a factor σ, followed by a translation of μ unit to the right.

In Python, we can sample from non-standard probability distributions by specifying the optional arguments ```scale``` and ```loc``` which perform the scaling and translation of standard pdfs. Here are some examples:

Sampling from scaled/shifted probability distributions	
	`from scipy.stats import uniform, norm`
i) Sample a number $x \in [1, 3]$ from the uniform distribution $U(1, 3)$ Note: $x \in$`[loc, loc + scale]`	`x = uniform.rvs(scale = 2, loc = 1)`
ii) Sample a number $y \in \mathbb{R}$ from the normal distribution $\mathcal{N}(\mu = -1, \sigma^2 = 4)$ Note: $\mu =$ `loc`, $\sigma =$ `scale`	`y = norm.rvs(scale = 2, loc = -1)`

Iid random variables

We will be discussing statistical properties of a combination of multiple random variables. For example, if $X_1 \sim \mathcal{N}(0, 1)$ and $X_2 \sim \mathcal{N}(1, 2)$, we might ask what distribution the sum $X_1 + X_2$ follows.

If X_1 and X_2 have same probability distribution, then they are said to be *identically distributed* random variables. More precisely, let F_1 and F_2 be the cdfs of X_1 and X_2. Then X_1 and X_2 are *identically distributed* if $F_1(x) = F_2(x)$ for all $x \in \mathbb{R}$.

X_1 and X_2 are said to be *independent* random variables if

$$\Pr(X_1 \in A \text{ and } X_2 \in B) = \Pr(X_1 \in A) \Pr(X_2 \in B),$$

(we discussed this condition in eq. 7.3).

When X_1 and X_2 are *independent, identically distributed* random variables, we usually abbreviate this to "X_1 *and* X_2 *are iid random variables*".

Sample statistics

In statistics, we are often presented with a small sample from which we deduce information about the population. We use different symbols to distinguish the mean and variance of a sample from those of the population:

	Population	Sample
Mean	μ	$\bar{x}$
Variance	σ^2	s^2

Given n numbers, $x_1, x_2, \ldots, x_n$ sampled from a population, we can calculate the *sample mean*, $\bar{x}$, as

$$\bar{x} = \frac{1}{n} \sum_{i=1}^{n} x_i. \tag{8.4}$$

Suppose that we repeat drawing n samples and calculating the sample mean many times. Consider the random variable $\bar{X}$ which can take all possible values of the sample mean:

$$\bar{X} = \frac{1}{n} \sum_{i=1}^{n} X_i, \tag{8.5}$$

where X_i are iid random variables with $E[X_i] = \mu$ and $\text{Var}(X_i) = \sigma^2$.

What does $\bar{X}$ tell us about the population mean μ? Naturally, we would expect the expectation value of $\bar{X}$ to approach μ as we repeat drawing n samples many times. This is indeed the case since:

$$E[\bar{X}] = E\left[\frac{1}{n} \sum_{i=1}^{n} X_i\right] = \frac{1}{n} \sum_{i=1}^{n} E[X_i] = \frac{1}{n} n\mu = \mu, \tag{8.6}$$

where we have used the linearity of the operator E (see eq. 7.7).

When the expectation value of a sample statistic T coincides with the population parameter θ, we say that T is an *unbiased estimator* of θ. In our case, we have just shown that $\bar{X}$ is an unbiased estimator of μ.

It is also worth noting that

$$\text{Var}(\bar{X}) = \text{Var}\left(\frac{1}{n} \sum_{i=1}^{n} X_i\right) = \frac{1}{n^2} \sum_{i=1}^{n} \text{Var}(X_i) = \frac{1}{n^2} n\sigma^2 = \frac{1}{n}\sigma^2, \tag{8.7}$$

where we have used the property (7.11) of the Var operator and the fact that X_i are independent variables.

What does an unbiased estimator of σ^2 look like? One might guess that the random variable

$$\hat{S}^2 = \frac{1}{n} \sum_{i=1}^{n} (X_i - \bar{X})^2$$

is a good estimate of the population variance (since this mirrors the definition of population variance (7.8)). However, the working below shows that $\hat{S}^2$ is not an unbiased estimator of σ^2. First, note the simplification

$$\hat{S}^2 = \frac{1}{n} \sum_{i=1}^{n} \left(X_i^2 - 2X_i\bar{X} + \bar{X}^2\right)$$

$$= \frac{1}{n} \sum_{i=1}^{n} X_i^2 - 2\bar{X}\left(\frac{1}{n} \sum_{i=1}^{n} X_i\right) + \frac{\bar{X}^2}{n} \sum_{i=1}^{n} 1$$

$$= \frac{1}{n} \sum_{i=1}^{n} X_i^2 - 2\bar{X}^2 + \bar{X}^2$$

$$= \frac{1}{n} \sum_{i=1}^{n} X_i^2 - \bar{X}^2.$$

Taking the expectation value and using eq. 8.6-8.7, we have

$$E[\hat{S}^2] = \sigma^2 + \mu^2 - E[\bar{X}^2]$$

$$= \sigma^2 + \mu^2 - (\text{Var}(\bar{X}) + \mu^2)$$

$$= \frac{n-1}{n}\sigma^2.$$

Since the expectation does not equal σ^2, we say that $\hat{S}^2$ is a *biased estimator*. This also implies that the unbiased estimator, S^2, of σ^2 is given by a simple adjustment.

$$S^2 := \frac{n}{n-1}\hat{S}^2 \implies E[S^2] = \sigma^2. \tag{8.8}$$

This means that given a sample of n numbers $x_1, x_2, \ldots, x_n$, then we should calculate the *adjusted sample variance* using the formula

$$s^2 = \frac{1}{n-1}\sum_{i=1}^{n}(x_i - \bar{x})^2. \tag{8.9}$$

We will always take the term 'sample variance' to mean the adjusted variance. For large n, however, the adjustment only has a tiny effect.

Here are two more useful sample statistics: the *median* and the *mode*.

Given a sample of n numbers, $x_1, x_2, \ldots, x_n$, we arrange them in increasing order, and relabel the ordered list as $y_1, y_2, \ldots, y_n$. The *median* is then defined as

$$\text{Median} = \begin{cases} y_{(n+1)/2} & \text{if } n \text{ is odd,} \\ \frac{1}{2}\left(y_{n/2} + y_{(n/2)+1}\right) & \text{if } n \text{ is even.} \end{cases}$$

The intuition is that the median splits the ordered sample into equal halves.

Finally, the *mode* is the value that occurs most frequently in the sample.

Here are some useful Python functions for calculating sample statistics. Try calculating them by hand before checking with Python.

Calculating sample statistics	
	`import numpy as np` `from scipy import stats`
Our sample	`A = [3, 1, 4, 1, 5, 9]`
Mean	`np.mean(A)`
Median	`np.median(A)`
Mode	`stats.mode(A, keepdims=False)`
Sample variance (eq. 8.9)	`np.var(A, ddof =1)`
Unadjusted variance	`np.var(A)`

The argument `ddof` stands for 'delta degrees of freedom'. Let's clarify this terminology.

Our sample contains 6 observations ($x_i, i = 1, 2, \ldots, 6$). The ingredients for calculating the sample variance s^2 are: x_1 to x_6, plus the sample mean $\bar{x}$. But the knowledge of $\bar{x}$ means that not all 6 numbers are independent: we can work out the 6th number from knowing 5 numbers and the sample mean. In this case, we say that there are only 5 degrees of freedom. This corresponds to the prefactor $\frac{1}{5}$ in the adjusted variance formula.

Scipy's `ddof` counts the reduction in the degrees of freedom due to extra known parameters. We will revisit the degrees of freedom again in §8.4.

8.2 *Basics of statistical packages*

In this chapter, we will be using 3 popular statistical libraries, namely, *Pandas*, *Seaborn* and *Scikit-learn*. This is quite a break from the previous chapters, where we have tried to minimise the use of specialist libraries. However, we have included them here due to their popularity in data science.

Pandas

Pandas[1] is a Python library for easy data handling and analysis. Instead of NumPy arrays, in Pandas data are stored as *series* (for 1D data) or *dataframes* (2D). Working with dataframes is analogous to working with Excel spreadsheets, but with more mathematical logic and, dare we say, less frustration.

Pandas is traditionally imported with the following line:

```
import pandas as pd
```

In §8.6 (on linear regression and Simpson's paradox) we will use Pandas to work with data files. This is an important skill in real data analysis where data are typically stored in multiple large files which may have to be merged or trimmed.

Seaborn

Seaborn[2] is a data visualisation library created by Michael Waskom in 2012. Seaborn simplifies the process of creating beautiful data visualisations that would have otherwise been difficult and time consuming to make using Matplotlib alone.

Seaborn is traditionally imported with the line:

```
import seaborn as sns
```

In §8.3 (on the Central Limit Theorem), we will use Seaborn to plot histograms (much more easily than how we previously plotted histograms with Matplotlib in §7.5). More Seaborn functionalities are explored in the exercises.

Scikit-learn

The last section of this chapter explores topics in machine learning using *scikit-learn*[3] (often shortened to *sklearn*).

Sklearn is a Python library originally authored by David Cournapeau and publicly available since 2013. You will need sklearn version 1.2.2 or later. To check which version

---

[1] https://pandas.pydata.org

[2] https://seaborn.pydata.org

[3] https://scikit-learn.org

you may already have, run these two lines:

```
import sklearn
sklearn.__version__
```

And finally: although we will be exploring statistics with Python, the statistician's programming language of choice is not Python but R[4]. For good introduction to R, see [193].

---

[4] https://www.r-project.org

8.3 Central Limit Theorem

Consider n numbers $x_1, x_2, \ldots, x_n$ randomly drawn from the uniform distribution $U(0, 1)$. Let $\bar{x} := \frac{1}{n} \sum_{i=1}^{n} x_i$ be the sample mean.
When $n = 100$, find the probability that $\bar{x} > 0.55$.

We are interested in the probability distribution of the sample mean where the samples are drawn from some underlying distribution such as the $U(0, 1)$ in this case. Let the random variable $\bar{X}$ denote all possible mean values of n samples.

Let's experiment with some values of n. When $n = 1$, the sample means are the samples themselves, so the distribution of $\bar{X}$ is simply $U(0, 1)$.

When $n = 2$, pairs of numbers x_1 and x_2 are drawn from $U(0, 1)$ and an average is taken. We note that there are many more possible pairs of numbers that will average to 0.5 compared with pairs of numbers that will average to, say, 0.99 or 0.01. Thus, we expect that the probability distribution of $\bar{X}$ will peak at 0.5, and drop off to 0 at $\bar{X} = 0$ and $\bar{X} = 1$.

Using Python to simulate this problem for more values of n, we obtain fig. 8.2. Here, n samples ($n = 1, 2, 3, 10$) are drawn from $U(0, 1)$ and repeated 10^4 times. The histograms shown (plotted with Seaborn) are the normalised frequency plotted with 25 bins on the interval $(0, 1)$. These are good approximations of the probability distributions of $\bar{X}$.

As n increases, the distribution of $\bar{X}$ approaches a familiar shape, *i.e.* the normal distribution. Taking this observation further, when $n = 100$, the distribution of $\bar{X}$ (shown on the top panel of fig. 8.3) can in fact be approximated by $N(\mu_{\bar{X}}, \sigma_{\bar{X}})$ (pdf shown as a red curve) where

$$\mu_{\bar{X}} \approx 0.5, \qquad \sigma_{\bar{X}} \approx 0.029. \tag{8.10}$$

In the code CLT.ipynb, we approximate $\mu_{\bar{X}}$ and $\sigma_{\bar{X}}$ directly from the list of simulated $\bar{x}$ values. Keep in mind that at this point, the normal approximation is only a conjecture – we will justify this shortly.

Let's now address the question: what is the probability that $\bar{x} > 0.55$? In one run of the simulation, out of 10^4 sample means, the code reported that 420 numbers were greater than 0.55. Hence, we could say that

$$\Pr(\bar{X} > 0.55) \approx 420/10^4 \approx 0.042. \tag{8.11}$$

Alternatively, we could use the normal approximation $N(0.5, 0.029)$, and calculate the probability by integrating the pdf using SciPy's quad. We can also transform the integral to the standard normal distribution as follows.

$$\Pr(\bar{X} > 0.55) = \frac{1}{\sigma_{\bar{X}}\sqrt{2\pi}} \int_{0.55}^{\infty} e^{-\frac{1}{2}\left(\frac{\bar{X}-\mu_{\bar{X}}}{\sigma_{\bar{X}}}\right)^2} d\bar{X}$$

$$= \frac{1}{\sqrt{2\pi}} \int_{z=1.724}^{\infty} e^{-z^2/2} dz \qquad \left(\text{let } z = \frac{\bar{X}-\mu_{\bar{X}}}{\sigma_{\bar{X}}}\right)$$

$$= 1 - \Phi(1.724)$$

$$\approx 0.04235 \quad (4 \text{ S.F.}) \tag{8.12}$$

in agreement with the simulation result (8.11). (To evaluate $\Phi(1.724)$, use stats.norm.cdf.)

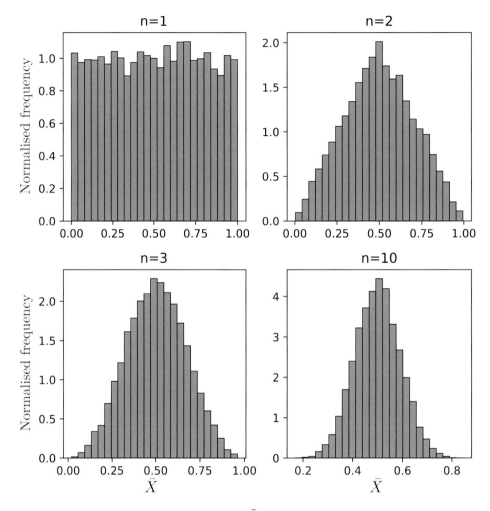

Fig. 8.2: Distribution of the sample mean $\bar{X}$ for $n = 1, 2, 3$ and 10 when n numbers are sampled from the uniform distribution $U(0, 1)$, repeated 10^4 times. The distribution approaches a normal distribution.

The Central Limit Theorem

The fact that the distribution of the sample mean $\bar{X}$ approaches the normal distribution is due to the following fundamental theorem of probability and statistics:

Theorem 8.1 *(Central Limit Theorem) Let $X_1, X_2, \ldots, X_n$ be iid random variables sampled from a distribution with mean μ and variance σ^2 where $0 < \sigma^2 < \infty$. Then, for each fixed x, the sample mean $\bar{X}$ satisfies*

$$\lim_{n \to \infty} \Pr\left(\frac{\bar{X} - \mu}{\sigma/\sqrt{n}} \leq x\right) = \Phi(x),$$

where Φ is the cdf of the standard normal distribution (8.2).

Equivalently, the Central Limit Theorem (CLT) says that for large n, the pdf of the sample mean approaches $N(\mu_{\bar{X}}, \sigma_{\bar{X}})$ where

$$\mu_{\bar{X}} = \mu, \qquad \sigma_{\bar{X}} = \frac{\sigma}{\sqrt{n}},$$

where μ and σ^2 are the mean and variance of the underlying distribution from which the samples are drawn.

We can now see why the normal approximation with $\mu_{\bar{X}}, \sigma_{\bar{X}}$ given in eq. 8.10 holds. For $U(0, 1)$, we have $\mu = \frac{1}{2}$ and $\sigma^2 = \frac{1}{12}$. Therefore, according to the CLT, we have

$$\mu_{\bar{X}} = 0.5, \qquad \sigma_{\bar{X}} = \frac{\sigma}{\sqrt{n}} = \frac{1}{20\sqrt{3}} = 0.0288675\ldots$$

in agreement with the values found in eq. 8.10.

Amazingly, the CLT holds for any discrete or continuous underlying distribution with a finite nonzero variance. Fig. 8.3 shows the distributions of the sample mean when the underlying distributions are:

- The uniform distribution $U(0, 1)$
- The binomial distribution with $n = 12$ and $p = 0.3$ (see eq. 7.16)
- The exponential distribution with $\beta = 2$ (see eq. 7.27)
- The arcsine distribution with pdf

$$f(x) = \frac{1}{\pi\sqrt{x(1-x)}}, \tag{8.13}$$

(so called because its cdf $F(x) = \frac{2}{\pi}\sin^{-1}\sqrt{x}$).

In each case, the normal distribution (shown in red) is a good fit to the sample-mean distribution.

We end with a little history of the CLT, the proof of which spans centuries. We already mentioned in §7.8 that de Moivre discovered the special case of the CLT for the binomial distribution in 1733, and this was later generalised by Laplace in ca. 1810. A definitive proof for an arbitrary distribution was published by A. Lyapunov in 1900 and further refinements were made by J. Lindeberg in 1920 and P. Lévy in 1935, resulting in the version of the theorem that we know today. Reference [64] presents a comprehensive discussion of the historical development of the CLT.

> **DISCUSSION**

- **Bates distribution**. The probability distributions seen in fig. 8.2 can be calculated exactly: the mean of n random variables drawn from $U(0, 1)$ follow the *Bates distribution* with pdf

$$f_n(x) = \frac{n}{2(n-1)!}\sum_{k=0}^{n}(-1)^k\binom{n}{k}(nx-k)^{n-1}\operatorname{sign}(nx-k), \tag{8.14}$$

where $\operatorname{sign}(t) = 1, 0, -1$ when t is positive, zero or negative. You should verify that when $n = 1$, we recover $U(0, 1)$, and when $n = 2$, we obtain two straight line segments joining $(0, 0)$, $(0.5, 2)$ and $(1, 0)$ (a *triangular distribution*).

Grace Bates (1914–1996) was an American mathematician who worked on algebra and probability.

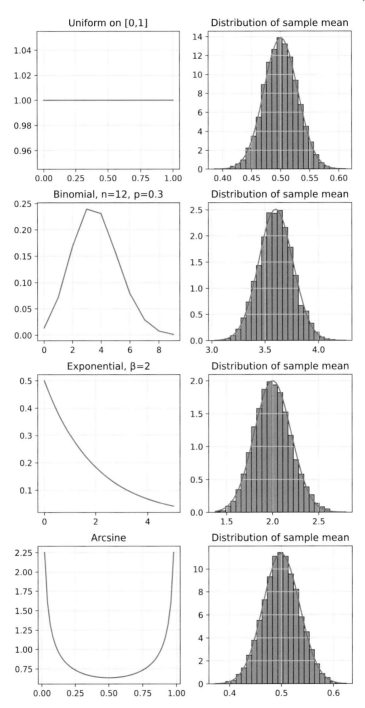

Fig. 8.3: Demonstration of the Central Limit Theorem for four underlying probability distributions shown on the left. From each distribution, 100 numbers are sampled and the sample mean is calculated. Repeating this 10^4 times, we obtain the probability distributions of the sample mean shown on the right. These are consistent with the normal distribution, shown as red curves.

- **Cauchy distribution**. The CLT only applies when $0 < \sigma < \infty$. An example of a pathological distribution which violates this condition is the (standard) *Cauchy distribution*

$$f(x) = \frac{1}{\pi(1 + x^2)}. \tag{8.15}$$

It is straightforward to show that the mean and variance are undefined. In exercise 1, you will show that the CLT fails when samples are drawn from this distribution.

- **Generalised CLT**. The sample mean of n independent variables (not necessary identically distributed) also approaches the normal distribution for large n, provided certain generic conditions are satisfied - see [163] for details.

The graph below shows the distribution of the sample mean $\bar{X}$ of 100 numbers, consisting of 25 numbers drawn from each of the four distributions shown in fig. 8.3. The normal approximation is shown as the red curve, hence verifying the generalised CLT.

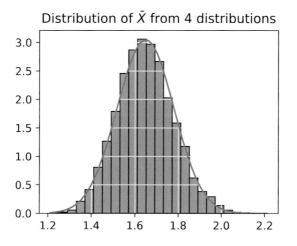

Fig. 8.4: Demonstration of the generalised CLT: when 25 samples are drawn from each of the 4 different distributions shown in fig. 8.3, the distribution of the sample mean $\bar{X}$ is still approximately normal.

CLT.ipynb (for plotting the first pair of figures in fig. 8.3)	
For uniform and normal distributions	```import numpy as np
import matplotlib.pyplot as plt
from scipy.stats import uniform, norm
import seaborn as sns
%matplotlib```
List of sample means How many sample means?	```Means = []
Musize = int(1e4)```
Batch size for calculating each mean $\bar{X}$	```size = 100```
Collect lots of sample means by 1. Generating a batch of uniform variates 2. Calculating the sample mean 3. Storing it	```for i in range(Musize):
samples = uniform.rvs(size=size)
mu = np.mean(samples)
Means.append(mu)```
Two graphs side-by-side 1st graph = uniform pdf	```fig,(ax1, ax2)=plt.subplots(1,2,figsize=(7,3))
x1 = np.linspace(0,1)
pdf = uniform.pdf(x1)
ax1.plot(x1, pdf)
ax1.set_title(f'Uniform on [0,1]')
ax1.grid('on')```
2nd graph = seaborn histogram of sample mean (normalised frequency) $\mu_{\bar{X}}$ $\sigma_{\bar{X}}$ Normal approximation (due to the CLT) Add red normal curve on top of histogram	```sns.histplot(Means, bins=25,
            stat = 'density', ax=ax2)
ax2.set_title(f'Distribution of sample mean')
x2 = np.linspace(min(Means), max(Means))
ax2.set_ylabel('')
MU = np.mean(Means)
SIG = np.sqrt(np.var(Means, ddof=1))
y2 = norm.pdf(x2, MU, SIG)
ax2.plot(x2,y2, 'r')
ax2.grid('on')

plt.show()``` |
| Count of samples > 0.55 | ```Mcount = sum(i > 0.55 for i in Means)
Ans = Mcount/Musize
print(f'{Mcount} samples were >0.55. '
      f'P(X>0.55)={Ans}')``` |

8.4 Student's *t* distribution and hypothesis testing

a) Sample n random numbers from the standard normal distribution $\mathcal{N}(0, 1)$. Let $\bar{x}$ be the sample mean and s^2 be the sample variance. Calculate $t = \dfrac{\bar{x}}{s/\sqrt{n}}$.

Let the random variable T denote the values of t. Plot the distribution of T and investigate the limit as $n \to \infty$.

b) The following 5 numbers were sampled from a normal probability distribution whose mean μ and variance σ^2 are unknown.

$$0.6517 \qquad -0.5602 \qquad 0.8649 \qquad 1.8570 \qquad 1.9028$$

Estimate the population mean μ.

c) A student believes that the population from which the numbers in part (b) were drawn from has mean $\mu_0 = 1.5$. Does the data support this statement?

This section deals with statistical information that can be extracted from a small dataset. In particular, we will see what we can learn about the mean of an infinite population from the mean of a small sample. We will encounter the *Student's t distribution*, and learn about *hypothesis testing*, *p values* and *confidence intervals*.

$\boxed{a)}$ Let's go straight to Python to sample n random variates from $\mathcal{N}(0, 1)$ and find the sample mean $\bar{x}$ and the sample variance s^2. Repeating this many times (say 10^4), we obtain a histogram showing the normalised frequency of the variable $t = \frac{\bar{x}}{s/\sqrt{n}}$. The code is similar to `CLT.ipynb` in the previous section. The results are shown in fig. 8.5 for $n = 3, 5, 8$, and 10.

Each histogram corresponds approximately to a symmetric distribution whose shape resembles the normal distribution. However, the red curve plotted over each histogram is *not* the normal distribution, but rather, *Student's t distribution*, due to the following theorem.

Theorem 8.2 *Let* $X_1, X_2, \ldots, X_n$ *be* n *random iid variables drawn from the standard normal distribution* $\mathcal{N}(0, 1)$. *Let the random variables* $\bar{X}$ *and* S^2 *be the sample mean and sample variance. Then, the random variable* $T = \frac{\bar{X}}{S/\sqrt{n}}$ *follows Student's t distribution with pdf*

$$f(t) = \frac{1}{\sqrt{\pi\nu}} \frac{\Gamma(\frac{\nu+1}{2})}{\Gamma(\frac{\nu}{2})} \left(1 + \frac{t^2}{\nu}\right)^{-\frac{\nu+1}{2}}, \qquad t \in \mathbb{R}, \tag{8.16}$$

where Γ *is the gamma function and the parameter* $\nu = n - 1 > 0$.

(See eq. 6.29 for the definition of the gamma function.) This result was discovered by *Student*, the pseudonym of *William Sealy Gosset* (1876–1937), an English statistician who studied small-sample statistics whilst working at the Guinness brewery in Ireland.

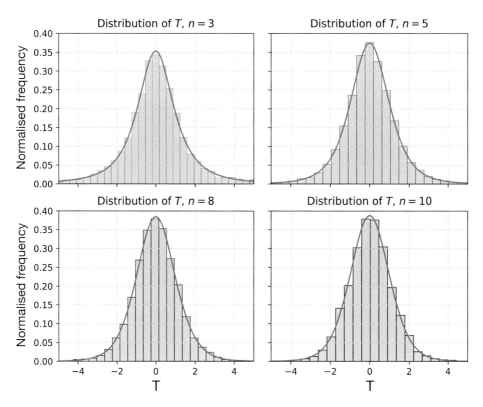

Fig. 8.5: Histograms showing the distributions of the variable $T = \sqrt{n}\bar{X}/S$ for $n = 3, 5, 8$ and 10. The red curves show Student's t distribution (8.16) with $\nu = n - 1$.

The parameter ν is called the *degree of freedom*. With n samples, we say that there are $n - 1$ degrees of freedom for the distribution of the sample mean $\bar{X}$. This is because, for a given mean $\bar{x}$ of n numbers, we are free to choose only $n - 1$ numbers, $x_1, x_2, \ldots, x_{n-1}$, as the final remaining number is fixed by the relation $x_n = n\bar{x} - x_1 - x_2 - \cdots - x_{n-1}$. Although this explanation suggests that the degree of freedom ν should be an integer, the pdf admits any positive real values of ν.

Fig. 8.6 shows a family of the t distribution for $\nu = 0.5 - 10$ in steps of 0.5. Generally the pdf of the t distribution has a lower peak and fatter tails than the normal distribution $N(0, 1)$. As ν becomes large, the limit approaches $N(0, 1)$. The proof requires the limit $\lim_{n \to \infty}(1 + \frac{x}{n})^n = e^x$ and Stirling's approximation (6.37).

Interestingly, when $\nu = 1$, the t distribution reduces to the Cauchy distribution (eq. 8.15).

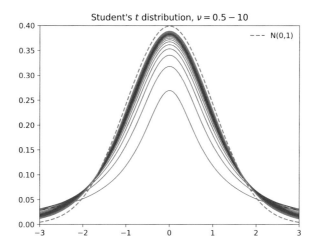

Fig. 8.6: The pdf of Student's t distribution (8.16) where the degrees-of-freedom parameter ν goes from 0.5 to 10 (solid lines going from blue to red) in steps of 0.5. The dashed red curve is the pdf of the standard normal distribution $\mathcal{N}(0, 1)$.

Confidence interval of μ from a small sample

$\boxed{b)}$ We are given a small sample of 5 numbers with sample mean and sample variance

$$\bar{x} = 0.9732, \qquad s^2 = 1.0268 \quad \text{(4 dec. pl.)}.$$

We know that the population is normal with mean μ and variance σ^2 (both are unknown). We wish to estimate μ using seemingly limited information in the small sample.

The mean and variance of the random variable $\bar{X}$ are given by

$$\mu_{\bar{X}} = E\left[\frac{X_1 + X_2 + \cdots + X_n}{n}\right] = \frac{1}{n}\sum_{i=1}^{n} E[X_i] = \mu \tag{8.17}$$

$$\sigma_{\bar{X}}^2 = \text{Var}\left(\frac{X_1 + X_2 + \cdots + X_n}{n}\right) = \frac{1}{n^2}\sum_{i=1}^{n} \text{Var}(X_i) = \frac{\sigma^2}{n}. \tag{8.18}$$

The standardised variable (z score) for $\bar{X}$ is therefore $\frac{\bar{X}-\mu}{\sigma/\sqrt{n}}$. Since we do not know σ, we can replace it with the sample variance S, giving precisely the T random variable

$$T = \frac{\bar{X} - \mu}{S/\sqrt{n}}, \tag{8.19}$$

which, according to theorem 8.2, follows the t distribution (8.16).

Another way to interpret eq. 8.19 is to reverse the roles of $\bar{X}$ and μ, and conclude that the unknown population mean μ follows the t distribution which has undergone the scaling transformation $\mu \rightarrow \mu/(S/\sqrt{n})$ and then shifted so that its peak is at $\mu = \bar{X}$. The top panel of fig. 8.9 shows this probability distribution (produced by the code `ttest.ipynb`).

The purplish shaded area (which is symmetric about the peak) equals 0.95, spanning the interval

$$-0.3149 < \mu < 2.2014$$

We call this the 95% *confidence interval* for the estimate of μ.

Hypothesis testing

$\boxed{c)}$ In the context of statistics, a *hypothesis* is a statement about a population (whose data we have). A hypothesis is usually stated in terms of a population parameter (*e.g.* its mean, variance or the type of distribution) reflecting the belief of an observer.

Part (c) of this question phrases the estimate of the parameter μ in terms of a hypothesis (in this case, that $\mu = 1.5$). In light of the data (5 numbers drawn from the population), how confident can we be that this hypothesis holds? This is the art of statistical *hypothesis testing*.

Hypothesis testing is a crucial concept in statistics with many variations and specialist terminology that can be daunting for new learners. The essential concept is simpler to understand graphically rather than algorithmically. Of course, one can simply follow a hypothesis-testing recipe to get the right answer (unfortunately many students are taught this way). However, without understanding the reason behind each step, the whole process becomes an opaque, robotic routine that sheds no light on the underlying mathematics.

Let's go through hypothesis testing in this question step-by-step. Along the way, key terminology will be introduced, along with relevant graphical interpretations.

- **Step 1: State the hypotheses**
  Basic hypothesis testing involves two statements: the *null hypothesis*, H_0; and the *alternative hypothesis*, H_1.
  In general, the null hypothesis represents a viewpoint that is expected, status quo or uninteresting about the population. The alternative hypothesis is a statement that contradicts the null hypothesis.
  In the process of hypothesis testing, we initially assume H_0 to be true. The goal is simple: can we reject H_0 given the data? If so, we then conclude that H_1 is true. If we fail to reject H_0, then it continues to hold true.
  The null hypothesis usually includes one of the symbols $=, \leq, \geq$. The alternative hypothesis includes $\neq, >, <$.
  Thus, in our case, we have

  $$H_0 : \mu = 1.5$$
  $$H_1 : \mu \neq 1.5$$

- **Step 2: Decide on the rejection region**
  Consider the top panel of fig. 8.9, which shows the distribution of the population mean with a 95% confidence interval. We could say that if the hypothetical value of μ_0 lies within this region (as indeed it does), then we do not reject H_0. If μ_0 falls outside the shaded region, it is said to lie in the *rejection region* – meaning that H_0 should be rejected. (The rejection region is also called the *critical region*.)
  Instead of 95%, we might have chosen 90%, or 99%. It all depends on our desired level of confidence in rejecting H_0. In this question, let's go with 95%.

Another way to put this is that we have chosen to take a 5% chance that H_0 might be wrongly rejected. In statistics speak, we say that the *significance level* $\alpha = 0.05$.

Also note that the rejection region in our case is split between both tails of the pdf. This is because our alternative hypothesis $H_1 : \mu \neq 1.5$ allows μ to be at either tails of the pdf. In technical terms, we say that our test is a *two-tailed* test.

If, instead we had chosen $H_1 : \mu < 1.5$. Then the rejection region is only on the left tail of the pdf, and we say that the test is *one-tailed* (see exercise 2).

- **Step 3a: Calculate the test statistic**

  Consider the middle panel of 8.9, which shows the standard *t* distribution $f(t)$ with $\nu = 4$ degrees of freedom. The rejection region with total area 0.05 (corresponding to the α value) can be transformed to the corresponding area on this distribution as shown below.

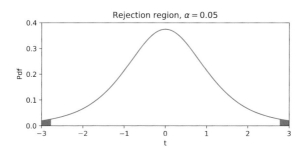

Fig. 8.7: The standard *t* distribution with $\nu = 4$, showing the shaded rejection region with area $\alpha = 0.05$. Critical values are at $t_{\text{crit}} \approx \pm 2.7764$.

The boundary of the rejection region on this plot works out to be at $t_{\text{crit}} \approx \pm 2.7764$ (these are called *critical values*). We can check this with SciPy using the following syntax.

```
from scipy import stats
rv = stats.t(4)
A, B = rv.ppf(0.025), rv.ppf(0.975)
print(f'({A:.4f}, {B:.4f})')
```

Output: `(-2.7764, 2.7764)`

Here `ppf` (*percent point function*) is simply the inverse of the cdf.

A *test statistic*, t, is the value on the (standard) distribution which determines whether H_0 is rejected, depending whether or not it falls within the rejection region. In our case, according to H_0, $\mu = \mu_0 = 1.5$. Thus, using eq. 8.19, the test statistic is

$$t = \frac{\bar{x} - \mu_0}{s/\sqrt{n}}$$

$$= -1.2286 \text{ (4 dec. pl.)}$$

Since $|t| < t_{\text{crit}}$, the test statistic falls outside the rejection region, meaning that H_0 fails to be rejected.

Note the sign of t. Negative t means that the hypothetical value of μ_0 is greater than the sample average.

- **Step 3b: Calculate the p-value**

  Another way to test the hypothesis is to use the p-value. To do this, consider the area shaded (in yellow) shown in the middle panel of fig. 8.9. This area, called the *p value*, represents the critical level of significance which determines whether H_0 is rejected or not. In other words, we have the following criteria:

  $$p \leq \alpha \implies \text{Reject } H_0, \qquad p > \alpha \implies \text{Do not reject } H_0.$$

  Some students remember this condition this way: "*if the p-value is a wee value, reject the null hypothesis*".

  The probabilistic interpretation of the p-value can be deduced from its area interpretation: assuming the null hypothesis, the p-value represents the probability of observing our samples (or an even more extreme set of samples with a larger $|t|$).

  To calculate the p-value, we find the area under the pdf with the test statistic t in the integration limit. Using its symmetry, we have

  $$p = 2 \int_{-\infty}^{t} f(x)\, dx$$
  $$= 2F(t) \qquad \text{(where } F \text{ is the cdf)}$$
  $$= 0.2866 \ (4 \text{ dec. pl.})$$

Since the p is *not* a wee value (0.05 or less), we conclude that the data is consistent with H_0, which cannot be rejected.

We can now reveal that the sample of 5 numbers were actually drawn from $N(1, 1)$. Although $\mu = 1$ in actuality, our data does not allow us to dismiss the claim that $\mu = 1.5$.

t-test with SciPy

The particular type of hypothesis testing we have discussed is known in the trade as a *one-sample t-test*.

Conveniently, SciPy can help us calculate the test statistic t and the p-value associated with this test, as well as the confidence interval of μ. The syntax are shown in the code `ttest.ipynb`. The key syntax is the line

```
res = stats.ttest_1samp(samples, mu0)
```

We then report the t-value using `res.statistic` and the p-value using `res.pvalue`. As reported in the code output shown at the bottom of fig. 8.9, both SciPy's values agree with our previous calculations. The syntax `res.confidence_interval(0.95)` gives the 95% confidence interval.

> ### DISCUSSION

- **Statistical tables**. In exam situations where computers and calculators are not allowed, students are usually supplied with statistical tables. These tables are also found at the back of most textbooks on statistics.

  Here is an example of a table of critical t-values for a given α for the Student's t distribution (from [56]). There are also tables of p-values for various distributions. With Python on hand, however, there is no need to consult any tables.

Perhaps these tables will be phased out in the near future, just as log and trig tables have disappeared decades ago.

t Distribution: Critical Values of t

		Significance level					
Degrees of	*Two-tailed test:*	10%	5%	2%	1%	0.2%	0.1%
freedom	*One-tailed test:*	5%	2.5%	1%	0.5%	0.1%	0.05%
1		6.314	12.706	31.821	63.657	318.309	636.619
2		2.920	4.303	6.965	9.925	22.327	31.599
3		2.353	3.182	4.541	5.841	10.215	12.924
4		2.132	2.776	3.747	4.604	7.173	8.610
5		2.015	2.571	3.365	4.032	5.893	6.869

Fig. 8.8: Excerpt from a statistical table [56]. Note the critical *t*-value we found earlier using Python (highlighted).

- **Type I and type II errors**. We discussed earlier that the significance value $\alpha = 0.05$ means that there is a small probability of 5% that the null hypothesis H_0 is rejected when it is actually true. This is called a *type I error*, associated with false positives in medical tests, or a guilty verdict when a defendant is innocent.

  Type II error, on the other hand, is the probability β that the null hypothesis fails to be rejected when it is actually false. This is associated with false negatives in medical tests, or a not-guilty verdict when a defendant is guilty.

  Typically, α and β are set (somewhat arbitrarily) between 0.05 and 0.2.

  In our example, the samples were drawn from a normal distribution with $\mu = 1$, yet we were unable to reject the hypothesis that $\mu = 1.5$. It appears that a type II error has occurred. This may have been avoided with a bigger sample – exercise 2e quantifies how much β is reduced as the sample size increases.

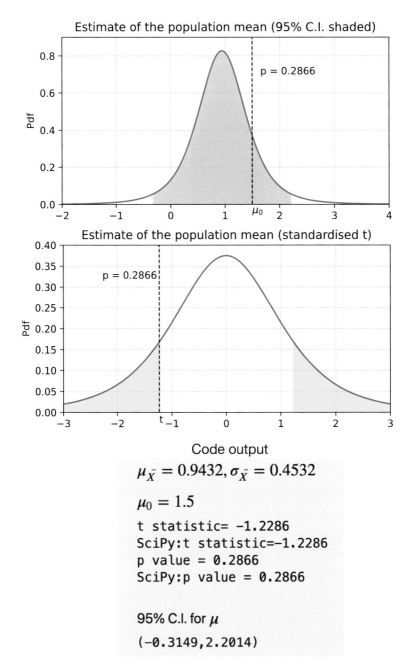

Fig. 8.9: *Top*: The pdf of the estimate of the population mean μ obtained from the given 5 numbers. The pdf is a t distribution with $v = 4$, scaled by $\sigma_{\bar{X}}$ and centred at $\mu_{\bar{X}}$. The domain of the shaded region is the 95% confidence interval of μ. The dashed line shows the hypothetical value $\mu_0 = 1.5$. *Middle*: The standard t distribution with test statistic t (eq. 8.19). The area of the shaded region equals the p-value. *Bottom*: Statistical information displayed by the code `ttest.ipynb`.

ttest.ipynb (for plotting the top panel of fig. 8.9 and displaying results relevant to the t-test)	
	```import numpy as np```
	```import matplotlib.pyplot as plt```
	```from scipy import stats```
For displaying maths in the code output	```from IPython.display import Math, Latex```
	```%matplotlib```
The data	```samples =[0.6517,-0.5602,0.8649,1.8570,1.9028]```
	```N = len(samples)```
Sample mean ($\bar{x}$)	```Smean = np.mean(samples)```
Sample variance (s)	```Svar = np.var(samples, ddof=1)```
Estimate $\sigma_{\bar{X}}$	```scale = np.sqrt(Svar/N)```
Show output on Jupyter notebook	```display(Math(r'\mu_\bar{X} = ' f"{Smean:.4}, "```
	```r'\sigma_{\bar{X}} = 'f"{scale:.4}"))```
Plotting top panel of fig. 8.9	```fig, ax = plt.subplots(1,1, figsize=(6,3))```
SciPy's t distribution (scaled and shifted)	```rv = stats.t(N-1, loc = Smean, scale = scale)```
Domain for plotting	```xmin, xmax = -2, 4```
	```x2 = np.linspace(xmin, xmax, 150)```
The pdf	```y2 = rv.pdf(x2)```
Plot the pdf in red	```ax.plot(x2, y2, 'r')```
	```ax.set_xlim((xmin,xmax))```
	```ax.set_ylim((0,0.9))```
	```ax.set_ylabel('Pdf')```
	```ax.set_title('Estimate of the population'```
	```' mean (95% C.I. shaded)')```
Null hypothesis: $\mu = \mu_0 = 1.5$	```mu0 = 1.5```
	```display(Math(r'\mu_0 = ' f"{mu0}"))```
Plot dotted vertical line at $\mu = \mu_0$	```ax.axvline(mu0, c='k', ls='--', lw=1)```
	```ax.text(mu0, -0.05 , r'μ_0')```
SciPy's one-sample t-test	```res = stats.ttest_1samp(samples, mu0)```
Test statistic (manual)	```t = (Smean-mu0)/scale```
	```print(f't statistic= {t:.4f}')```
Test statistic (SciPy)	```print(f'SciPy:t statistic={res.statistic:.4f}')```
p-value (manual-ish)	```p = 2*(1-rv.cdf(mu0))```
	```print(f'p-value = {p:.4f}')```
p-value (SciPy)	```print(f'SciPy:p-value = {res.pvalue:.4f}')```
	```print('')```
Display p-value on figure	```ax.text(1.1*mu0,0.7, f'p = {p:.4}')```
Report confidence interval	```display(Latex('95% C.I. for μ'))```
	```CL = res.confidence_interval(0.95)```
	```print(f'({CL.low:.4f},{CL.high:.4f})')```
The confidence interval	```Xcl = np.linspace(CL.low, CL.high)```
Shade the area under the pdf over the C.I.	```ax.fill_between(Xcl, 0, rv.pdf(Xcl),```
	```color='plum', alpha=0.7)```
	```plt.grid('on')```
	```plt.show()```

8.5 χ^2 distribution and goodness of fit

a) Let $X_1, X_2, \ldots, X_n$ be n iid random variables drawn from the standard normal distribution $N(0, 1)$.

Plot the distribution of the random variable $\sum_{i=1}^{n} X_i^2$. Investigate the limit as $n \to \infty$.

b) I toss four coins simultaneously. Let N be the number of heads shown in each throw. I repeat this 100 times. The following data for the number of heads was obtained.

N	0	1	2	3	4
Observed frequency	12	33	36	16	3

Are the coins fair?

In section 8.4, we tested a hypothesis on the parameter of a known distribution (it is recommended that you work through the previous section first to understand the framework of hypothesis testing). In this section, we will test a hypothesis concerning the nature of the distribution itself. In particular, we want to test whether some given data comes from a certain distribution. We will see that the *goodness of fit* can be measured by a quantity called χ^2 (chi-squared).

$a)$ We use Python to generate n random variates drawn from $N(0, 1)$ and compute the sum of their squares. When we repeat this many times (say 10^4), a distribution of the random variable $X = \sum_{i=1}^{n} X_n^2$ is obtained. Fig. 8.10 shows the histograms of normalised frequencies for $n = 1, 2, 3$ and 5.

The following theorem gives the identity of the red curve plotted over each histogram.

Theorem 8.3 *Let $X_1, X_2, \ldots, X_n$ be n random iid variables drawn from the standard normal distribution $N(0, 1)$. Then, the random variable $X = \sum_{i=1}^{n} X_i^2$ follows the χ^2 distribution with pdf*

$$f(x) = \begin{cases} \frac{1}{2^{n/2}\Gamma(n/2)} x^{\frac{n}{2}-1} e^{-x/2}, & x > 0, \\ 0, & x \leq 0, \end{cases} \tag{8.20}$$

where Γ is the gamma function and the degrees-of-freedom parameter n is positive.

(See [52] for proof.) Note from the power of x that the f is continuous at 0 if and only if $n > 2$.

Looking at fig 8.10, we might guess that as n increases the distribution looks more and more like a normal distribution, with the peak shifting further and further to large x. Indeed, this follows from the Central Limit Theorem.

$b)$ Let's recast the question "*are the coins fair?*" in the form a hypothesis test. If the coins are fair, then the number of heads, k, in each 4-coin throw would follow the binomial distribution $B(4, k)$ with $p_{\text{success}} = 0.5$ (we reserve the symbol p for the p-value to be calculated later). Here are the null and alternative hypotheses:

H_0 : The number of heads follows the binomial distribution with $p_{\text{success}} = 0.5$

H_1 : The number of heads does not follow the binomial distribution with $p_{\text{success}} = 0.5$.

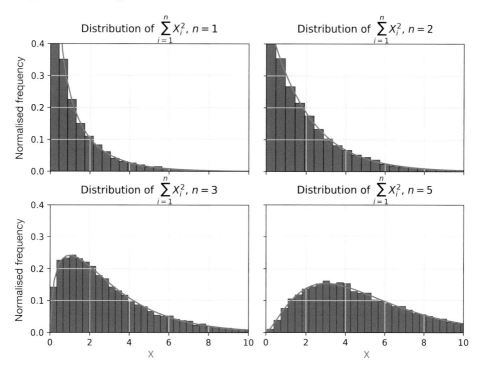

Fig. 8.10: Histograms showing the distributions of the sum of squares $X = \sum_{i=1}^{n} X_i^2$ for $n = 1, 2, 3$ and 5. Each red curve is a χ^2 distribution (8.20).

In particular, H_1 would suggest that the coins are not fair (assuming that no other factors are systematically affecting the outcome).

We are going to perform a *goodness-of-fit test* by first creating another row called "expected frequency" in the given table. Assuming H_0, we expect that out of 100 throws, the number of throws showing k heads equals

$$100 \binom{4}{k} (0.5)^k (0.5)^{4-k} = \frac{25}{4} \binom{4}{k}.$$

N	0	1	2	3	4
Observed frequency	12	33	36	16	3
Expected frequency	6.25	25	37.5	25	6.25

Fig. 8.12 shows the observed and expected (normalised) frequencies of the number of heads. The figure is produced using the code `chi2test.ipynb`. There are significant discrepancies between the two distributions, and this might make you suspect that H_0 should be rejected. Let's see how this can be done using the goodness-of-fit test.

The procedure is similar to the t-test in the previous section. We will calculate the test statistic and the p-value using the significance level $\alpha = 0.05$ (*i.e.* 95% confidence).

Let's begins by calculating the test statistic given by

$$\chi^2 = \sum_1^c \frac{(\text{observed} - \text{expected})^2}{\text{expected}}, \tag{8.21}$$

where c is the number of categories in the data ($c = 5$ in our case). The idea is simple: if χ^2 is large, then the observed and expected frequencies are very different, meaning that the hypothetical distribution is no good. We reject H_0.

In our example, we find

$$\chi^2 = \frac{(12 - 6.25)^2}{6.25} + \frac{(33 - 25)^2}{25} + \frac{(36 - 37.5)^2}{37.5} + \frac{(16 - 25)^2}{25} + \frac{(3 - 6.25)^2}{6.25}$$

$$= 12.84.$$

The reason we have used the same symbol χ^2 for both the test statistic (8.21) and the probability distribution (8.20) is due to the following theorem.

Theorem 8.4 (Convergence to χ^2) *Consider a random sample of size n drawn from a population. Let O_i and E_i be the observed and expected numbers of samples that are in category i (where i = 1, 2, . . . , c). As n → ∞, the distribution of the statistic*

$$\sum_{i=1}^c \frac{(O_i - E_i)^2}{E_i}$$

converges to the χ^2 distribution (8.20) with c − 1 degrees of freedom.

This theorem is due to *Karl Pearson* (1857–1936), a British polymath and influential statistician whose work has shaped the way statistics is applied in modern research. Sadly, like his mentor Francis Galton, Pearson saw scientific justification in eugenics and worked fervently to promote it.

But back to our problem. With 100 samples, we can use Theorem 8.4 to determine how large the test statistic χ^2 would have to be for H_0 to be rejected with significance $\alpha = 0.05$. We simply determine the critical value of χ^2 such that the area under the distribution (with 4 degrees of freedom) from χ^2 to ∞ equals 0.05. This rejection region is shown below.

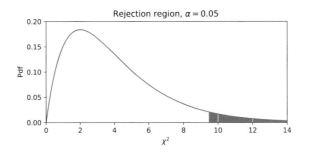

Fig. 8.11: The χ^2 distribution with 4 degrees of freedom. The shaded rejection region (with infinite tail) has area $\alpha = 0.05$, and the critical value is $\chi^2_{\text{crit}} \approx 9.4877$.

The critical value is at $\chi^2_{\text{crit}} \approx 9.4877$. We can check this with SciPy as follows:

```
from scipy import stats
rv = stats.chi2(4)
crit = rv.ppf(0.95)
print(f'{crit:.4f}')
```

Output: 9.4877

Since our test statistic $\chi^2 > \chi^2_{\text{crit}}$, it lies in the rejection region, and therefore H_0 is rejected at 95% confidence.

Note the difference between the rejection regions in figs. 8.7 (the t-test) and 8.11 (the χ^2- test). The former is a two-tailed test whilst the latter is a one-tailed test.

Finally, we may wish to calculate the p-value associated with the null hypothesis. The p-value equals the area in the tail of the pdf bounded by the test statistic χ^2:

$$p = \int_{\chi^2}^{\infty} f(x)\, \mathrm{d}x$$
$$= F(\chi^2) \quad \text{(where } F \text{ is the cdf)}$$
$$= 0.0121 \text{ (4 dec. pl.)}$$

Since the p-value *is* a wee value ($p < 0.05$), we reject H_0. The p-value tells us that under the null hypothesis, the probability of observing our data (or a more extreme dataset with an even larger χ^2) is only 1.21%. We conclude that, at 95% confidence, the coins are not all fair.

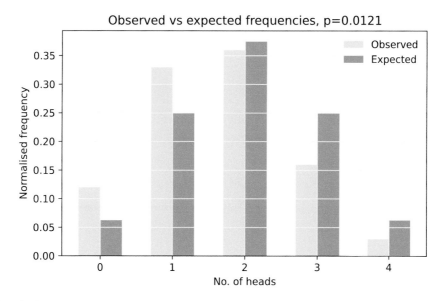

Fig. 8.12: Histogram comparing the observed and expected (normalised) frequencies assuming the binomial distribution with $n = 4$ and Pr(heads) = 0.5. The figure is produced by chi2test.ipynb.

χ^2 test in SciPy

SciPy can help us quickly calculate the test statistic χ^2 and the p-value using the syntax shown in the code `chi2test.ipynb`. The key syntax is the line

$$\text{res = stats.chisquare(f\_obs, f\_exp)}$$

where `f_obs` and `f_exp` are the observed and expected frequencies. We then report the χ^2 value using `res.statistic` and the p-value using `res.pvalue`. Be careful to distinguish between SciPy's syntax `chi2` (the distribution) and `chisquare` (the test). The code output shown below shows that SciPy's results agree with our calculations.

Code output

```
chi2 statistic = 12.8400
SciPy: chi2 statistic = 12.8400

p-value = 0.0121
SciPy:p-value = 0.0121
```

DISCUSSION

- **Practical advice and tea-tasting**. Since theorem 8.4 only holds in the large sample limit, many authors recommend not using the χ^2 test if the sample size is less than 50. In addition, if the expected frequencies are too small (< 5), it is recommended that the data in different categories be combined to increase the frequencies.

  For a test than can deal with small samples, see *Fisher's exact test*, devised by the eminent British statistician *Sir Ronald Fisher* (1890–1962) (who unfortunately also worked on eugenics). Fisher was motivated to statistically test whether one Dr Muriel Bristol was really able to taste if tea or milk was added first to a cup of tea. See [181] for a readable historical review, or [77] for a more mathematical account of this experiment.

- **The Gamma distribution**. The χ^2 distribution is a special case of the *gamma distribution* with pdf

$$f(x) = \frac{\beta^\alpha}{\Gamma(\alpha)} x^{\alpha-1} e^{-\beta x}, \qquad x > 0,$$

and $f(x) = 0$ for $x \leq 0$. The gamma distribution has two parameters $\alpha, \beta > 0$. When $(\alpha, \beta) = (n/2, 1/2)$, we recover the χ^2 distribution. The gamma distribution is a large family of distributions, finding applications in various areas from biology (*e.g.* modelling the distribution of protein concentration in cells [69]) to astronomy (modelling the luminosity distribution of galaxies [61]).

chi2test.ipynb (for plotting fig. 8.12 and reporting the results relevant to the χ^2 test)	
	``` import numpy as np import matplotlib.pyplot as plt from scipy import stats %matplotlib ```
Observed frequencies	``` f_obs = np.array([12, 33, 36, 16, 3]) ```
Number of categories Sample size (no. of throws)	``` C = len(f_obs) Nsmpl = sum(f_obs) ```
Expected frequencies	``` X = np.arange(C) f_exp = Nsmpl*stats.binom.pmf(X, C-1, 0.5) ```
Test statistic χ^2 (manual calculation)	``` chisq = sum((f_obs - f_exp)**2/f_exp) ```
SciPy's goodness-of-fit test	``` res = stats.chisquare(f_obs, f_exp) ```
Report test statistic Report SciPy's test statistic	``` print(f'chi2 stat. = {chisq:.4f}') print(f'SciPy: chi2 stat.={res.statistic:.4f}') print('') ```
χ^2 distribution Read p-value from cdf	``` rv = stats.chi2(df = C-1) p = 1 - rv.cdf(chisq) ```
Report p-value Report SciPy's p-value	``` print(f'p-value = {p:.4f}') print(f'SciPy:p-value = {res.pvalue:.4f}') ```
Let's plot pairs of bars side-by-side	``` fig, ax = plt.subplots(1,1, figsize=(6,4)) ```

Width of each histogram bar
Normalised observed frequencies
(blue bars, offset left)
Normalised expected frequencies
(orange-ish bars, offset right)
	``
	width = 0.3
	ax.bar(X - width/2, f_obs/Nsmpl,
	        width = width, color = 'skyblue')
	ax.bar(X + width/2, f_exp/Nsmpl,
	        width = width, color = 'coral')
	ax.set_title('Observed vs expected '
	            f'frequencies, p = {p:.4f}')
	ax.set_xlabel('No. of heads')
	ax.set_ylabel('Normalised frequency')
	
	plt.legend(['Observed', 'Expected'])
	plt.grid('on')
	plt.show()
	``

8.6 Linear regression and Simpson's paradox

Download the files `datasetA.csv` and `datasetB.csv` from the book's website[a]. In each data file, there are 500 rows of two comma-separated numbers x and y. Here are the first few entries in each file.

datasetA.csv	datasetB.csv
3.0468, 0.8157	−0.2851, 5.0319
4.6406, −2.0899	−4.0064, 1.1191
2.8146, −1.5743	3.1098, 4.9286
2.7879, −2.8717	−1.4257, 1.8992
$\vdots$	$\vdots$

Find the equation of the regression line $y = \alpha x + \beta$ for each of the following datasets.
a) dataset A,   b) dataset B,   c) the combined dataset.

[a] https://github.com/siriwarwick/book/tree/main/Chapter8

Linear regression involves fitting a curve through scatter points in order to understand the trend in the data. For example, suppose we have data points (x_i, y_i) in $\mathbb{R}^2$. We might try to fit a straight line $y = \alpha x + \beta$ through the data as 'best' as we can (*e.g.* by minimising some function measuring the mismatch between data and model).

The *linear* part of linear regression refers to the fact that the model can be regarded as a linear combination of the parameters (α, β in the case of a straight line). In this sense, fitting a polynomial $y = \sum \alpha_k x^k$ to the data is also regarded as linear regression.

The *regression* part refers to the phenomenon of *regression to the mean* (as coined by Galton). This will explained later in the discussion section.

In this section, we try our hand at handling data contained in files using Pandas, and performing linear regression on the datasets. We will also discuss Simpson's paradox in which an apparent trend seen in various datasets vanishes or reverses when the datasets are combined.

Least-square method

Let's consider the *method of least squares* which will give us the equation of the regression line. Suppose the line

$$y = \alpha x + \beta$$

is drawn through N data points (x_i, y_i), where $1 \leq i \leq N$. In the ideal situation where the line goes through all data points, we would be able to solve the following system of linear equations *exactly* for (α, β)

$$\alpha x_1 + \beta = y_1$$
$$\alpha x_2 + \beta = y_2$$
$$\vdots$$
$$\alpha x_n + \beta = y_n$$

However, in most situations, the data and the line will not match exactly. Let $\varepsilon_i :=$ $(\alpha x_i + \beta - y_i)$ be the difference between the y coordinates of the ith data point and the line – see fig. 8.13.

We define the regression line as one which minimises the *residual sum of squares*, SS_{res}, defined as:

$$\text{Minimise } SS_{\text{res}} := \sum_{i=1}^{n} \varepsilon_i^2 = \sum_{i=1}^{n} (\alpha x_i + \beta - y_i)^2. \tag{8.22}$$

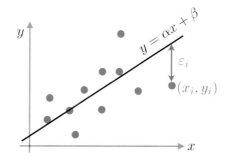

Fig. 8.13: The least-squares method. The coefficients α, β minimise the residual sum of squares $SS_{\text{res}} := \sum_i \varepsilon_i^2$.

Eq. 8.22 can be expressed in matrix form as

$$\text{Minimise } SS_{\text{res}} := |A\mathbf{x} - \mathbf{b}|^2, \tag{8.23}$$

$$\text{where } A = \begin{pmatrix} x_1 & 1 \\ x_2 & 1 \\ \vdots & \\ x_n & 1 \end{pmatrix}, \quad \mathbf{x} = \begin{pmatrix} \alpha \\ \beta \end{pmatrix}, \quad \mathbf{b} = \begin{pmatrix} y_1 \\ y_2 \\ \vdots \\ y_n \end{pmatrix}.$$

Although we have a situation where the linear system $A\mathbf{x} = \mathbf{b}$ cannot be solved exactly, we can still find the best approximation $\hat{\mathbf{x}}$ (the least-squares solution) which minimises $|A\mathbf{x} - \mathbf{b}|^2$.

Linear algebra gives us an elegant solution (see [194] for proof).

Theorem 8.5 *(Least-squares solution) Suppose we have a system $A\mathbf{x} = \mathbf{b}$ where A is an $m \times n$ matrix with $m \geq n$ and rank $A = n$. The least-squares solution $\hat{\mathbf{x}}$ is unique and satisfies*

$$A^T A\hat{\mathbf{x}} = A^T \mathbf{b}. \tag{8.24}$$

The condition rank $A = n$ ensures that the matrix $A^T A$ is invertible. Note that if A is invertible, then A^T is also invertible, and the least-squares solution is simply the exact solution $A^{-1}\mathbf{b}$.

Fitting a higher-degree polynomial through data points can be done using exactly the same method (exercise 4).

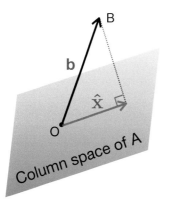

There is a nice geometric interpretation of the least-squares solution $\hat{\mathbf{x}}$ in the language of linear algebra. If vector $\mathbf{b}$ is not in the column space of A, the approximate solution $\hat{\mathbf{x}}$ represents the projection of $\mathbf{b}$ onto the column space. The picture on the left represents this situation in 3D. The least-squares solution $\hat{\mathbf{x}}$ is the vector joining O to the point on the plane closest to the point B, and the minimum error (min SS_{res}) is the corresponding shortest distance.

The code below reads in data from a file and stores it as a dataframe. The least-squares solution is then computed by using SciPy's `linalg.solve` to solve the linear system (8.24). See §5.4 for our earlier discussion on solving linear systems.

Important: make sure the `csv` file is in the same folder as the `ipynb` file which reads it.

Reading data file and computing the least-squares solution	

	`import numpy as np`
	`from scipy import linalg as LA`
	`import pandas as pd`

Read in data in csv file as a dataframe	`df = pd.read_csv('datasetA.csv', sep=',',` `                                header=None)`

x and y coords of the data points	`x = df.values[:,0]`
	`y = df.values[:,1]`
	`ones = np.ones_like(x)`

Construct matrix A	`A = np.column_stack((x , ones))`
Solve eq. 8.24	`alpha, beta = LA.solve(A.T@A, A.T@y)`
Display result to 5 dec. pl.	`print(f'alpha={alpha:.5f}, beta={beta:.5f}')`

Here are the code outputs for the values of the least-squares coefficients for dataset A, dataset B and the combined dataset.

For dataset A : `alpha = 0.64874, beta = -4.90148`

For dataset B : `alpha = 0.59491, beta = 3.88723`

For combined dataset : `alpha = -0.28755, beta = 0.84805`

Simpson's paradox

The code `regression.ipynb` produces fig. 8.14 which shows the datasets in the *x-y* plane along with the regression lines. The latter are conveniently plotted using the `fit` function in NumPy's `polynomial` library which we previously used in §7.10. You should verify that NumPy's values for α and β agree with the solutions we obtained using matrices.

The red and blue clusters of points both show a trend in which *y* increases as *x* increases (*i.e. x* and *y* are positively correlated). However, when the dataset are combined, the data shows negative correlation. This is *Simpson's paradox*, named after the British statistician *Edward Hugh Simpson* (1922–2019). The 'paradox' here is that a trend seen in several datasets can disappear or reverse when the datasets are combined.

This phenomenon has led to misunderstanding and misuse of data. For instance, the reduced death rate in various age groups due to vaccination against Covid can be reversed in the data for the total population[5]. Simpson's paradox cautions us to be wary of inferring information from a large datasets in which hidden variables have been ignored.

In exercise 5, we will express Simpson's paradox as a mathematical statement involving inequalities.

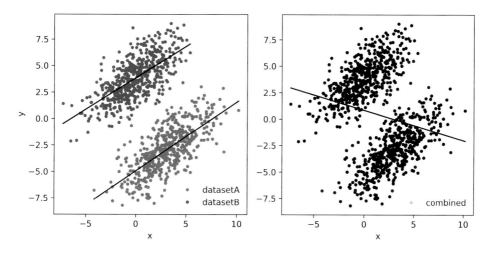

Fig. 8.14: *Left*: (x, y) data from the files `datasetA.csv` (red cluster) and `datasetB.csv` (blue), together with the regression lines. *Right*: Simpson's paradox: Combining the datasets gives a regression line with the opposite trend. The figure is produced by `regression.ipynb`

Finally, you may be interested to know that the data points were generated by sampling 500 pairs of numbers from a *bivariate normal distribution* with different covariance matrices. This will all be investigated in the next section.

---

[5] https://covidactuaries.org/2021/11/22/simpsons-paradox-and-vaccines/

DISCUSSION

- **Correlation coefficient** r. Note that in fig. 8.14, the combined dataset has a larger spread and shows a 'weaker' linear trend than that in the individual datasets.

  The strength of conformity of a dataset to a linear trend can be measured by the *correlation coefficient* r (sometimes called *Pearson's correlation*), defined as

  $$r := \frac{\sum_i (x_i - \bar{x})(y_i - \bar{y})}{\sqrt{\sum_i (x_i - \bar{x})^2 \sum_i (y_i - \bar{y})^2}}, \tag{8.25}$$

  where $\bar{x}$ and $\bar{y}$ are the sample means of x_i and y_i. The sign of r tells us whether x and y are correlated or anti-correlated, and the magnitude measures the strength of a linear relationship. It can be shown that $-1 \leq r \leq 1$, with $r = \pm 1$ if the datapoints line up perfectly in a line.

  A bit of algebra shows that the slope α and intercept β of the regression line are related to r by

  $$\alpha = r \frac{s_y}{s_x}, \qquad \beta = \bar{y} - \alpha \bar{x}, \tag{8.26}$$

  where s_x and s_y are the sample standard deviations. Keep in mind though that r can be calculated without having to draw any line through the data points.

- **Regression to the mean**. Substituting eq. 8.26 into the equation of the regression line, we have

  $$\frac{y - \bar{y}}{s_y} = r \left(\frac{x - \bar{x}}{s_x} \right).$$

  Suppose that x is one standard deviation away from the mean ($x = \bar{x} + s_x$). The equation of the regression line tells us that y will be r standard deviations away from $\bar{y}$. Unless the data points line up a perfect line, r will be a small number. Thus, this observation shows that the predicted value of y is closer (in unit of standard deviations) to the mean $\bar{y}$ than x is to $\bar{x}$. This *regression to the mean* (as observed by Galton) explains the origin of the term regression.

- **Coefficient of determination** R^2. Once we have a line through the data points, we might want to measure how well the line fits the data. More precisely, we ask: does introducing a linear relationship between x and y help to explain the variation in the y values in the data?

  If we hadn't introduced a linear relationship between x and y, we might measure the variation in y by the *total sum of squares* around the sample mean $\bar{y}$ of the n y-coordinates y_i.

  $$SS_{\text{tot}} := \sum_{i=1}^{n} (y_i - \bar{y})^2.$$

  However, with a linear relationship $\hat{y} = mx + c$, the variation in y about the line can be measured by the residual sum of squares.

  $$SS_{\text{res}} := \sum_{i=1}^{n} (y_i - \hat{y})^2.$$

  The *coefficient of determination*, R^2, is the fractional difference in the variation in y due to introduction of the linear relationship.

$$R^2 := \frac{SS_{\text{tot}} - SS_{\text{res}}}{SS_{\text{tot}}} = 1 - \frac{SS_{\text{res}}}{SS_{\text{tot}}}. \tag{8.27}$$

Note that R^2 can be measured using any line through the data, not just the regression line. A perfect fit gives $R^2 = 1$. A horizontal line drawn at $y = \bar{y}$ gives $R^2 = 0$. In fact, R^2 can be negative for a particularly terrible fit to the data (*e.g.* when a line with a negative gradient is drawn through dataset A).

Exercises 6 and 7 ask you to measure r and R^2 in the data shown in fig. 8.14.

- **Anscombe's quartet**. Given a dataset, the equation of the regression line, and additional measures like μ, σ, r and R^2, still give us an incomplete understanding of data in general. *Visualisation* is an indispensable part of data analysis, as demonstrated by the famous *Anscombe's quartet*, which comprises 4 very different datasets with identical statistical measures. You can explore these intriguing datasets yourself with Seaborn[6].

---

[6] https://seaborn.pydata.org/examples/anscombes_quartet.html

regression.ipynb (for plotting fig. 8.14)	
| For plotting regression lines | ```
import numpy as np
import matplotlib.pyplot as plt
from numpy.polynomial import Polynomial as P
import pandas as pd
%matplotlib
``` |
| Create 2 plots side by side | ```
fig,(ax1,ax2) = plt.subplots(1,2,figsize=(8,4))
``` |
| Read in csv file as a dataframe | ```
df1 = pd.read_csv('datasetA.csv', sep=',',
                  header=None)
``` |
| *x* and *y* coords of dataset A | ```
x1 = df1.values[:,0]
y1 = df1.values[:,1]
``` |
| Left panel: Scatter plot (red dots) | ```
ax1.plot(x1,y1,'.r')
``` |
| Repeat for dataset B | ```
df2 = pd.read_csv('datasetB.csv', sep=',',
                  header=None)
x2 = df2.values[:,0]
y2 = df2.values[:,1]
``` |
| Plot as blue dots | ```
ax1.plot(x2,y2,'.b')
``` |
| Insert legend | ```
ax1.legend(['datasetA','datasetB'],
           loc='lower right')
``` |
| Gradient and *y*-intercept of line | ```
poly1 = P.fit(x1, y1, 1).convert()
xfit1 = np.linspace(min(x1),max(x1))
yfit1 = poly1(xfit1)
``` |
| Overlay the regression line in black | ```
ax1.plot(xfit1, yfit1, 'k')
``` |
| Repeat for dataset B | ```
poly2 = P.fit(x2, y2, 1).convert()
xfit2 = np.linspace(min(x2),max(x2))
yfit2 = poly2(xfit2)
ax1.plot(xfit2, yfit2, 'k')

ax1.set_xlabel('x')
ax1.set_ylabel('y')
ax1.grid('on')
``` |
| *x* and *y* coords of combined dataset | ```
X = np.concatenate((x1,x2))
Y = np.concatenate((y1,y2))
``` |
| Right panel: plot as grey dots Insert legend | ```
ax2.plot(X,Y,'.', c='gray')
ax2.legend(['combined'], loc='lower right')
``` |
| Plot regression line for combined dataset | ```
Poly = P.fit(X, Y, 1).convert()
Xfit = np.linspace(min(X),max(X))
Yfit = Poly(Xfit)
ax2.plot(Xfit, Yfit, 'k')

ax2.set_xlabel('x')
ax2.grid('on')

plt.show()
``` |

8.7 Bivariate normal distribution

Let X_i $(i = 1, 2, \ldots, n)$ be random variables with zero mean. We say that the random variables X_i follow the *multivariate normal distribution* if their joint pdf is given by

$$f(\mathbf{X}) = \frac{1}{\sqrt{(2\pi)^n \det \Sigma}} \exp\left(-\frac{1}{2}\mathbf{X}^T \Sigma^{-1} \mathbf{X}\right), \qquad (8.28)$$

where $\mathbf{X} = \begin{pmatrix} X_1 & X_2 & \ldots & X_n \end{pmatrix}^T$. The covariance matrix Σ is defined as

$$\Sigma_{ij} = \text{Cov}(X_i, X_j) := E[(X_i - \mu_{X_i})(X_j - \mu_{X_j})].$$

Since the variables have zero mean, $\Sigma_{ij} = E[X_i X_j]$.

In 2 dimensions $(n = 2)$, f is said to be the pdf of the *bivariate normal distribution*.

Let $\mathbf{X} = \begin{pmatrix} X & Y \end{pmatrix}$. The covariance matrix can be expressed as

$$\Sigma = \begin{pmatrix} \sigma_X^2 & \rho\sigma_X\sigma_Y \\ \rho\sigma_X\sigma_Y & \sigma_Y^2 \end{pmatrix},$$

where σ_X^2 and σ_Y^2 are the variances of X and Y. The *correlation* ρ is given by

$$\rho := \frac{\text{Cov}(X, Y)}{\sqrt{\text{Var}(X)\text{Var}(Y)}}$$

and satisfies $-1 < \rho < 1$.

a) If $\sigma_X = \sigma_Y = 1$, investigate the effect of ρ on the shape of the pdf.

b) It is known that the file `datasetA.csv` are random numbers drawn from a bivariate normal distribution. Estimate μ_X, μ_Y and Σ.

Two random variables X and Y are said to be *jointly continuous* if there exists a non-negative function $f : \mathbb{R}^2 \to \mathbb{R}$ such that the probability that $(X, Y) \in A \subseteq \mathbb{R}^2$ is given by

$$\Pr\left((X, Y) \in A\right) = \iint_A f(x, y)\, dx\, dy.$$

f is called the *joint* or *bivariate probability distribution*.

The bivariate normal distribution (8.28) is applicable when X and Y have zero mean. If X and Y have nonzero mean μ_X and μ_Y, we simply apply the translation $X \mapsto X - \mu_X$ and $Y \mapsto Y - \mu_Y$ to the pdf (8.28). The resulting pdf can be written explicitly as

$$f(X, Y) = \frac{1}{2\pi\sigma_X\sigma_Y\sqrt{1 - \rho^2}} \exp\left(-\frac{1}{2(1 - \rho^2)}\left[Z_X^2 - 2\rho Z_X Z_Y + Z_Y^2\right]\right) \qquad (8.29)$$

where $\quad Z_X = \dfrac{X - \mu_X}{\sigma_X}, \quad Z_Y = \dfrac{Y - \mu_Y}{\sigma_Y}.$

Compare this with the pdf for the univariate normal pdf given by

$$g(X) = \frac{1}{\sigma\sqrt{2\pi}} \exp\left(-\frac{1}{2}Z^2\right), \quad \text{where} \quad Z = \frac{x - \mu}{\sigma}. \qquad (8.30)$$

We see that when X and Y are *uncorrelated* ($\rho = 0$), the bivariate normal pdf is given by $f(X,Y) = g(X)g(Y)$. This also means that X and Y are independent[7] random variables (we defined this in eq. 7.33).

Here is a comparison between the univariate and bivariate normal pdfs:

	Univariate normal	Bivariate normal
Peak position	μ	(μ_X, μ_Y)
'Shape' parameters	σ	σ_X, σ_Y, ρ
Normalisation	$\int_{\mathbb{R}} f(x)\,\mathrm{d}x = 1$	$\iint_{\mathbb{R}^2} f(x,y)\,\mathrm{d}x\,\mathrm{d}y = 1$

The bivariate normal distribution is a very useful tool for data analysis. For instance, if X and Y are the height and weight of 40-year old women in England, each variable would very likely follow a normal distribution with its own mean and variance. However, modelling them jointly as a bivariate normal distribution gives us extra information on how much height and weight are correlated. This information is captured by the correlation coefficient ρ.

You might have noticed that the definition for the correlation ρ looks very similar to the coefficient of correlation r (eq. 8.25) for two sets of samples x and y. Indeed they are the same concept: we use ρ to for measuring correlation between two populations, and r for two sets of samples.

$\boxed{a}$ Let $\mu_X = \mu_Y = 0$ and $\sigma_X = \sigma_Y = 1$. The effect of ρ on the shape of the bivariate pdf is explored using the code `bivariate.ipynb`. By varying ρ from -1 to 1 using a slider, we see the changes in the pdf shown in fig. 8.15. In the figure, we also show the contours of the pdf on the x-y plane.

Some observations from the figure:

- When $\rho = 0$, the pdf has a rotational symmetry about the z-axis. This is because when $\rho = 0$, we have $z = f(X,Y) \propto \exp(X^2 + Y^2)$, which can be transformed using cylindrical coordinates (r, θ, z) to $z \propto \exp(r^2)$. There is no θ dependence, meaning that the pdf looks the same in all directions. The contours (curves of constant z) are therefore concentric circles $X^2 + Y^2 = $ constant.
- When $\rho > 0$, the pdf is stretched in the direction $y = x$ and squashed in the direction $y = -x$. The contours are ellipses with principal axes along $y = \pm x$ direction. Higher values of ρ stretch/squash the ellipse further.
  When $\rho < 0$, the stretching and squashing effects are reversed. The pdf and contour ellipses are elongated in the direction $y = -x$.

Here is a useful result to help us understand why the stretching and squashing occur in these directions.

Lemma (*Quadratic form of an ellipse*) Let M be a real symmetric 2×2 matrix with eigenvalues λ_1, λ_2 and the corresponding eigenvectors $\mathbf{e}_1, \mathbf{e}_2$. Let $\mathbf{x} = \begin{pmatrix} x \\ y \end{pmatrix}$. Then, the equation $\mathbf{x}^T M \mathbf{x} = $ constant describes an ellipse centred at the origin with principal axes along $\mathbf{e}_1, \mathbf{e}_2$. Furthermore, the ratio of the principal axes is $1/\sqrt{\lambda_1} : 1/\sqrt{\lambda_2}$.

---

[7] In general, however, two random variables can be uncorrelated but not independent. For example, $X = \mathcal{N}(0, 1)$ and $Y = X^2$.

The expression $\mathbf{x}^T M \mathbf{x}$ is called a *quadratic form*. We give the proof of the above lemma in the discussion.

Note that the contours of the bivariate normal pdf is $z = $ constant, *i.e.* $\exp\left(-\frac{1}{2}\mathbf{x}^T \Sigma^{-1} \mathbf{x}\right) = $ constant, which implies that $\mathbf{x}^T \Sigma^{-1} \mathbf{x} = $ constant.

Assuming that $\sigma_X = \sigma_Y = 1$, the covariance matrix Σ is

$$\Sigma = \begin{pmatrix} 1 & \rho \\ \rho & 1 \end{pmatrix} \implies \Sigma^{-1} \frac{1}{1 - \rho^2} \begin{pmatrix} 1 & -\rho \\ -\rho & 1 \end{pmatrix}.$$

Note that Σ^{-1} is a real symmetric matrix. Its eigenvalues and eigenvectors are

$$\lambda_1 = \frac{1}{1 + \rho}, \quad \mathbf{e}_1 = \begin{pmatrix} 1 \\ 1 \end{pmatrix}, \qquad \lambda_2 = \frac{1}{1 - \rho}, \quad \mathbf{e}_2 = \begin{pmatrix} -1 \\ 1 \end{pmatrix}.$$

According to the Lemma, $\mathbf{x}^T \Sigma^{-1} \mathbf{x} = $ constant describes an ellipse with principal axes $\mathbf{e}_1$ and $\mathbf{e}_2$ (*i.e.* $y = \pm x$). Furthermore, the ratio of the principal axes is $\sqrt{1 + \rho} : \sqrt{1 - \rho}$.

For example, with $\rho = 0.5$, the contours are ellipses with axes ratio $\sqrt{3} : 1$. With $\rho = -0.8$, the ratio is $1 : 3$.

b) From `datasetA.csv`, we find the following sample statistics.

$$\bar{x} = 2.9994 \qquad s_x^2 = 5.2191$$
$$\bar{y} = -2.9557 \qquad s_y^2 = 4.5409$$
$$r = 0.6955$$

We then insert these values into the estimator, S, for the covariance matrix Σ.

$$S := \begin{pmatrix} s_x^2 & r s_x s_y \\ r s_x s_y & s_y^2 \end{pmatrix} = \begin{pmatrix} 5.2191 & 3.3859 \\ 3.3859 & 4.5409 \end{pmatrix}.$$

Alternatively, we can also obtain the matrix S using `scipy.stats.Covariance` or `numpy.cov`.

Fig. 8.16 shows the data from `datasetA.csv` along with the 1, 2, and 3σ contours. For each ellipse, the lengths of the semi major/minor axes are $n/\sqrt{\lambda_1}$ and $n/\sqrt{\lambda_2}$, aligned along the directions of the corresponding eigenvectors $\mathbf{e}_1$ and $\mathbf{e}_2$. Unlike the univariate normal distribution, the 1,2,3σ contours do not correspond to the 68-95-99.7% confidence level. In exercise 8, you will calculate the corresponding percentages for the bivariate normal distribution.

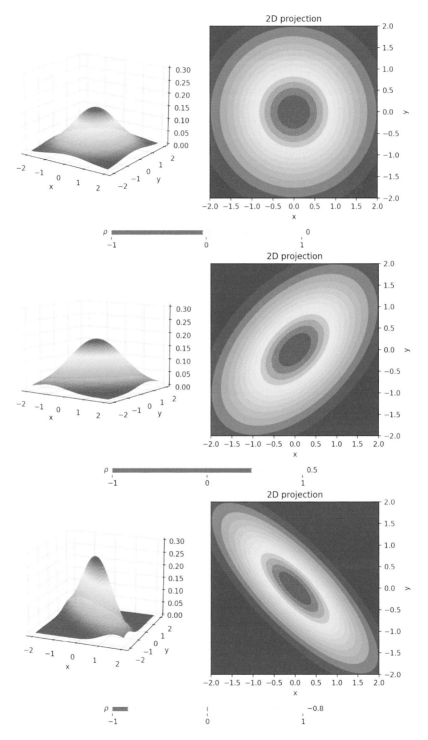

Fig. 8.15: Bivariate normal distribution with $\rho = 0, 0.5$ and -0.8 (top to bottom) with $\sigma_x^2 = \sigma_y^2 = 1$. Each pair of figures (with ρ slider) is produced by the code `bivariate.ipynb`. The pdfs are shown on the left, and their projections onto the x-y plane on the right.

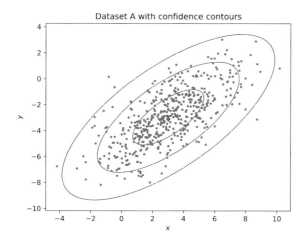

Fig. 8.16: Data from `datasetA.csv` with 1, 2 and 3σ confidence ellipses.

We can now reveal that the code used to generate `datasetA.csv` (and writing the data to file) is given below.

```
Writing pandas dataframe to csv file
import pandas as pd
from scipy import stats

mean  = [3,-3]
Sigma = [[5,3],[3,4]]
rv  = stats.multivariate_normal(mean, Sigma)
S = rv.rvs(500)
pd.DataFrame(S).to_csv("filename.csv", index=False,
                    header=None, float_format='%.4f')
# (choose your own filename)
```

The main takeaway message in this section is that for bivariate data, the covariance matrix Σ encodes the 'shape' of the data. For a more complete mathematical treatment of the multivariate normal distribution, see [203].

DISCUSSION

• **Diagonalisation of quadratic form**. Let M be a real, symmetric 2×2 matrix with nonzero eigenvalues λ_1, λ_2, and let $\mathbf{x} = \begin{pmatrix} x \\ y \end{pmatrix}$. Here is the proof that $\mathbf{x}^T M \mathbf{x} = 1$ is an ellipse with axes ratio $1/\sqrt{\lambda_1} : 1/\sqrt{\lambda_2}$.

Since M is a real, symmetric matrix, the spectral theorem (thm. 5.5 in §5.7) tells us that $M = PDP^T$, where the columns of P are the eigenvectors of M, and $D = \begin{pmatrix} \lambda_1 & 0 \\ 0 & \lambda_2 \end{pmatrix}$.

The key observation is that, using the eigenvectors as the basis, the quadratic form has no cross terms.

$$\mathbf{x}^T M \mathbf{x} = 1 \xrightarrow{\mathbf{u} = P^T \mathbf{x}} \mathbf{u}^T D \mathbf{u} = 1 \iff \lambda_1 u_1^2 + \lambda_2 u_2^2 = 1.$$

Thus, in the eigenbasis, we see an ellipse whose ratio of the principal axes is $1/\sqrt{\lambda_1} : 1/\sqrt{\lambda_2}$.

- **Marginal distribution**. Suppose that random variables X, Y follow a bivariate distribution $f(X, Y)$. A *marginal distribution* is obtained when a variable is integrated out of the joint pdf. For instance, the pdf of the marginal distribution of X is obtained by integrating out the y variable, *i.e.*

$$f_X(x) = \int_{\mathbb{R}} f(x, y) \, \mathrm{d}y.$$

(We say that Y is *marginalised*.) It can be shown that if X, Y follow that bivariate normal distribution with $\mu_X = \mu_Y = 0$ and $\Sigma = \begin{pmatrix} \sigma_X^2 & \rho\sigma_X\sigma_Y \\ \rho\sigma_X\sigma_Y & \sigma_Y^2 \end{pmatrix}$, then the marginal distributions of X and Y are univariate normal distributions, namely, $\mathcal{N}(0, \sigma_X^2)$ and $\mathcal{N}(0, \sigma_Y^2)$ respectively (the correlation information is lost).

- **Correlation and causation**. Much has now been discussed about the correlation between two variables, but it is worth remembering the old adage that correlation does not imply causation. Two variables could be positively correlated by sheer coincidence, or the correlation may have been caused by a hidden 'confounding' variable. Ref. [159] gives and in-depth yet accessible discussion on establishing causation statistically. See this link[8] for a light-hearted take.

---

[8] https://xkcd.com/552/

bivariate.ipynb (for plotting fig. 8.15)	
| | ```python
import numpy as np
import matplotlib.pyplot as plt
from matplotlib.widgets import Slider
from scipy.stats import multivariate_normal
%matplotlib
``` |
| Specify the correlation coefficient $\rho$ . . . and the variances $\sigma_X^2, \sigma_Y^2$ | ```python
rho = 0.5
S1, S2 = 1, 1
``` |
| $\sigma_X \sigma_Y$ | ```python
S12 = np.sqrt(S1*S2)
``` |
| We'll plot 2 figures side by side<br>A little space at the bottom for a slider | ```python
fig = plt.figure(figsize=(10,6))
plt.subplots_adjust(bottom=0.15)
``` |
| Left plot: 3D surface of the pdf<br>Right plot: contours on the $x$-$y$ plane<br>Equal aspect ratio for the right plot | ```python
ax1 = fig.add_subplot(121, projection='3d')
ax2 = fig.add_subplot(122)
ax2.set_box_aspect(1)
ax2.set_title('2D projection')
ax2.yaxis.tick_right();
ax2.yaxis.set_label_position('right')
ax2.set_xlabel('x'); ax2.set_ylabel('y')
``` |
| Set domain for the 3D plot | ```python
xplot = np.linspace(-2,2)
yplot = np.linspace(-2,2)
X, Y = np.meshgrid(xplot,yplot)
``` |
| Plotting 2 graphs given $\rho$<br>Covariance matrix $\Sigma$<br><br>The bivariate normal pdf given $(X, Y)$<br>Specify colormap for both plots<br>Plot the surface | ```python
def plots(rho):
    Cov = [(S1, rho*S12), (rho*S12, S2)]
    rv = multivariate_normal(cov = Cov)
    Z = rv.pdf(np.dstack([X,Y]))
    cmap = 'rainbow'
    plot1 = ax1.plot_surface(X, Y, Z,
                         alpha=0.9, cmap=cmap)
    ax1.set_zlim(0, 0.3)
    ax1.set_xlabel('x')
    ax1.set_ylabel('y')
``` |
| Set number of contours<br>Plot the contour ellipses | ```python
    levels = np.linspace(0,np.max(Z),15)
    plot2 = ax2.contourf(X, Y, Z,
                   levels = levels, cmap=cmap)
    return plot1, plot2
``` |
| Slider dimensions and position<br>Create $\rho$ slider<br>$-1 < \rho < 1$<br><br>Add ticks to slider manually | ```python
axt = plt.axes([0.34, 0.1, 0.4, 0.03])
rho_slide = Slider(axt, r'$\rho$',
                 -1, 1, valstep=0.01, valinit=0)
axt.add_artist(axt.xaxis)
axt.set_xticks([-1,0,1])
``` |
| Initialise plots at $\rho = 0$ | ```python
plots(0)
``` |
| Update plots if slider is moved<br><br>Get new $\rho$ value from slider<br>Replot | ```python
def update(val):
    ax1.clear()
    rho = rho_slide.val
    plots(rho)

rho_slide.on_changed(update)
plt.show()
``` |
```

8.8 Random walks

A particle is initially at $x_0 = 0$. In each time step, it moves left or right by a distance 1 unit with equal probability. Find the expected value of the distance from the origin after N steps.

Generalise the problem to two dimensions.

A *random walk* is a sequence (X_N) where $X_N = \sum_{i=1}^{N} \varepsilon_i$ and $\{\varepsilon_i\}$ are iid variables. Random walks are a type of stochastic process which occurs in everyday life - from diffusion of fluids to the fluctuation of stock prices. For introductions to stochastic processes, see [21, 158, 178].

In this example, we will study a simple stochastic process called the *symmetric random walk*, defined by

$$X_N = \sum_{i=1}^{N} \varepsilon_i, \quad \text{where } \varepsilon_i = \pm 1 \text{ with equal probability.}$$

In other words, $\{\varepsilon_i\}$ are iid random variables drawn from a Bernoulli distribution with $\Pr(\varepsilon_i = 1) = \Pr(\varepsilon_i = -1) = 0.5$ (*i.e.* a coin toss). We can think of X_N is the displacement of the particle with respect to the origin at step N.

This setup might remind you of the Galton board (§7.8) where X_n represents the horizontal displacement of the ball after encountering n pegs. Here we have reduced the problem to 1 dimension, which allows a clearer visualisation and statistical analysis. We will also see how this approach allows the problem to be generalised to any dimensions.

Firstly, as in the Galton board setup, we can apply the Central Limit Theorem (§8.3) to ε_i, and deduce that the random variable $X_N / \sqrt{N}$, in the large N limit, follows the standard normal distribution $\mathcal{N}(0, 1)$. Equivalently, $X_N \sim \mathcal{N}(0, \sqrt{N})$ in the large N limit.

However, the question asks for the distribution of the distance $|X_N|$ rather than the displacement. It is not immediately clear what this distribution is. Let's see what Python can show us.

Simulating random walks

The code `randomwalk.ipynb` simulates the 1D random walk problem and produces trajectories and the mean-distance plot shown in fig. 8.17. The key syntax in the code is the function `np.cumsum` which gives the cumulative sum of Bernoulli random variables chosen from $\{1, -1\}$. This gives us an array X_N of the particle's displacement at the Nth step.

10^3 trajectories, evolved up to 500 steps, are shown in the left panel of fig. 8.17. They appear to form a cluster with a parabolic shape. To see why, let's calculate the mean and variance of X_N.

$$E[X_N] = \sum_{i=1}^{N} E[\varepsilon_i] = 0.$$

$$\text{Var}(X_N) = \sum_{i=1}^{N} \text{Var}(\varepsilon_i) = \sum_{i=1}^{N} E[\varepsilon^2] = \sum_{i=1}^{N} 1 = N.$$

We expect most of the trajectories to stay within the 3σ boundary, which is the parabola $f(N) = \pm 3\sqrt{N}$, shown as dashed lines. A little modification to the code shows that the trajectories stay within the 3σ boundary $\sim 99.7\%$ of the time (exercise 9).

The jagged blue curve in the right-hand panel shows the mean of $|X_N|$ taken over the 10^3 trajectories. The curve is well approximated by the function $\sqrt{2N/\pi}$. We will see where this comes from shortly. It is interesting that such a simple, predictable average should arise out of a bunch of seemingly random trajectories.

The χ distribution

The following theorem tells us about the distribution of the mean distance at step N of a random walk.

Theorem 8.6 (*The χ distribution*) *Let $Z_1, Z_2, \ldots, Z_k$ be iid random variables drawn from $N(0, 1)$. Then, the random variable $S_k = \sqrt{Z_1^2 + Z_2^2 + \cdots + Z_k^2}$ follows the χ (chi) distribution with k degrees of freedom. Its pdf given by*

$$f(x) = \frac{x^{k-1}e^{-x^2/2}}{2^{(k/2)-1}\Gamma(k/2)}, \quad (x \geq 0) \tag{8.31}$$

and $f(x) = 0$ for $x < 0$.

A little algebra shows that the mean and variance of the variable S_k is

$$E[S_k] = \sqrt{2}\frac{\Gamma((k+1)/2)}{\Gamma(k/2)}, \quad \text{Var}(S_k) = k - (E[S_k])^2. \tag{8.32}$$

Note that with 1 degree of freedom, $E[S_1] = E[|Z_1|] = \sqrt{2}/\Gamma(1/2) = \sqrt{2/\pi}$.

Using this observation, we deduce that the mean of the variable $S_1 = |X_N/\sqrt{N}|$ approaches $\sqrt{2/\pi}$ in the large N limit; that is, $E[|X_N|] \approx \sqrt{2N/\pi}$. This proves the behaviour seen on the right of fig. 8.17.

Another interesting distribution emerges when we consider the time that the particle spends in the region $x > 0$. This leads to the astonishing *arcsine law* explored in exercise 9.

2D random walk

Let's generalise the problem to 2 dimensions. Starting at the origin $(0, 0)$, the particle chooses the displacement randomly from $\{\pm 1\}$ in both the x and y directions. This means that at the next step, the particle will be at either $(1, 1)$, $(-1, 1)$, $(1, -1)$ or $(-1, -1)$. These trajectories are shown in fig. 8.18 in which we simulated 10^3 trajectories up to $N = 500$ steps.

Let $X_N = Y_N = \sum_{i=1}^{N} \varepsilon_i$. The distance D from the origin at step N is therefore given by $D = \sqrt{X_N^2 + Y_N^2}$.

Theorem 8.6 tells us that mean of the variable $S_2 = \sqrt{(X_N^2 + Y_N^2)/N} = D/\sqrt{N}$ approaches that of the χ distribution with 2 degrees of freedom, namely, $E[S_2] = \sqrt{2}\Gamma(3/2) = \sqrt{\pi/2}$.

In other words, $E[D] \approx \sqrt{\pi N/2}$. This is corroborated by Python: the mean distance plot on the right of fig. 8.18 agrees well with our theoretical estimate.

In addition, from eq. 8.32 the variance of S_2 is given by

$$\text{Var}(S_2) = 2 - \frac{\pi}{4}.$$

This implies that the standard deviation of the distance D is given by $\sigma \approx 1.1\sqrt{N}$. We draw the 3σ circle around the cluster of trajectories on the LHS of fig. 8.18. As expected, this encompasses almost the entire cluster of trajectories. In fact, Python tells us that the trajectories stay within the 3σ boundary $\sim 99\%$ of the time (exercise 9).

A nice takeaway from all these observations about random walks is that, although the trajectories appear random, statistics reveals the hidden order and predictability to us.

DISCUSSION

- **A drunk bird**. What is the probability that a random walk returns to the origin? It turns out that the 1D and 2D random walks are *recurrent*, meaning that they return to the origin with probability 1. However, in 3 dimensions and higher, random walks are *transient*, meaning that the probability of returning to the origin is strictly less than 1. This inspired the mathematician Shizuo Kakutani to quip that "*a drunk man will find his way home, but a drunk bird may get lost forever*".

- **Wiener processes**. The random walks that we studied in this section are discrete in time and space. In the continuous limit, the phenomenon is known as *Brownian motion*, or, a *Wiener process*, $W(t)$, defined by the following properties:

  1. $W(0) = 0$;
  2. W is continuous in t with probability 1;
  3. $W(t+s) - W(t) \sim \mathcal{N}(0, s)$ for any $t \geq 0$ and $s > 0$.

  The Wiener process plays an important role in physics [141] and mathematical finance [43].
  Norbert Wiener (1894–1964) was an American mathematician who published pioneering work on stochastic processes and analysis. His work was partly inspired by the observation of pollen motion in water by the Scottish botanist *Robert Brown* (1773–1858).

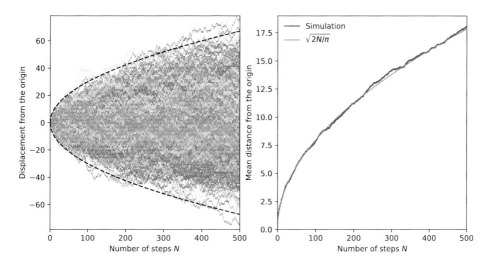

Fig. 8.17: Random walk in 1 dimension generated by `randomwalk.ipynb`. *Left*: 10^3 trajectories, all starting from the origin, evolved up to $N = 500$ steps. The dotted line shows the contour $\pm 3\sqrt{N}$. *Right*: The mean distance (absolute value of the displacement) at step N. The blue simulation points are well approximated by $E[|X_N|] = \sqrt{2N/\pi}$ shown in red.

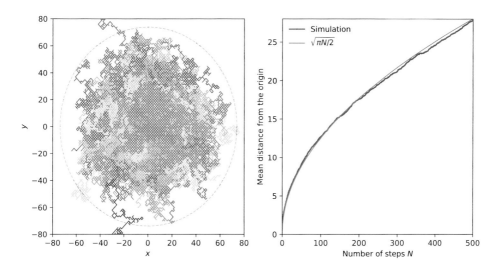

Fig. 8.18: Random walk in 2 dimensions. *Left*: 10^3 trajectories, all starting from the origin, evolved up to $N = 500$ steps. The dotted line shows the circle with radius 3σ. *Right*: The mean distance from the origin at step N. The blue simulation points are well approximated by $E\left[\sqrt{X_N^2 + Y_N^2}\right] = \sqrt{\pi N/2}$ shown in red.

randomwalk.ipynb (for plotting fig. 8.17)	
| | ```python
import numpy as np
import matplotlib.pyplot as plt
import random
%matplotlib
``` |
| Number of trajectories
Number of time steps per trajectory | ```python
Nsims = int(1e3)
Nsteps = 500
``` |
| Initialise mean distance in each step | ```python
Sims = np.arange(Nsims)
Steps= np.arange(Nsteps+1)
SumDist = np.zeros_like(Steps)
``` |
| Plot 2 figures side-by-side
Left plot: Colourful random walks

Assign trajectories different colours | ```python
fig,(ax1, ax2)=plt.subplots(1,2,figsize=(10,5))
ax1.set_xlim(0, Nsteps)
ax1.set_ylim(-3.5*np.sqrt(Nsteps),
             3.5*np.sqrt(Nsteps))
ax1.set_xlabel('Number of steps $N$')
ax1.set_ylabel('Displacement from the origin')
ax1.grid('on')
color=plt.cm.prism_r(np.linspace(0, 1, Nsims))
``` |
| For each trajectory,
choose `Nsteps` random integers (0 or 1)

Re-scale them to -1 and 1
Summing cumulatively
Pre-pend the 0th step
Plot the multicoloured trajectories

Record the distance in each step | ```python
for i in Sims:
    R = [random.randint(0,1) for
         i in range(1,Nsteps+1)]
    moves = 2*np.array(R)-1
    traj = np.cumsum(moves)
    traj = np.insert(traj, 0, 0)
    ax1.plot(Steps, traj, '-',
             lw = 0.5, c = color[i])
    SumDist += np.abs(traj)
``` |
| The $3\sigma$ boundary | ```python
bound = 3*np.sqrt(Steps)
``` |
| Plot the boundary (dashed black) | ```python
ax1.plot(Steps, bound, 'k--',
         Steps, -bound, 'k--')
``` |
| Average distance in each time step | ```python
AveDist = SumDist/Nsims
``` |
| **Right plot**: mean distance in blue
Theoretical prediction in red | ```python
ax2.plot(Steps, AveDist, 'bo-', ms=1)
ax2.plot(Steps,np.sqrt(2*Steps/np.pi),'r',lw=1)
ax2.set_xlim(0, Nsteps)
ax2.set_ylim(0, np.ceil(np.max(AveDist)))
ax2.set_xlabel(r'Number of steps $N$')
ax2.set_ylabel('Mean distance from the origin')
ax2.legend(['Simulation', r'$\sqrt{2N/\pi}$'])
ax2.grid('on')
``` |
| | ```python
plt.show()
``` |

8.9 Bayesian inference

> Some students suggested to the maths department at a certain university that they
> would like a new water fountain (for dispensing drinking water) to be installed in
> the building for students' use.
>
> The department decided to do a survey in which some students were asked if they
> would be using the new water fountain if one were to be installed. Out of 20 students
> surveyed, 9 students said they would.
>
> Should the department install a new water fountain?

At first glance this doesn't seem to be a well-defined mathematical problem. However, this is typical of many real-world problems that require mathematical modelling in order to make a well-informed decision.

The simplest approach is to simply say that since only 9 out of 20 (45%) of those surveyed would use the water fountain, then, extrapolating to the larger population, we might expect that only 45% of the maths department will be using the water fountain. Perhaps the department administrators were hoping that at least 60% of the department would use a new water fountain. So they may decide that it is not worth installing a new water fountain after all. Problem solved - or is it?

In this section, we will show how the problem could be modelled in a more sophisticated way by involving probability in our assumptions, modelling and decision making. The process we will be studying is called *Bayesian inference* which plays a key role in data science and complex decision making (e.g. in machine learning).

Here are two ways in which we can introduce probability into our modelling.

- Let's assume that a randomly chosen student has a fixed probability p of saying that they will use the water fountain if one is built.
  If we think of p as the probability of success in a Bernoulli trial, then it follows that the number of students who respond positively to a new water fountain follows the binomial distribution (see §7.5).
- We don't know exactly what p is, but we might have some initial belief on the ballpark values that it can take. Say, there are already a number of water fountains in the department, or if the survey was done soon after a Covid outbreak, then p would likely to be low (most likely < 0.5). On the other hand, if there are no water outlets around the department, or if the survey was done in a hot summer month, then p would probably be high. This belief can be modelled mathematically as a probability distribution which can be used to inform our decision.
  For now, let's assume that we don't know much about the students in the department, so our initial belief might be that p can take any value between 0 and 1 with equal probability, *i.e.* p is drawn from the uniform distribution $U(0, 1)$. This is called a *non-informative prior*.

Our goal is to use the data (informed by our belief) to infer a statistical constraint on the value of p. The answer will be in the form of a probability distribution of p. This process is called *Bayesian inference*.

Prior, likelihood, posterior

The goal of Bayesian inference is to obtain a statistical constraint on the model parameters θ (which may be a vector of many parameters) using data. At the heart of Bayesian inference is Bayes' Theorem from §7.7.

$$\Pr(\theta|\text{data}) = \frac{\Pr(\text{data}|\theta)\,\Pr(\theta)}{\Pr(\text{data})}. \tag{8.33}$$

It is worth learning the technical name for each of the terms in the above equation.

- $\Pr(\theta)$ on the RHS is called the *prior*. It reflects our initial belief about the values of the model parameters.
  In our scenario, we will start by taking the prior to be $U(0, 1)$, in which case $\Pr(\theta)$ is constant.
- $\Pr(\text{data}|\theta)$ is called the *likelihood*. It is the probability that the underlying generative model takes the parameters θ and produces the observed data. The generative model depends on the situation at hand.
  In the water-fountain scenario, $\theta = p$ and the generative model is the random variable drawn from the binomial distribution $B(20, p)$. The likelihood is the probability that the number drawn equals 9 (matching the observed data), *i.e.*

$$\Pr(\text{data}|\theta) = \Pr(X = 9 | n = 20, p) = \binom{20}{9} p^9 (1 - p)^{11}.$$

- $\Pr(\text{data})$ is called the *evidence*. It can simply be regarded as the normalisation factor which will make the RHS of eq. 8.33 a probability distribution. (In more advanced applications, the evidence can be used to compare the predictive powers of different models.)
- $\Pr(\theta|\text{data})$ on the LHS is called the *posterior* probability distribution (or simply the posterior). This represents our updated belief about θ upon seeing the data.

In short, one might summarise the Bayesian inference process as follows: starting with an initial belief (the prior[9]), we use the data and a probability model (the likelihood) to update our belief (the posterior).

Coding

The code `bayesian.ipynb` performs the Bayesian inference pipeline via the following steps

- **Step 1: Prior** – Draw a large number, say 10^5, of p-values from the prior distribution.
- **Step 2: Likelihood** – For each value of p, draw a number from $B(20, p)$.
- **Step 3: Posterior** – If then result in Step 2 produces the number 9, store the value of p, and reject otherwise. Normalising this distribution of p gives the posterior.

---

[9] Interesting examples of how the the prior can lead to blatantly biased decisions are described in the classic psychology paper by Tversky and Kahneman [204].

To understand why Step 3 corresponds to the RHS of Bayes' Theorem (8.33), think about the filtering process as the multiplication between the prior and the likelihood. For a value p to be viable, it has to be drawn from the prior distribution *and*, in addition, drawing from $B(20, p)$ gives 9. The posterior is the probability of both events happening.

Fig. 8.19 (top panel) shows the prior and posterior probabilities using the uniform prior. We represent these probabilities as histograms, and therefore require the data to be binned (here we used 20 bins in the p-values, but you can change that in the code). To put numbers in a given array into bins, we use the NumPy function `np.histogram` which bins the data without producing a histogram. We then normalise the height of the bars to show the probability values on the y-axis (rather than the values of the pmf).

We see that the posterior is a distribution which peaks when p is approximately 0.4–0.5, with a significant spread. In fact, the code also reports some statistical properties of the posterior:

```
Posterior mean = 0.455
Mode at p = 0.475
90% credible interval from p = 0.275 to 0.625
```

There is a 90% probability that the parameter value p falls within the 90% *credible interval*. Note the philosophical difference between a Bayesian credible interval and a frequentist confidence interval. A 90% confidence interval would capture the true value of the parameter 90% of the time.

The strategy for calculating the credible interval is to consider the cumulative probability distribution and determine in which bin the probability equals 0.05 or 0.95. The same strategy was used in the Birthday problem §7.6 to determine where probability exceeds 0.5.

The beta distribution

Let's investigate how sensitive the posterior is to the choice of prior.

A commonly used family of probability distribution for the prior is the *beta distribution* $\beta(a, b)$, where $a, b > 0$ are shape parameters of the distribution. Its pdf is given by

$$f(x) = \frac{\Gamma(a + b)}{\Gamma(a)\Gamma(b)} x^{a-1}(1 - x)^{b-1}. \qquad (8.34)$$

The pdf is continuous on $(0, 1)$ (or $[0, 1]$ if $a, b \geq 1$). The beta distribution captures a wide range of trends on $[0, 1]$ and encompasses many famililar distributions. Exercise 12 explores the range of pdf shapes as a and b vary.

We note that when $a = b = 1$, the pdf reduces to $f(x) = 1$. Hence, $\beta(1, 1) = U(0, 1)$, which is the prior seen in the top panel of fig. 8.19.

In the middle panel, we use the prior $\beta(1, 3)$, which reduces to a quadratic function $\propto (x - 1)^2$. One can imagine using this prior if, say, we believe that most students wouldn't use the new water fountain (perhaps there are already lots of water fountains).

In the lower panel, we use the prior $\beta(4, 2)$, which gives a skewed distribution peaking at around $p = 0.75$. Can you think of a situation in which this prior could reflect our initial belief?

Despite the stark difference in the priors, the shape of the posteriors is similar. Here are the comparison of the posterior distributions resulting from the 3 priors.

Prior	Posterior mean	Mode	90% credible interval
$\beta(1,1)$	0.455	0.475	$[0.275, 0.625]$
$\beta(1,3)$	0.416	0.425	$[0.275, 0.575]$
$\beta(4,2)$	0.500	0.525	$[0.325, 0.675]$

Table 8.1: Analysis of the posteriors for three different priors

(Note that you might get slightly different results in your own run of the code.) We see that the posterior is largely influenced by the data, whilst the prior has a much smaller effect. In fact, *with more data, the prior matters less and less*.

Analysing the posterior will inform the decision whether the department should install another water fountain. For example, if the department has 100 students, and the department will only install a new water fountain if at least 60 students are likely to use it. Then under the prior $\beta(1,1)$ or $\beta(4,2)$, the credible interval supports the installation initiative.

In conclusion, a huge amount of mathematics lies hidden in this problem which contains only two numbers: 9 out of 20. Real-life modelling is a complex process with assumptions, beliefs and uncertainties. Bayesian statistics helps us articulate some of these uncertainties and use them to make mathematically informed decisions.

Bayesian statistics is a vast topic that this small section cannot do justice. For excellent introductions to the subject, see [123] and [139].

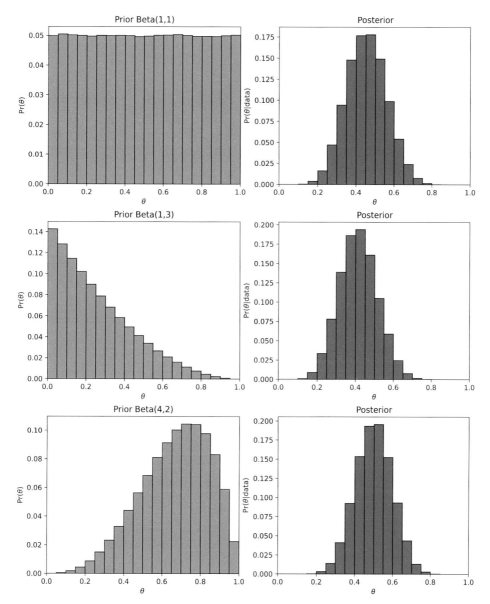

Fig. 8.19: The prior (left) and posterior (right) probabilities for the Bayesian analysis of the water-fountain problem.

DISCUSSION

- **Frequentist vs. Bayesian**. It is very likely that all the probability and statistics you have studied so far are based on the *frequentist* interpretation of probability.

  Here are some key conceptual differences between the frequentist and Bayesian interpretations.

Frequentist interpretation	Bayesian interpretation
Data comes from some probability distribution.	Data is fixed.
Model parameters are fixed.	Model parameters follow some probability distributions.
Beliefs do not matter.	Initial belief (prior) is an essential ingredient.
The confidence interval contains the true parameter value $x\%$ of the time.	The credible interval contains the true parameter value with $x\%$ probability.

- **Markov-Chain Monte Carlo (MCMC) methods**. In our code, we manually accepted/rejected the value p drawn from the prior by testing whether or not drawing from $B(20, p)$ gives 9.

  Now imagine that we were to constrain n parameters $\theta = (\theta_1, \theta_2, \ldots, \theta_n)$. Manually searching for acceptable values of θ in an n-dimensional parameter space quickly becomes computationally unfeasible – this is the curse of dimensionality.

  A more efficient way to explore the parameter space is to perform a special type of random walk in the parameter space called *Markov-Chain Monte Carlo* (MCMC) sampling. This is explored in detail in exercise 14. MCMC algorithms are an essential part of Bayesian inference in real data analysis.

- **PyMC** is a popular Python library for performing Bayesian inference. With PyMC, the code for analysing our problem (using a range of sophisticated MCMC samplers) is reduced to just a handful of lines. See `https://www.pymc.io`.

bayesian.ipynb (for plotting fig. 8.19)

```
import numpy as np
import matplotlib.pyplot as plt
from scipy.stats import binom, beta
%matplotlib
```

The data
Number of realisations
Number of bins in the histograms
We'll collect θ values for the posterior

```
Nsample, Nsuccess = 20, 9
Nsim = int(1e5)
Nbins = 20
post = []
```

Parameters for the prior $\beta(a, b)$

```
a, b = 1, 1
```

Sample from the prior

```
pri = beta.rvs(a, b, size=Nsim)
```

For each parameter value
Feed into the likelihood
Filtering: accept only parameter values that produce the required data

```
for p in pri:
    like = binom.rvs(Nsample, p)
    if like == Nsuccess:
        post.append(p)
```

Bin edges for plotting the results

```
bins = np.linspace(0, 1, Nbins+1)
```

Prior (binned), _=redundant output
Prior (normalised)

```
pri_b, _ = np.histogram(pri, bins= bins)
pri_prob = pri_b/sum(pri_b)
```

Posterior (binned)
Posterior (normalised)

```
post_b, _ = np.histogram(post, bins= bins)
post_prob = post_b/sum(post_b)
```

Plot two figures side by side

```
fig,(ax1, ax2)=plt.subplots(1,2,figsize=(11,4))
```

Width of each histogram bar
Centres of bins

```
width = 1/Nbins
X = np.arange(0, 1 ,width) + width/2
```

Plot the prior as histogram

```
ax1.bar(X, pri_prob, width = width,
        color='coral', edgecolor='k')
ax1.set_title(f'Prior Beta({a},{b})')
ax1.set_xlabel(r'$\theta$')
ax1.set_ylabel(r'Pr($\theta$)')
ax1.set_xlim(0,1)
```

Plot the posterior as histogram

```
ax2.bar(X, post_prob, width = width,
        color='royalblue', edgecolor='k')
ax2.set_title('Posterior')
ax2.set_xlabel(r'$\theta$')
ax2.set_ylabel((r'Pr($\theta|$data)'))
ax2.set_xlim(0,1)

plt.show()

# Code continues on the next page
```

bayesian.ipynb (continued)	
Report the posterior mean	``` post_mean = sum(X*post_b)/sum(post_b) print(f'Posterior mean = {post_mean:.3}') ```
Report the mode	``` post_mode = np.argmax(post_prob) print(f'Mode at p = {X[post_mode]:.3}') ```
Cumulative probability distribution Find the 5th and 95th percentiles	``` C = np.cumsum(post_b)/np.sum(post_b) p05 = np.searchsorted(C, 0.05) p95 = np.searchsorted(C, 0.95) ```
Report the 90% credible interval	``` print('90% credible interval from ' f'p = {X[p05]:.3} to {X[p95]:.3}') ```

8.10 Machine learning: clustering and classification

> Refer to datasets A and B in §8.6.
> a) If we combine them into a single dataset (and perhaps even shuffle the rows), how do we disentangle them again?
> b) A data point (previously overlooked) reads $(0, 0)$. Which dataset is it in?

In statistics, we use mathematical tools to analyse data to help us make inferences and decisions. *Machine learning* takes ideas in statistics (and probability, calculus, linear algebra *etc.*) and uses those ideas to make decisions and predictions based on some data and (if applicable) training. What statistics and machine learning have in common is the use of *data*.

Two main disciplines of machine learning are *supervised* and *unsupervised* machine learning[10].

In supervised learning, an algorithm uses a training dataset to learn how various characteristics determine the labels of objects in the data. Objects with some shared characteristics share the same label. Once trained, the algorithm is then presented with new unlabelled data, and determines the label based on the training. If the labels are discrete (*e.g.* cat or dog, 1 or 0), we call this task *classification*. (If the labels are continuous, the task is *regression*.)

In unsupervised learning, we do not know the 'true' labels of objects in the data, and there are no training datasets. One possible task is to find subgroups of objects that are similar (with respect to some metric) and give each subgroup a label. This is the task of *clustering*.

In this section, we will investigate how simple classification and clustering algorithms can be performed in Python.

k-means clustering

Let's first consider a clustering task. Start with the combined datasets A and B from §8.6 as shown in the left panel of fig. 8.20. Suppose that we don't know which data point comes from which file. The goal is to disentangle them into two clusters. Since the goal is to assign labels to an unlabelled dataset, this is an unsupervised machine-learning task.

The *k-means algorithm* is a simple clustering algorithm consisting of the following steps.

1. Specify k, the number of clusters in the data[11].
2. Assign k random data points as cluster centres c_i (also called *centroids*), for $i = 1, 2, \ldots, k$.
3. For each data point x_i, calculate its distance to the centroids $|x_i - c_j|$.
   If c_j is the nearest centroid to x_i, we assign x_i to cluster j. In other words, x_i is in cluster j if[12]
   $$j \text{ minimises } D(n) = |x_i - c_n|, \quad n = 1, 2, \ldots, k.$$

---

[10] There is also *reinforcement learning* – an important part of artificial intelligence.

[11] We judge by eye that the combined dataset contains two clusters, but sometimes the number of clusters in the data may not be so obvious. The k-means algorithm can be modified so that the most suitable value of k is determined from the data. This is known in the industry as the 'elbow method'.

[12] If a point is equidistant to more than one cluster, one can be chosen at random.

4. Update the centroids to the mean of each clustered data. In other words, let $\mathbf{x}_i$ be all data points in cluster j in the previous step. Then, the new centroid is the mean $\langle \mathbf{x}_i \rangle$.
5. Repeat steps 3 and 4 until there are no changes in the centroids.

The results of the algorithm are 1) a label (cluster ID) for each data point, 2) the coordinates of the k centroids.

In practice, the distance metric $|\cdot|$ does not have to be the physical distance. For example, given a colour image, one might want to cluster pixels with a similar colour together, in which case the distance between two points would be the difference in their RGB values. In this way, by retaining only k colours, k-means clustering to be used for image compression and segmentation (exercise 16).

$\boxed{a)}$ Let's apply k-means clustering to add a label (0 or 1) to each point in the combined dataset. We will colour all points in clusters 0 red, and all points in cluster 1 blue. Since the labels were removed, we cannot tell which colour belongs to which dataset. Our only goal is to sort them into two clusters.

The code `clustering.ipynb` shows how *sklearn* can be used to perform k-means clustering. The code produces the right panel of fig. 8.20, showing the two clusters in different colours and their centroids as black crosses. The coordinates of these centroids are reported by the code as

```
After 3 iterations, the centroids are at
[ 3.00014779 -2.97393273] and
[-0.09937072  3.81232351]
```

In fact, the data was generated by sampling from two bivariate normal distributions with means at $(3, -3)$ and $(0, 4)$. The k-means centroids are consistent with the actual means of the clusters.

$\boxed{b)}$ We can ask *sklearn* to predict which cluster the data point $(0, 0)$ belongs to.

```
point = [0,0]
kmeans.predict([point])[0]
```

The output is 1, meaning that it belongs to the blue cluster.

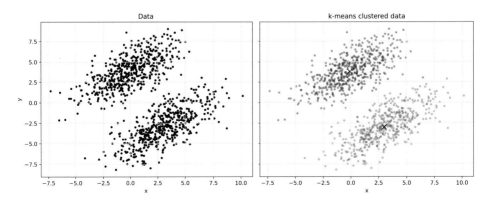

Fig. 8.20: The data before (left) and after k-means clustering (right). The black crosses are the centroids of the clusters.

clustering.ipynb (for plotting the right panel of fig. 8.20)	

Use sklearn to perform k-means clustering	```import pandas as pd```
	```import matplotlib.pyplot as plt```
	```from sklearn.cluster import KMeans```
	```%matplotlib```

Import data from two csv files (make sure they are in the same directory as this ipynb file)	```data0 = pd.read_csv('datasetA.csv',```
	``` sep=',', header=None)```
	```data1 = pd.read_csv('datasetB.csv',```
	``` sep=',', header=None)```

Combine (concatenate) the datasets	```X = pd.concat([data0, data1])```

Perform k-means clustering with $k = 2$
`random_state` allows reproducibility
Fetch the centroids
	```kmeans = KMeans(n_clusters=2, random_state=4,```
	``` n_init='auto').fit(X)```
	```centr = kmeans.cluster_centers_```

	```fig, ax = plt.subplots(1,1)```

Plot the datapoints, using the labels (0 or 1) as dot colours (red or blue)	```ax.scatter(X.values[:,0], X.values[:,1],```
	``` c=kmeans.labels_, cmap='bwr_r',```
	``` s=10, alpha=0.5)```

Mark the centroids with big black crosses	```ax.scatter(centr[:,0], centr[:,1], c='k',```
	``` s=80, marker = 'x')```
	```ax.set_title('k-means clustered data')```
	```ax.set_xlabel('x')```
	```ax.set_ylabel('y')```
	```plt.grid('on')```
	```plt.show()```

Report number of iterations and centroid coordinates	```print(f'After {kmeans.n_iter_} iterations, '```
	``` 'the centroids are at')```
	```print(f'{centr[0,:]} and \n{centr[1,:]}')```

k-nearest neighbour (kNN) classification

In clustering, we started with unlabelled data and assigned a label to each data point. What if we were to start with *some* labelled data, and use them to deduce the labels of unlabelled data points? This is the goal of *classification*.

Let's study a type of classification called the *k-nearest neighbours (kNN) algorithm*. Follow these steps.

1. Create the training dataset $X^t = \{\mathbf{x}_i^t, \ell_i\}$ where each $\mathbf{x}_i^t$ is an unlabelled data point and ℓ_i its label.
2. Let $\mathbf{x}$ be an unlabelled data point to which we wish to assign a label. Compute the distance of $\mathbf{x}$ to all points in the training data, *i.e.* for all $\mathbf{x}_i^t$, compute

$$|\mathbf{x} - \mathbf{x}_i^t|.$$

3. Choose an integer $k \in \mathbb{N}$ and select the k-nearest neighbours (in the training set) of $\mathbf{x}$ with respect to the distance $|\cdot|$. Let the set L be the labels for the k-nearest neighbours of $\mathbf{x}$.
4. The label of $\mathbf{x}$ is determined by the mode of L, *i.e.* $\mathbf{x}$ is assigned the most common label amongst those of its k nearest neighbours.

The cartoon below illustrates kNN classification with $k = 4$. The labels of the training set shown are either *apple*, *pear* or *avocado*. According to the kNN classification algorithm, the unlabelled point marked ? will be labelled *apple*.

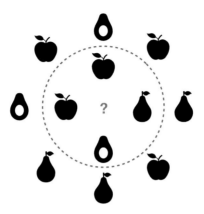

Fig. 8.21: The 4-nearest neighbours classification algorithm will assign the unlabelled point ? the *apple* label.

If there is a tie, a label can be left unassigned, or one can be chosen from amongst the modal values, either randomly, or (as is done in *sklearn*) choose whichever modal label appears first in the ordered training data.

Just like clustering, the distance metric $| \cdot |$ does not necessarily have to be the physical distance, but can be defined to fit the purpose. This allows classification techniques to be used for image processing, for example, to identify objects shown in an image, or to fill in missing details (*inpainting*).

Now let's experiment with classification using datasets A and B.

The code `classification.ipynb` will demonstrate *training*, *testing* and *classification* using *sklearn*. We now discuss the key concepts and highlights from the code.

1. Training First, we create a list of the correct label for each point in the combined dataset.

Then, we randomly pick 10% of the combined data (with their correct labels) to use as the training set X^t. This is accomplished using the *sklearn* function

```
train_test_split(X, label, train_size=0.1, random_state=4)
```

The argument `random_state` allows the same result to be reproduced in every run.

The training set is shown in fig. 8.22 in which training data points are marked with crosses. The three panels show the classification results using $k = 1, 2, 3$ nearest neighbours. The shaded regions are coloured blue or red according to the classification – points in the red region would be labelled 0, and blue would be 1. The key syntax for the classification (if $k = 3$) is

```
clf = KNeighborsClassifier(n_neighbors = 3)
```

The shading is produced using the function `DecisionBoundaryDisplay`.

Note the different boundaries between the red and blue regions in fig. 8.22 for different values of k. For $k = 1$, the boundary (white jagged line) consists of points whose nearest neighbour is any of two training points of different colours, one on either side of the boundary. This means that along this boundary, the $1NN$ classification results in a tie, although in practice it would be highly unlikely to have to deal with this ambiguity since the boundary is an infinitely thin line.

For $k = 2$, any point in the (now bigger) white region has 2 nearest neighbours of different colours. The shading algorithm does not try to break the tie and leaves this region as white.

Finally, for $k = 3$, we see light blue and light red regions straddling the $k = 1$ line. For a point in the light blue region, its 3 nearest neighbours are 2 blue and 1 red training points (and similar for points in the light red region).

We deduce that it is best to choose k to be an odd integer to avoid tie-breaks as much as possible.

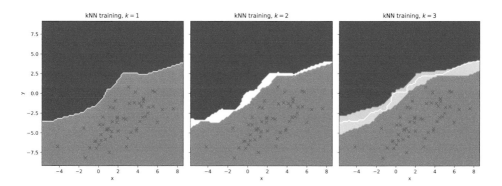

Fig. 8.22: The training dataset for the kNN classification with $k = 1, 2$ and 3 nearest neighbours.

2. Testing Let's see how well kNN performs given that only 10% of the data was used for training. The remaining data (without labels) are fed into the kNN algorithm using the syntax

$$\texttt{clf.predict(X\_test)}$$

The resulting classification predictions are then tested against the actual classification. The measure of accuracy in the testing phase of classification is normally presented as a *confusion matrix*, whose diagonal are the counts of correct classifications, and off-diagonal entries count the mis-classifications. In our case, with two classes $\{0, 1\}$, the confusion matrix is the 2×2 matrix defined as:

$$\text{Confusion matrix} = \begin{pmatrix} \text{Predicted} = 0, \text{ true} = 0 & \text{Predicted} = 0, \text{ true} = 1 \\ \text{Predicted} = 1, \text{ true} = 0 & \text{Predicted} = 1, \text{ true} = 1 \end{pmatrix}.$$

The code `classification.ipynb` reports the following output for the testing phase.

```
Confusion matrix =
[[444    1]
 [  1 454]]
Accuracy = 0.998
(-0.8404,0.1561) is classified as 1 but should be 0
(2.2737,1.5929) is classified as 0 but should be 1
```

The accuracy score is simply the fraction of correct classifications. The score is very high indeed, despite using only 10% of the data for testing. This is due to the fact that the data is quite clearly segregated into 2 classes with only a few ambiguous points along the interface.

Fig. 8.23 summarises the testing phase. Comparing the left (true classification) and right (test result), we see that the classification results are excellent! In fact, you can pick out by eye where the two mis-classified points are. This gives us high confidence that the classifier can be used for predicting the classification of unseen data.

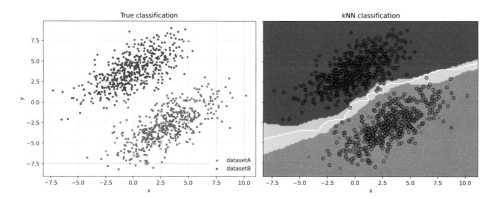

Fig. 8.23: Testing the kNN classification. The true classified data is shown on the left, and the result of 3 nearest-neighbour classification is on the right, where 10% of the data was used for training (shown as crosses). The data used for testing are shown as circles. There were 2 mistakes in the classification (can you spot them?), giving an accuracy score of 99.8%

3. Classification Finally, we use the *k*NN classifier to predict the label for the data point (0,0) as follows.

```
point  = [0,0]
clf.predict([point])[0]
```

The output is 0, meaning that it belongs to the red cluster. This suggests that the data point came from dataset A. Note that using a different set of training data can result in a different prediction.

There are many other clustering[13] and classification[14] algorithms built into scikit-learn. Some of these algorithms are explored in exercise 17.

DISCUSSION

- **Much, much more on machine learning**. It is of course impossible to capture the vast and rapidly advancing field of machine learning within a short section of this book. Here are some good introductory references on the topic. A good all-round introduction to machine learning can be found in [117]. A gentle mathematical introduction can be found in [53]. Ref. [136, 186] offer very readable introductions with Python code, and cover key topics such as neural networks and deep learning.

---

[13] https://scikit-learn.org/stable/modules/clustering.html

[14] https://scikit-learn.org/stable/auto_examples/classification/plot_classifier_comparison.html

classification.ipynb (for plotting the right panel of fig. 8.23)

```
                                            import pandas as pd
                                            import matplotlib.pyplot as plt
The kNN classifier                          from sklearn.neighbors \
                                                import KNeighborsClassifier
For splitting data into training and        from sklearn.model_selection \
testing datasets                                import train_test_split
For shading classification regions and      from sklearn.inspection \
plotting the boundary                           import DecisionBoundaryDisplay
For producing the confusion matrix during   from sklearn.metrics \
testing                                         import confusion_matrix
                                            %matplotlib

Import data from csv files in the           data0 = pd.read_csv('datasetA.csv',
same directory as this ipynb file                              sep=',', header=None)
                                            data1 = pd.read_csv('datasetB.csv',
                                                               sep=',', header=None)
Combine (concatenate) the datasets          X = pd.concat([data0, data1])
Create a list of correct labels             label = len(data0)*[0] + len(data1)*[1]

Training: Keep 10% of data for              X_train, X_test, y_train, y_test = \
training and the rest for testing               train_test_split(X, label,
random_state allows reproducibility              train_size=0.1, random_state=4)

Initialise the kNN classifier               clf = KNeighborsClassifier(n_neighbors = 3)

Perform kNN on training set                 clf.fit(X_train, y_train)

                                            fig, ax = plt.subplots(1,1)
This colourmap maps 0 to red and 1 to blue  cm = 'bwr_r'

Shade red/blue regions                      DecisionBoundaryDisplay.from_estimator(
(using full data)                               clf, X, cmap=cm, alpha=0.8, ax=ax)

Testing: Try kNN on testing set             predi = clf.predict(X_test)
Calculate and report accuracy score         score = clf.score(X_test, y_test)
and confusion matrix                        print('Confusion matrix = ')
                                            print(confusion_matrix(predi, y_test))
                                            print(f'Accuracy = {score:.3}')

Report mis-classifications                  for i,(c1,c2) in enumerate(zip(y_test,predi)):
                                                if c1 != c2:
(iloc locates a dataframe's compo-              x = X_test.iloc[i, 0]
nent)                                           y = X_test.iloc[i, 1]
                                                print(f'({x},{y}) is classified as '
                                                    f'{c2} but should be {c1}')

Plot the training set as crosses of         ax.scatter(X_train.iloc[:,0],X_train.iloc[:,1],
the correct colours                             c=y_train, cmap=cm, marker ='x')
Plot the testing set as black-edged circles ax.scatter(X_test.iloc[:,0], X_test.iloc[:,1],
of the correct colours                          c=predi, cmap=cm, edgecolors="k",
                                                alpha=0.5,)

                                            ax.set_title(r'kNN classification')
                                            ax.set_xlabel('x')
                                            ax.set_ylabel('y')
                                            plt.show()
```

8.11 Exercises

1 (*The Central Limit Theorem*) Use `CLT.ipynb` to help you answer these problems.

    a. Use the code to reproduce each row of fig. 8.3.
Now consider the last row (the arcsine distribution 8.13). Verify that Python's answers for the mean and variance of the red curve are close to the values suggested by the CLT.
Suggestion: First, show by hand that for the arcsine distribution 8.13, we have

$$\mu = \frac{1}{2}, \qquad \sigma^2 = \frac{1}{8}.$$

    b. (*Bates distribution*) Reproduce each panel in fig. 8.2. In addition, overlay the Bates distribution (8.14) on the histogram.

    c. Reproduce each panel in fig. 8.2 when the underlying distribution is the arcsine distribution. How large would you say n has to be for the normal approximation to be accurate? (This is slightly subjective.)

    d. (*Cauchy distribution*) Verify that when random variates are drawn from the Cauchy distribution (8.15), the sample mean follows the same Cauchy distribution (rather than a normal distribution). This is an example of when the CLT is not applicable. Suggestion: Plot everything over the interval $[-5, 5]$ say, and tell Seaborn to restrict the histogram to this range using the argument `binrange = [-5,5]`.

    e. (*Generalised CLT*) Reproduce fig. 8.4 (*i.e.* when random variates are drawn from 4 different distributions). Experiment further using your favourite pdfs.
Look up the *Lindeberg condition* and try to break the convergence predicted by the generalised CLT.

2 (*Hypothesis testing*) Use `ttest.ipynb` as a starting point for these questions.

    a. Reproduce fig. 8.5 (histogram for the *t*-distribution). Suggestion: Use Seaborn and adjust the number of bins.

    b. Plot fig. 8.6 (pdf of Student's *t* distribution for varying degrees of freedom ν). In addition to changing the colour of the curves as ν increases, try increasing the opacity (`alpha` value) of the curves at the same time.

    c. Plot the middle panel of fig. 8.9 (showing shaded area representing the *p*-value).

    d. (*One-tailed t-test*) Hypothesis testing! t-testModify the code to perform the following one-tail hypothesis test using the same sample as described in §8.4.

$$(A) \qquad H_0 : \mu = 1.5 \qquad H_1 : \mu < 1.5$$
$$(B) \qquad H_0 : \mu = 1.5 \qquad H_1 : \mu > 1.5$$

The code should report the *p*-value in each case, with and without using SciPy's `ttest_1samp`, and produce a plot the pdf for the estimate of μ (analogous to the top panel of fig. 8.9). The 95% confidence interval should be shaded.
Tip: Use the `alternative` argument in `ttest_1samp`. See SciPy documentation.
[Ans: For (A) $p = 0.1433$. $\mu < 1.9093$ at 95% confidence.]

e. (*Type II error*) Perform the *t*-test again as described in §8.4 with the same null and alternative hypotheses ($H_0 : \mu = 1.5$ and $H_1 : \mu \neq 1.5$) with a different set of $n = 5$ numbers sampled from $\mathcal{N}(1, 1)$. Repeat 1000 times. Keep track of how many tests are performed and how many fail to be rejected. Hence calculate the value of β (the probability of committing type II error).
[Ans: β equals the *p*-value, which is uncomfortably high. We really want a small number.]
Repeat the exercise, this time increasing the sample size to $n = 10$. Show that β is drastically reduced to just around 1%.

3 (χ^2 *distribution*) Use `chi2test.ipynb` to help answer the following questions.

a. The following statistical table is taken from [56].

χ^2 **(Chi-Squared) Distribution: Critical Values of** χ^2

Degrees of freedom	5%	1%	?
		Significance level	
1	3.841	6.635	10.828
2	?	9.210	13.816
3	7.815	11.345	16.266
4	9.488	?	18.467

Use Python to fill in the 3 missing entries in the table.

b. Reproduce each panel of fig. 8.10. Suggestion: Use Seaborn.

c. (*Minimum χ^2 parameter estimation*) Using the dataset in §8.5, we showed that when $p_{\text{success}} = 0.5$, the test statistic $\chi^2 \approx 12.84$
Now calculate the test statistic χ^2 for a range of value of p_{success} in $(0.1, 0.9)$. Hence, plot χ^2 against p_{success}. Suggestion: Use log y scale for dramatic effect.
Estimate to 2 decimal places the value of p_{success} that minimises χ^2.
Interpret the meaning of this value of p_{success} that minimises χ^2.

d. (*Degrees of freedom*) Using the dataset in §8.5, we can estimate p_{success} using the simple formula

$$\hat{p} = \frac{\text{Total number of heads observed}}{\text{Total number of coins thrown}}.$$

Show that $\hat{p} = 0.4125$.
Now consider the following null and alternative hypotheses.

H_0 : The number of heads follows the binomial distribution

H_1 : The number of heads does not follow the binomial distribution

Perform hypothesis testing using the χ^2 test statistic and $\alpha = 0.05$.
Suggestion: The degree of freedom $= C - N_{\text{param}} - 1$ where C is the number of categories in the data, and N_{param} is the number of parameters in the pdf. In this context, the binomial distribution has one parameter p_{success}. Therefore, you will need to set the degree of freedom to $C - 2$.

4 (*Regression*) Use `regression.ipynb` to help you with these questions.

    a. Modify the code so that the regression lines are plotted using Seaborn's `regplot` function instead of NumPy's `polynomial` library.

    b. Here are some easy least-squares problems:

       i. Consider 50 evenly-spaced values of $x \in [0, 1]$. Randomly select y from the normal distribution $N(x, 1)$. Produce a scatter plot of (x, y) and find the equation of the regression line. Add this line to your plot.
       Suggestion: Use the matrix method discussed in §8.6 to obtain the regression line. Then check that it matches NumPy's or Seaborn's answer.

      ii. Modify the code in part (i) so that y is instead drawn from the normal distribution $N(x^2, 1)$. Produce a scatter plot of (x, y) and find the equation of the least-squares parabola. Add this parabola to your plot. (Same suggestions as above).

      iii. Consider 100 evenly-spaced grid points on $[0, 1] \times [0, 1]$, let $z = 1 - \rho$ where ρ is drawn from the normal distribution $N(0, 1)$. Produce a scatter plot of (x, y, z) and find the equation of the least-squares plane. Add this plane to your plot.
      Suggestion: Use the matrix method (eq. 8.24). Useful commands: `np.meshgrid` and `flatten()`.

5 (*Quantifying Simpson's paradox*) Consider the following data which gives the number of success and failures when students and staff were asked to take a certain test.

University A	Pass	Fail	Total
Student	a	$b - a$	b
Staff	A	$B - A$	B

University B	Pass	Fail	Total
Student	c	$d - c$	d
Staff	C	$D - C$	D

    a. Deduce that Simpson's paradox occurs if

$$\frac{a}{b} < \frac{A}{B} \quad \text{and} \quad \frac{c}{d} < \frac{C}{D}, \qquad \text{but} \qquad \frac{a + c}{b + d} > \frac{A + C}{B + D}.$$

    This means that in each university, students are less likely to pass the test compared to staff. However, for the combined population, students appear to be more likely to pass compared to staff.

    b. Give an example of non-negative integers (a, b, c, d) and (A, B, C, D) such that Simpson's paradox occurs. Make sure $a < b$, $A < B$, $c < d$ and $C < D$.

    c. Sample integers b, B, d, D from $U(1, 100)$ and sample integers a, A, c, C from, respectively, $U(0, b)$, $U(0, B)$, $U(0, d)$ and $U(0, D)$.
    Find probability that Simpson's paradox occurs. (Ans: $\approx 1\%$.)

6 (*Correlation coefficient*) Using eq. 8.26, show that the correlation coefficients r for the dataset shown in fig. 8.14 are given by

- Dataset A: $r = 0.6955$
- Dataset B: $r = 0.6652$
- Combined: $r = -0.1981$

We conclude that datasets A and B show moderate positive correlations between x and y. The combined data shows a weaker negative correlation.

Suggestion: There are several ways to calculate r in Python, such as:

- SciPy's `stats.pearsonr`
- SciPy's `stats.linregress`
- NumPy's `corrcoef`
- Pandas's `print(df.corr())`

The last two methods print a symmetric matrix showing the r values between all pairs of columns.

7 (*Coefficient of determination*) Use eq. 8.27 to show that in dataset A, the coefficient of determination $R^2 \approx 0.48$. Note that this is simply the square of the correlation coefficient.

Now draw another line through the data (it can be as bad a fit as you like). Calculate R^2, which you should find to always be lower than that of the regression line.

Now a challenge. Suppose we use the line $y = mx + c$ to fit dataset A. Plot the value of R^2 as a heatmap as (m, c) vary near the least-square parameters (α, β), which you should locate with a dot.

Suggestion: Adjust `vmin` and `vmax` in `pcolormesh` until you see an elliptical contour whose centre is (α, β).

8 (*Bivariate normal distribution*) Use the code `bivariate.ipynb` as a starting point for these questions.

a. The pairs of random numbers (x, y) in the file `datasetB.csv` came from a bivariate normal distribution. Estimate $\mu_x, \mu_y, \sigma_x, \sigma_y$ and ρ.

b. Suppose that $\mu_X = \mu_Y = 0$, and that the covariance matrix is given by $\Sigma = \begin{pmatrix} 5 & 2 \\ 2 & 4 \end{pmatrix}$.

Calculate (by hand) the dimensions and orientation of the contour ellipses for the bivariate normal pdf. Verify that your answers are consistent with the code output.

c. Reproduce fig. 8.16 (confidence ellipses for dataset A). For instructions, see https://matplotlib.org/stable/gallery/statistics/confidence_ellipse.html.

d. (*68-95-99.7 rule for bivariate normal*) Consider the standard bivariate normal distribution ($\mu_x = \mu_y = 0$ and $\sigma_x = \sigma_y = 1$). Show (e.g. with Monte Carlo simulation) that when $\rho = 0$, the probability that a randomly chosen pair of numbers $(x, y) \in \mathbb{R}^2$ lies in the 1σ ellipse is approximately 39%.

(Compare this with 68% for the 1D normal distribution.)

(Suggestion: The code for doing this is really short. If $\rho = 0$, and X and Y are drawn from $N(0, 1)$. Their sum therefore follows the χ^2 distribution with 2 degrees of freedom. The 1σ ellipse are all points such that $X^2 + Y^2 \leq 1$.)

Hence, show numerically that the 68-95-99.7 rule for the standard univariate normal becomes 39-86-99 for the standard bivariate normal.

Investigate how changing ρ affect these numbers.

e. (*Marginal distribution*) Use Seaborn's `jointpoint` function to visualise the marginal pdfs for X and Y in `datasetA.csv`.

9 (*Random walks*) Use `randomwalk.ipynb` as a starting point for these questions.

a. In the 1D random walk model, plot a histogram showing the distribution of
(i) X_{1000}, the displacement from the origin after 1000 steps,
(ii) $|X_{1000}|$, the distance from the origin after 1000 steps.
In each case, overlay a suitable curve which approximates the distribution.
Suggestion: From the discussion in the text, (i) is approximately normal, (ii) is approximately a χ distribution (8.31).

b. Verify that the percentage of the time in which the 1D random walk spends inside the 3σ boundary shown in fig. 8.17 is 99.7%. Predict the answers for 1σ and 2σ boundaries and verify your answer.
Repeat for the 2D random walk.

c. Reproduce fig. 8.18 for the 2D random walk.
Now modify the code so that at each step, the particle can only make one of the following moves: up, down, left or right, with equal probability. Show that the mean distance can be approximated by $E\left[\sqrt{X_N^2 + Y_N^2}\right] = \sqrt{\pi N}/2$.
Modify the 3σ circle accordingly. Verify numerically that trajectories stay within the 3σ circle $\sim$ 99% of the time.

d. Plot the χ distribution with N degrees of freedom, where $N = 1$ to 10, on the same figure. Vary the colour of the curves in a sensible way.

10 (*Gambler's ruin*) Imagine a scenario in which a gambler starts off with Âč10, and with each game wins Âč1 with probability p, or loses Âč1 with probability $1 - p$

a. First consider the case $p = 0.5$. This could be a game in which tossing a fair coin determines if the gambler wins or loses. The gambler stops playing when either of the following scenarios occurs.
A) The gambler has Âč20,   B) The gambler has Âč0.
Simulate this scenario. Plot the distribution of the number of games played and find the mean.
(Ans: 100 games)

b. Repeat the experiment with $p = 0.4$. Suggestion: Use `scipy.stats` to draw random variates from a Bernoulli distribution.
(Ans: 48 games)

For analytic solutions, further references and the historical significance of this problem (dating back to Pascal), see ch. 5 of [77].

11 (*Arcsine law*) Consider the 1D symmetric random walk starting at $X_0 = 0$. Let N_+ be the total number of steps during which a random walk spends in the region $X > 0$
Simulate a large number of random walks, each with N_{steps} timesteps.
Plot the distribution of N_+/N_{steps} (this ratio quantifies the time that the trajectory spends in the region $X > 0$). Show that the distribution is well approximated by the arcsine distribution with pdf

$$f(x) = \frac{1}{\pi\sqrt{x(1 - x)}},$$

(see eq. 8.13). You should obtain something similar to the figure below (the red curve is the arcsine distribution).

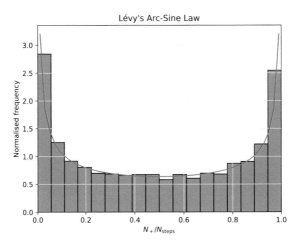

This result is known as *Lévy's first arcsine law*. For more details and a counterintuitive interpretation of this result, see ch. 29 of [77].

12 (*The beta distribution*) The beta distribution has pdf

$$f(x) = \frac{\Gamma(a + b)}{\Gamma(a)\Gamma(b)} x^{a-1}(1 - x)^{b-1}, \tag{8.35}$$

where $a, b > 0$. On domain $[0, 1]$, plot the following graphs:

a. $\beta(a, a)$ for various values of $a > 0$
b. $\beta(a, 2)$ for various a
c. $\beta(1, b)$ for various b

In each case, describe the effect of varying the free parameter.

13 Use the code `bayesian.ipynb` to help you answer these questions.

a. Experiment with different beta priors to see the response of the posterior. One suggestion is to try the prior $\beta(0.5, 0.5)$. Does this function look familiar? (see previous question). Can you think of a situation where someone might believe that this prior is suitable?

b. (*Conjugate priors*) Suppose that the prior is $\beta(a, b)$ and the likelihood is the binomial distribution. Then, given the data of k successes out of N observations, the posterior can be shown to be another beta distribution:

$$\beta(a + k, b + N - k).$$

Verify this observation by overlaying the graph of a suitable beta distribution on the posterior plot in each panel of fig. 8.19.

For a given likelihood function, if the prior and posterior belong to the same family of distribution, we say that the prior is a *conjugate prior* for the likelihood. More details and proof in ch. 9 of [123].

14 (*MCMC sampling with Metropolis-Hastings algorithm*) Let's revisit the water-fountain problem try the following *Metropolis-Hastings* algorithm. Follow this recipe.

- **Step 1**: Choose any initial parameter value θ_i
- **Step 2**: Draw a sample θ_{i+1} from the normal distribution $N(\theta_i, 0.1)$ (this is called the *proposal distribution*).
- **Step 3**: Calculate the *acceptance ratio* α given by

$$\alpha := \frac{\Pr(\theta_{i+1}|\text{data})}{\Pr(\theta_i|\text{data})} = \frac{\text{Likelihood pmf}(\theta_{i+1}) \times \text{Prior pdf}(\theta_{i+1})}{\text{Likelihood pmf}(\theta_i) \times \text{Prior pdf}(\theta_i)}.$$

  In the water-fountain problem, the likelihood is the binomial distribution $B(20, 9)$. Use the prior $\beta(1, 1)$.
- **Step 4**: Decide whether to accept or reject θ_{i+1}. To do this:
  - Draw a random number u from $U[0, 1]$
  - If $u \leq \alpha$, accept θ_{i+1}. Store it, and assign $\theta_i = \theta_{i+1}$. Go to Step 2.
  - If $u > \alpha$, reject θ_{i+1}. Redo Step 2.

a. Code the algorithm, looping over it say, 10^5 times. You should then obtain a sequence of accepted parameter values θ. You should find that the acceptance rate is about 72%, resulting in a sequence of around 72000 θ values.
b. Plot the sequence of accepted θ values. You should see a 1D random walk concentrating around the mean which you should find to be around 0.455. This is called a *trace* plot.
c. From the random-walk sequence, generate a histogram. This is the posterior. You should obtain a figure similar to the following (your data might be slightly different). Note the similarity between the resulting posterior and that at the top of fig. 8.19.

Note: An algorithm which samples parameter values θ from a proposed distribution and accepting/rejecting them probabilistically is broadly known as an *Markov-Chain Monte Carlo* (MCMC) algorithm. The precise recipe for accepting/rejecting as described above is called the *Metropolis-Hastings* algorithm, which explores the parameter space by performing a specialised random walk. The real benefit of the MCMC method for Bayesian inference becomes apparent when the parameter space is multi-dimensional. For a detailed treatment of the MCMC method and the Metropolis-Hastings algorithm, see [123].

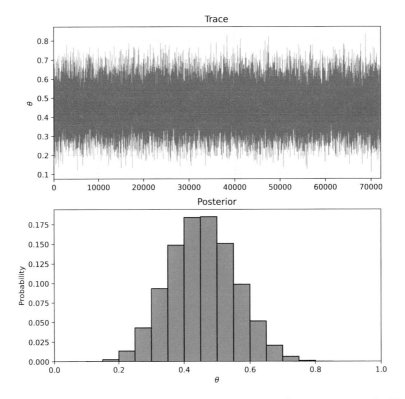

Fig. 8.24: *Top*: The trace plot showing a random walk resulting from the Metropolis-Hastings algorithm. *Bottom*: The posterior probability distribution plotted using the random walk data.

15 (*Clustering*) Predict what will happen if you ask *sklearn* to sort the combined dataset (shown on the left of fig. 8.20) into 3, 4 or 5 clusters. Do you expect to get the same result every time?

Modify the code `clustering.ipynb` to allow the data to be sorted into into k clusters where k is any integer greater than 1. The code should also report their centroids.

Suggestion: Choose a colourmap which does not display points in white. Remove `random_state` to see that the result is probabilistic.

16 (*Image segmentation using k-means clustering*)

Image segmentation is the process of decomposing an image into simple components. Use k-means clustering to reduce the greyscale image *flowers.jpg* (or your own image) to k shades. Fig. 8.25 shows the segmented image for $k = 3$.

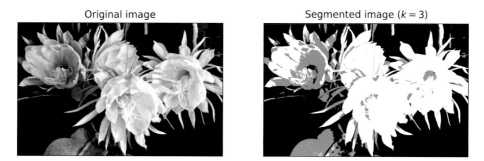

Fig. 8.25: Image segmentation using k-means clustering. The image on the right contains 3 shades.

Suggestions:

- See §5.8 on how to load an image from file. Use |np.array| to convert the image into a matrix.
- Reshape the matrix into an $(n \times 1)$ array X containing the greyscale values, *i.e.* integers between 0 (black) and 255 (white).
- Use *sklearn* to perform k-means clustering on X (see `clustering.ipynb`). Take note of the cluster centroids, which are the greyscale values of the k components.
- Reshape the clustered array into the original shape and display the matrix using `imshow`.

17 (*Exploring Scikit-learn further*)

   a. Perform k-means clustering and kNN classification using appropriate toy datasets that are included in the *sklearn* library[15].

   b. Research into the following popular algorithms.
- Clustering: i) *Mean-shift*   ii) *Gaussian mixture models*
- Classification: i) *Support vector machine (SVM)*   ii) *Neural network (multi-layer perceptron)*

Revisit the clustering and classification tasks in part (a) using these algorithms.

---

[15] https://scikit-learn.org/stable/datasets/toy_dataset.html

Appendix A: Python 101

Fig. A.1: *Guido van Rossum* (b.1956), creator of Python. (Image credit: Michael Cavotta CC BY-NC-ND 4.0)

Since the inception of Python in 1989 by the Dutch computer scientist *Guido van Rossum*, Python has now grown to become arguably the world's most popular programming language (as of 2023). Python is easy and intuitive to learn, with a strong community support. It is therefore ideal for learners with little or no background in computing, and as such it is now taught in schools all over the world.

A.1 Installation

If you are totally new to Python, the following installation method is recommended.

- **Anaconda and JupyterLab.** We recommend installing *Anaconda* on your computer from

$$\texttt{https://www.anaconda.com}$$

  Anaconda is a popular software suite containing, amongst other applications. *JupyterLab* which we recommend for using with this book. The complete guide to using JupyterLab can be found at

$$\texttt{https://jupyterlab.readthedocs.io}$$

- **Pip** is an essential tool for installing and updating Python libraries that are not part of the standard distribution.

S. Chongchitnan, *Exploring University Mathematics with Python*,

Pip uses command lines. This means that you need to have a *Terminal* open (this applies to both Windows and Mac).

Check if you already have pip on your machine. In your terminal, type

```
pip --version
```

If pip has not yet been installed, follow the instructions at

```
https://pip.pypa.io
```

A.2 Learning Python

There is no shortage of learning resources for beginners. Here are some free resources for beginners that you may find useful (as of 2023). I have selected resources that focus on scientific computing.

Of course, Python evolves all the time, so you may like to start by searching for the latest resources out there.

There is no need to a complete an entire Python course to start using this book. You can pick things up along the way by modifying the code given in this book.

- **Online resources**
  - Aalto university in Finland has produced a wonderful, free course with the focus on scientific computing. There are lots of videos and easy-to-follow tutorials. See:

    ```
    https://aaltoscicomp.github.io/python-for-scicomp/
    ```

  - Scientific Python lectures (Gaël Varoquax *et. al.*) is a comprehensive set of tutorials, guiding beginners through NumPy, SciPy, SymPy *etc.* on all the way up to machine learning.

    ```
    https://lectures.scientific-python.org
    ```

- **Books**
  - Linge and Langtangen, *Programming for computations – Python*, Springer (2020) [131]
  - Kong, Siauw and Bayen, *Python programming and numerical methods – a guide for engineers and scientists*, Academic Press (2020) [114]
  - Lin, Aizenman, Espinel, Gunnerson and Liu, *An introduction to Python programming for scientists and engineers*, Cambridge University Press (2022) [130]
  - Lynch, *Python for scientific computing and artificial intelligence*, CRC Press (2023) [136]

A.3 Python data types

Here we summarise key facts about data types in Python.

Basic data types

The basic data types used in this book are:

- **Integer** *e.g.* `0`, `1`, `-99`
- **Float** *e.g.* `1.0`, `2/3`
- **String** *e.g.* `'Hello'`, `'2'`
- **Boolean** namely, `True`, `False`
- **Complex** *e.g.* `1j`, `5-2j`

Composite data types

Composite data types and their properties are given in table (A.1) below.

Data type	Ordered?	Mutable?	Duplicate elements allowed?
List and **NumPy array**	Ordered	Mutable	Duplicate elements allowed
Tuple	Ordered	Immutable	Duplicate elements allowed
Set	Unordered	Immutable	No duplicate elements
Dictionary	Ordered	Mutable	No duplicate elements

Table A.1: Properties of composite data types in Python 3.9+

Examples:

- **Lists** and **arrays** are used in most programs in this book.
- **Tuple**: `s` in ellipse.ipynb (§3.3).
- **Set**: `uniq` in birthday.ipynb (§7.6)
- **Dictionary**: `step` in planes.ipynb (§5.3)

Let's now discuss the three properties in the heading of table A.1.

- **Ordered.** A composite data type is said to be *ordered* if its elements are arranged in a fixed, well, order. This means that it is possible to pinpoint the first or second object in the data type. For example, suppose we define a list L as:

$$L = ['goat', 999, True]$$

Then, the first element `L[0]` is reported as the string `'goat'`. Similarly, define a tuple as

$$T = ('goat', 999, True)$$

Then, T[0] is reported as the string 'goat'.

In contrast, suppose we define a set S as:

$$S = \{'goat', 999, True\}$$

Then, calling S[0] gives an error.

Finally, a dictionary is ordered (since version 3.7), but Python does not index its elements (which are *keys*).

- **Mutable.** A composite data type is said to be *mutable* if its elements can be changed after it is defined. For example, using the list L above, the command

$$L[1] = L[1] + 1$$

changes L to ['goat', 1000, True]. This happens because lists are mutable. However, the command

$$T[1] = T[1] + 1$$

produces an error, since the elements of T cannot be changed.

- **Duplicate elements.** The list

$$L1 = ['goat', 999, True, True]$$

is different from L (they have different numbers of elements). However, the set S1 = {'goat', 999, True, True} equals the set S (you can test this by performing the equality test S==S1).

Whilst duplicate elements are ignored by a set, a dictionary will override previous duplicate elements. For example, if we define a dictionary of animal types with 2 duplicate keys:

$$D = \{'animal': 'goat', 'animal': 'pig', 'animal': 'bear'\}$$

then D is simply a dictionary with one key, namely, D = {'animal': 'bear'}.

Lists vs arrays

	Python list	**Numpy array**
Mixed data types?	Allowed	Not allowed
+ means	Concatenation	Addition elementwise
* means	Duplication	Multiplication elementwise
Storage	Less efficient	More efficient
Computational speed	Slower	Faster (due to vectorisation)

Table A.2: Comparing properties of Python lists and NumPy arrays.

Let's now discuss table A.2 line-by-line.

- **Mixed data types.** A list can hold mixed data types. For example, the list:

$$\text{Lmixed} = [1 \text{ , } 2.3 \text{ , } 4+5\text{j}]$$

contains an integer, a float and a complex number.

Now let's convert the list to an array using the command `A=np.array(Lmixed)`. We now find that the array `A` reads

$$\text{array}([1. +0.\text{j}, 2.3+0.\text{j}, 4. +5.\text{j}])$$

This shows that the list has been converted to an array of a single data type (*i.e.* complex). This is because NumPy arrays are *homogeneous*, meaning that every element in an array is of the same data type. NumPy converts the elements in the list to the data type that best satisfies all of the list elements involved.

- **+ operator.** Adding two NumPy arrays together element-wise is what we often need to do as mathematicians (*e.g.* adding vectors and matrices). But take note that for two lists, L1 and L2, the operation L1+L2 creates a new list by merging (also known as concatenating) the two lists.

  In the mathematical tasks discussed in this book, we sometimes find ourselves adding an array of numbers to a list of numbers, in which case the operator + thankfully acts like element-wise addition.

  In short, the operator + acts differently depending on the data types involved. In technical terms, this is called *operator overloading*.

  Run the following code which demonstrates overloading of the operator +.

```
import numpy as np
A = np.array([7, 8, 5])
X = [0, -1, 2]
L = ['goat', 999, True]
Sum1 = A + A
Sum2 = A + X
Sum3 = X + L
#Sum4 = A + L    #This line will produce an error
print('A+A =', Sum1, type(Sum1),
      '\nA+X =', Sum2, type(Sum2),
      '\nX+L =', Sum3, type(Sum3))
```

Output:

```
A+A = [14 16 10] <class 'numpy.ndarray'>
A+X = [7 7 7] <class 'numpy.ndarray'>
X+L = [0, -1, 2, 'goat', 999, True] <class 'list'>
```

- **∗ operator.** Let L be a list and A be an array. Let c be a constant. Consider the following 'multiplications' involving the operator ∗. The results are not always what you might expect due to overloading.

  1. If c is a positive integer, then c*L is a list comprising c concatenated copies of the list L. In other words, c*L= L + L + ... + L (c copies).
     If c is a negative integer or zero, then c*L is an empty list.
     If c is not an integer, c*L produces an error.
  2. c*A is an array whose elements are those A multiplied by c.
  3. L*A is an array whose ith element is the product of the ith element of L and the ith element of A.

4. A*A is an array whose *i*th element is the square of the *i*th element of A. This is
   equivalent to A**2.

The following code demonstrates the above points.

```
import numpy as np
L = [0, -1, 2]
A = np.array([7, 8, 5])
Prod1 = 3*L; Prod2 = 3*A
Prod3 = L*A; Prod4 = A*A
print('3*L =', Prod1, type(Prod1),
      '\n3*A =', Prod2, type(Prod2),
      '\nL*A =', Prod3, type(Prod3),
      '\nA*A =', Prod4, type(Prod4))
```

Output:

```
3*L = [0, -1, 2, 0, -1, 2, 0, -1, 2] <class 'list'>
3*A = [21 24 15] <class 'numpy.ndarray'>
L*A = [ 0 -8 10] <class 'numpy.ndarray'>
A*A = [49 64 25] <class 'numpy.ndarray'>
```

Here is another example: in the code classification.ipynb (§8.10), we find the
following line:

$$label = len(data0)*[0] + len(data1)*[1]$$

This line uses the * operator to duplicate the singleton lists [0] and [1] and concatenate
them using the + operator.

- **Storage.** In broad terms, a large array takes up less storage (in terms of bytes) than a
  list of the same length. Let's quantify this statement.
  In the code ratio-size.ipynb, we store the sequence

$$S = (0, 1, 2, \ldots, n - 1) \tag{A.1}$$

in two ways: as a list and as an array. We then find out how many bytes are required to
store each representation, and calculate the ratio

$$\frac{\text{Number of bytes used to store } S \text{ as a list}}{\text{Number of bytes used to store } S \text{ as an array}}.$$

This ratio is plotted in fig. A.2 for sequence length n up to 10^6. We see that for a
long sequence ($n \geq 10^3$), storing it as a list can take up as much as 10% more space
compared to an array. On the other hand, there are no real space-saving advantages for
short sequences ($n \lesssim 100$).

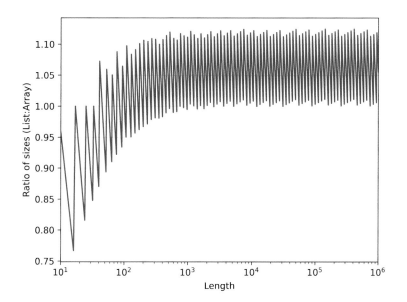

Fig. A.2: Ratio of the numbers of bytes needed to store the sequence $(0, 1, 2, \ldots, n-1)$ as a list vs as an array. For a long sequence $(n \gtrsim 10^3)$, storing it as a list can require as much as 10% more space compared to an array. This graph is produced by `ratio-size.ipynb`.

ratio-size.ipynb (for plotting fig. A.2)	
	```import numpy as np``` ```import matplotlib.pyplot as plt```
getsizeof = size of an object in bytes	```from sys import getsizeof```

Sequence lengths (up to 10^6)
For storing list sizes. . .
and array sizes
	```N = np.round(np.logspace(1,6,1000))```
	```sizeL = []```
	```sizeA = []```

For the sequence $(0, 1, 2, \ldots, n-1)$
Create the corresponding list. . .
and the corresponding array
Store their sizes
	```for n in N:```
	    ```L = [x for x in range(int(n))]```
	    ```A = np.arange(n)```
	    ```sizeL.append(getsizeof(L))```
	    ```sizeA.append(getsizeof(A))```

Size ratio	```ratio = np.array(sizeL)/np.array(sizeA)```

	```plt.semilogx(N, ratio, 'b')```
	```plt.xlabel('Length')```
	```plt.ylabel('Ratio of sizes (List:Array)')```
	```plt.xlim(10, max(N))```
	```plt.grid('on')```
	```plt.show()```

- **Computational speed.** Broadly, computations using arrays are faster than those using lists. This is because many array operations can be *vectorised*, meaning that the operations are performed in parallel by the CPU. This is much faster than, say, using a *for* loop to perform the same operations on each element of a list one at a time. Let's quantify this speed boost.

  In the code `ratio-time.ipynb`, we measure how long it takes to add one to each element of the list and array representations of the sequence S (eq. A.1). Using a list L, we time how long it takes to perform the list comprehension

  $$[1+1 \text{ for } 1 \text{ in } L]$$

  In contrast, using an array A, the operation $A + 1$ is vectorised, where 1 is understood by NumPy to be the array $(1,1,\ldots, 1)$ of the same size as A (this shape-matching is called *broadcasting*[1]).

  Fig. A.3 shows the ratio of the runtimes for the list and the array calculations. We see that the list computation is generally slower. On my computer, the worst cases occur for sequences of length $\approx 3 \times 10^4$, where the list computation is up to 70 times slower than the array computation.

  Your graph will be slightly different, depending on your hardware and Python distribution. But your graph should support the conclusion that list-based calculations are generally slower than those using arrays.

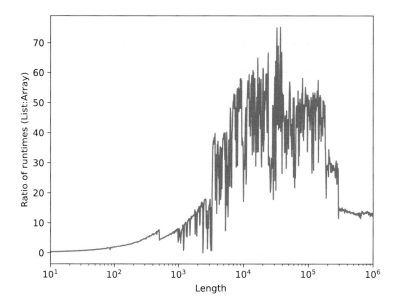

Fig. A.3: Ratio of the runtimes taken to add 1 to the sequence $(0, 1, 2, \ldots, n-1)$, using a list vs using an array. In the worst case (when $n \approx 3 \times 10^4$), the list computation is around 70 times slower than the array computation. This graph is produced by `ratio-time.ipynb`. Your graph will be slightly different.

---

[1] https://numpy.org/doc/stable/user/basics.broadcasting.html

ratio-time.ipynb (for plotting fig. A.3)	
| | ```python
import numpy as np
import matplotlib.pyplot as plt
from time import perf_counter as timer
``` |
| For measuring operation runtime | |
| Array of sequence lengths (up to $10^6$) | ```python
N = np.around(np.logspace(1,6,1000))
timeL = []
timeA = []
``` |
| For storing runtime using a list... | |
| and using an array | |
| For the sequence $(0, 1, 2, \ldots, n-1)$ | ```python
for n in N:
    L = [x for x in range(int(n))]
    A = np.arange(n)
``` |
| Create the corresponding list... | |
| and the corresponding array | |
| *Start the clock!* | ```python
    tic = timer()
    [l+1 for l in L]
    toc = timer()
    timeL.append(toc-tic)
``` |
Add 1 to every element in the list	
Stop the clock!	
Store the runtime	
| *Start the clock!* | ```python
    tic = timer()
    A+1
    toc = timer()
    timeA.append(toc-tic)
``` |
Repeat for the array (vectorised method)	
Stop the clock!	
Store the runtime	
| Ratio of runtimes | ```python
ratio = np.array(timeL)/np.array(timeA)
``` |
| | ```python
plt.semilogx(N, ratio, 'r')
plt.xlabel('Length')
plt.ylabel('Ratio of runtimes (List:Array)')
plt.xlim(10, max(N))
plt.grid('on')
plt.show()
``` |

A.4 Random musings

Free visualisation tools

In this book, we have used *Matplotlib*, *Pandas*, *Plotly* and *Seaborn* to create visualisations in Python. Here are other useful (and free) visualisation tools.

- **Visualising curves and surfaces.** *Desmos*[2] is a powerful visualisation tool for plotting curves and exploring 2D geometry. One of the best features is that sliders are instantly and intuitively created for you. For 3D geometry, *math3d*[3] offers an easy-to-use, Desmos-like interface for creating 3D surfaces and vector fields. *GeoGebra*[4] gives advanced functionalities and is my go-to app for creating beautiful 3D figures like fig. 3.9.

---

[2] https://www.desmos.com

[3] https://www.math3d.org

[4] https://www.geogebra.org
```

498 A Appendix A: Python 101

- **Data visualisation.** *R* is a powerful programming language used by the statistics community. The R library *htmlwidgets*[5] makes it easy to create interactive data visualisation with only basic R knowledge.

Parallelisation

We have come across tasks that can be done in parallel, *e.g.* performing multiple Monte Carlo simulations, solving an ODE with a range of initial conditions, and machine-learning tasks. Your computer will most likely contain multiple computing cores, and we can speed up our code by manually distributing tasks over multiple cores. To see how many computing cores your computer has, run the following lines:

```
import multiprocessing
multiprocessing.cpu_count()
```

If you have not tried parallel programming before, a good starting point is the documentation[6] for the module `multiprocessing`.

Python pitfalls and oddities

Overall, Python is easy for beginners to pick up. But there are some pitfalls that can trip up many beginners.

We have already mentioned that when using lists, the + and * operators are not really addition and multiplication.

Here are more pitfalls and some oddities that Python learners should watch out for.

1. **The last element.** One of the most common beginner's mistakes is forgetting that `range` and `np.arange` do not include the last element, but `np.linspace` does. This applies to slicing of arrays and lists too. For example, `A[-1]` is the last element of array `A`, but `A[0:-1]` is the array `A` *excluding* the last element.

2. **Tranposing 1D arrays.** Sometimes you may want to transpose a one-dimensional array (*e.g.* to turn a row vector into a column vector). For example:

```
u = np.array([0,1,2])
v = u.T
```

However, you will find that then `v` is still precisely `u`. Nothing happens when you transpose a 1D array!
If you really need the transpose explicitly, try adding another pair of brackets and transpose `v=np.array([[0,1,2]])` instead. In practice, such a transpose can often be avoided altogether.

[5] http://www.htmlwidgets.org
[6] https://docs.python.org/3/library/multiprocessing.html

3. **Modifying a duplicate list/array** can be dangerous. Consider the following lines of code:

```
A = [1,1,1]
B = A
B[0] = 9
```

Clearly B=[9,1,1]. But you may be surprised to find that A is also [9,1,1]! The original list has been modified, perhaps inadvertently.

To modify B whilst preserving the original list, use B=A.copy() instead of B=A. Alternatively, if using arrays, use B=np.copy(A).

This caution applies to all mutable objects in Python. In short, be careful when duplicating mutable objects[7].

4. **Iterators can only be used once.** What do you think you will see if you run this code?

```
A = [1,2]
B = [5,9]
Z = zip(A,B)
for i,j in Z:
    print(i,j)
for i,j in Z:
    print('Can you see this?')
```

You will find that the string 'Can you see this?' is never printed, because zip is an *iterator*, and each iterator can only be used once. When it comes to the second for loop, Z is now empty.

To avoid this pitfall, we should replace each Z in the for loops by zip(A,B).

5. **Mutable default argument.** Probably the most infamous Python pitfall of all is illustrated in the following function

```
def F(x=[]):
    x.append(9)
    return x
print(F(), F())
```

The function F takes an argument x, which, if not supplied, is set to the empty list by default. Therefore, calling F() gives [9]. So we might expect that the output of the code is [9],[9]. However, you may be surprised to see the output

$$[9, 9] [9, 9]$$

See this discussion thread[8] for explanation. The commonly prescribed remedy is to use the (immutable) keyword x=None in the argument instead.

```
def F(x=None):
    if x==None:
        x=[]
    x.append(9)
    return x
```

---

[7] You will need deepcopy rather than copy when duplicating, say, a list containing lists. Read more about deepcopy at https://docs.python.org/3/library/copy.html

[8] https://stackoverflow.com/questions/1132941

Concluding remarks

Mathematics and programming are both lifelong pursuits for me. The main difference is that whilst mathematics is a body of universal truths that will never change, programming changes constantly. Even in the space of almost two years of writing this book, Python has constantly evolved, and things that used to work now produce warnings or errors.

Naturally this means that this book will not have an indefinite shelf-life, but at least I hope that I have demonstrated how maths and programming can work in synergy (in the non-contrived sense of the word). I hope that the techniques demonstrated in this book have given readers plenty of inspirations to explore mathematics more deeply for themselves. I am looking forward to sharing more ideas for exploring university mathematics with Python in future updates.

REFERENCES

1. Agarwal, R.P., Hodis, S., O'Regan, D.: 500 examples and problems of applied differential equations. Springer, Cham (2019)
2. Aggarwal, C.: Linear algebra and optimization for machine learning. Springer, Cham (2020)
3. Ahlfors, L.V.: Complex analysis, third edn. McGraw-Hill, New York (1979)
4. Aigner, M., Ziegler, G.M.: Proofs from THE BOOK, 6th edn. Springer, Berlin (2018)
5. Alcock, L.: How to think about abstract algebra. Oxford University Press, Oxford (2021)
6. Alligood, K.T., Sauer, T.D., Yorke, J.A.: Chaos: an introduction to dynamical systems. Springer, New York (1997)
7. Altmann, S.L.: Hamilton, rodrigues, and the quaternion scandal. Mathematics Magazine **62**(5), 291 (1989)
8. Anderson, D.F., Seppäläinen, T., Valkó, B.: Introduction to Probability. Cambridge University Press, Cambridge (2017)
9. Andreescu, T., Andrica, D.: Complex numbers from A to –Z, 2nd edn. Birkhäuser, New York (2014)
10. Apostol, T.M.: Mathematical analysis, 2nd edn. Addison-Wesley, London (1974)
11. Apostol, T.M.: Introduction to analytic number theory. Springer, New York (2010)
12. Armstrong, M.A.: Groups and Symmetry. Springer, New York (1988)
13. Atkinson, K.E., Han, W., Stewart, D.: Numerical solution of ordinary differential equations. Wiley, New Jersey (2009)
14. Axler, S.: Linear algebra done right, 3rd edn. Springer (2015)
15. Ayoub, R.: Euler and the zeta function. The American Mathematical Monthly **81**(10), 1067 (1974)
16. Bak, J., Newman, D.J.: Complex analysis, 3rd edn. Springer, New York (2010)
17. Baker, G.L., Blackburn, J.A.: The pendulum : a case study in physics. Oxford University Press, Oxford (2006)
18. Bannink, T., Buhrman, H.: Quantum Pascal's Triangle and Sierpinski's carpet. arXiv e-prints arXiv:1708.07429 (2017)
19. Barnard, T., Neill, H.: Discovering group theory: a transition to advanced mathematics. CRC Press, Boca Raton (2017)
20. Bartle, R.G., R., S.D.: Introduction to Real Analysis, 4th edn. Wiley, New Jersey (2011)
21. Bas, E.: Basics of probability and stochastic processes. Springer, Cham (2019)
22. Bays, C., Hudson, R.H.: A new bound for the smallest x with $\pi(x) > \text{li}(x)$. Mathematics of Computation **69**(231), 1285 (1999)
23. Beléndez, A., Pascual, C., Méndez, D.I., Beléndez, T., Neipp, C.: Exact solution for the nonlinear pendulum. Revista Brasiliera de Ensino de Física **29**(4), 645 (2007)
24. Beltrametti, M.C., Carletti, E., Gallarati, D., Bragadin, G.M.: Lectures on curves, surfaces and projective varieties. European Mathematical Society, Zürich (2009)
25. Berndt, B.C., Robert, A.R.: Ramanujan: Letters and Commentary. American Mathematical Society, Providence (1995)
26. Birkhoff, G., Mac Lane, S.: A survey of modern algebra. Macmillan, New York (1941)
27. Birkhoff, G., Mac Lane, S.: A survey of modern algebra, 4th edn. Macmillan, London; New York; (1977)
28. Boas, M.L.: Mathematical Methods in the Physical Sciences, 3rd edn. Wiley (2005)
29. Bork, A.M.: "vectors versus quaternions"—the letters in nature. American Journal of Physics **34**(3), 202 (1966)
30. Borwein, J.M., Bradley, D.M., Crandall, R.E.: Computational strategies for the riemann zeta function. Journal of Computational and Applied Mathematics **121**(1), 247 (2000)

S. Chongchitnan, *Exploring University Mathematics with Python*,

31. Bower, O.K.: Note concerning two problems in geometrical probability. The American Mathematical Monthly **41**(8), 506 (1934)
32. Brauer, F., Castillo-Chavez, C., Feng, Z.: Mathematical models in epidemiology. Springer, New York (2019)
33. Bronson, R., Costa, G.B.: Schaum's outline of differential equations, 5th edn. McGraw Hill (2021)
34. Brunton, S.L., Kutz, J.N.: Data-Driven Science and Engineering: Machine Learning, Dynamical Systems, and Control, 2nd edn. Cambridge University Press, Cambridge (2022)
35. Burton, D.M.: Elementary number theory, 7th edn. McGraw Hill, New York (2011)
36. Butcher, J.C.: Numerical methods for ordinary differential equations, 3rd edn. Wiley, Chichester (2016)
37. do Carmo, M.P.: Differential geometry of curves and surfaces. Prentice Hall (1976)
38. Carter, N.C.: Visual group theory. Mathematical Association of America, Washington, D.C. (2009)
39. Chao, K.F., Plymen, R.: A new bound for the smallest x with $\pi(x) > \text{li}(x)$ (2005). URL https://arxiv.org/abs/math/0509312
40. Cheney, W., Kincaid, D.: Numerical mathematics and computing, 7th edn. Cengage (2012)
41. Chihara, T.S.: An introduction to orthogonal polynomials. Dover, New York (2011)
42. Choquet-Bruhat, Y., de Witt-Morette, C., Dillard-Bleick, M.: Analysis, manifolds and physics. North Holland, Amsterdam (1983)
43. Chung, K.L., AitSahlia, F.: Elementary probability theory: with stochastic processes and an introduction to mathematical finance, 4th edn. Springer, New York (2003)
44. Clarke, R.D.: An application of the poisson distribution. Journal of the Institute of Actuaries **72**(3), 481 (1946)
45. Collins, P.J.: Differential and integral equations. Oxford University Press, Oxford (2006)
46. Comninos, P.: Mathematical and computer programming techniques for computer graphics. Springer, London (2006)
47. Conrey, J.B.: The riemann hypothesis. Notices of the AMS **50**(3), 341 (2003)
48. Cramér, H.: On the order of magnitude of the difference between consecutive prime numbers. Acta Arithmetica **2**(1), 23 (1936)
49. Crandall, R., Pomerance, C.: Prime numbers: a computational perspective, 2nd edn. Springer, New York (2005)
50. Crowe, M.J.: A history of vector analysis: the evolution of the idea of a vectorial system. University of Notre Dame Press, London;Notre Dame (Illinois) (1967)
51. Crowe, W.D., Hasson, R., Rippon, P.J., Strain-Clark, P.E.D.: On the structure of the mandelbar set. Nonlinearity **2**(4), 541 (1989)
52. DeGroot, M.H., Schervish, M.J.: Probability and statistics, 4th edn. Pearson Education, London (2012)
53. Deisenroth, M.P., Faisal, A.A., Ong, C.S.: Mathematics for machine learning. Cambridge University Press, Cambridge (2020)
54. Dolotin, V., Morozov, A.: The Universal Mandelbrot Set: The beginning of the story. World Scientific, Singapore (2006)
55. Douady, A., Hubbard, J.H.: Étude dynamique des polynômes complexes. Publications Mathématiques d'Orsay **84** (1984)
56. Dougherty, C.: Introduction to econometrics, fifth edn. Oxford University Press, Oxford (2016)
57. Dudley, U.: Elementary number theory, 2nd edn. W. H. Freeman, San Francisco (1978)
58. Elaydi, S.: An introduction to difference equations, 3rd edn. Springer, New York (2005)
59. Evans, G., Blackledge, J., Yardley, P.: Numerical methods for partial differential equations. Springer-Verlag, London (2000)
60. Farlow, S.J.: Partial differential equations for scientists and engineers. Dover, New York (1982)
61. Feigelson, E.D., Babu, G.J.: Modern statistical methods for astronomy: with R applications. Cambridge University Press, Cambridge (2012)
62. Feller, W.: An introduction to probability theory and its applications, vol. I, 3rd edn. Wiley, London (1968)
63. Fine, B., Rosenberger, G.: Number Theory: an introduction via the density of primes, 2nd edn. Birkhäuser, Cham (2016)
64. Fischer, H.: A History of the Central Limit Theorem. Springer, New York (2011)
65. Folland, G.B.: Fourier analysis and its applications. American Mathematical Society, Providence (2009)
66. Forbes, C.S., Evans, M.: Statistical distributions, 4th edn. Wiley-Blackwell, Oxford (2010)
67. Fortney, J.P.: A Visual Introduction to Differential Forms and Calculus on Manifolds. Birkhäuser, Cham (2018)

68. Fraleigh, J.B., Brand, N.E.: A first course in abstract algebra, 8th edn. Pearson (2020)
69. Friedman, N., Cai, L., Xie, X.S.: Linking stochastic dynamics to population distribution: An analytical framework of gene expression. Phys. Rev. Lett. **97**, 168302 (2006)
70. Gallian, J.A.: Contemporary abstract algebra, 10th edn. Chapman and Hall /CRC, Boca Raton (2020)
71. Gardner, M.: Mathematical games. Scientific American **231**(4), 120 (1974)
72. Gelbaum, B.R., Olmsted, J.M.H.: Counterexamples in Analysis. Dover, New York (1964)
73. Gerver, J.: The differentiability of the Riemann function at certain rational multiples of π. Proceedings of the National Academy of Sciences of the United States of America **62**(3), 668–670 (1969). URL `http://www.jstor.org/stable/59156`
74. Gezerlis, A.: Numerical methods in physics with Python. Cambridge University Press, Cambridge (2020)
75. Glendinning, P.: Stability, instability, and chaos: an introduction to the theory of nonlinear differential equations. Cambridge University Press, Cambridge (1994)
76. Goldstein, H., Poole, C., Safko, J.: Classical mechanics, 3rd edn. Addison Wesley, San Francisco (2002)
77. Gorroochurn, P.: Classic problems of probability. John Wiley, Hoboken (2012)
78. Gradshteyn, I.S., Ryzhik, I.M.: Table of integrals, series, and products, 8th edn. Academic Press, Amsterdam (2014)
79. Granville, A.: Zaphod beeblebrox's brain and the fifty-ninth row of pascal's triangle. The American Mathematical Monthly **99**(4), 318 (1992)
80. Granville, A., Martin, G.: Prime number races. The American Mathematical Monthly **113**(1), 1 (2006)
81. Gray, J.: A history of abstract algebra. Springer, Cham (2018)
82. Grieser, D.: Exploring mathematics. Springer, Cham (2018)
83. Griffiths, D.J.: Introduction to Electrodynamics, 4th edn. Cambridge University Press, Cambridge (2017)
84. Griffiths, D.J., Schroeter, D.F.: Introduction to quantum mechanics, 3rd edn. Cambridge University Press, Cambridge (2018)
85. Griffiths, M., Brown, C., Penrose, J.: From pascal to fibonacci via a coin-tossing scenario. Mathematics in School **43**(2), 25–27 (2014)
86. Grimmett, G., Stirzaker, D.: One thousand exercises in probability, 3rd edn. Oxford University Press, Oxford (2020)
87. Hall, L., Wagon, S.: Roads and wheels. Mathematics Magazine **65**(5), 283–301 (1992)
88. Hamill, P.: A student's guide to Lagrangian and Hamiltonians. Cambridge University Press, Cambridge (2014)
89. Hanley, J.A., Bhatnagar, S.: The "poisson" distribution: History, reenactments, adaptations. The American Statistician **76**(4), 363 (2022)
90. Hansen, J., Sato, M.: Regional climate change and national responsibilities. Environmental Research Letters **11**(3), 034009 (2016)
91. Hart, M.: Guide to Analysis, 2nd edn. Palgrave, Basingstoke (2001)
92. Haslwanter, T.: An introduction to statistics with Python: with applications in the life sciences. Springer, Switzerland (2016)
93. Hass, J., Heil, C., Weir, M.: Thomas' Calculus, 14th edn. Pearson (2019)
94. Herman, R.L.: An introduction to Fourier analysis. Chapman and Hall /CRC, New York (2016)
95. Hiary, G.A.: Fast methods to compute the riemann zeta function (2007). URL `https://arxiv.org/abs/0711.5005`
96. Hirsch, M.W., Smale, S., Devaney, R.L.: Differential equations, dynamical systems, and an introduction to chaos, 3rd edn. Academic Press, Amsterdam (2013)
97. Howell, K.B.: Ordinary differential equations: an introduction to the fundamentals, 2nd edn. Chapman and Hall /CRC, Abingdon (2020)
98. Jarnicki, M., Pflug, P.: Continuous Nowhere Differentiable Functions: The Monsters of Analysis. Springer, Cham (2015)
99. Jaynes, E.T.: The well-posed problem. Foundations of Physics **3**(4), 477 (1973)
100. Johansson, R.: Numerical Python. Apress, Berkeley (2019)
101. Johnson, P.B.: Leaning Tower of Lire. American Journal of Physics **23**(4), 240 (1955)
102. Johnston, D.: Random Number Generators—Principles and Practices. De Gruyter Press, Berlin (2018)
103. Johnston, N.: Advanced linear and matrix algebra. Springer, Cham (2021)
104. Johnston, N.: Introduction to linear and matrix algebra. Springer, Cham (2021)
105. Jones, G.A., Jones, J.M.: Elementary number theory. Springer, London (1998)
106. Jones, H.F.: Groups, representations and physics, 2nd edn. Taylor & Francis, New York (1988)
107. Kajiya, J.T.: The rendering equation. SIGGRAPH Comput. Graph. **20**(4), 143–150 (1986)

108. Katz, V.J.: The history of stokes' theorem. Mathematics Magazine **52**(3), 146 (1979)
109. Kay, S.M.: Intuitive probability and random processes using MATLAB. Springer, New York (2006)
110. Kenett, R., Zacks, S., Gedeck, P.: Modern statistics: a computer-based approach with Python. Birkhäuser, Cham (2022)
111. Kettle, S.F.A.: Symmetry and structure: readable group theory for chemists, 3rd edn. John Wiley, Chichester (2007)
112. Kibble, T.W.B., Berkshire, F.H.: Classical mechanics, 5th edn. Imperial College Press, London (2004)
113. Kifowit, S.J., Stamps, T.A.: The Harmonic Series diverges again and again. AMATYC Review **27**(2), 31–43 (2006)
114. Kong, Q., Siauw, T., Bayen, A.: Python programming and numerical methods – a guide for engineers and scientists. Academic Press (2020)
115. Kortemeyer, J.: Complex numbers: an introduction for first year students. Springer, Wiesbaden (2021)
116. Kosinski, A.A.: Cramer's rule is due to cramer. Mathematics Magazine **74**(4), 310–312 (2001)
117. Kubat, M.: An introduction to machine learning, third edn. Springer, Cham (2021)
118. Kucharski, A.: Math's beautiful monsters; how a destructive idea paved the way to modern math. Nautilus Q.(11) (2014)
119. Kuczmarski, F.: Roads and wheels, roulettes and pedals. The American Mathematical Monthly **118**(6), 479–496 (2011)
120. Kuhl, E.: Computational epidemiology: data-driven modelling of COVID-19. Springer, Cham (2021)
121. Lagarias, J.C.: Euler's constant: Euler's work and modern developments. Bulletin of the American Mathematical Society **50**(4), 527–628 (2013)
122. Lam, L.Y.: Jiu zhang suanshu (nine chapters on the mathematical art): An overview. Archive for History of Exact Sciences **47**(1), 1 (1994)
123. Lambert, B.: A student's guide to Bayesian statistics. SAGE Publications, London (2018)
124. Langtangen, H.P., Linge, S.: Finite difference computing with PDEs. Springer, Cham (2017). URL `https://link.springer.com/book/10.1007/978-3-319-55456-3`
125. Lay, D.C., Lay, S.R., McDonald, J.: Linear algebra and its applications, 5th edn. Pearson, Boston (2016)
126. Lemmermeyer, F.: Reciprocity laws. Springer, Berlin (2000)
127. Lengyel, E.: Mathematics for 3D game programming and computer graphics. Cengage (2011)
128. Leon, S.J., Björck, A., Gander, W.: Gram-schmidt orthogonalization: 100 years and more. Numerical Linear Algebra with Applications **20**(3), 492 (2013)
129. Li, T.Y., Yorke, J.A.: Period three implies chaos. The American Mathematical Monthly **82**(10), 985 (1975)
130. Lin, J.W., Aizenman, H., Espinel, E.M.C., Gunnerson, K.N., Liu, J.: An introduction to Python programming for scientists and engineers. Cambridge University Press, Cambridge (2022)
131. Linge, S., Langtangen, H.P.: Programming for Computations - Python, 2nd edn. Springer (2020)
132. Liu, Y.: First semester in numerical analysis with Python (2020). URL `http://digital.auraria.edu/IR00000195/00001`
133. Lorenz, E.N.: Deterministic nonperiodic flow. Journal of Atmospheric Sciences **20**(2), 130 (1963)
134. Lyche, T.: Numerical linear algebra and matrix factorizations. Springer, Cham (2020)
135. Lynch, S.: Dynamical systems with applications using Python. Birkhäuser, Cham (2018)
136. Lynch, S.: Python for scientific computing and artificial intelligence. CRC Press (2023)
137. MacTutor History of Mathematics Archive: URL `https://mathshistory.st-andrews.ac.uk/`
138. MacTutor History of Mathematics Archive: URL `https://mathshistory.st-andrews.ac.uk/Curves/Cycloid/`
139. Matsuura, K.: Bayesian Statistical Modeling with Stan, R, and Python. Springer, Singapore (2022)
140. May, R.M.: Simple mathematical models with very complicated dynamics. Nature **261**(5560), 459 (1976)
141. Mazo, R.M.: Brownian motion: fluctuations, dynamics, and applications, vol. 112. Clarendon Press, Oxford (2002)
142. Mazur, B., Stein, W.A.: Prime numbers and the Riemann hypothesis. Cambridge University Press, Cambridge (2016)
143. McCluskey, A., B., M.: Undergraduate Analysis. Oxford University Press, Oxford (2018)
144. McMullen, C.T.: The Mandelbrot set is universal. In The Mandelbrot Set, Theme and variations, p. 1. Cambridge University Press, Cambridge (2007)
145. Michelitsch, M., Rössler, O.E.: The "burning ship" and its quasi-julia sets. Computers & Graphics **16**(4), 435 (1992)
146. Michon, G.P.: Surface area of an ellipsoid (2004). URL `http://www.numericana.com/answer/ellipsoid.htm`

147. Mickens, R.E.: Difference equations: Theory, applications and advanced topics, 3rd edn. Chapman and Hall /CRC (2015)
148. Misner, C.W., Thorne, K.S., A., W.J.: Gravitation. W. H. Freeman, San Francisco (1973)
149. Mullen, G.L., Sellers, J.A.: Abstract algebra: a gentle introduction. Chapman and Hall /CRC, Boca Raton (2017)
150. Muller, N., Magaia, L., Herbst, B.M.: Singular value decomposition, eigenfaces, and 3d reconstructions. SIAM review **46**(3), 518 (2004)
151. Nahin, P.J.: Duelling Idiots and Other Probability Puzzlers. Princeton University Press, Princeton (2012)
152. Nahin, P.J.: Digital Dice: Computational Solutions to Practical Probability Problems. Princeton University Press, New Jersey (2013)
153. Nahin, P.J.: Inside Interesting Integrals. Springer-Verlag, New York (2015)
154. Needham, T.: Visual complex analysis. Clarendon, Oxford (1997)
155. Nickerson, R.: Penney ante: counterintuitive probabilities in coin tossing. UMAP journal **27**(4), 503 (2007)
156. Paolella, M.S.: Fundamental probability: a computational approach. John Wiley, Chichester, England (2006)
157. Patarroyo, K.Y.: A digression on hermite polynomials (2019). URL `https://arxiv.org/abs/1901.01648`
158. Paul, W., Baschnagel, J.: Stochastic processes: from physics to finance, 2nd edn. Springer, New York (2013)
159. Pearl, J., Glymour, M., Jewell N, P.: Causal Inference in Statistics: A Primer. John Wiley and Sons, Newark (2016)
160. Peck, R., Short, T.: Statistics: learning from data, 2nd edn. Cengage, Australia (2019)
161. Pedersen, S.: From calculus to analysis. Springer, Cham (2015)
162. Peitgen, H.O., Jürgens, H., Saupe, D.: Chaos and fractals: new frontiers of science, 2nd edn. Springer (2004)
163. Petrov, V.V.: Limit theorems of probability theory: sequences of independent random variables. Clarendon, Oxford (1995)
164. Pharr, M., Humphreys, G.: Physically based rendering: from theory to implementation, 2nd edn. Morgan Kaufmann, San Francisco (2010)
165. Piessens, R., Doncker-Kapenga, E.d., Überhuber, C., Kahaner, D.: QUADPACK: A subroutine package for automatic integration. Springer-Verlag, Berlin (1983)
166. Platt, D., Trudgian, T.: The Riemann hypothesis is true up to $3 \cdot 10^{12}$. Bulletin of the London Mathematical Society **53**(3), 792 (2021)
167. Pollack, P., Roy, A.S.: Steps into analytic number theory: a problem-based introduction. Springer, Cham (2021)
168. Polyanin, A.D., Zaitsev, V.F.: Handbook of ordinary differential fquations, 3rd edn. Chapman and Hall /CRC, New York (2017)
169. Posamentier, A.S., Lehmann, I.: The (Fabulous) Fibonacci Numbers. Prometheus Books, New York (2007)
170. Press, W.H., Tekolsky, S.A., Vetterling, W.T., Flannery, B.P.: Numerical Recipes: the art of scientific computing, 3rd edn. Cambridge University Press, Cambridge (2007)
171. Pressley, A.N.: Elementary Differential Geometry. Springer-Verlag, London (2010)
172. Priestley, H.A.: Introduction to complex analysis, second edn. Oxford University Press, Oxford (2005)
173. Reid, M.: Undergraduate algebraic geometry. Cambridge University Press, Cambridge (1988)
174. Riley, K.F., Hobson, M.P., Bence, S.J.: Mathematical Methods for Physics and Engineering, 3rd edn. Cambridge University Press (2006)
175. Robinson, J.C.: An introduction to ordinary differential equations. Cambridge University Press, Cambridge (2004)
176. Rosenhouse, J.: The Monty Hall Problem. Oxford University Press, Oxford (2009)
177. Roser, M., Appel, C., Ritchie, H.: Human height. Our World in Data (2013). URL `https://ourworldindata.org/human-height`
178. Ross, S.M.: Stochastic processes, 2nd edn. Wiley, Chichester (1996)
179. Ross, S.M.: A first course in probability, 10th edn. Pearson, Harlow (2020)
180. Salinelli, E., Tomarelli, F.: Discrete dynamical models, vol. 76. Springer, Wien (2014)
181. Salsburg, D.: The Lady Tasting Tea: How Statistics Revolutionized Science in the Twentieth Century. Holt Paperbacks (2002)
182. Sauer, T.: Numerical analysis, 2nd edn. Pearson, Boston (2012)
183. Selvin, S.: Letters to the editor. The American Statistician **29**(1), 67 (1975)

184. Serway, R.A., Jewett, J.W.: Physics for scientists and engineers, 10th edn. Cengage (2011)
185. Shafarevich, I.R.: Basic algebraic geometry I, 3rd edn. Springer, Berlin (2013)
186. Shah, C.: A Hands-On Introduction to Machine Learning. Cambridge University Press (2022)
187. Shinbrot, T., Grebogi, C., Wisdom, J., Yorke, J.: Chaos in a double pendulum. American Journal of Physics **60**(6), 491–499 (1992)
188. Skegvoll, E., Lindqvist, B.H.: Modeling the occurrence of cardiac arrest as a poisson process. Annals of emergency medicine **33**(4), 409 (1999)
189. Spivak, M.: Calculus, 3rd edn. Cambridge University Press, Cambridge (2006)
190. Stein, E.M.: Fourier analysis: an introduction. Princeton University Press, Princeton (2003)
191. Stewart, I., Tall, D.: Complex analysis: the hitch hiker's guide to the plane, second edn. Cambridge University Press, Cambridge (2018)
192. Stewart, J., Watson, S., Clegg, D.: Multivariable Calculus, 9th edn. Cengage (2020)
193. Stinerock, R.: Statistics with R: A Beginner's Guide. SAGE Publishing.1 (2018)
194. Strang, G.: Introduction to linear algebra, 5th edn. Wellesley-Cambridge Press (2016)
195. Strang, G.: Linear algebra and learning from data. Wellesley-Cambridge Press (2019)
196. Strauss, W.A.: Partial differential equations : an introduction, 2nd edn. Wiley, New Jersey (2008)
197. Strogatz, S.H.: Nonlinear dynamics and chaos, with applications to physics, biology, chemistry, and engineering, 2nd edn. Westview Press (2015)
198. Sutton, E.C.: Observational Astronomy: Techniques and Instrumentation. Cambridge University Press (2011)
199. Sýkora, S.: Approximations of ellipse perimeters and of the complete elliptic integral (2005). URL `http://dx.doi.org/10.3247/SL1Math05.004`
200. Tall, D.: The blancmange function continuous everywhere but differentiable nowhere. The Mathematical Gazette **66**(435), 11–22 (1982)
201. Tapp, K.: Differential Geometry of Curves and Surfaces. Springer (2016)
202. Thim, J.: Continuous nowhere differentiable functions. Masters Thesis, Luleå University of Technology (2003)
203. Tong, Y.L.: The Multivariate Normal Distribution. Springer New York, New York (1990)
204. Tversky, A., Kahneman, D.: Judgment under uncertainty: Heuristics and biases. Science **185**(4157), 1124 (1974)
205. Unpingco, J.: Python for Probability, Statistics, and Machine Learning, 2nd edn. Springer International Publishing, Cham (2019)
206. Vince, J.: Quaternions for computer graphics, 2nd edn. Springer, London (2021)
207. Vince, J.: Mathematics for computer graphics, 6th edn. Springer, London (2022)
208. Vălean, C.I.: (Almost) Impossible Integrals, Sums, and Series. Springer, Cham (2019)
209. Watson, G.N.: Three triple integrals. The Quarterly Journal of Mathematics **os-10**(1), 266 (1939)
210. Wikimedia: URL `https://commons.wikimedia.org/wiki/File:Blaise_Pascal_Versailles.JPG`
211. Wikimedia: URL `https://en.wikipedia.org/wiki/Jia_Xian#/media/File:Yanghui_triangle.gif`
212. Wikimedia: URL `https://en.wikipedia.org/wiki/Florence_Nightingale#/media/File:Florence_Nightingale_(H_Hering_NPG_x82368).jpg`
213. Wikimedia / Mario Biondi: URL `https://commons.wikimedia.org/wiki/File:Al_Khwarizmi%27s_Monument_in_Khiva.png`
214. Wikipedia: URL `https://en.wikipedia.org/wiki/Seki_Takakazu`
215. Wilcox, A.J.: On the importance—and the unimportance— of birthweight. International Journal of Epidemiology **30**(6), 1233 (2001)
216. Witte, R.S., Witte, J.S.: Statistics, 4th edn. Wiley, Hoboken (2021)
217. Wolfram, S.: Statistical mechanics of cellular automata. Rev. Mod. Phys. **55**, 601 (1983)
218. Wolfram, S.: Geometry of binomial coefficients. The American Mathematical Monthly **91**(9), 566 (1984)
219. Yesilyurt, B.: Equations of Motion Formulation of a Pendulum Containing N-point Masses. arXiv e-prints arXiv:1910.12610
220. Young, H.D., Freedman, R.A.: University physics with modern physics, 15th edn. Pearson (2020)
221. Yuan, Y.: Jiu zhang suan shu and the gauss algorithm for linear equations. Documenta Mathematica (Extra volume: optimization stories) p. 9 (2012)

INDEX

A

Algebra   283
Anaconda   489
Analysis
  Complex   342
  Real   5
Animation (Matplotlib)   150
Ansombe's quartet   449
Apéry's constant $\zeta(3)$ 338
arange (syntax)   6
Arc length   104
Archimedian property   31
Arcsine law (random walk)   483
Array (NumPy)   6

B

Bayes' Theorem   381
Bayesian statistics
  Conjugate priors   484
  Credible interval   465
  Evidence   464
  Inference   463
  Likelihood   464
  Posterior   464
  Prior   464
  vs. Frequentist statistics   468
Bernoulli numbers   338
Bernoulli trials   369
Bertrand paradox   413
Bifurcation   179
Big O notation   58
Binet's formula   22
Binomial coefficient   364
Birthday problem   377

Generalisations   408
Bisection method   34
Blancmange function   88
Broadcasting (Python concept)   496
Buffon's needle   401
Buffon's noodles   404
Bézout's identity   320

C

Cardioid   187
Cartographic projections   122
Catalan's constant   396
Cayley's Theorem   310
Cayley-Hamilton Theorem   248, 279
Ceiling   13
Cellular automata   368
Central Limit Theorem   424
  Generalised   427
Chaos   167, 172
Chinese Remainder Theorem   323
Circulation   135
Classification   471
  k-nearest neighbour algorithm   473
Clustering   471
  k-means algorithm   471
Coefficient of determination R^2 448, 482
Coin toss (simulation techniques)   369
Combination   356
Comma (Python operator)   29
Comparison test   16
Completeness Axiom   32
Confidence interval   432
Conic sections   118
Continuity (ε-δ definition)   23

BIOGRAPHICAL INDEX